ELECTROSTATIC PRECIPITATOR MANUAL

ELECTROSTATIC PRECIPITATOR MANUAL

JACK R. McDONALD
ALAN H. DEAN

Southern Research Institute

NOYES DATA CORPORATION
Park Ridge, New Jersey, U.S.A.
1982

Copyright © 1982 by Noyes Data Corporation
Library of Congress Catalog Card Number: 82-3449
ISBN: 0-8155-0895-6
ISSN: 0090-516X
Printed in the United States

Published in the United States of America by
Noyes Data Corporation
Mill Road, Park Ridge, New Jersey 07656

10 9 8 7 6 5 4 3 2 1

Library of Congress Cataloging in Publication Data

McDonald, J. R. (Jack Raymond), 1944-
   Electrostatic precipitator manual.

   (Pollution technology review, ISSN 0090-516X ;
no. 91)
   Bibliography: p.
   Includes index.
   1. Electrostatic precipitation--Handbooks,
manuals, etc. I. Dean, Alan H. II. Title.
III. Series.
TH7695.E4M33   628.5'32   82-3449
ISBN 0-8155-0895-6   AACR2

Foreword

This manual, based on research by Southern Research Institute, is a summary of the results of studies performed by various individuals and organizations on the application of electrostatic precipitators to the collection of fly ash particles produced in the combustion of pulverized coal. These studies include comprehensive performance evaluations of full-scale precipitators, in-situ and laboratory measurement of fly ash resistivity, rapping reentrainment investigations, tests to evaluate the effects of flue gas conditioning agents on precipitator performance, investigations into the fundamental operation of hot-side precipitators, basic laboratory experiments, and development of a mathematical model of electrostatic precipitation.

The manual covers the fundamentals of electrostatic precipitation; mechanical and electrical components of electrostatic precipitators; factors influencing precipitator performance; measurement of important parameters; advantages and disadvantages of cold-side, hot-side, and flue gas conditioned electrostatic precipitators; safety aspects; maintenance procedures; troubleshooting procedures; the usage of a computer model for electrostatic precipitation; and features of a well-equipped electrostatic precipitator.

As a result of these studies, new sources of information are available that can be used by power plant personnel as an aid in selecting, sizing, maintaining, and troubleshooting electrostatic precipitators.

The information in the book is from:

> *A Manual for the Use of Electrostatic Precipitators to Collect Fly Ash Particles,* prepared by Jack R. McDonald and Alan H. Dean of Southern Research Institute for the U.S. Environmental Protection Agency, May 1980.

Foreword

The expanded table of contents is organized in such a way as to serve as a subject index and provides easy access to the information contained in the book.

Advanced composition and production methods developed by Noyes Data are employed to bring this durably bound book to you in a minimum of time. Special techniques are used to close the gap between "manuscript" and "completed book." In order to keep the price of this book to a reasonable level, it has been partially reproduced by photo-offset directly from the original report and the cost savings passed on to the reader. Due to this method of publishing, certain portions of the book may be less legible than desired.

NOTICE

The material in this book was prepared as an account of work sponsored by the U.S. Environmental Protection Agency. Publication does not signify that the contents necessarily reflect the views and policies of the contracting agencies or the publisher, nor does mention of trade names or commercial products constitute endorsement or recommendation for use.

Contents and Subject Index

1. INTRODUCTION 1

2. TERMINOLOGY AND GENERAL DESIGN FEATURES ASSOCIATED WITH
   ELECTROSTATIC PRECIPITATORS USED TO COLLECT FLY ASH PARTICLES 4

3. FUNDAMENTAL PRINCIPLES OF ELECTROSTATIC PRECIPITATION 7
   General Considerations 7
   Creation of an Electric Field and Corona Current 7
   Particle Charging 10
   Particle Collection 15
   Removal of Collected Material 17

4. LIMITING FACTORS AFFECTING PRECIPITATOR PERFORMANCE 19
   Allowable Voltage and Current Density 19
   Nonideal Effects 20
      Nonuniform Gas Velocity Distribution 20
      Gas Sneakage 21
      Particle Reentrainment 21

5. USE OF ELECTROSTATIC PRECIPITATORS FOR THE COLLECTION OF FLY ASH 23
   Reasons for Using Electrostatic Precipitators to Collect Fly Ash .23
   Design of Precipitators Used to Collect Fly Ash 23
      General Description 23
      Precipitator Shell 24
      Electrical Sections 25
         Electrical Energization 25
            Historical Development 25
            Power Supplies 26
            High Voltage Rectifiers 27
            Spark Rate 27
            Design and Operating Requirements 27
            Sources of High Voltage Electrical Equipment 29
         Discharge Electrode System 29
            Geometries of Discharge Electrodes 30
            Types of Discharge Electrodes 31
            New Designs 35
            Discharge Electrode Support 36
         Collecting Electrode System 37
            Geometries of Collecting Electrodes 37
      Ash Removal Designs 39
         General 39
         Rappers 39
         New Technology in Rapper Control 42
         Hoppers 43
         Removal from Hoppers 50
         Dust Removal Systems 51
            Container Removal 51
            Dry Vacuum Systems 51

viii   Contents and Subject Index

        Wet Vacuum Systems 51
        Screw Conveyors 51
        Scraper Bottom 51
    Gas Flow Devices 51
        General 51
        Straighteners 53
        Splitters 54
        Transformation Splitters 54
        Vanes 56
        Diffusion Plates 57
    Types of Precipitators Used to Collect Fly Ash 58
        Cold-Side 58
        Hot-Side 59
    Compilation of Installations Using Electrostatic Precipitators to Collect Fly Ash ... 61

6. ANALYSIS OF FACTORS INFLUENCING ESP PERFORMANCE 63
    Particle Size Distribution 63
        General Discussion 63
        Characterization of Particle Size Distributions 63
        Field Methods for Measuring Particle Size Distributions 71
            General Considerations in Making Field Measurements 71
            Inertial (Aerodynamic) Methods 72
            Optical Methods 84
            Diffusional and Condensation Nuclei Methods 86
            Electrical Mobility Method 91
        Other Specialized Partical Sizing Systems for Field Use 96
            Respirable Particle Classifier (RPC) Impactor 96
            Large Particle Sizing System (LPSS) 101
        Laboratory Methods for Measuring Particle Size Distributions 103
            Sedimentation and Elutriation 103
            Centrifuges 104
            Microscopy 107
            Sieves 108
            Coulter Counter 110
        Effect of Particle Size Distribution on ESP Performance 110
        Measured Size Distributions from Various Installations 113
            Plant Number One 113
            Plant Number Two 113
            Plant Number Three 120
            Plant Number Four 120
            Plant Number Five 120
            Plant Number Six 126
            Plant Number Seven 126
            Plant Number Eight 126
            Plant Number Nine 126
            Plant Number Ten 126
            Plant Number Eleven 132
            Plant Number Twelve 132
            Plant Number Thirteen 134
            Plant Number Fourteen 134
            Plant Number Fifteen 135
        Summary of Inlet Particle Size Distributions 135
    Specific Collection Area 142
    Voltage-Current Characteristics 145
        Electrical Circuitry for a Precipitator 145
        Measurement of Voltage-Current Characteristics 147
        Effect of Electrode Geometry 150
        Effect Due to Gas Properties 160
        Effects Due to Particles 172
        Effects Due to Chemical Conditioning Agents 177
        Effects of Voltage-Current Characteristics on Precipitator Performance 185
        Measured Secondary Voltage-Current Data from Full-Scale Precipitators Collecting
            Fly Ash 189

Measured Cold-Side Curves	191
Plant 1—Cold-Side ESPs Collecting Ash from Low Sulfur Western Coal	191
Plant 2—Cold-Side ESPs Collecting Ash from High Sulfur Eastern Coal	193
Plant 3—Cold-Side ESPs Collecting Ash from High Sulfur Eastern Coal	196
Plant 4—Cold-Side ESPs Collecting Ash from Low Sulfur Western Coal	200
Plant 5—Cold-Side ESP Collecting Ash from Medium Sulfur Southeastern Coal	201
Plant 6—Cold-Side ESP Collecting Ash from Midwestern Coal	203
Plant 7—Cold-Side ESP Collecting Ash from Low Sulfur Western Coal	204
Measured Hot-Side Curves	205
Plant 8—Hot-Side ESP Collecting Ash from Low Sulfur Eastern Coal	205
Plant 9—Hot-Side ESP Collecting Ash from Low Sulfur Western Coal	207
Plant 10—Hot-Side ESP Collecting Ash from a Western Power Plant Burning Low Sulfur Coal	214
Resistivity of Collected Fly Ash	**215**
Effect of Ash Resistivity on Precipitator Performance	215
Measured Voltage-Current Curves Demonstrating Back Corona	219
Factors Influencing Ash Resistivity	221
Volume and Surface Conduction	221
Factors Influencing Volume Resistivity	222
Factors Influencing Surface Resistivity	226
Combined Effects of Volume and Surface Conduction	231
Prediction of Fly Ash Resistivity	232
Measurement of Ash Resistivity	236
Factors Influencing Measurement of Resistivity	236
Particle Size Distribution and Porosity	236
Electric Field	237
Method of Depositing Ash Layer	237
Thickness of Ash Layer	238
Time of Current Flow	238
Source Variability	238
Methods for Measuring Ash Resistivity	238
General Considerations	238
Laboratory versus In Situ Measurements	239
Laboratory Measurements—Standard Technique	240
Apparatus for Standard Technique	240
Experimental Procedure for Standard Technique	242
Variations for the Standard Technique Used in Laboratory Studies	243
Laboratory Studies Simulating Flue Gases Containing SO_x—Experimental Apparatus Utilizing ASME, PTC-28, Test Cells	243
Experimental Procedure	246
Problems Encountered Using SO_x	247
Experiments to Develop Apparatus and Procedure to Utilize Environments Containing SO_x	248
Development of a Radial Flow Test Cell and Procedure—Equipment	249
Test Procedure	252
In Situ Measurements—General Considerations	253
In Situ Resistivity Probes—Point-to-Plane Probe	255
Description of SoRI Point-Plane Probe	258
General Maintenance of SoRI Point-Plane Probe	259
Operation of the SoRI Point-Plane Probe—Pre-Field Trip Preparation	260
Operating Instructions	260
Operating Outline	264
Calculations	264
Cyclone Resistivity Probes	268
Kevatron Electrostatic Precipitator Analyzer	270
Lurgi Electrostatic Collection Resistivity Device	271
Comparison of In Situ Resistivity Probes	272
Limitations Due to Non-Ideal Effects	**274**
Gas Velocity Distribution	274
General Discussion	274
Criteria for a Good Gas Flow Distribution	275
Field Experience with Gas Flow Distribution	275
Correlation of Collection Efficiency with Gas Velocity Distribution	293

Gas Sneakage..297
Air Flow Model Studies..300
  Basis for Model Studies...300
  Similarity of Fluid Flows...301
  Flow Model Construction...304
  Instrumentation...305
Particle Reentrainment..305
Rapping Reentrainment...305
  Background..305
  Emissions Due to Rapping..309
  Summary of the Results of Rapping Studies.........................329
Reentrainment from Factors Other than Rapping.......................330
Nonuniform Temperature and Dust Concentration.......................331

7. EMISSIONS FROM ELECTROSTATIC PRECIPITATORS 332
Particulate Emissions...332
Methods for Determination of Overall Mass Efficiency................332
  EPA Test Method 5...332
Description of Components..334
ASTM—Test Method..335
ASME Performance Test Code 27.......................................335
Status of Rules and Regulations Governing Particulate Matter, Sulfur Oxide, Nitrogen Oxide, and Opacity for Coal-Fired Power Boilers in the United States 337
Background..337
Current Status of Emission Regulations..............................337
Performance Evaluation..338
Discussion and Definition of Opacity................................339
Relationship Between Opacity and Mass Concentration and Particle Size 341
  Theoretical Relationship..341
  Observed Relationship...344
Example of Modeling of Opacity Versus Mass at the Exit of an Electrostatic
  Precipitator..345
Measurement of Relative Stack Emission Levels and Opacity...........346

8. CHOOSING AN ELECTROSTATIC PRECIPITATOR: COLD-SIDE VERSUS HOT-SIDE CONDITIONING AGENTS...351
Advantages and Disadvantages of the Different Precipitator Options.................351
General Discussion..351
Cold-Side Electrostatic Precipitator................................351
Hot-Side Electrostatic Precipitator.................................352
Cold-Side Electrostatic Precipitator with Chemical Flue Gas Conditioning...........353
  Possible Advantages of Chemical Flue Gas Conditioning.............353
  Properties and Utilization of Well-Known Conditioning Agents......355
  Utility Utilization and Capital and Operating Costs of Conditioning Systems.......358
  Possible Disadvantages of Chemical Flue Gas Conditioning..........358
  Precipitator Requirements and Economic Comparisons................359

9. SAFETY ASPECTS OF WORKING WITH ELECTROSTATIC PRECIPITATORS........365
Rules and Regulations..365
Hazards..365
Fire and Explosion Hazards..365
Electrical Shock Hazards..366
Toxic Gas Hazard..367
Other More Minor Hazards..367

10. MAINTENANCE PROCEDURES...368

11. TROUBLESHOOTING..375
Diagnosis of ESP Problems..375
Available Instrumentation for Electrostatic Precipitators........378
Spark Rate Meters...378
Secondary Voltage and Current Meters................................383
Opacity Meters..384
Hopper Level Meters...385

Contents and Subject Index   xi

12. AN ELECTROSTATIC PRECIPITATOR COMPUTER MODEL .... .386
    Introduction .... .386
    Capabilities of the Model .... .387
    Basic Framework of the Model .... .388
    Latest Improvements to the Model .... .390
        Calculation of Voltage-Current Characteristics .... .390
        Method for Predicting Trends Due to Particulate Space Charge .... .390
        Method for Estimating Effects Due to Rapping Reentrainment .... .391
        Empirical Corrections to No-Rap Migration Velocities .... .392
        User-Oriented Improvements .... .393
    Applications and Usefulness of the Model .... .394
        Use of the Model for Troubleshooting .... .394
        Use of the Model for Sizing of Precipitators .... .398

13. FEATURES OF A WELL-EQUIPPED ELECTROSTATIC PRECIPITATOR .... .403

REFERENCES .... .405

APPENDIX A—POWER PLANT AND AIR QUALITY DATA FOR PLANTS WITH ESP ...... .426

APPENDIX B—CASCADE IMPACTOR STAGE PARAMETERS .... .448

APPENDIX C—PARTICULATE MATTER, SO_x, AND NO_x EMISSION LIMITS .... .450

APPENDIX D—LOW TEMPERATURE CORROSION AND FOULING .... .455
    Introduction .... .455
    Sulfuric Acid Occurrence in Flue Gas .... .455
        SO_x, H_2O, and H_2SO_4 Equilibria .... .455
        Determination of the Sulfuric Acid Dew Point .... .457
        Condensation Characteristics .... .462
    Factors Influencing Corrosion Rates .... .463
        Acid Strength .... .463
        Acid Deposition Rate .... .466
        Fly Ash Alkalinity .... .468
        Hydrochloric Acid .... .470
    Fouling of Low Temperature Surfaces .... .473
    Laboratory Corrosion Studies .... .474
    Summary of Field Experience and Plant Data .... .477
    Methods of Assessing Corrosion Tendencies of Flue Gases .... .482
        Introduction .... .482
        Corrosion Probes .... .482
        Acid Deposition Probes .... .483
        Gas and Ash Analysis .... .483
    Summary and Conclusions .... .483

Section 1

Introduction

Recent studies performed by various individuals and organizations have been directed toward obtaining a fuller and better understanding of the application of electrostatic precipitators to collect fly ash particles produced in the combustion of pulverized coal. These studies include comprehensive performance evaluations of full-scale precipitators, in situ and laboratory measurement of fly ash resistivity, rapping reentrainment investigations, tests to evaluate the effects of flue gas conditioning agents on precipitator performance, investigations into the fundamental operation of hot-side precipitators, basic laboratory experiments, and development of a mathematical model of electrostatic precipitation. As a result of these studies, new sources of information are available that can be used by power plant personnel as an aid in selecting, sizing, maintaining, and troubleshooting electrostatic precipitators.

The purpose of the present work is to bring together the results of these and previous studies and to incorporate them into a document which is oriented toward the collection of fly ash particles by electrostatic precipitation. Since the scope and detail of this document are rather extensive, an expanded table of contents has been provided for use in retrieving information on specific topics contained in the text. It is suggested that the user familiarize himself with the table of contents so that he can use the text in the most effective manner when addressing specific needs. An attempt has been made to present concepts, measurement techniques, factors influencing precipitator performance, data, and data analysis from a practical standpoint. Theoretical developments and equations have been avoided where possible. Therefore, discussions, descriptions, and data from small-scale and full-scale precipitators have been stressed in illustrating many of the important considerations associated with electrostatic precipitators. The extensive use of data from full-scale precipitators should familiarize the user with what to expect in actual field applications.

In the text, Sections 2-5 deal primarily with the basic components of electrostatic precipitators and with the fundamental principles of electrostatic precipitation in order to establish the framework for ensuing discussions. The basic mechanical and electrical components associated with electrostatic precipitators are discussed with respect to their functions and various designs. The fundamental steps in electrostatic precipitation involving the maintainence of an electric field and corona current, particle charging, particle transport to the collection electrodes, and removal of particles from the collection electrodes are discussed in sufficient detail to provide an understanding of the importance of the various physical mechanisms and of the factors affecting these mechanisms. Limiting factors affecting electrostatic precipitator performance are discussed in order to familiarize the reader with effects that result in less than optimal performance. The types of electrostatic precipitators presently used to collect fly ash particles are described briefly. These include cold-side, hot-side, and flue gas conditioned electrostatic precipitators. A compilation of installations in the U.S. using electrostatic precipitators to collect fly ash particles has been prepared. This compilation includes coal, boiler, and electrostatic precipitator data for each installation.

In Section 6, factors influencing electrostatic precipitator performance, along with measurement techniques and experimental data, are discussed extensively. These factors include particle size distribution, specific collection area, voltage-current characteristics, resistivity of the collected fly ash, and nonideal effects such as nonuniform gas velocity distribution, gas bypassage of electrified regions (sneakage), and particle reentrainment. Methods and instrumentation for measuring particle size distributions, voltage-current characteristics, fly ash resistivity, gas velocity distribution, gas sneakage, and rapping reentrainment are described in detail. Methods of interpretation and analysis of the data obtained from the various types of measurements are discussed.

Since the particulate emissions from an electrostatic precipitator must meet mass and opacity standards, it is important to be familiar with methods for measuring these quantities. Section 7 deals with the different methods for measuring mass and opacity. The dependence of opacity on mass and particle size distribution is discussed.

The material in Section 8 is intended to be used as a guide in selecting the type of electrostatic precipitator which is best suited from a cost and reliability standpoint for a particular application. The advantages and disadvantages of cold-side, hot-side, and flue gas conditioned electrostatic precipitators are discussed. Estimates are made of the costs for the different options when treating ashes with low, moderate, and high resistivities.

Sections 9, 10, and 11 deal with effective utilization of electrostatic precipitators by discussing safety considerations, maintenance procedures, and troubleshooting of problems, respectively. Since serious accidents can occur when working with electrostatic precipitators, it is important to be aware of the hazards involved and to take the proper precautions. Following proper maintenance procedures will result in better precipitator performance over the long term, fewer operating problems, less down time, and longer life of certain components. Many electrostatic precipitator problems can be diagnosed and corrected by using appropriate troubleshooting procedures. The equipping of an electrostatic precipitator with instrumentation which is helpful in troubleshooting of problems is discussed.

Since it has been shown that a computer model, which has been developed under the sponsorship of the U.S. Environmental Protection Agency, can be used to advantage in predicting electrostatic precipitator performance as a function of the operating parameters, Section 12 discusses this approach. The capabilities of the modeling approach are stressed. The applications and usefulness of the model are discussed extensively. Applications include predictions of efficiency as a function of particle size distribution, specific collection area, electrical operating conditions, and nonideal conditions. These applications are incorporated into useful procedures for troubleshooting and sizing electrostatic precipitators.

Section 13 points out features that a well-equipped electrostatic precipitator should possess. These features are a natural consequence of the preceding material in the manual. These features are intended to yield flexibility, reliability, ease in analysis of precipitator performance, and, ultimately, the best possible precipitator performance.

Section 2

Terminology and General Design Features Associated with Electrostatic Precipitators Used to Collect Fly Ash Particles

An electrostatic precipitator (ESP) is a device which is used to remove suspended particulate matter from industrial process streams. Dry electrode, parallel plate electrostatic precipitators are used by the electric utility industry to remove fly ash particles from the effluent gas produced in the combustion of coal. Figure 1 shows a schematic diagram of a wire-plate electrostatic precipitator.[1] Although the details of construction will vary from one manufacturer to another, the basic features are the same.

Since uniform, low turbulence gas flow is desirable in the collection regions of a precipitator, several devices may be employed to achieve good gas flow quality before the gas is treated. Turning or guide vanes are used in the duct work prior to the precipitator in order to preserve gas-flow patterns following a sharp turn or sudden transition. This prevents the introduction of undue turbulence into the gas flow. Plenum chambers and/or diffusion screens (plates) are used to achieve reduced turbulence and improved uniformity of the gas flow in expansion turns or transitions prior to the gas treatment regions of the precipitator.

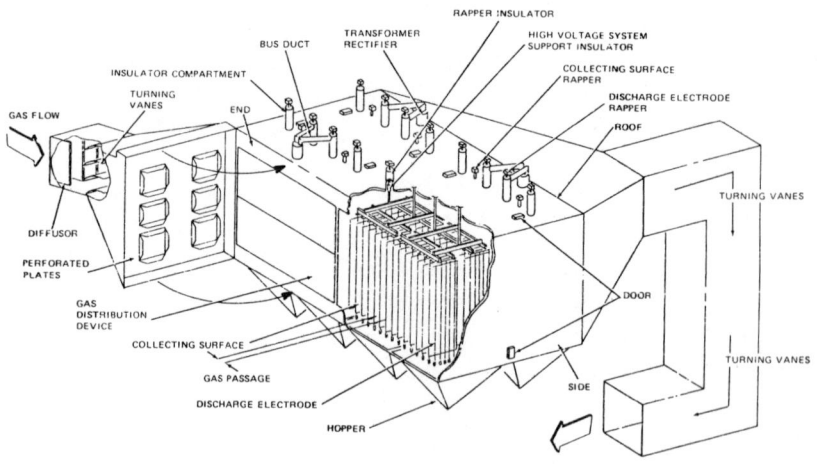

Figure 1. General precipitator layout and nomenclature.[1]

The gas entering the treatment regions of the precipitator flows through several passage ways (<u>gas passages</u>) formed by plates (<u>collection electrodes</u>) which are parallel to one another. A series of <u>discharge electrodes</u> is located midway between the plates in each gas passage. High voltage electrical <u>power supplies</u> provide the voltage and current which are needed to separate the particles from the gas stream. The discharge electrodes are held at a high negative potential with the collection electrodes grounded.

A precipitator may be both physically and electrically sectionalized. Figure 2 shows two possible precipitator layouts with the terminology concerning sectionalization.[2]

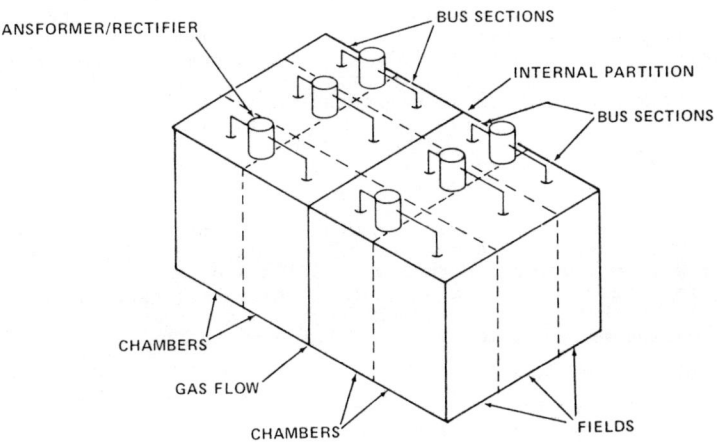

CASE I: 1 PRECIPITATOR, 2 CHAMBERS, 12 BUS SECTIONS, 6 POWER SUPPLIES, 3 FIELDS

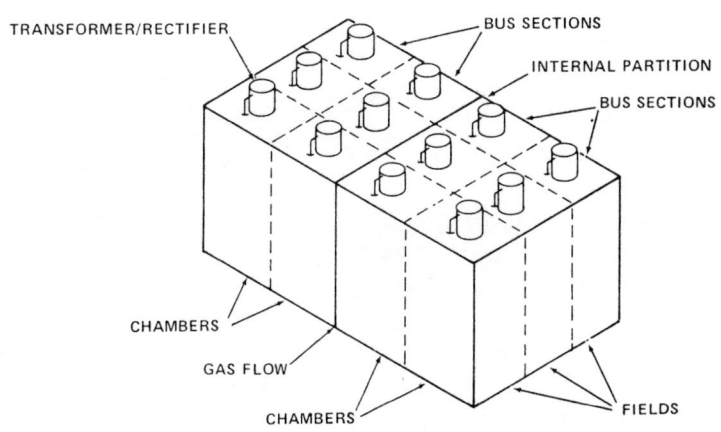

CASE II: 1 PRECIPITATOR, 2 CHAMBERS, 12 BUS SECTIONS, 12 POWER SUPPLIES, 3 FIELDS

Figure 2. Typical precipitator electrical arrangements and terminology.[2]

A _chamber_ is a gas-tight longitudinal subdivision of a precipitator. A precipitator without any internal dividing wall is a single chamber precipitator. A precipitator with one dividing wall is a two-chamber precipitator, etc.

An _electrical field_ is a physical portion of a precipitator that is energized by a single power supply. A _bus section_ is the smallest portion of an electrostatic precipitator which can be deenergized independently. An electrical field may contain two or more bus sections. Electrical fields in the direction of gas flow may be physically separated in order to provide internal access to the precipitator.

The material which is collected on the collection and discharge electrodes is removed by mechanical jarring (or _rapping_). Devices called _rappers_ are used to provide the force necessary to dislodge the collected material from the electrode surfaces. Rappers may provide the rapping force through impact or vibration of the electrodes. The material which is dislodged during rapping falls under the influence of gravity. A certain amount of the material dislodged during rapping falls into _hoppers_ which are located below the electrified regions. Material collected in the hoppers is transported away from the precipitator in some type of disposal process.

Portions of the gas flowing through a precipitator may pass through regions below and above the collection electrodes where treatment will not occur. Normally, _baffles_ are located in the region below the collection electrodes. These baffles redirect the gas flow back into the treatment region and prevent the disturbance of the material collected in the hoppers.

Section 3

Fundamental Principles of Electrostatic Precipitation

GENERAL CONSIDERATIONS

The electrostatic precipitation process involves several complicated and interrelated physical mechanisms: the creation of a nonuniform electric field and ionic current in a corona discharge; the ionic and electronic charging of particles moving in combined electro- and hydro-dynamic fields; and the turbulent transport of charged particles to a collection surface. In many practical applications, the removal of the collected particulate layer from the collection surface presents a serious problem since the removal procedures introduce collected material back into the gas stream and cause a reduction in collection efficiency. Other practical considerations which reduce the collection efficiency are nonuniform gas velocity distribution, bypassage of the electrified regions by particle-laden gas, and particle reentrainment during periods when no attempt is being made to remove the collected material. In certain applications, the flue gas environment and fly ash composition are such that the collected particulate layer limits the maximum values of useful voltage and current.

CREATION OF AN ELECTRIC FIELD AND CORONA CURRENT

The first step in the precipitation process is the creation of an electric field and corona current. This is accomplished by applying a large potential difference between a small-radius electrode and a much larger radius electrode, where the two electrodes are separated by a region of space containing an insulating gas. For industrial applications, a large negative potential is applied at the small-radius electrode and the large-radius electrode is grounded.

At any applied voltage, an electric field exists in the interelectrode space. For applied voltages less than a value referred to as the "corona starting voltage", a purely electrostatic field is present. At applied voltages above the corona starting voltage, the electric field in the vicinity of the small-radius electrode is large enough to produce ionization by electron impact. Between collisions with neutral molecules, free electrons are accelerated to high velocities and, upon collision with a neutral molecule, their energies are sufficiently high to cause an electron to be separated from a neutral molecule. Then, as the increased number of electrons moves out from the vicinity of the small-radius electrode, further collisions between electrons and neutral molecules occur. In a limited high electric field region near the small-radius electrode, each collision between an electron and a neutral molecule has a certain probability of forming a positive molecular ion and another electron, and an electron avalanche is established.

The positive ions migrate to the small-radius electrode and the electrons migrate into the lower electric field regions toward the large-radius electrode. These electrons quickly lose much of their energy and, when one of them collides with a neutral electronegative molecule, there is a probability that attachment will occur and a negative ion will be formed. Thus, negative ions, along with any electrons which do not attach to a neutral molecule, migrate under the influence of the electric field to the large-radius electrode and provide the current necessary for the precipitation process.

Figure 3-a is a schematic diagram showing the region very near the small-radius electrode where the current-carrying negative ions are formed.[3] As these negative ions migrate to the large-radius electrode, they constitute a steady-state charge distribution in the interelectrode space which is referred to as an "ionic space charge". This "ionic space charge" establishes an electric field which adds to the electrostatic field to give the total electric field. As the applied voltage is increased, more ionizing sequences result and the "ionic space charge" increases. This leads to a higher average electric field and current density in the interelectrode space.

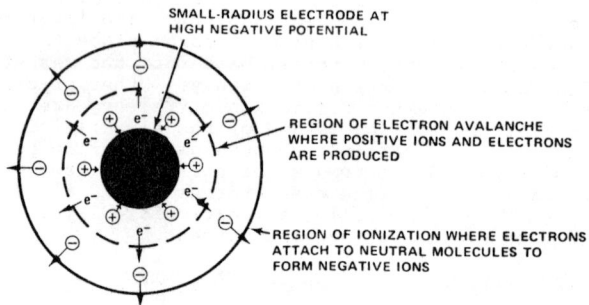

Figure 3-a. Region near small-radius electrode.[3]

Figure 3-b gives a qualitative representation of the electric field distribution and equipotential surfaces in a wire-plate geometry which is commonly used.[3] Although the electric field is very nonuniform near the wire, it becomes essentially uniform near the collection plates. The current density is very nonuniform throughout the interelectrode space and is maximum along a line from the wire to the plate. Figure 4 contains experimental data showing the positional dependence of the current density and electric field at the plate.[4] The data were taken under laboratory conditions with positive corona in ambient air at an applied voltage of 26 kV. The geometry consisted of a wire radius of 0.15 mm, plate-to-plate spacing of 23 cm, and a wire-to-wire spacing of 10 cm. In Figure 4, corona wires are located directly across from the points X = -0.1, 0, and 0.1 m at the plate. Positions x = -0.05 and 0.05 m correspond to positions at the plate, midway between corona wires. The data show both the current density and electric field at the plate to be maximum directly across from a corona wire. Although the degree of uniformity of the electric field and current density distributions will vary for different electrode geometries, the general features will be the same as those of a wire-plate geometry.

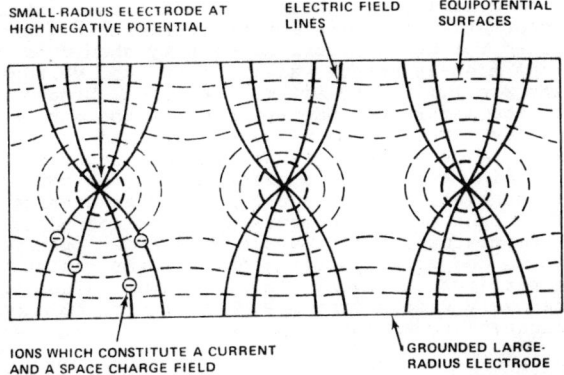

Figure 3-b. Electric field configuration for wire-plate geometry.[3]

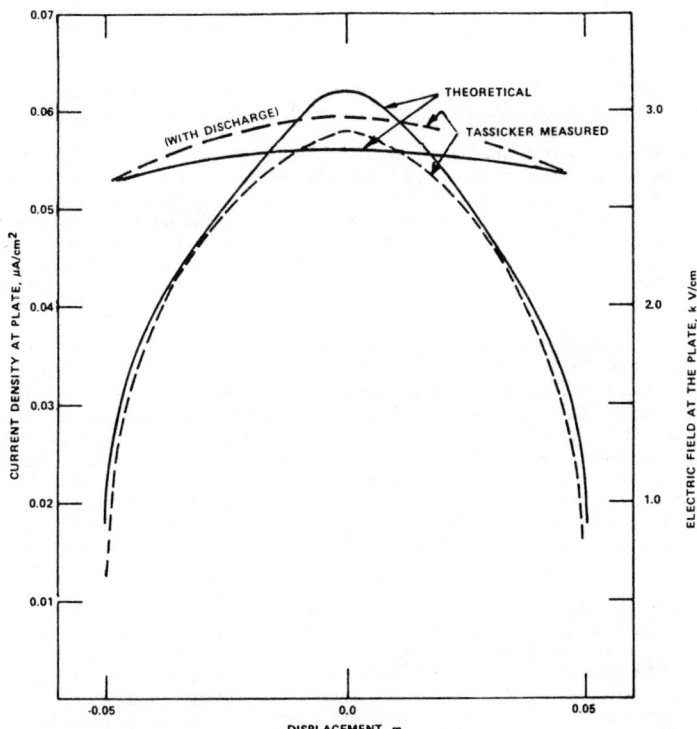

Figure 4. Experimental data showing the dependence of the current density and electric field at the plate.[4]

In order to maximize the collection efficiency obtainable from the electrostatic precipitation process, the highest possible values of applied voltage and current density should be employed. In practice, the highest useful values of applied voltage and current density are limited by either electrical breakdown of the gas throughout the interelectrode space or of the gas in the collected particulate layer. High values of applied voltage and current density are desirable because of their beneficial effect on particle charging and particle transport to the collection electrode. In general, the voltage-current characteristics of a precipitator depend on the geometry of the electrodes, the composition, temperature, and pressure of the gas, the particulate mass loading and size distribution, and the resistivity of the collected particulate layer. Thus, maximum values of voltage and current can vary widely from one precipitator to another and from one application to another.

PARTICLE CHARGING

Once an electric field and current density are established, particle charging can take place. Particle charging is essential to the precipitation process because the electrical force which causes a particle to migrate toward the collection electrode is directly proportional to the charge on the particle. The most significant factors influencing particle charging are particle diameter, applied electric field, current density, and exposure time.

The particle charging process can be attributed mainly to two physical mechanisms, field charging and thermal charging:[5,6,7]

(1) At any instant in time and location in space near a particle, the total electric field is the sum of the electric field due to the charge on the particle and the applied electric field. In the field charging mechanism, molecular ions are visualized as drifting along electric field lines. Those ions moving toward the particle along electric field lines which intersect the particle surface impinge upon the particle surface and place charge on the particle.

Figure 5 depicts the field charging mechanism during the time it is effective in charging a particle.[3] In this mechanism, only a limited portion of the particle surface ($0 \leq \theta < \frac{\pi}{2}$) can suffer an impact with an ion and collisions of ions with other portions of the particle surface are neglected. Field charging takes place very rapidly and terminates when sufficient charge (the saturation charge) is accumulated to repel additional ions. Figure 6-b depicts the electric field configuration once the particle has attained the saturation charge.[3] In this case, the electric field lines are such that the ions move along them around the particle.

Theories based on the mechanism of field charging agree reasonably well with experiments whenever particle diameters exceed about 0.5 μm and the applied electric field is moderate to high. In these theories, the amount of charge accumulated by a particle depends on the particle diameter, applied electric field, ion density, exposure time, ion mobility, and dielectric constant of the particle.

(2) The thermal charging mechanism depends on collisions between particles and ions which have random motion due to their

thermal kinetic energy. In this mechanism, the particle charging rate is determined by the probability of collisions between a particle and ions. If a supply of ions is available, particle charging occurs even in the absence of an applied electric field. Although the charging rate becomes negligible after a long period of time, it never has a zero value as is the case with the field charging mechanism. Charging by this mechanism takes place over the entire surface of the particle and requires a relatively long time to produce a limiting value of charge.

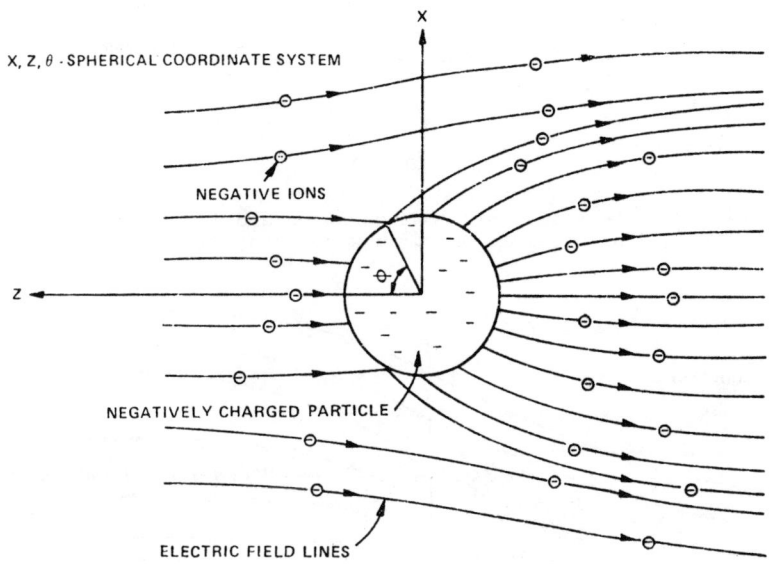

Figure 5. Electric field configuration during field charging.[3]

Figure 6-a depicts the thermal charging process in the absence of an applied electric field.[3] In this case, the ion distribution is uniform around the surface of the particle and each element of surface area has an equal probability of experiencing an ion collision. Thermal charging theories which neglect the effect of the applied electric field adequately describe the charging rate over a fairly broad range of particle sizes where the applied electric field is low or equal to zero. In addition, they work well for particles less than 0.1 μm in diameter regardless of the magnitude of the applied electric field.

Figure 6-b depicts the thermal charging process in the presence of an applied electric field after the particle has attained the saturation charge determined from field charging theory.[3] The effect of the applied electric field is to cause a large increase in ion concentration on one side of the particle while causing only a relatively small decrease on the other side. Although the ion concentration near the surface of the particle becomes very nonuniform, the net effect is to increase the average ion concentration, the probability of collisions between ions and the particle, and the particle charging rate.

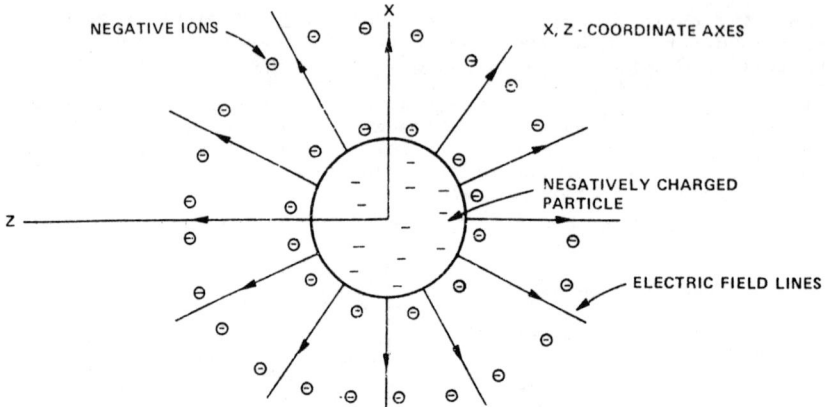

Figure 6-a. Electric field configuration and ion distribution for particle charging with no applied field.³

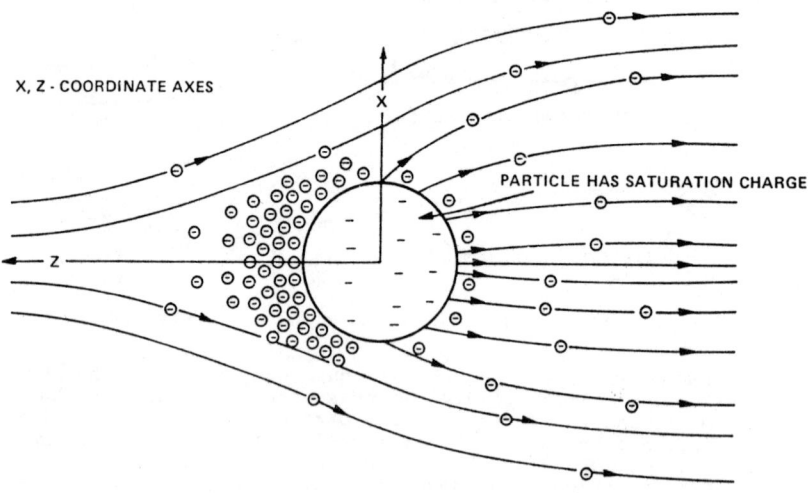

Figure 6-b. Electric field configuration and ion distribution for particle charging in an applied field after saturation charge is reached.³

In thermal charging theories, the amount of charge accumulated by a particle depends on the particle diameter, ion density, mean thermal velocity of the ions, absolute temperature of the gas, particle dielectric constant, residence time, and the applied electric field. The effect of the applied electric field on the thermal charging process must be taken into account for fine particles having diameters between 0.1 and 2.0 µm. Depending most importantly on the applied electric field and to a lesser extent on certain other variables, particles in this size range can acquire values of charge which are 2-3 times larger than that predicted from either the field or the thermal charging theories. For these particles, neither field nor thermal charging predominates and both mechanisms must be taken into account simultaneously.

Figures 7, 8, and 9 contain experimental data[8,9] showing the dependence of particle charge on the variables which are most important in the charging process. These variables are the particle diameter (d), charging electric field strength (E), and ion density-residence time product (Nt). The data were obtained under laboratory conditions using dioctyl phthalate (DOP), polyvinyltoluene latex (PVTL), and polystyrene latex (PSL) particles ranging in diameter from 0.109 to 7 μm. In the data shown here, the particles were charged by positive ions formed in a corona discharge in ambient air. In electrostatic precipitators used to collect fly ash particles, the average values of E and Nt are approximately in the ranges of 1.5-4.5 kV/cm and 0.1-1.0×10^{14} sec/m^3, respectively. The data clearly show that particle charge can be increased by increasing d, E, and Nt. However, for a fixed value of E, increasing Nt beyond a certain value will not result in a significant increase in charge on a particle with a given diameter.

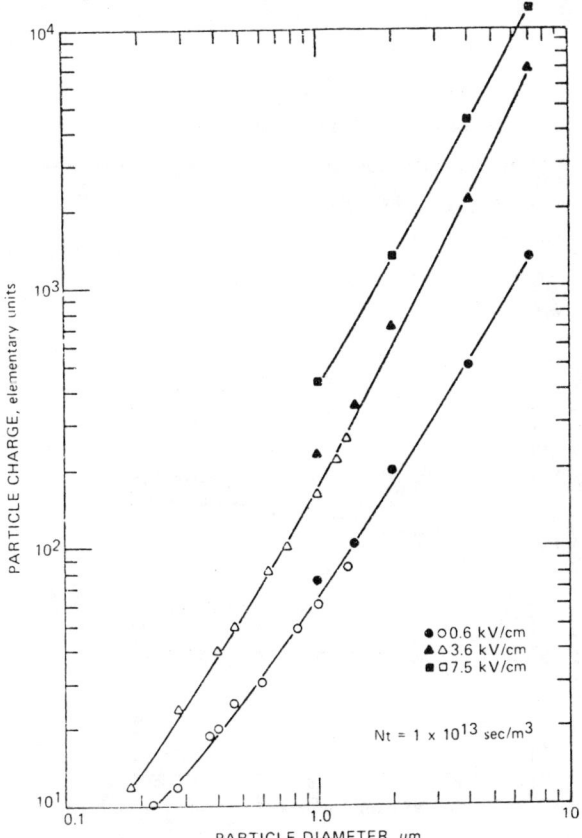

Figure 7. Particle charge vs. dia. for DOP aerosols. The open symbols are Hewitt's (1957) data.[8,9]

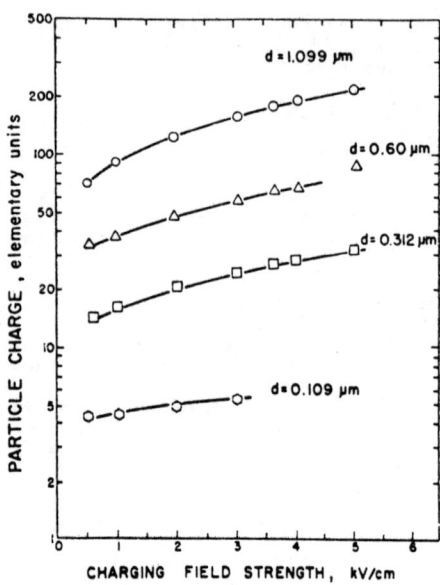

Figure 8. Number of charges per particle vs. the charging field strength for PSL and PVTL particles, with $Nt = 1.5 \times 10^{13}$ sec/m^3.[8,9]

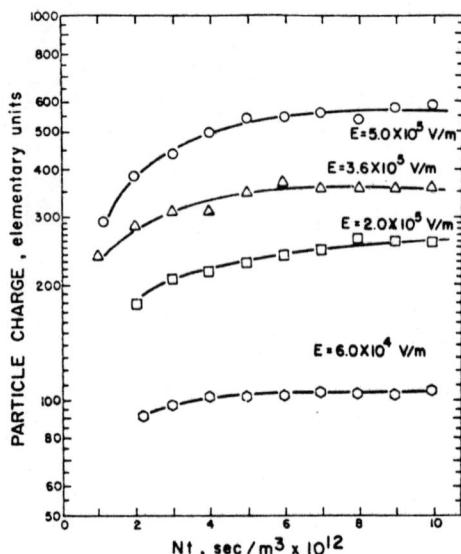

Figure 9. Number of charges per particle vs. the Nt product for a 1.4 μm dia. DOP aerosol. Four different values of the charging field strength were used.[8,9]

In most cases, particle charging has a noticeable effect on the electrical conditions in a precipitator. The introduction of a significant number of fine particles or a heavy concentration of large particles into an electrostatic precipitator significantly influences the voltage-current characteristic. Qualitatively, the effect is seen by an increased voltage for a given current compared to the particle-free situation. As the particles acquire charge, they must carry part of the current but they are much less mobile than the ions. This results in a lower "effective mobility" for the charge carriers and, in order to obtain a given particle-free current, higher voltages must be applied to increase the drift velocities of the charge carriers and the ion densities.

The charged particles, which move very slowly, establish a particulate space charge in the interelectrode space. The distribution of the particulate space charge results in an electric field distribution which adds to the electric fields due to the electrostatic field and the ionic field to give the total electric field distribution. It is important to consider the space charge resulting from particles because of its influence on the electric field distribution, especially the electric field near the collection plate. The electric field at the plate for a given current is higher in the particle containing case than in the particle-free case. The particulate space charge is a function of position along the length of the precipitator since particle charging and collection are a function of length.

PARTICLE COLLECTION

As the particle-laden gas moves through a precipitator, each charged particle has a component of velocity directed towards the collection electrode. This component of velocity is called the electrical drift velocity, or electrical migration velocity, and results from the electrical and viscous drag forces acting upon a suspended charged particle. For particle sizes of practical interest, the time required for a particle to achieve a steady-state value of electrical migration velocity is negligible. Near the collection electrode,[10]

$$w_p = \frac{qE_p C}{6\pi a \mu} \quad , \qquad (1)$$

where w_p = electrical migration velocity near the collection electrode of a particle of radius a (m/sec),

q = charge on particle (coul),

E_p = electric field near the collection electrode (volt/m),

a = particle radius (m),

μ = gas viscosity (kg/m-sec),

C = Cunningham correction factor, or slip correction factor[11] = $(1 + A\lambda/a)$,

where $A = 1.257 + 0.400 \exp(-1.10\, a/\lambda)$, and

λ = mean free path of gas molecules (m).

If the gas flow in a precipitator were laminar, then each charged particle would have a trajectory which could be determined from the velocity of the gas and the electrical migration velocity. In this case, the collection length required for 100% collection of particles with a known migration velocity can be calculated. For cases where turbulence exists, a laminar flow calculation is of interest only from the standpoint that it establishes the best possible collection efficiency for a given collection length.

In industrial precipitators, laminar flow never occurs and, in any collection mechanism, the effect of turbulent gas flow must be considered. The turbulence is due to the complex motion of the gas itself, electric wind effects of the corona, and transfer of momentum to the gas by the movement of the particles. Average gas flow velocities in most cases of practical interest are between 0.6 and 2.0 m/sec. Due to eddy formation, electric wind, and other possible effects, the instantaneous velocity of a small volume of gas surrounding a particle may reach peak values which are much higher than the average gas velocity. In contrast, migration velocities for particles smaller than 0.6 μm in diameter are usually less than 0.3 m/sec. Therefore, the motion of these smaller particles tends to be dominated by the turbulent motion of the gas stream. Under these conditions, the paths taken by the particles are random and the determination of the collection efficiency of a given particle becomes, in effect, the problem of determining the probability that a particle will enter a laminar boundary zone adjacent to the collection electrode in which capture is assured.

Using probability concepts and the statistical nature of the large number of particles in a precipitator, White[12] derived an expression for the collection efficiency in the form

$$\eta = 1 - \exp(-A_p w_p/Q), \qquad (2)$$

where η = collection fraction for a monodisperse aerosol,

A_p = collection area (m^2),

w_p = electrical migration velocity near the collection electrode of the particles in the monodisperse aerosol (m/sec), and

Q = gas volume flow rate (m^3/sec).

The simplifying assumptions on which the derivation of equation (2) is based are:

(1) The gas is flowing in a turbulent pattern at a constant, mean foward-velocity.

(2) Turbulence is small scale (eddies are small compared to the dimensions of the duct), fully developed, and completely random.

(3) The particle electrical migration velocity near the collecting surface is constant for all particles and is small compared with the average gas velocity.

(4) There is an absence of disturbing effects, such as particle reentrainment, back corona, particle agglomeration, or uneven corona. Experimental data[13] under conditions which are consistent with the above assumptions demonstrate that equation (2) adequately describes the collection of monodisperse aerosols in an electrostatic precipitator under certain idealized conditions.

In industrial precipitators, the above assumptions are never completely satisfied but they can be approached closely for fine particles. With proper design, the ratio of the standard deviation of the gas velocity distribution to the average gas velocity can be made to be 0.25 or less so that an essentially uniform, mean forward-velocity would exist. Although turbulence is not generally a completely random process, a theoretical determination of the degree of correlation between successive states of flow and between adjacent regions of the flow pattern is a difficult problem and simple descriptive equations do not presently exist for typical precipitator geometries. At the present, for purposes of discussion, it appears practical and plausible to assume that the turbulence is highly random. The turbulence does not dominate the motion of particles larger than about 10 μm diameter due to their relatively high electrical migration velocities. Under these conditions, equation (2) would be expected to underpredict collection efficiencies. The practical effect in determining precipitator performance will be slight, however, since even equation (2) predicts collection efficiencies greater than 99.6% for 10 μm diameter particles at relatively low values of current density and collection area [i.e., a current density of 10 nA/cm^2 and a collection area to volume flow ratio of 39.4 m^2/(m^3/sec)].

It should be kept in mind that in the real situation the particles inside a precipitator are not uniformly mixed in cross-sections perpendicular to the direction of gas flow and that particle concentration gradients do exist. These concentration gradients are not predicted from equation (2). The concentration profiles for the finer particles will deviate only slightly from a uniform distribution with the deviation increasing with increasing particle diameter. Thus, although equation (2) represents a simple and most times adequate calculational tool for practical purposes, it does not provide for all particle diameters a faithful representation of the physical mechanisms which occur in the precipitation process.

According to equation (2), the collection efficiency for a given particle diameter can be increased by increasing A_p and/or w_p or by decreasing Q. Increasing w_p involves increasing q and/or E_p. In order to increase w_p, the applied voltage and current density must be increased. This increases both q and E_p.

REMOVAL OF COLLECTED MATERIAL

In dry collection, the removal of the precipitated material from the collection plates and subsequent conveyance of the material away from the precipitator represent fundamental steps in the collection process. These steps are fundamental because collected material must be removed from the precipitator and because the buildup of excessively thick layers on the plates must be prevented in order to ensure optimum electrical operating conditions. Material which has been precipitated on the collection plates is usually dislodged by mechanical jarring or vibration of the plates, a process called rapping. The dislodged material falls under the influence of gravity into hoppers located below the plates and is subsequently removed from the precipitator.

The effect of rapping on the collection process is determined primarily by the intensity and frequency of the force applied to the plates. Ideally, the rapping intensity must be large enough to remove a significant fraction of the collected material but not so large as to propel material back into the main gas stream. The rapping frequency must be adjusted so that a larger thickness which is easy to remove and does not significantly degrade the electrical conditions is reached between raps. In practice, the optimum rapping intensity and frequency must be determined by experimentation. With perfect rapping, the sheet of collected material would not reentrain, but would migrate down the collection plate in a stick-slip mode, sticking by the electrical holding forces and slipping when released by the rapping forces.

Section 4

Limiting Factors Affecting Precipitator Performance

ALLOWABLE VOLTAGE AND CURRENT DENSITY

The performance of a precipitator which has good mechanical and structural features will be determined primarily by the electrical operating conditions. Any limitations on applied voltage and current density will be reflected in the optimum collection efficiency which can be obtained. A precipitator should be operated at the highest <u>useful</u> values of applied voltage and current density for the following reasons: (1) high applied voltages produce high electric fields; (2) high electric fields produce high values of the saturation and limiting charge that a particle may obtain; (3) high current densities produce high rates at which particles charge to the saturation or limiting values of charge; (4) high current densities produce an increased electric field near the collection electrode due to the "ionic space charge" contribution to the field; and (5) high values of electric field and particle charge produce high migration velocities and increased transport of particles to the collection electrode.

Electrical conditions in a precipitator are limited by either electrical breakdown of the gas in the interelectrode space or by electrical breakdown of the gas in the collected particulate layer. In a clean-gas, clean-plate environment, gas breakdown can originate at the collection electrode due to surface irregularities and edge effects which result in localized regions of high electric field. If the electric field in the interelectrode space is high enough, the gas breakdown will be evidenced by a spark which propagates across the interelectrode space. The operating applied voltage and current density will be limited by these sparking conditions.

If a particulate layer is deposited on the collection electrode, then the corona current must pass through the particulate layer to the grounded, collection electrode. The voltage drop (V_L) across the particulate layer is

$$V_L = j\rho t, \qquad (3)$$

where j = current density (A/cm^2),

ρ = resistivity of particulate layer (ohm-cm), and

t = thickness of the layer (cm).

The average electric field in the particulate layer (E_L) is given by

$$E_L = j\rho. \qquad (4)$$

The average electric field in the particulate layer can be increased to the point that the gas in the interstitial space breaks down electrically. This breakdown results from the acceleration of free electrons to ionization velocity to produce an avalanche condition similar to that at the corona electrode. When this breakdown occurs, one of two possible situations will ensue. If the electrical resistivity of the particulate layer is moderate (~ 0.1-1.0×10^{11} ohm-cm), then the applied voltage may be sufficiently high so that a spark will propagate across the interelectrode space. The rate of sparking for a given precipitator geometry will determine the operating electrical conditions in such a circumstance. If the electrical resistivity of the particulate layer is high ($>10^{11}$ ohm-cm), then the applied voltage may not be high enough to cause a spark to propagate across the interelectrode space. In this case, the particulate layer will be continuously broken down electrically and will discharge positive ions into the interelectrode space. This condition is called back corona. The effect of these positive ions is to reduce the amount of negative charge on a particle due to bipolar charging and reduce the electric field associated with the "ionic space charge". Both the magnitude of particle charge and rate of particle charging are affected by back corona. Useful precipitator current is therefore limited to values which occur prior to electrical breakdown whether the breakdown occurs as sparkover or back corona.

Field experience shows that current densities for cold side precipitators are limited to approximately 50-70 nA/cm² due to electrical breakdown of the gases in the interelectrode space. Consequently, this constitutes a current limit under conditions where breakdown of the particulate layer does not occur.

Electrical breakdown of the particulate layer has been studied extensively by Penney and Craig[14] and Pottinger[15] and can be influenced by many factors. Experimental measurements show that particulate layers experience electrical breakdown at average electric field strengths across the layers of approximately 5 kV/cm. Since it takes an electric field strength of approximately 30 kV/cm to cause electrical breakdown of air, this suggests that high localized fields exist in the particulate layer and produce the breakdown of the gas in the layer. The presence of dielectric or conducting particles can cause localized regions of high electric field which constitute a negligible contribution to the average electric field across the layer. The size distribution of the collected particles also influences the electrical breakdown strength by changing the volume of interstices.[16] It has also been found that breakdown strength varies with particulate resistivity with the higher breakdown strength being associated with the higher resistivity.

NONIDEAL EFFECTS

The nonidealities which exist in full-scale electrostatic precipitators will reduce the ideal collection efficiency that may be achieved with a given specific collection area. The nonideal effects of major importance are (1) nonuniform gas velocity distribution, (2) gas sneakage, and (3) particle reentrainment. These nonideal effects must be minimized by proper design and optimization of a precipitator in order to avoid serious degradation in performance.

Nonuniform Gas Velocity Distribution

Uniform, low-turbulence gas flow is essential for optimum

precipitator performance. Nonuniform gas flow through a precipitator lowers performance due to two effects. First, due to the exponential nature of the collection mechanism, it can be shown mathematically that uneven treatment of the gas lowers collection efficiency in the high velocity zones to an extent not compensated for in the low velocity zones. Secondly, high velocity regions near collection plates and in hopper areas can sweep particles back into the main gas stream.

Although it is known that a poor gas velocity distribution results in reduced collection efficiency, it is difficult to formulate a mathematical description for gas flow quality. White[17] discusses nonuniform gas flow and suggests corrective actions. Preszler and Lajos[18] assign a figure-of-merit based upon the relative kinetic energy of the actual velocity distribution compared to the kinetic energy of the average velocity. This figure-of-merit provides a measure of how difficult it may be to rectify the velocity distribution but not necessarily a measure of how much the precipitator performance would be degraded. At the inlet of a precipitator, a value of 0.25 or less for the ratio of the standard deviation of the gas velocity distribution to the average gas velocity is generally recommended. However, it must be noted that the gas velocity distribution can change significantly throughout the length of a precipitator and, depending upon the design of the precipitator and the manner in which it is interfaced with other plant equipment, the gas velocity distribution may improve or degrade along the length of a precipitator.

Gas Sneakage

Gas sneakage occurs when gas bypasses the electrified regions of an electrostatic precipitator by flowing through the hoppers or through the high voltage insulation space. Gas sneakage can be reduced by the use of frequent baffles which force the gas to return to the main gas passages between the collection plates. If there were no baffles, the percent gas sneakage would establish the maximum possible collection efficiency because it would be the percent volume having zero collection efficiency. With baffles, the sneakage gas remixes with part of the main gas flow and then another fraction of the main gas flow re-bypasses in the next unbaffled region. The upper limit on collection efficiency due to gas sneakage will therefore depend on the amount of sneakage gas per baffled section, the degree of remixing, and the number of baffled sections. Gas sneakage becomes increasingly important for precipitators designed for high collection efficiencies where only a small amount of gas sneakage per section can result in a severe limitation on collection efficiency.

Particle Reentrainment

Particle reentrainment occurs when collected material reenters the main gas stream. This can be caused by several different effects and, in certain cases, can severely reduce the collection efficiency of a precipitator. Causes of particle reentrainment include (1) rapping which propels collected material into the interelectrode space, (2) the action of the flowing gas stream on the collected particulate layer, (3) sweepage of material from hoppers due to poor gas flow conditions, air inleakage into the hoppers, failure to empty hoppers when required, or the boiling effect of rapped material falling into the hoppers, and (4) excessive sparking which dislodges collected material by electrical impulses and disruptions in the current which is necessary to provide the electrical force which holds the material to the collection plates.

Recent studies[19,20] have been made to determine the effect of particle reentrainment on precipitator performance. In studies where the rappers were not employed, real-time measurements of outlet emissions at some installations showed that significant reentrainment of mass was occurring due to factors other than rapping. These studies also showed that for high-efficiency, full-scale precipitators approximately 30-85% of the outlet particulate emissions could be attributed to rapping reentrainment. The results of these studies show that particle reentrainment is a significant factor in limiting precipitator performance.

Section 5

Use of Electrostatic Precipitators for the Collection of Fly Ash

REASONS FOR USING ELECTROSTATIC PRECIPITATORS TO COLLECT FLY ASH[21]

In 1975 utilities burned over 3.63×10^{11} kg (400 million tons) of coal which would produce about 2.72×10^{10} kg (30 million tons) of fly ash annually, assuming an average ash content of 10% and an average ash retention of 30% in the furnace. An illustration of the magnitude of the ash problem can be shown by the output of one 600 MW power plant which typically exhausts 7×10^4 am^3/min (2.5 million acfm) of flue gas. With a typical ash loading of 5 grains/scf at the air preheater outlet, the ash emitted could be about 7.3×10^5 kg (800 tons) per day. To achieve an efficiency of collection of at least 99% and to dispose of almost 7.3×10^5 kg (800 tons) per day of fly ash is very demanding of a collection system. The best reasons for using electrostatic precipitators for the gas cleaning problem described above are:

(1) Electrostatic precipitators can be designed to provide high collection efficiency for all sizes of particles from submicroscopic to the largest present in the gas stream.

(2) They are economical in operation because of low internal power requirements and inherently low draft loss. Gas pressure drop through a precipitator may be of the order of 2-3 cm of water or less as compared with pressures of 8-36 cm of water for filters and 25-254 cm of water for scrubbers.[22]

(3) They can treat very large gas flows.

(4) They are very flexible in gas temperatures used, ranging in the power field from as low as 93°C (200°F) to as high as 427°C (800°F).

(5) They have long useful life.

Today a total of more than 1300 fly ash electrostatic precipitator installations having a rated gas flow of over 1.4×10^7 am^3/min (500 million acfm) have been made in the United States. Future expansion of the power industry due to ever greater energy consumption by the public and an increased dependence on coal as the major energy source for power production are factors favoring continued growth of fly ash precipitation.

DESIGN OF PRECIPITATORS USED TO COLLECT FLY ASH

General Description

The design of an electrostatic precipitator for a particular installation involves many parameters that can influence both cost and performance. The most significant variables (besides fly ash character) involved in the design are:[23]

    Area and type of collection electrodes,
    Dimensions of the precipitator shell,
    Size, spacing, and type of discharge electrodes,
    Size and type of power supply units,
    Degree of sectionalization,
    Layout of the precipitator in accordance with physical space limitations,
    Design of the gas handling system,
    Size and shape of the hoppers,
    Type and number of electrode rappers,
    Type of ash removal equipment.

Stringent air pollution control standards require low stack emissions. Also, new regulations have enforcement provisions which can curtail or even shut down entire production units in order to comply with emission standards. Optimum precipitator design, therefore, is of paramount importance for economic reasons as well as aesthetic and health reasons. Some of the factors involved in designing electrostatic precipitators are described in detail below to allow an understanding of the importance of each factor in the total design. Different manufacturers sometimes have different recommendations as to the type of discharge electrode, power supply, rappers, etc. to use in a fly ash precipitator, so an attempt has been made to discuss and show different types of each component.

Precipitator Shell[24,22]

The purpose of the shell is to confine the gas flow for proper exposure to the electrodes, to avoid excessive heat loss, and to provide structural support for the electrodes and rapping equipment. The shell is normally rectangular, where plate electrodes are used, or cylindrical if tube electrodes are used. Cylindrical shells may also be used with plate-type precipitators where relatively high or low gas pressures are encountered. Shell material is usually steel, but because of some corrosion problems, it may be lined with or made of tile, brick, concrete, or special corrosion-resistant steels. Insulation is usually required to maintain the shell at a temperature above the dew point if the gases contain corrosive materials. Access doors and stairways and safety provisions are provided as auxiliary equipment.

Gas diffuser plates are sometimes provided as part of the shell in order to improve gas flow. Roof, wall, and hopper baffles are used to minimize the amount of gas which may by-pass the electrodes. Hopper baffles generally extend below the dust level in the hopper to provide a seal and keep gas from flowing through the hopper. Design of diffuser plates and baffles will be covered in more detail in the section on gas flow.

Cross bracing is generally provided by diagonal members across the inlet and/or outlet of the shell. Horizonal structural members are built-up trusses, box beams, or various other configurations which vary with manufacturer. The structural members must be capable of supporting the electrodes and maintaining them in alignment over the range of temperatures and external load conditions encountered in operation. Some manufacturers contend that thermal expansion of the shell constitutes a major alignment problem unless provision is made to allow for expansion. Otherwise, buckling of the shell and subsequent distortion of the electrode system can occur. Since the assembly occurs at ambient temperature, expansion stresses of the structural members will obviously occur as the shell is heated to flue gas temperatures. One method utilized to overcome the expansion problem is to provide bearings at the base of the support columns to permit the shell to move without buckling. Figure 10 shows the details of such a bearing which generally is just two flat plates without lubrication.[24] Many installations have been built without such bearings, the claim being that expansion takes place uniformly and that distortion due to thermal expansion is inconsequential. However, expansion is generally more of a problem for hot-side units than for cold-side units.

There are other causes of shell distortion, principally inadequate foundations. When this occurs, electrode spacings can change from the designed four to five inches to two inches or less. Such a shift in spacing limits the operating voltage and seriously impairs precipitator performance.

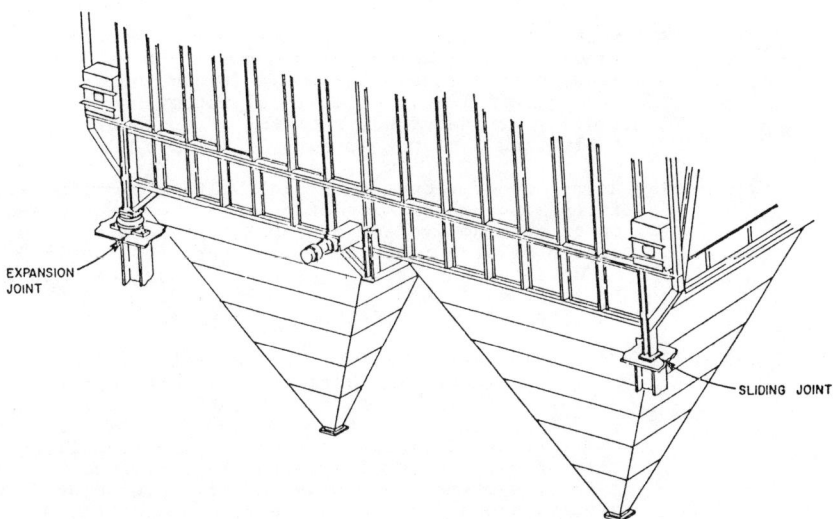

Figure 10. Illustration of thermal expansion bearing surface for precipitator installation.[24]

Electrical Sections

Electrical Energization--

Historical development[25]--The purpose of high-voltage equipment in electrostatic precipitation is to provide the intense electric fields and corona currents needed for particle charging and collection. In addition, the electrical sets must be highly stable in operation and have useful operating lives of twenty years or more. Proper voltage waveform, protection and stability against precipitator sparkover, proper voltage and current output ratings, and sturdy electrical and mechanical design are necessary requirements of the equipment. Automatic control of rectifier output is essential for most fly ash precipitators because of changing load conditions and characteristics of the flue gas and of the fly ash produced by boilers which are operated at varying loads and fuel conditions.

Development of high voltage equipment to energize precipitators has been an evolutionary process. Mechanical rectifier sets in the earliest precipitators were succeeded by more reliable and long lasting tube rectifier sets. Solid-state rectifiers have made the mechanical type virtually obsolete. Selenium solid-state rectifiers provide reliable service and long life but are subject to damage from high temperatures. Universal adoption of the solid-state rectifier began with the recent development of the silicon type. To maintain optimum energization levels, modern equipment uses silicon diode rectifiers, oil or askerel filled high-voltage transformers, thyristor control elements, and automatic feedback control.

A good summary of the general periods of development of the various components of high voltage electrical equipment is given by Hall:[26]

1906 - 1950 Mechanical rectifiers, generally with simple rheostat manual control; low power 250 mA dc sets, either double half-wave (beginning in 1932) or single full-wave electrical output; generally small size individual sections (four 10,000 ft^2 collecting area per set).

1950 - 1960 Major use of vacuum tube high voltage rectifiers; increasing use of automatic voltage control based on an optimum average precipitator sparking rate; growth of high power rectifier sets to 1000-1500 mA dc sizes and use of very large sections energized by individual sets. First silicon diode rectifier set designed in 1955-1956 and applied in 1958.

1960 - 1970 Commercial use of modern, solid-state silicon-controlled rectifier automatic voltage control system with linear reactor in 1965; universal use of silicon diode high voltage rectifiers; application of linear reactors to the stabilization of certain unreliable saturable reactor control systems (1963); use of high pressure cleaned process gas as dielectric medium for high voltage transformer design to eliminate high pressure feed through bushings; development of more sophisticated automatic voltage control techniques using fast computer-type logic circuitry and printed circuit boards capable of stable rectifier set operation at threshold or very low sparking rates over a wide range of loads.

Power supplies[22,27]--Each power supply consists of four components as shown in Figure 11: a step-up transformer, a high voltage rectifier, a control element, and a sensor for the control system.[27] The step-up transformer increases the voltage from the line voltage to that required by the precipitator. The high voltage rectifier converts the high voltage ac power to dc to be compatible with precipitator requirements.

One function of the control system is to vary the amplitude

of the dc voltage applied to the electrode system. This control
is usually located on the primary or low voltage side of the transformer. The control system can be operated either manually or in
one of several automatic modes, but automatic systems are typically
installed in modern installations. A well-designed automatic control system serves to maintain the voltage level at the optimum
value, even when the dust characteristics and concentration
fluctuate.

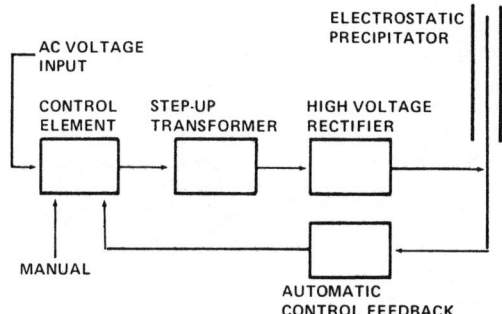

Figure 11. Power supply system for modern precipitators.[27]

High voltage rectifiers--The rectifiers change the ac to dc,
either full-or half-wave. In general, half-wave power supplies
allow a greater degree of sectionalization. Full-wave may be used
in situations where large dust loading or extremely fine particles
lead to a large space charge which limits the maximum current.

Spark-rate--The spark-rate is the number of times per minute
that electrical breakdown occurs between the corona wire and the
collection electrode. A spark-rate controller establishes the
applied voltage at a point where a fixed number of sparks per
minute occur (typically 50-150 per corona section). As the spark-
rate increases, a greater percentage of the input power is wasted.
One commonly used type of control device utilizes spark-rate as
the primary control. Another type of control circuit utilizes a
thyristor control element.[22,28] An explanation of automatic SCR
voltage control is given by Piulle[29] and a new precipitator voltage control using analog electronic networks is described by
Gelfand.[30]

Design and operating requirements--Table 1 summarizes the
design and operating requirements for modern high voltage electrical equipment of the conventional type.[26] Figure 12 shows
the schematic circuit diagram of a modern high voltage rectifier
set with SCR (silicon-controlled-rectifier) type automatic control.[26]
Multiple signal feedback loops are provided to obtain good regulation and fast response to transient spark disturbances. The
linear inductance reactor, although sometimes omitted because of
its added cost, is nevertheless an important factor in obtaining
good current waveform control and ability to operate the rectifier
set at or near rated output. A properly designed linear inductor
also eliminates spark bursting and arcing tendencies which contribute to instability and can also cause corona wire burning.
Metering should include kilovolt meters, milliammeters, and spark-
rate meters.

TABLE 1. DESIGN AND OPERATING REQUIREMENTS FOR MODERN HV ELECTRICAL EQUIPMENT IN ELECTROSTATIC PRECIPITATION[26]

Item	Specification
1. Of first importance	High reliability & stability under transient sparking conditions and occasional short-circuit load.
2. Precipitator operating voltage, kV_p	30-100+ (40-65 kV_p most common)
3. Precipitator current density, $mA/1000\ ft^2$	10-100+
4. Precipitator voltage waveform	Pulsating, negative polarity full-wave or double half-wave
5. Precipitator load	Capacitive - 0.02 to 0.125 μFD/section
6. Line input	460/480 V, 1 Ph, 60 Hz most common variation $\pm$ 5% line voltage
7. Rectifier circuit, standard sets	Single phase, FW bridge, silicon diodes
8. Rated output voltage, R load	70 kV peak, 45 kV dc average - most common, 105 kV peak, 67.5 kV dc average
9. HV transformer	400 V/53 kV rms or 400 V/78 kV rms
10. Individual set capacity, kVa	15 to 100
11. Rated dc output current, mA	250 to 1500+ per set (R load)
12. Transformer-rectifier insulation	Oil/askarel convection cooling
13. Duty	Continuous - outdoor or indoor installation
14. Ambient temperature	50°C max for TR oil-filled tank 55°C max for control cabinet
15. Voltage control	Automatic control essential based either on optimum avg sparking rate (adjustable), or nominally at spark threshold
16. Voltage control range	Essentially zero to 100% rated output. Modern systems use SCR type control with linear reactor in HV transformer primary.
17. Current limit - no sparking	Automatic limit at rated primary current. Full rated current abailability independent of voltage
18. Peak current limit during sparking	2 to 2.5 times normal peak current in the best systems

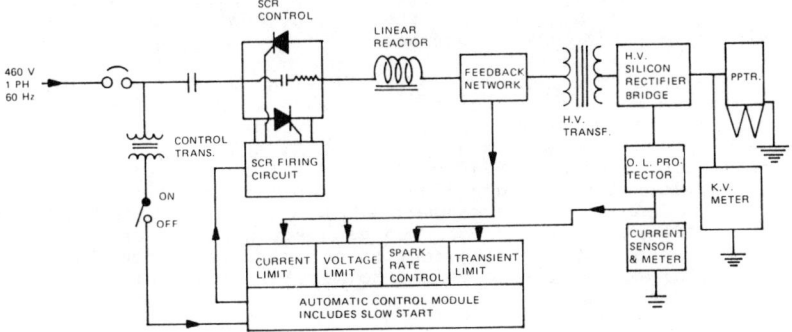

Figure 12. Schematic diagram - modern HV rectifier set with SCR type automatic control for electrostatic precipitators.[26]

Sources of high voltage electrical equipment[26]--The sources of high voltage electrical equipment for precipitators are somewhat limited. The following categories summarize the locations of these sources:

In-house - all electrical equipment designed and made by the precipitator supplier. These include Research-Cottrell, Inc. and CE-Walther (vis Helena Corp.). Buell is also essentially in this category since only the HV transformer core and coil are purchased.

Hybrid - purchase of HV transformer (usually General Electric or Westinghouse) to specification with the control unit being made to one's own design in-house or at a separate local company.

Industrial - HV transformer and controls made by General Electric or Westinghouse.

Commercial - several suppliers of ordinary high voltage power supplies offer equipment for electrostatic precipitation.

Specialty Companies - very few companies specialize in selling high voltage equipment and controls for electrostatic precipitators. An example of one which does is Environecs in Costa Mesa, California.

Typical oil-filled transformers weigh 1090 kg (2400 lbs) at 16 KVA to about 1816 kg (4000 lbs) at 100 KVA. Askarel nonflammable fluids increase the weight about 454 kg (1000 lbs). Modern control cabinets are typically about 454 kg (1000 lbs) or less, front access.

Discharge Electrode System--

The discharge electrode system is designed in conjunction with the collection electrode system to maximize the electric current and field strength. The discharge electrode is also referred to as the corona electrode, cathode, high voltage electrode, or corona wire. The shape and size of the discharge electrodes are governed by the corona current and mechanical requirements of the system. Where high concentrations of fine dusts are encountered, space charge limits the current flow, especially in the inlet sections. In such cases, special electrodes which give higher currents may be used to achieve a high power density within the inlet sections. Variation in the current flow and electric field within limits is possible by controlling the type and size of the discharge electrode.

Geometries of discharge electrodes--The shape of the electrodes may be in the form of cylindrical or square wires, barbed wire, or stamped from formed strips of metal of various shapes. Some discharge electrode geometries are shown in Figure 13.[27] A "square twisted" wire is usually 0.48 cm (3/16 inch) or 0.64 cm (1/4 inch) square and is twisted longitudinally to help straighten the rod and to increase the length of the sharp edge, which increases the corona current. "Spiral" wires are formed as a spring and then pulled for installation. The spring tension helps restrict the lateral motion. "Barbed" wires are merely commercial grade barbed wire.

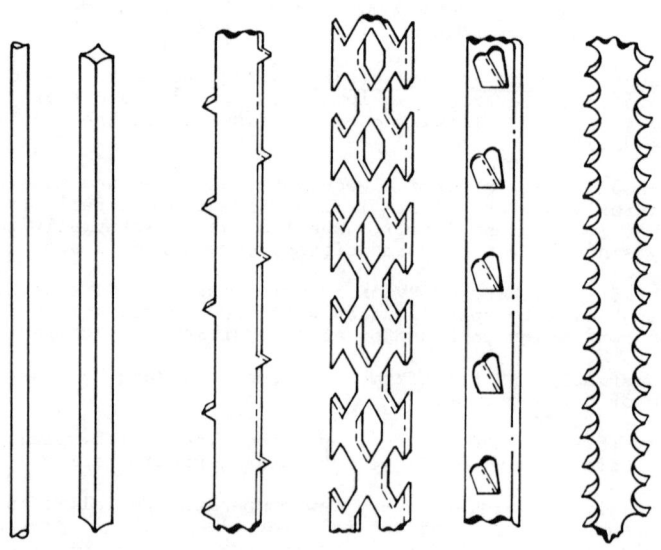

Figure 13. Typical forms of discharge or corona electrodes.[27]

A wire considered basic, though not restricted to, the "European" design explained below has a star-shaped cross section as shown in Figure 14-a.[31] A wire known as Isodyn wire is described by Engelbrecht[31] as being advantageous on a rigid frame system when a higher current at a given voltage is desired compared to the star-shaped discharge wire (Figure 14-b). As stated earlier, mechanical and electrical requirements usually determine the shape and size of the discharge electrodes.

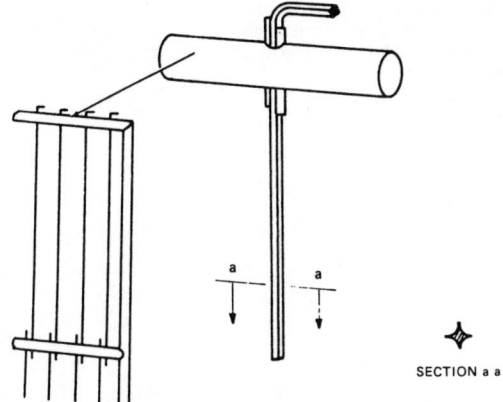

Figure 14-a. Rigid discharge electrode star wire.[31]

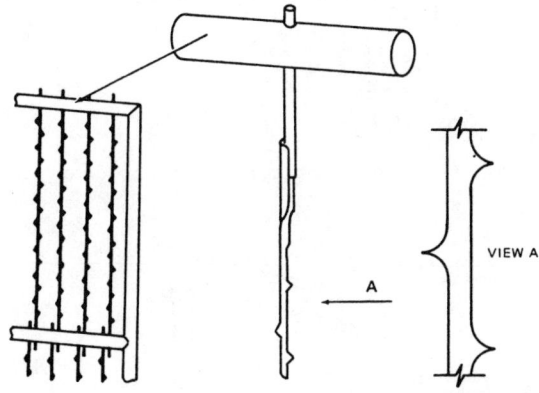

Figure 14-b. Rigid discharge electrode isodyn wire.[31]

Types of discharge electrodes--Various types of discharge electrodes are used in electrostatic precipitators, but one of the major differences between manufacturers is in the method of supporting the discharge electrodes. One approach, typical of European practice, is to provide a frame or tubular support for the electrodes. The other approach which has been used by most American manufacturers is to suspend the electrodes from a support and maintain them in position by weights and guides at the bottom.

A typical weighted-wire electrode is illustrated in Figure 15,[32] and a complete weighted-wire electrode system illustrating the method of fastening the wire at the top and maintaining the wire in place by bottom weight guides is shown in Figure 16.[33] There is considerable variation among manufacturers as to the method of supporting the discharge wire from the support frame. Since the discharge wires tend to move under the influence of both aerodynamic and electrical forces, mechanical fatigue failure can occur. Various methods of allowing some movement of the support have been attempted to minimize the fatigue problem. Wires are also subjected to localized sparking in the regions of high field strength and shrouds are sometimes used to give a larger diameter, and hence low field strength, in critical regions near the ends of the electrodes.

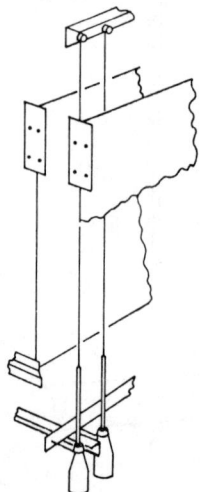

Figure 15. Weighted wire corona electrodes.

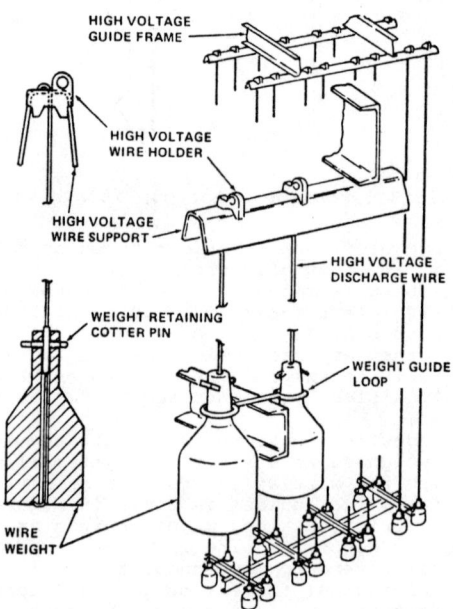

Figure 16. Example of weighted wire electrode system.[33]

Various types of the "European" discharge wire system or rigid discharge electrode system are available, and a classification may be attempted based on the following criteria:[31] (1) two dimensional frames with rigid discharge wires (Figure 17), (2) three dimensional frames with discharge wires strung between horizontal supports (Figure 18), (3) discharge wires supported off masts (Figure 19), and (4) self-supporting rigid discharge electrodes (Figure 20).

Rigid discharge electrode systems are now being offered by manufacturers previously identified only with the weighted-wire design. The rigid or supported wire electrode system has the advantage of minimizing the wire breakage problem since the electrodes are supported by rigid members and remain in position and energized even if breakage of the electrode occurs.

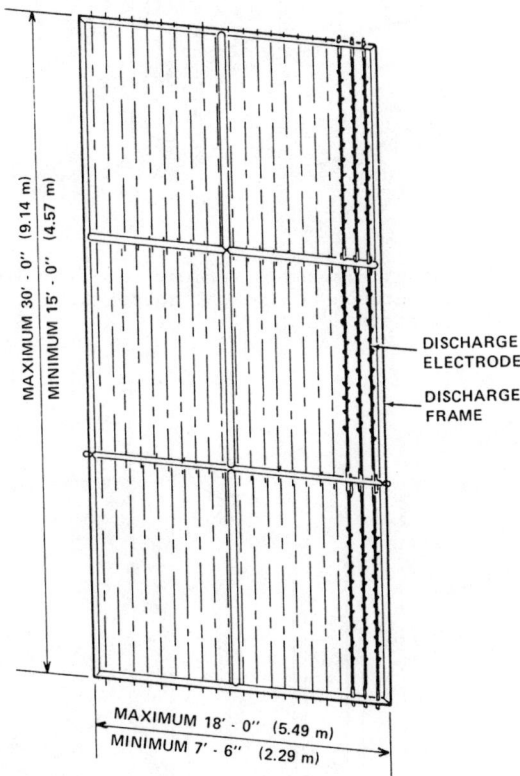

Figure 17. "European" discharge wire system with rigid discharge wires on a two dimensional frame.[31]

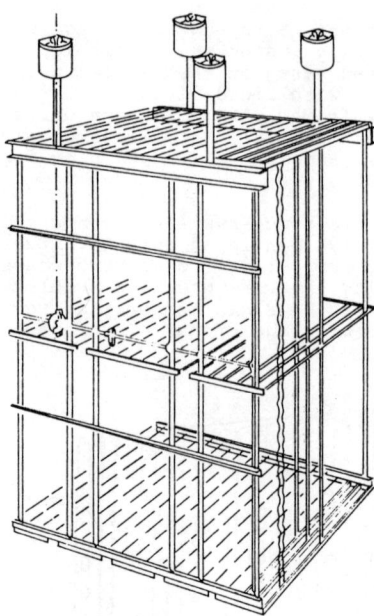

Figure 18. "European" discharge wire system with discharge wires strung between horizontal supports on a three dimensional frame.[31]

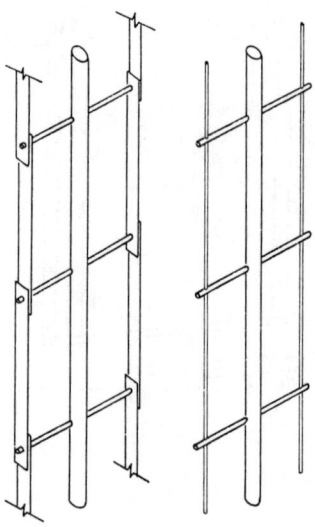

Figure 19. "European" discharge wire system with discharge wires supported off a mast.[31]

Use of Electrostatic Precipitators   35

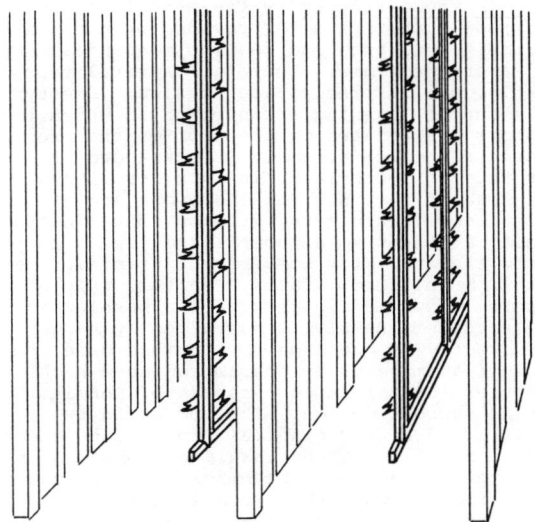

Figure 20. "European" discharge wire system with self-supporting rigid discharge electrodes.[31]

New designs--Research-Cottrell has developed a new rigid discharge electrode, called the Dura-Trode, which is said to offer longer life and better performance than earlier weighted-wire or rigid-type electrodes.[34] In cross section, the system is similar to a hollow airfoil. Corona discharge is delivered primarily by thin scalloped blades at leading and trailing edges. Research Cottrell estimates that maintenance costs are projected at zero over the unit's 30 year plus lifetime. Units have been tested successfully at five coal-fired generating stations and four industrial plants since 1975. Figure 21 shows the Dura-Trode rigid-type electrode.

Figure 21. Unitized Dura-trade rigid-type electrode.[34]

Discharge electrode support[22]--The main functions of the discharge electrode support are to provide the necessary high voltage electrical insulation and to give mechanical support to the discharge electrode frame. There are several types of support systems currently used in precipitator design. One type, shown in Figure 22, is a support bushing arrangement in which the high voltage insulators are located on the roof of the precipitator, and the discharge electrode assembly is suspended from the bus beam by hanger rods.[35] Porcelain pin-type insulators support the mechanical load of the internal framework and are located in a relatively low temperature zone with low contamination. These bushings are not gas tight so a common practice is to provide a flow of air into the insulator compartment to prevent entrance of dust-laden air from the precipitator. Another type of discharge electrode support, shown in Figure 23, is a bushing arrangement in which the electrode assembly is suspended by hanger rods which are supported directly by bushings.[35] In this case, the bushings are constructed of alumina or Pyroceram and have higher mechanical strength and better thermal shock resistance, permitting a much simpler electrode support design. The low porosity of the insulation materials and better gas seal provided by the gasket minimize the gas inleakage to the insulator compartment. However, for some applications, the bushings are continuously purged with air, either induced when the precipitator is under suction or forced by blowers. The bushings are housed in either individual roof tunnels or in a common housing on top of the precipitator.

Other types of electrode supports are in use and the type varies between manufacturers. For some applications, the flue gases are near the dew point and condensation on cooler parts of the insulator will cause localized arcing. When arcing occurs, a low resistance path can be formed which will partially short circuit the power supply or the heat generated by the arc can fracture the insulator. When such conditions can occur, special precautions must be taken to heat the insulators to prevent condensation of moisture or acid.

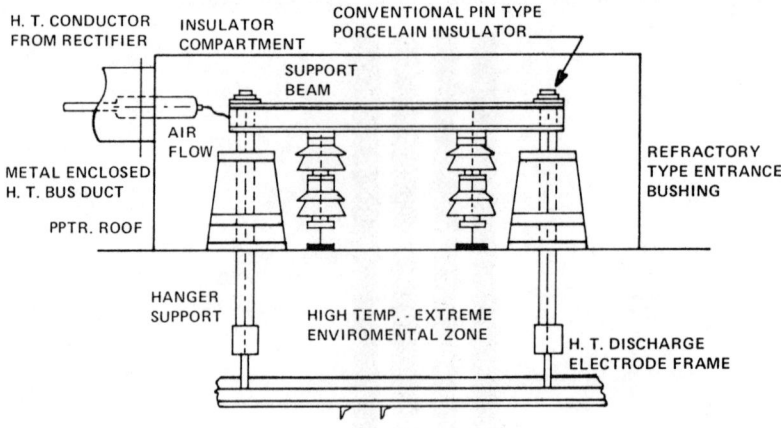

Figure 22. Example of high-temperature support bushings.[35]

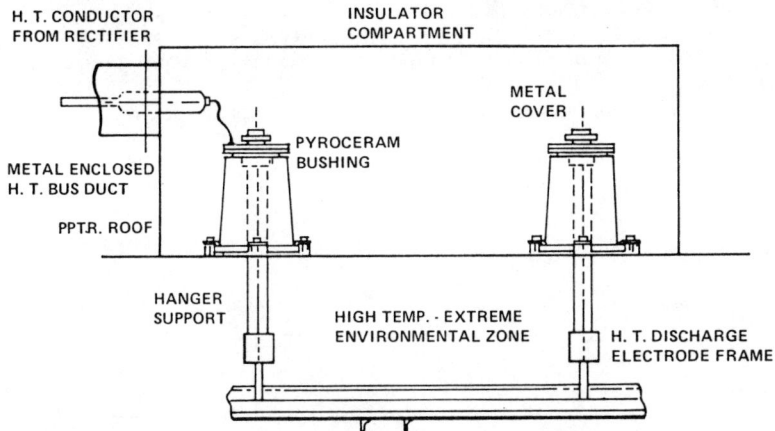

Figure 23. Example of high-temperature support bushings.[35]

Collecting Electrode System--

The collecting electrodes are the individual grounded surfaces on which particulate matter is collected. Generally, in spite of the many elaborate concepts given in the patent literature, most collection electrodes are simple configurations, the main considerations being stiffness of the collection plates and shielding of the collected dust layer to prevent reentrainment. An additional requirement is that the edge of the collecting plates be free of sharp edges or protrusions which can provide localized high field regions, resulting in sparking at low voltage. If welded structures are utilized, the weld must be smooth to minimize localized sparking. These considerations are especially important in the wire-and weight-type discharge electrode since the wire extends beyond the edge of the collection plate. A further requirement of the collection electrode is that the rapping impact should be transmitted to all parts of the plate as uniformly as possible to facilitate uniform dust removal. The plates should be heavy enough to prevent damage due to rapping, especially where high impact rapping is used.

Geometries of collecting electrodes--A few types of collecting electrodes, representative of those most often used by precipitator manufacturers today, are illustrated in Figure 24.[23] This list is by no means exhaustive since the patent literature contains numerous other electrode types which have been designed to shield the collected dust and minimize reentrainment. Many of these are unacceptable because of excessive weight or cost, or because of high reentrainment losses.

The shielded flat plate collecting electrode, used chiefly in horizontal flow, dust-type precipitators, is the most popular in present day use in the United States. In order to shield the precipitated dust from the gas passing across the plate, baffles are mounted along the plate. The baffles are fabricated as formed shapes and welded to the ends and surfaces of the collecting plate. Baffle shapes vary from flat strips perpendicular to the collecting surface to aerodynamic designs to minimize gas turbulence. The size of the collecting electrodes in an electrical section ranges from twenty feet to fifty feet in height, from three feet to twelve feet in the direction of gas flow, and is usually about 18 gauge thickness.

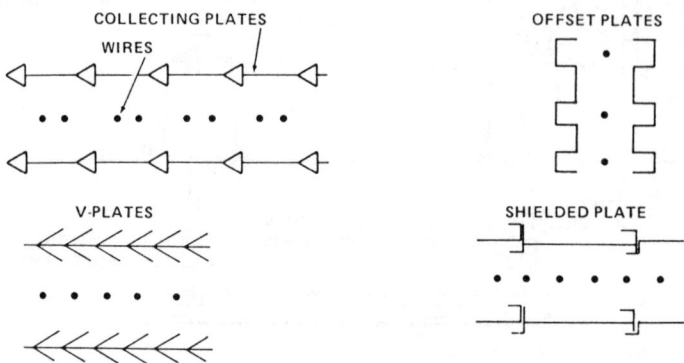

Figure 24. Various types of collection electrodes.[23]

Offset plates are made by bending a flat sheet into a square or angular zig-zag or a corrugated pattern. The dust precipitated in the troughs is shielded from the main gas stream to minimize reentrainment. The plates are usually from ten to thirty feet in height and from three to nine feet in the direction of gas flow. Two design variations of this arrangement used by Wheelabrator Lurgi are shown in Figure 25.[36]

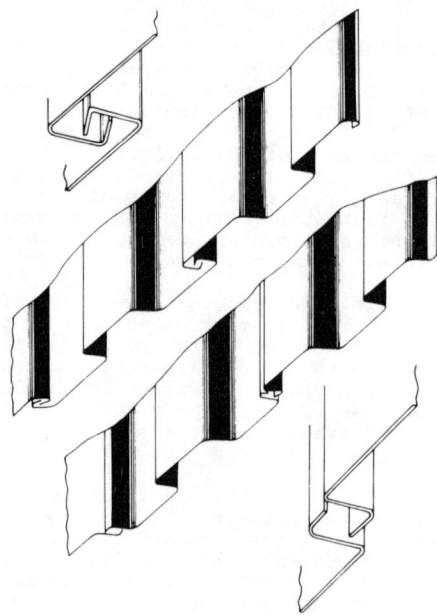

Figure 25. Exclusive Wheelabrator Lurgi collecting electrodes. The CSW, with single overlap, and the double overlap CSH design.[36]

A vee plate is a composite assembly of metal strips bent into the shape of a vee or chevron. The vees are spaced about 3.18 cm (1 1/4 inch) apart for the full length of the plate which is about two inches thick. The points of the vees face upstream and the spaces between the vee members act as quiescent zones in which the dust is precipitated with minimum reentrainment. The collecting plates in use today range from about eighteen to thirty-five feet in height, and an individual plate is usually three feet in the direction of gas flow. Two plates are customarily fastened together in order to make up a six foot section.

Ash Removal Designs

General[37]--

Once fly ash has been collected on the collecting electrode, it must be removed to a hopper or storage facility, not only to remove the material from the precipitator per se, but also to maintain optimum electrical conditions in the precipitation zones. The deposits are dislodged by mechanical impulses or vibrations of the electrodes, a process known as rapping. Many of the problems associated with poor electrostatic precipitator performance can be related directly to degradation in rapper system performance. Because of the complex nature of the dust removal mechanics in an electrostatic precipitator, a number of factors should be considered when evaluating rapper system problems. These factors are related not only to hardward quality and manufacture, but also to charging process conditions, maintenance procedures, and initial application of the rapping hardware. Hardware malfunctions have been a problem in the past. Therefore, the latest technologies available in solid-state electronics are being incorporated into system designs to provide continuous on-line monitoring.

Rappers[22]--

Depending on individual vendor philosophy, rapping impulses are provided by either single impact or vibratory-type rappers. These in turn are activated either electrically or pneumatically, using accelerated or gravitational type impacts. Some commonly used methods of dry removal of fly ash from collecting and discharge electrodes are discussed below.

Single impact rapper (electromagnetic solenoid) - electromagnetic solenoid rappers consist of a plunger which is lifted by energizing the solenoid. On release of the plunger by deenergizing the coil, it falls under the influence of gravity against an anvil which transmits the rap through a rod to the electrodes to be cleaned (Figure 26).[38] Solenoid-type rappers are used for both discharge and collecting electrode cleaning and are usually located on top of the precipitator. Solenoid rappers can be spring actuated as well as gravity actuated. Control consists of varying the electrical energy, which changes the magnitude of the impulse or the frequency of rapping. The acceleration of the rap can be as low as 5 g, but raps from 30 g to 50 g are required for most fly ash precipitators.

Vibrators (electromagnetic) - electromagnetic vibrators consist of a balanced spring-loaded armature suspended between two synchronized electromagnetic coils. When energized, the armature vibrates at line frequency. This vibrating energy is transmitted through a rapper rod to the electrodes. When used for cleaning discharge electrodes, the rapper rod is provided with an electrical insulating section in order to isolate the high voltage electrode charge from ground. Control consists of varying the electrical energy input, which changes the amplitude of vibrations, the operation time duration, and the frequency of vibration. Figure 27 shows a typical electromagnetic vibrator installation.[35,40]

Vibrators (eccentrically unbalanced motors) - this system consists of mechanical vibrators with an electric motor equipped with adjustable cam weights mounted on a single shaft or on both shafts of a double ended motor. When operated, the eccentrically positioned cam weights set the entire assembly into vibration. The motor is mounted directly on the rapper shaft which transmits the generated vibration to the electrodes to be cleaned. Control consists of varying the degree of eccentricity by cam weight adjustment, the length of time operated, and the frequency of operation.

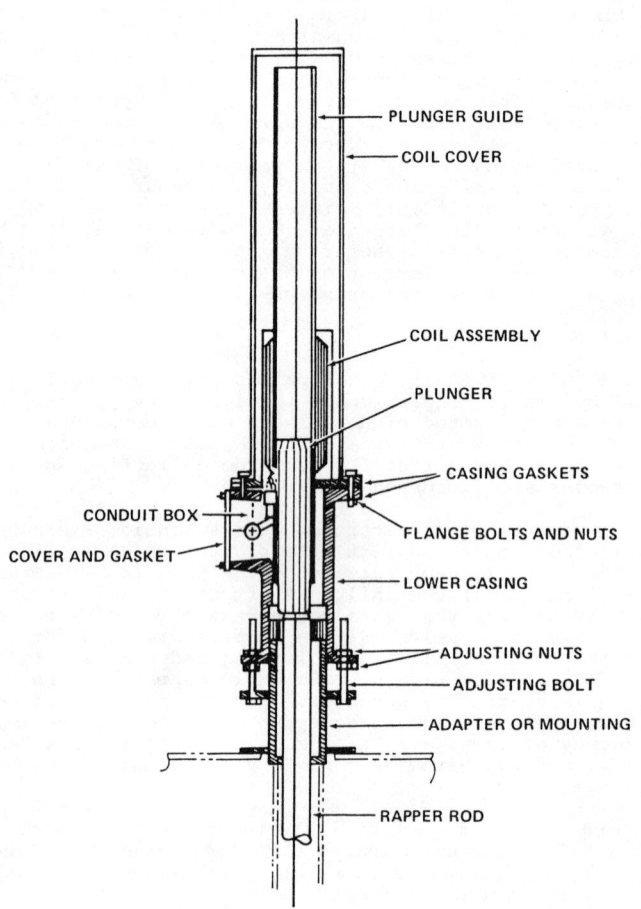

Figure 26. Typical electromagnetic rapper assembly.[38]

Use of Electrostatic Precipitators   41

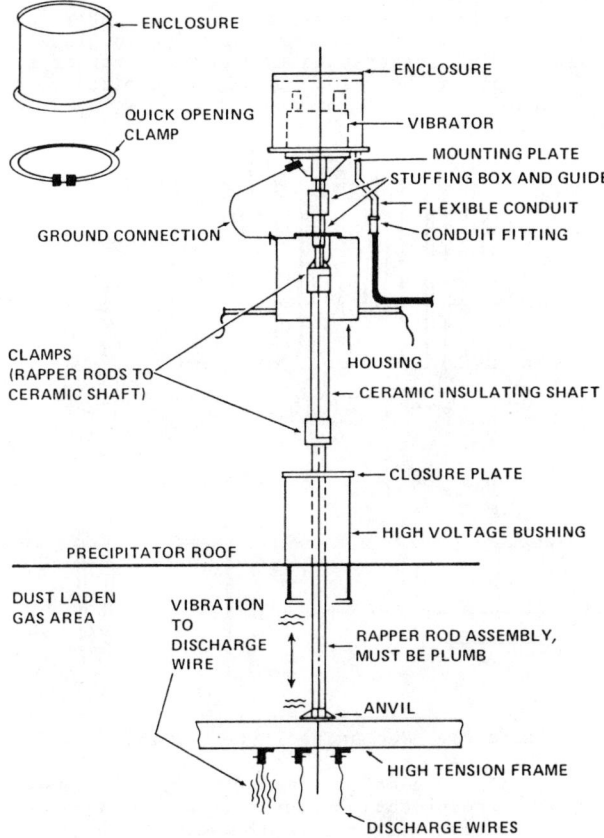

Figure 27. Typical vibratory rapper.[39,40]

Single impact (motor-driven cams) - this mechanism consists of a motor-driven shaft running horizontally across the precipitator. Cams are located along the shaft which raise small hammers by their handles. When the rotating cam reaches the end of its lobe, the hammer swings downward, striking an anvil located on the end of a single collecting electrode. Rapping control is limited to adjustment of operating time and shaft speed.

Single impact (motor-driven swing hammers) - this mechanism consists of a shaft running horizontally across the precipitator between banks of collecting plates. Hammer heads are connected to the shaft by spring leaf arms, and the shaft is oscillated by a motor-driven mechanical linkage. The hammers strike against anvils attached to the ends of all the collecting plates. Control is accomplished by varying operating time and the arc of the hammer swing.

Single impact mechanical rappers - Figure 28 shows this system which consists of a drive shaft running across the precipitator.[41] The shaft rotation carries the swing hammers around the shaft. When the hammer rods swing over the center cam disc and raise the hammer rods, the hammers fall due to gravity, striking an anvil which is attached to the discharge or collecting electrode structure. Rapping control is limited to operating time and shaft speed.

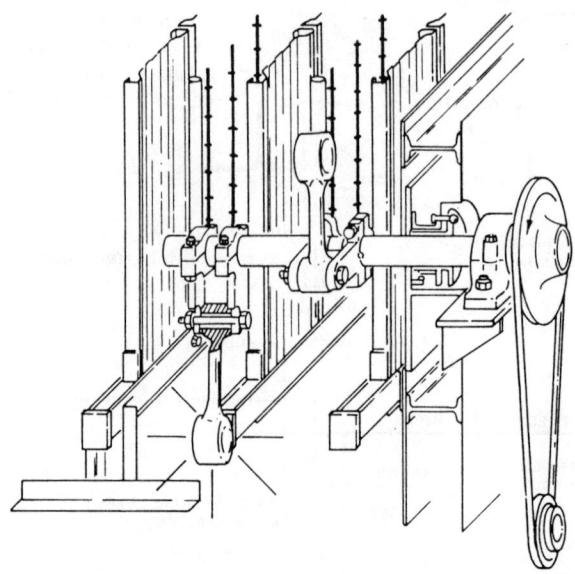

Figure 28. Mechanical-type rapper.[41]

Vibrators (air) - the major components of this system typically consist of a reciprocating piston in a sleeve-type cylinder. The assembly is fastened directly to the end of a rapper rod which transmits the rapping energy to the discharge or collecting electrode to be cleaned. Control consists of varying the air pressure, the duration of the rapping period, and the time elapsed between cleaning.

Failure to match rapping requirements to process characteristics can result in the need for higher rapping intensity than expected which in turn leads to accelerated degradation in system hardware. Generally, electric or pneumatic impulse rapped devices have been more successful in difficult rapping applications.[37]

New Technology in Rapper Control[37]--

Many of the problems involved with rapper control are associated with proper rap sequencing or individual rapper energization. Particularly vulnerable to this type of malfunction are those controls which incorporate mechanical switching and sequencing. Many solid-state devices are now being substituted for the mechanical-type devices. New technology has also made available rapper control systems that permit continual on-line monitoring of rapper system operation. The use of microprocessor-type control technology, previously uneconomical, has provided a high degree of rapper control flexibility and has reduced maintenance problems. Where rapper assembly malfunctions have previously caused control damage from ground fault currents, new control systems will test each individual rapper circuit prior to energization. Should that circuit prove defective, the control will automatically bypass the grounded rapper or circuit and indicate the defective unit in an LED display, thus permitting quick and easy location and repair. New technology has also been developed to incorporate precipitator power-off rapping techniques which increase rapping effectiveness for difficult dusts and reduce system wear.

Hoppers--

Hoppers are used to collect and store dry particulate which is removed from the electrodes. Insulation and heat tracing of the hoppers are very important in keeping the fly ash hot and dry, thereby facilitating removal from the hoppers to storage areas. If fly ash is allowed to cool, moisture condensation followed by caking of the ash may occur, making removal very difficult. Caking may be a potential problem especially when conditioning agents such as SO_3 are utilized to improve precipitator efficiency. Hopper heat tracing systems may be obtained from the Heat Tracing Division of Cooperheat, Inc., Rahway, New Jersey.

If the precipitator system is operated with internal pressures less than ambient atmospheric, air inleakage through the hopper can cause a reentrainment of the dust from the hoppers. Baffles are often placed in the hoppers and extend below the minimum dust level to minimize undesirable gas flows which may reentrain dusts. Good seals around hopper doors and around the connections of dust removal systems operating under a vacuum are a necessity.

Overflow of hoppers with fly ash can be a serious problem leading to an electrical shorted electrical system or reentrainment of fly ash, thereby reducing collection efficiency. Because of the importance of eliminating the overflow problem a number of hopper level detectors have been developed. These detectors use several different principles of operation and some of the more common detection methods and the companies which manufacture the detectors are given below.[42] This list is merely a representative sampling and should not be considered exhaustive. Also, the information contained is a condensation of promotional literature and does not reflect an opinion of Southern Research or the Environmental Protection Agency as to which system is superior.

   Kay-Ray, Inc.
   516 West Campus Drive
   Arlington Heights, Illinois 60004
   Phone: 312/259-5600

Kay-Ray, Inc. has placed over 1000 level switches on fly ash precipitators. This system is equivalent to an infinite number of probe sensors in that it detects ash at any location along the total width of the collection hopper. The system operates on a noncontacting radiation principle. A narrow beam of gamma rays is directed across the hopper (penetrating insulation, walls, and baffles) to a radiation detector located on the opposite wall. When ash builds up, the rays are absorbed, causing the detector to activate a relay for alarm or control purposes. The source of the gamma radiation is Cesium 137 with a half-life of 33 years. The detector consists of two Geiger-Müller detectors and associated electronics and produces an output current that is inversely proportional to the amount of material between source and detector. The major advantages of this system (Model 4810) are: (1) the sources and detectors are mounted outside the hoppers. None of the equipment comes in contact with the hot, abrasive fly ash; (2) the equipment can be mounted or repaired without having the fly ash hoppers down; (3) a high alarm is given whenever fly ash buildup intersects the path of the beam across the hopper. This assures an alarm condition whenever any fly ash builds up on either hopper wall or any of the baffles; (4) equipment operates at fly ash temperatures of 815°C (1500°F).

A typical system is shown in Figure 29.

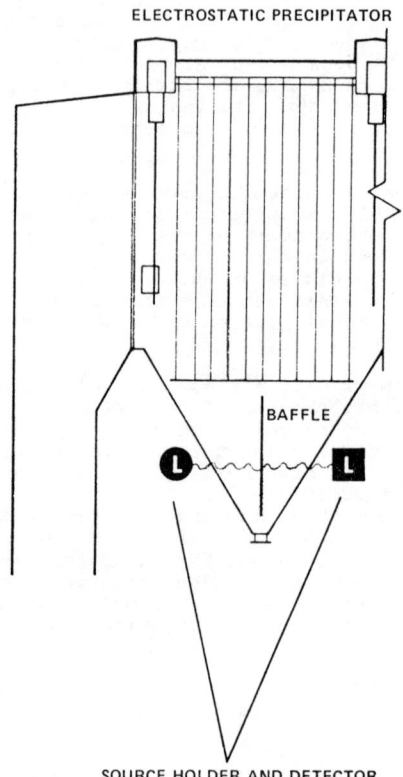

Figure 29. KAY-RAY fly ash control system.

More detailed drawings and specifications of a typical housing (Model 7063P) and a typical detector (Model 7316P) are given in Figures 30 and 31, respectively.

Use of Electrostatic Precipitators   45

SPECIFICATIONS

* Welded steel construction
* Rugged, simple mechanical design
* Lead filled, sealed in steel
* Lockable shutter mechanism
* Highly collimated radiation beam to provide inherent safety

* Low surface radiation level
* Wide range of source sizes and types
* Painted with chemically resistant epoxy
* Weight - 85 pounds **(38.6 kg)**

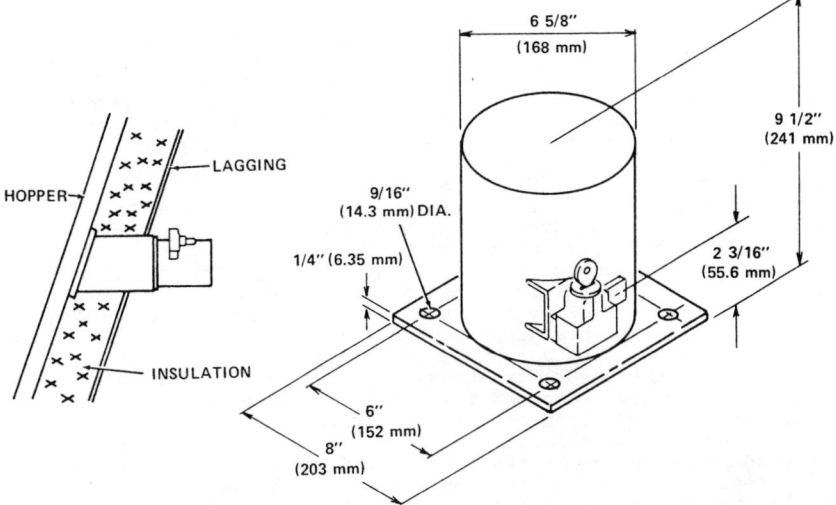

Figure 30. Level source housing - Model 7063P.

SPECIFICATIONS

* Reproducibility to ± 1/8"
* Solid state circuitry
* Two GM sensors
* High sensitivity, operates at less than 0.5 mr/hr
* Fail safe high or low
* Fast reponse - 1 second

* Output SPDT or DPDT relay contact 10A
* Input 115V, 50-60 HZ, 25 VA, 115 or 230VAC, 50-60 HZ
* Approx. weight 20 pounds **(9.1 kg)**
* Painted with chemically resistant epoxy
* Factory pre-calibrated

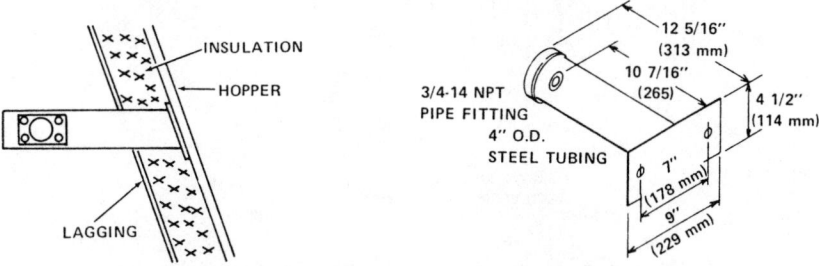

Figure 31. Fly ash level detector - Model 7316P.

A new system by Kay Ray, the Model 4400 Fly Ash Level Detection System features a remote electronics annunciator.

Texas Nuclear Division
Ramsey Engineering Company
Post Office Box 9267
Austin, Texas 78766
Phone: 512/836-0801

The Texas Nuclear Fly Ash Level System operates on the principle of gamma ray absorption. A radioactive source emits a narrow beam of gamma radiation from its protective housing. This beam passes through the hopper walls to the detector. When the fly ash level rises above the source, radiation is absorbed. When the number of gamma rays falls below a predetermined reference, the detector logic circuitry concludes that material is present and an output relay is switched from low to high level indication. Figure 32 shows a typical two hopper installation. The Texas Nuclear System takes advantage of the symmetry which usually exists in fly ash hopper installations by using a single source head with dual ports to illuminate opposing hoppers. Thus, most plants will require only half the number of sources and source heads as those using previous level switch system designs. The electron components are designed for continuous 93°C (200°F) operation. The system uses a Geiger-Müller radiation detector tube. One of the major advantages of the system is its 100% digitial circuitry. Another feature is a remote source actuator mechanism which provides positive opening, closing, and lock-out of the source head at locations convenient to the operator. Also, the mechanical design of the source head and detector eliminates the necessity of penetrating the insulation and welding mounting brackets to the hopper walls. Table 2 lists specifications for the detector and the source and source heads.

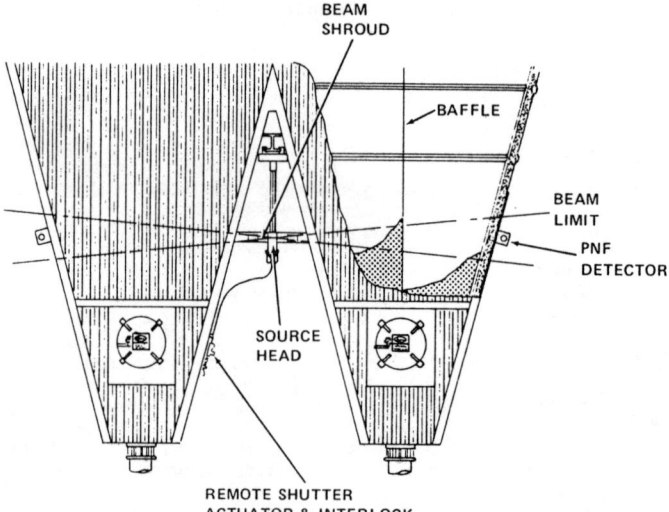

Figure 32. Typical hopper installation, Texas Nuclear Division, Ramsey Engineering Co.

TABLE 2. SPECIFICATIONS FOR TEXAS NUCLEAR
            DETECTOR AND SOURCE AND SOURCE HEADS

SPECIFICATIONS

DETECTOR-ELECTRONICS

Level reproducibility: $\pm$ .64 cm ($\pm$ 1/4 inch).
Sampling time: 1 to 3 minutes depending on application.
Radiation field required: 0.05 to 0.1 mR/h.
Minimum radiation change for operation: 25%.
Product temperature: Unlimited.
Ambient temperature: Designed for continuous operation at 93°C (200°F). Minimum operating temperature, -40°C (-40°F).
Detector: Single halogen-quenched Geiger-Müller tube.
Output: DPDT contacts, (10 ampers @ 115 VAC).
Circuitry: Total digital--all integrated circuits. Premium, high temperature components.
Controls: None--no adjustments in electronics required.
Power requirements: 115 VAC $\pm$ 15% @ 10 VA, 50-60 Hz.
Detector housing: Special lightweight aluminum construction. Dust and waterproof. Integral mounting bracket.
Size: 28.3 cm (11 1/8") long (plus conduit hub) x 15.2 cm (6") high x 15.2 cm (6") deep including mounting bracket (283mm x 152mm).
Weight: 2 kg (4 1/2 lb).

SOURCE & SOURCE HEADS

Source material: Cesium-137.
Source sizes: 5 to 100mCi depending on hopper size.
Head construction: Lead filled, steel encased. Special dual beam ports. Adaptable to Kirk or Superior key interlocks. Remote actuator with interlocks available.
Head weight: $\sim$ 16 kg (35 lbs).

>   Automation Products, Inc.
>   3030 Max Roy Street
>   Houston, Texas 77008
>   Phone: 713/869-0361

Automation Products, Inc. manufactures Dynatrol Detector Model CL-10DJ which consists of a rod which is installed through the wall of the collection hopper at the desired level detection.

When the probe is uncovered, the drive coil drives the rod into self-sustained mechanical oscillations at the natural resonant frequency of the rod. The pickup coil, located opposite to the drive coil, is excited by the mechanical oscillations of the rod and produces an ac signal voltage. The presence of this signal voltage indicates that the rod is uncovered or that a low level exists. When the fly ash covers the rod a dampening of the rod oscillations occurs. The magnitude of the rod oscillations are greatly reduced and the output from the pickup coil drops to a very low value, indicating that the rod is covered or that a high level exists.

A typical installation of the detector is shown in Figure 33.

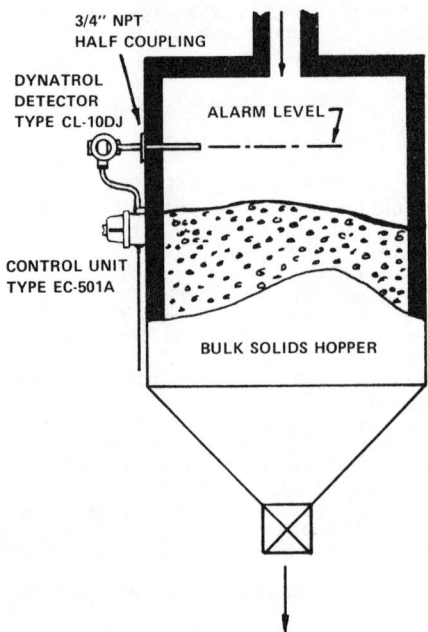

Figure 33. Typical installation of detector Type CL-10DJ.

United Conveyor Corporation
300 Wilmot Road
Deerfield, Illinois 60015
Phone: 312/948-0400

The United Conveyor Corporation Hopper Level Detector is a capacitance sensing on-off control instrument used for detecting or controlling product level in vessels or containers. Control action is provided by means of relay contact closure and alarm lamp indication in the display unit. The control is designed for mounting remotely from the level detector assembly. The detector assembly senses the change in product or material level as a function of the capacitance change between the detector and the vessel wall, and transmits this change to the control instrument. Specifications for this detector assembly are given in Table 3 and a diagram of a typical installation is shown in Figure 34.

Bindicator
1915 Dove Street
Port Huron, Michigan 48060
Phone: 1/800/521-6361

Bindicator manufactures several different types of level controls, but the one which appears most applicable for hopper level control is the PRTM-CO series 700 radio frequency level control. This control consists of a vessel, the vessel's components, a sensing probe, and an electronic unit. The electronic

TABLE 3. SPECIFICATIONS FOR UNITED CONVEYOR CORPORATION HOPPER LEVEL DETECTOR

Operating Temperature Limits:

    Control   -40°C (-40°F) to +71°C (+160°F)

    Detector   427°C (+800°F)

Vibration Limits   2 g's 10 to 100 Hz

Enclosure Classification   Weatherproof

Operating Humidity Range   0% to 90% RH

Supply Voltage   117 V AC $\pm$ 10% 60 Hz

Supply Power   12 VA Maximum

Output Relay   5A, 117 VAC/26.5 VDC, Noninductive;

                                       3A, 230 VAC Noninductive

Zero Adjustment Range   20 to 225 pf

Response Time   50 ms

Differential (Dead band)   0.1 pf Maximum

Connecting Cable   TRIAX Cable 61 m (200 ft) max.

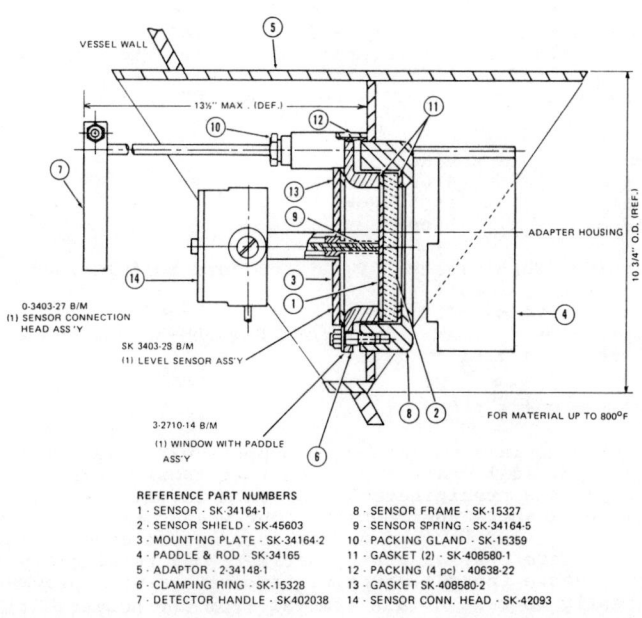

REFERENCE PART NUMBERS
1 - SENSOR - SK-34164-1
2 - SENSOR SHIELD - SK-45603
3 - MOUNTING PLATE - SK-34164-2
4 - PADDLE & ROD - SK-34165
5 - ADAPTOR - 2-34148-1
6 - CLAMPING RING - SK-15328
7 - DETECTOR HANDLE - SK402038
8 - SENSOR FRAME - SK-15327
9 - SENSOR SPRING - SK-34164-5
10 - PACKING GLAND - SK-15359
11 - GASKET (2) - SK-408580-1
12 - PACKING (4 pc) - 40638-22
13 - GASKET SK-408580-2
14 - SENSOR CONN. HEAD - SK-42093

Figure 34. Hopper level detector No. 3-3404-26.

unit provides a low power RF signal which is radiated from the sensing probe and measures any changes in the probe impedance caused by a change in the material level. Current changes not caused by change in the material level are eliminated by use of electronic compensation techniques. On a point control, current changes caused by the material cause activation of a relay. On a continuous control an electrical signal is given which is proportional to the level of material being measured. Table 4 gives the electrical specifications for both the point control and continuous control models of PRTMCO.

TABLE 4. ELECTRICAL SPECIFICATIONS FOR POINT CONTROL AND CONTINUOUS CONTROL MODELS OF PRTMCO

ELECTRICAL SPECIFICATIONS:

Point — Model 700

Line Voltage: 115 ± 20 VAC, 50/60 Hz
Power: 6 watts
Line Voltage Sensitivity: + 0.05 pf for ± 20 VAC
Ambient Temperature for electronic unit: -40°C to +71°C (-40°F to +160°F)
Temperature Sensitivity: ± 0.1 pf for -1°C (30°F)
Output: DPDT, 5A relay at 115 VAC non-inductive
Input Sensitivity: 0.1 pf
Stability: 0.1 pf for 6 months
Output Response Time: 0.02 sec.

Continuous -- Model 770

Line Voltage: 115 ± 20 VAC, 50/60 Hz
Power: 12 watts
Line Voltage Sensitivity: ± 0.5% for + 20 VAC
Ambient Temperature for electronic unit: -40°C to +71°C (-40°F to +160°F)
Temperature Sensitivity: 0.02% per °F or .08 pf (whichever is larger)
Span: .25 to 4000 pf
Output and Max. Load Resistance:
   1 - 5 ma   6000 ohms
   4 - 20 ma   1500 ohms
  10 - 50 ma   600 ohms

Load Resistance Sensitivity: 0.1% from zero to full load
Output Linearity: ± 0.5%
Output Response Time: 150 μ sec.
Probe Coating Error: max. error for 2000 ohm-cm coating 0.16 cm (1/16") thick is 3.81 cm (1.5")

Removal from Hoppers[43,22]--

Fly ash materials collected in hoppers will have different chemical and physical characteristics than those experienced at conditions inside the precipitator. For example, fly ash flows similar to a liquid well above the dewpoint, but when cooled below 121°C (250°F) to 149°C (300°F), its hygroscopic nature causes agglomeration and caking. Therefore, as stated earlier, maintaining fly ash sufficiently above the gas dewpoint temperature will prevent caking and add greatly to ease of ash removal from the hopper. Vibrators are sometimes used to prevent bridging.

Several types of systems exist for the removal of dusts accumulated in hoppers. These systems include container removal, dry vacuum, wet vacuum, screw conveyors, and scraper bottom. A brief description of these ash removal systems is given below.

Dust Removal Systems--

Container removal--This system is used on small installations collecting dry material in a hopper. The hoppers are usually of the conical or pyramidal type. The system consists of placing a transportable container below the hopper. The collected material stored in the hopper is transferred to the container through a simple manual valve or slide gate. When filled, the container is removed for emptying. In some instances the container is embodied as part of a truck.

Dry vacuum systems--In this system, dry bulk material is transferred from a precipitator hopper to a transport pipe system which is under vacuum. The material is metered from the hopper to the transport system through automatic rotary feeder valves or dump valves. The system vacuum is developed by an air pump. In order to maintain system fluidity, ambient air or hopper gas is induced as a carrier. The pump discharges the dust into a silo for storage.

Wet vacuum systems--In this system, dry dust is removed from a precipitator hopper into a transport pipe system which is maintained under vacuum by a water aspirator. The collected dust or ash is metered from the hopper into the transport system through automatic feeder valves or dump valves. In order to keep the dust suspended in the gas carrier, ambient air or additional hopper gas is induced into the transport line. The dry material being transported mixes with the water used for aspiration and forms a slurry. From this point the water-dust mixture is run to waste.

Screw conveyors--A screw conveyor system usually starts with an open screw in the bottom of a trough-type hopper which moves the dry dust to the outside. At the turns in the system each screw run passes the dust on to each successive screw by a gravity drop. The dust is moved on to a system silo or directly to some mobile conveyance. A screw conveyor system is also applicable with a conical or pyramidal type hopper. A rotary valve is required when the system is operating under vacuum.

Scraper bottom--The precipitator hopper is a flat bottom pan. An endless belt type scraper moves the collected dust to one end where a screw conveyor is located. The screw moves the dust out of the hopper. Once outside, the dust is conveyed to some remote point by any form of system such as container removal, vacuum, or screw.

Gas Flow Devices

General--

The best operating condition for an electrostatic precipitator will occur when the velocity distribution is uniform. Because of the logarithmic nature of the Deutsch efficiency formula, an increased velocity in one plate section will decrease the efficiency of that section more than the decreased velocity in a parallel section would increase the efficiency of that section.

As a consequence, for a precipitator with nonuniform gas velocity, the total plate area required to achieve a given efficiency will be greater than that for a precipitator with uniform gas velocity. One quantitative criterion describing the quality of gas distribution is that used by the Electrostatic Division of Industrial Gas Cleaning Institute. It states that acceptable gas distribution is achieved when 85% of all local gas velocities are within $\pm$ 25% of their average, and no single reading differs from the average by more than 40%. Quality gas distribution is especially important in the entrance to the gas cleaning device, but acceptable gas distribution in the transport system typically may be as low as 65% within $\pm$ 25% depending on the complexity of the system.[44]

Basically, the measurement technique consists of measuring gas velocities in a prescribed pattern. An imaginary plane is passed through the duct perpendicular to the gas flow at the location to be evaluated and this cross sectional area is broken down into smaller equal areas. Figure 35 gives an example of a rectangular system divided into equal areas.[44]

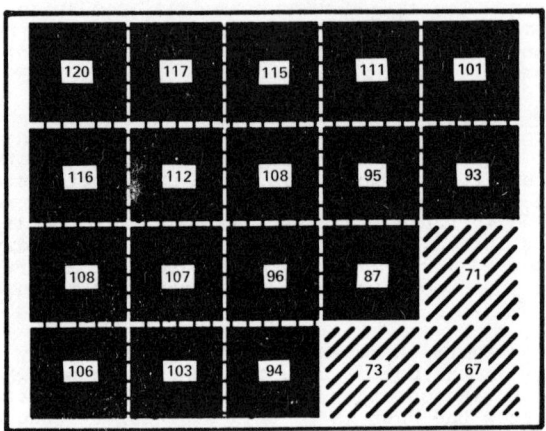

VELOCITY MEASUREMENTS AT EACH LOCATION IN DUCT

SUM OF VELOCITIES = 2,000
AVERAGE VELOCITY = $\frac{2,000}{20}$ = 100
ACCEPTABLE RANGE = ±25% OF AVG. VEL. = 75 TO 125

$\frac{17 \text{ (WITHIN RANGE)}}{20 \text{ (MEASUREMENTS)}}$ = 85% WITHIN RANGE

Figure 35. Computation of gas velocity distribution.

To achieve uniform flow in a duct according to ASME test procedures, one should have at least ten diameters of duct before and after any distrubance such as the elbow, expansion or contraction. In practice, such conditions cannot always be economically realized because of space limitations and the high cost of ducts in the sizes involved. However, given a reasonable space, it is possible to approach an acceptable quality of flow at the precipitator inlet by the use of straighteners, splitters, vanes, and diffusion plates.

Straighteners--

Partitions in a straight section of duct for the purpose of eliminating swirl are called straighteners. They may be "egg crate" dividers or nested tubes as shown in Figure 36-a.

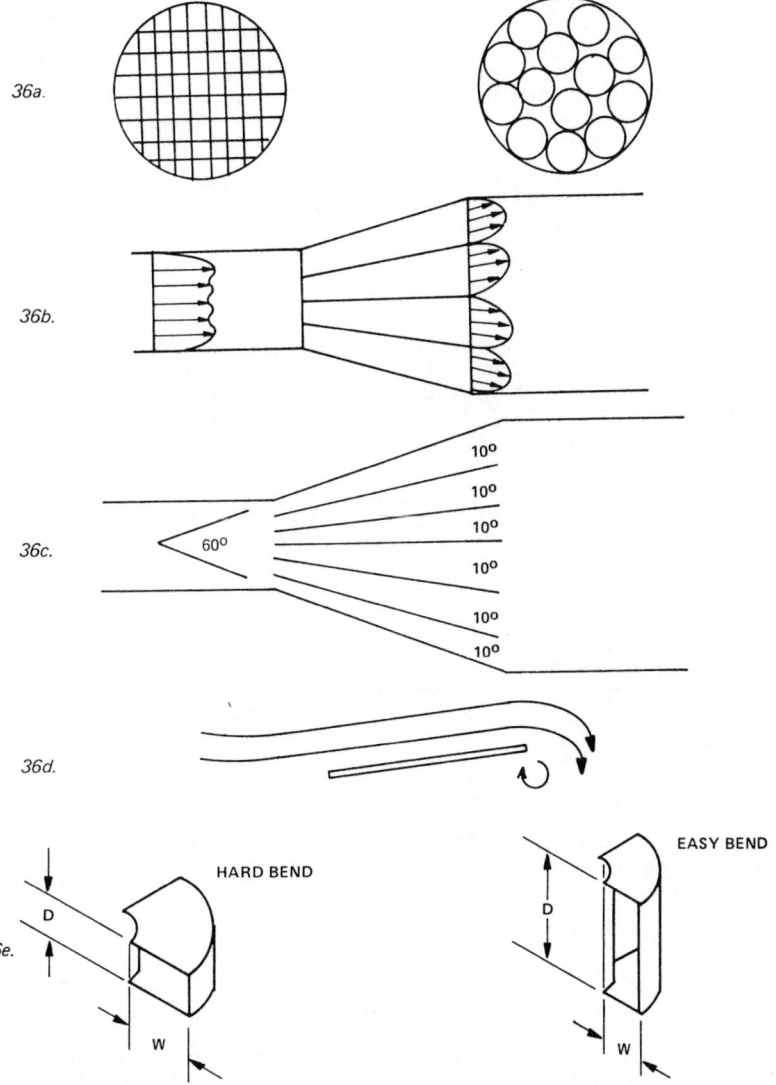

Figure 36. Flow devices.

A straightener will reduce the angle of a helical flow path to some angle less than that defined by arc tan = $\frac{\text{length}}{\text{spacing}}$.

The nonuniformity in the velocity in the axial direction will not be reduced. The scale of turbulence, or eddy diameter, will be temporarily reduced to the same size as the spacing of the straighteners, but because the Reynolds number is usually well above critical, the eddies will not die out, but will grow until they again reach the order of magnitude of the full duct. It is theoretically possible to make the spacing of straighteners small enough to obtain a Reynolds number less than critical and to obtain laminar flow through the straighteners, and to obtain nearly absolute uniformity. Unfortunately, such spacing would only be about the size of soda straws and the straightener would be expected to plug up with dust almost immediately.

A recommended straightener, according to AMCA Bulletin 210,[4,5] is an egg crate with a spacing of 7 1/2 to 15% of the diameter of a round duct or the average side of a rectangular duct, and with a length equal to three times the spacing. This reduces the swirl angle to arc tan 1/3 = 18 1/2°. An alternate straightener is a simple criss-cross at least one and a half diameters long.

The loss in these straighteners is equal to the loss in four plain duct diameters. If it is necessary to reduce the swirl angle to smaller values, the ratio of length to spacing must be increased, and the resulting friction will be higher.

Splitters--

A duct section that changes size or direction may be divided into smaller ducts over the full length of the change by partitions called splitters. Splitters may be used in elbows where direction is changed, or in transformations where velocity is changed. Splitters add wall friction, but can reduce total friction by optimizing velocity pressure losses.

Losses in elbows depend in part on how sharp the bend is. A square, or mitered, elbow will have a loss of 1.25 times the velocity pressure, but an elbow of optimum configuration could have a loss of only 0.11 to 0.14 velocity pressure.

As shown in Figure 37, the optimum configuration is an elbow with a ratio of inside radius to outside radius of about 0.66 for a square duct, or 0.7 for a round duct.[23]

To design a splitter elbow, therefore, it is only necessary to divide the given elbow into segments all having a radius ratio of about 2/3.

Transformation Splitters--

Splitters, shown in Figure 36-b, may also be used in a diverging duct transformation to divide the flow into nearly equal parts and then distribute the flow to the larger section.

The gas flow will not be uniform within each segment of the transformation section, but the volume through every segment can be made equal to that in every other segment if the splitters are manufactured to be field adjustable.

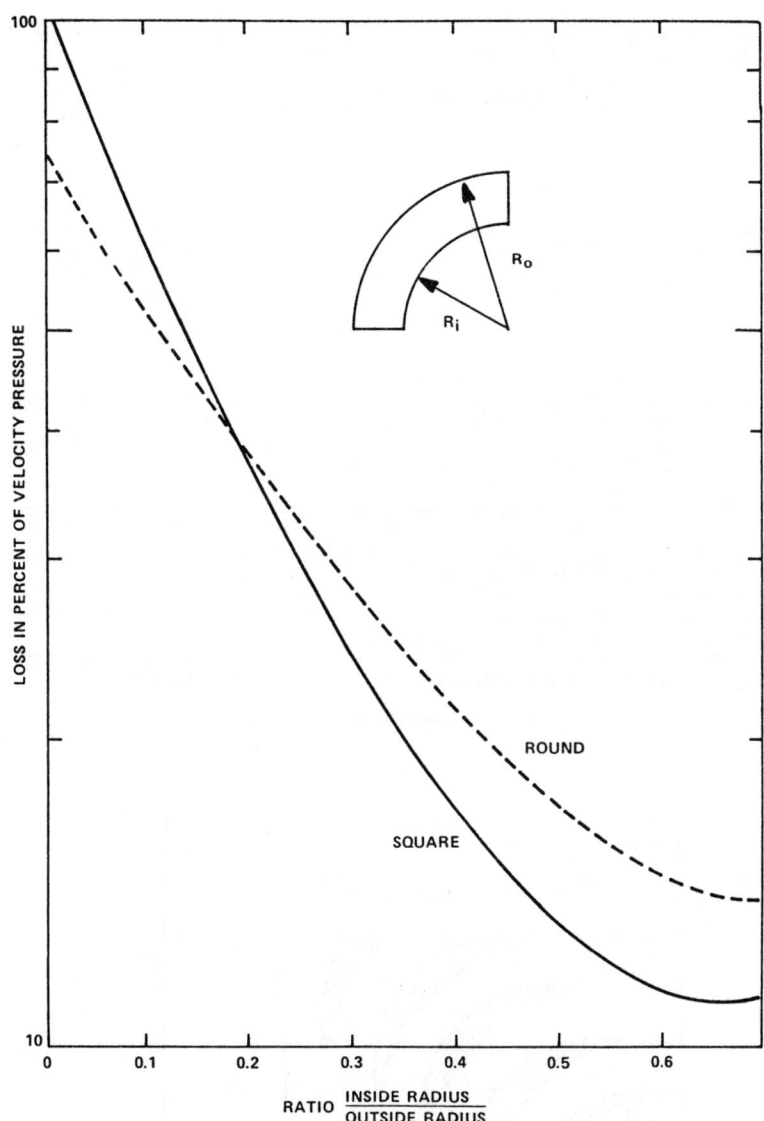

Figure 37. Elbow loss as a function of radius ratio.[23]

A sharp angle in the transformation causes the gas to separate from the walls of the duct, introducing turbulence and nonuniform flow. The maximum angle of divergence for no separation is about 7° included angle. Therefore, splitters in a transformation should be selected to have 7° to 19° included angle between successive splitters. For example, a transformation with 60°

included angle could be split into 8 channels with 7 1/2° spread, or 6 channels with 10° spread as shown in Figure 36-c.

Note that a transformation in one direction is the simpler. Transforming in two directions would require pyramidal splitters.

Vanes--

Another kind of deflector for redirecting gas flow is the turning vane. Turning vanes are flat, bent, or curved plates which are short relative to the duct section in which they are installed, as opposed to splitters which extend the full length. A plain flat plate vane used to deflect the air stream is partly effective, but it tends to increase turbulence as shown in Figure 36-d.

The low pressure area behind the plate also tends to pull the gas flow back toward its original path. Curved turning vanes in an elbow can be quite effective if they are spaced to give about a 6:1 aspect ratio and a 2/3 radius ratio, and are streamlined to give constant cross section through the turn.

The streamlined turning vanes shown in Figure 38 will preserve the flow pattern and will have a loss of about 10% of the velocity pressure.[23] A set of single thickness turning vanes will also preserve the flow pattern, but will have about 35% velocity pressure loss and may introduce some turbulence because of the unequal cross sections between them. Single thickness vanes should have a straight extension downstream with length about twice the spacing. In practice, single thickness turning vanes are generally used because of cost considerations.

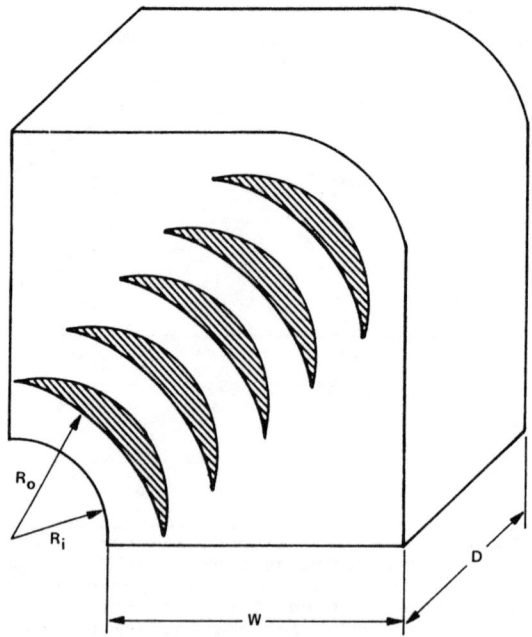

Figure 38. Streamlined turning vane elbow.[23]

For rectangular elbows, one parameter is the aspect ratio, or the ratio of the depth of the elbow measured parallel to the axis of the bend to the width of the elbow measured in the plane of the bend, as shown in Figure 36-e.

It is intuitively obvious from the sketches that a low aspect ratio elbow is a "hard" bend with high pressure loss which has very nonuniform flow caused by inertial forces. Any aspect ratio greater than unity will make a fair elbow, but aspect ratios from 4 to 6 are recommended.

Turning vanes are also used in transformation elbows; that is, elbows that change cross section between inlet and outlet. Although the combination of elbow and transformation is relatively poor design practice, severe space limitations may force it upon the designer. If a transformation elbow must be used, then turning vanes are essential, and they must be closely spaced to about the same spacing as the precipitator plates. They must also be followed by additional flow rectification means such as diffusion plates.

Diffusion Plates--

Diffusion plates, or screens, are simply perforated plates or wire screens which improve the uniformity of air flow by a combination of effects. First, they reduce the scale of turbulence from the order of magnitude of the duct to the order of magnitude of the holes. Of course, the kinetic energy that existed in the large scale eddies will reappear in the small scale eddies, but the large differences in velocity will be reduced. Second, there is a pressure drop across the screen and a reduction in area. The pressure drop will partly reappear as a velocity vector perpendicular to the plate. This vector added to the original velocity vector will give a resultant velocity always more nearly perpendicular to the plate. Thus, it might be possible to design a diffuser plate to turn the gas stream through a precise small angle. However, in practice, it is usually simpler and less costly to use two or more diffusion screens in series to achieve a fair degree of uniformity.

Perforated plate screens break the gas stream up into a multiplicity of small jets with high turbulent intensity and small scale of turbulence. These jets eventually coalesce downstream. The turbulent intensity reaches a peak at 2 to 3 mesh lengths (center to center of holes) downstream and declines exponentially thereafter. The scale of turbulence is of the order of the hole size at the screen and increases until it reaches the size of the duct. There is a critical parameter of 50% open area for diffusion screens. When the percentage of open area is less than 50%, the jets seem to be too far apart to coalesce uniformly and the screen introduces nonuniformity.

When the percentage of open area is between 50% and 65%, the jets appear to coalesce with 5 to 10 mesh lengths (center to center of holes) with improvement in uniformity.[46]

Screens may be used in series to provide greater uniformity at the cost of larger pressure drops. Dryden and Schubauer[47] developed the following relationship for the reduction in turbulent intensity:

$$r = (1 + k)^{-\frac{1}{2}n} \tag{5}$$

where

> r = reduction factor
>
> k = pressure drop coefficient = $\dfrac{\Delta p}{\rho v_0^2/2}$
>
> n = number of screens in series
>
> p = pressure drop
>
> ρ = density
>
> v_0 = average velocity.

All of the preceding duct work design techniques are available to the designer. None of the criteria are rigid, so there is considerable freedom in design. It is the designer's choice as to whether to use splitters, turning vanes, or screens to control the air distribution. On the inlet side of a precipitator, there may be a heavy dust loading of particle sizes large enough to settle out. Horizontal splitters or vanes form convenient shelves for the deposition of disastrous quantities of dust. Therefore, horizontal splitters and vanes are generally used only when the velocity is higher than the erosion velocity of deposited dust. Dust will collect by impaction on the diffusion screens, so some means of cleaning then is required, such as regular rapping or soot blowing.

TYPES OF PRECIPITATORS USED TO COLLECT FLY ASH

Cold-Side

Most electrostatic precipitators utilized to collect fly ash are of the cold-side type. That is, they are located downstream of the air preheater and operate at gas temperatures in the neighborhood of 150°C (300°F). The principal factor responsible for variations in performance of fly ash precipitators is the resistivity of the ash. Fly ash is composed largely of the oxides of aluminum, silicon, iron, and calcium, which at the operating temperatures of most precipitators, give it a very high electrical resistivity. However, moisture and sulfur trioxide present in the flue gases will be adsorbed on the fly ash particles and will reduce the resistivity. If the coal being burned has a sufficient sulfur content, the resistivity of the fly ash will be low enough ($\sim 2.0 \times 10^{10}$ Ω-cm) for good precipitator performance. However, if the sulfur content of the fuel is low, the amount of sulfur trioxide in the flue gas can be insufficient for proper conditioning of the ash. Owing to the increasing emphasis on the use of low-sulfur coals to minimize emission of sulfur oxides and the simultaneous demands for improvements in fly ash collection, increasing efforts are being made to find methods to overcome the problem of high resistivity. One method of improving the performance of fly ash precipitators collecting high resistivity dust is to reduce the gas temperature so that the resistivity is in a range more favorable for precipitation. However, most power plants burning coal with sulfur in the range of two to three percent, operate at gas temperatures from the air preheater in the vicinity of 150°C, primarily to minimize corrosion and fouling tendencies. Another method that is used to improve precipitator performance is to inject a chemical conditioning agent in the flue gas. The best known chemical conditioning agents are sulfur trioxide and ammonia. Of the two, sulfur tri-

oxide conditioning is the most familiar. Figure 39 is a diagram
of a liquid SO_2 system in which liquid sulfur dioxide is vaporized
and passed over a catalytic oxidizer in the presence of air.[48]

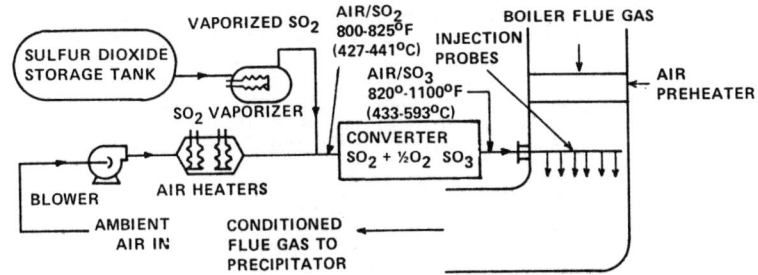

Figure 39. Liquid SO_2 system.[48]

Figure 40 is a diagram of a sulfur burning system in which molten sulfur is burned to produce gaseous sulfur dioxide and catalytically converted to sulfur trioxide.[48] Other conditioning agents that are potentially useful include sulfamic acid, ammonium sulfate, ammonium bisulfate, and triethylamine. Proprietary chemicals have also been used.

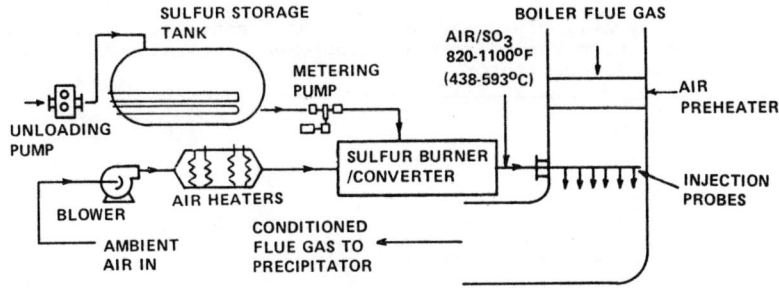

Figure 40. Sulfur burning SO_2 system.[48]

Hot-Side

The increasing use of low sulfur coal and the accompanying high ash resistivity at normal precipitator operating temperatures has led to the use of hot-side precipitators. Hot precipitators are located upstream of the air preheater and operate at temperatures generally in the range of 316 to 482°C (600 to 900°F). The resistivity of most fly ash is sufficiently low at these temperatures that current is not limited by fly ash resistivity. A schematic of an electrostatic precipitator system when a hot-side precipitator has been retrofitted to supplement an existing cold-side precipitator is shown in Figure 41.[27] Note the locations of the two collectors with respect to the air preheater. Besides the avoidance of resistivity problems, secondary advantages of hot precipitators include elimination of corrosion and hopper plugging problems, easier hopper emptying and ash transport, and better electrical stability and higher corona current densities than are possible with low temperature precipitators treating high resistivity ash. Some of the dis-

advantages of using hot precipitators are: operating voltages
are substantially reduced due to the lower densities of hot
gases, gas viscosity increases with temperature thus reducing
precipitation rate, structural and mechanical problems such as
precipitator shell failures and support structure distortions,
and the necessity for very long interconnecting flues needed
between the precipitator and the boiler, and gas flows are
about 50% higher because of expansion of the gases at the
higher temperatures (Figure 42 illustrates the relationship
between the sizes of a hot-side precipitator and a cold-side
precipitator for the same efficiency, as the dust resistivity
varies).[49]

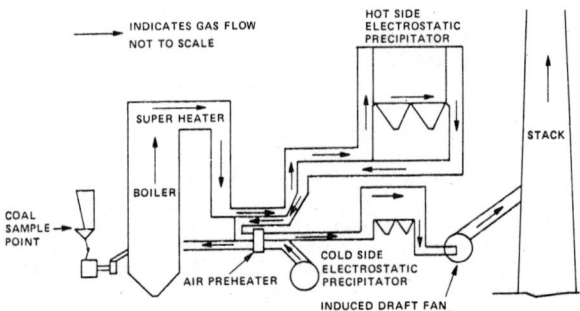

Figure 41. Schematic of an ESP system when a hot-side precipitator has been retrofitted to supplement the existing cold-side precipitator.[27]

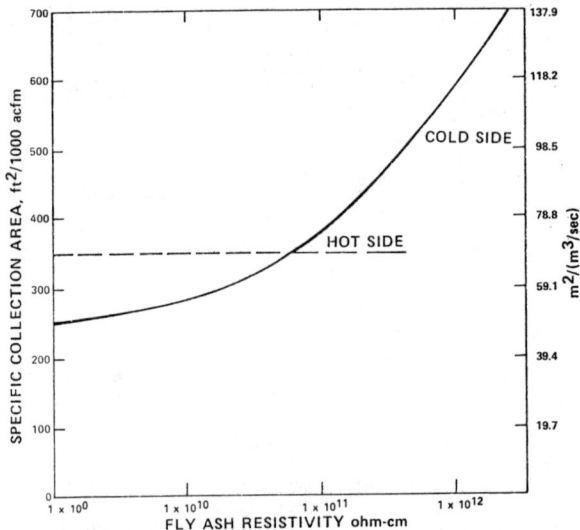

Figure 42. Illustration of the effects of fly ash resistivity on precipitator size for 99.5% collection efficiency. Curves are plotted on the basis of actual cubic feet per minute of gas flow. For 700°F hot-side and 300°F cold-side temperature, the ratio of gas flow for the same size boiler would be about 1.5. Hot-side resistivity is assumed to be not limiting.[49]

A number of successful hot precipitator installations exist, and Table 5 lists some design parameters for the hot-side units.[49] Recently, serious electrical problems associated with the collected ash layer have surfaced in many hot-side precipitators. This topic will be discussed in some detail later in the text.

COMPILATION OF INSTALLATIONS USING ELECTROSTATIC PRECIPITATORS TO COLLECT FLY ASH

There are a large number of electrostatic precipitator installations used for fly ash collection in the United States. Table A-1, located in the Appendix, contains a compilation of data from every precipitator installation used for collection of power plant fly ash in the U.S. These data were copied from Federal Power Commission files and are organized into three subject areas: (1) coal data, such as the heat content, sulfur content, and ash content, (2) boiler data, such as year placed in service, generating capacity, coal consumption, air flow, type of firing, manufacturer, efficiency, and percent excess air, and (3) precipitator data, such as manufacturer, year placed in service, design and tested efficiencies, mass emission rate, and installed costs.

This information is furnished by each power plant as part of FPC Form No. 67. Some of the data were not available for every plant and these cases are indicated by a hyphen. The most frequent omissions are for tested efficiencies of the electrostatic precipitators. There are other cases in which a range is given for tested efficiencies instead of exact numbers. These are instances in which there are more than one precipitator and it is unclear which efficiency belongs to which precipitator. These data are based on information furnished by individual electric utilities. Neither the Environmental Protection Agency nor Southern Research Institute were responsible for gathering the data and thus can not attest to its validity.

TABLE 5. HOT-SIDE PRECIPITATOR INSTALLATIONS[49]

Manufacturer	Collection Efficiency, %	Effective Migration Velocity, cm/sec	Specific Collecting Area, ft²/1000 ft³/min	Temperature, °F	Volume Flow Rate, 1000 acfm	Outlet Loading, gr/acf	Date Operational	Generating Rate, MW
Research Cottrell	99.2	9.03	270	650	1,250		72-73	870
Buell	99.0	9.75	240	800	337			
Buell	99.08	10.0	238	690	640			
Buell	99.0	9.75	240	800	340			
Research Cottrell	99.0	10.65	220	690	400			
	99.5	-		810	2,770		72	132
Research Cottrell	99.5	8.7	310	828	1,322		74	486
Research Cottrell	99.0	10.8	215	690	402		73	350
Research Cottrell	99.73	11.5	250	650	1,160	0.0163	71	150
Research Cottrell	99.0	10.65	220	650	690		72	250
Western	99.0	8.5	270	650	1,000		72	147
Research Cottrell	99.83	10.0	324	640	313	0.005	72	750
Buell	99.15	11.3	215	672	1,118	0.018	70	412
	99.5			690	825			210
Research Cottrell	99.57	11.1	245	550	487		73	200
Buell	99.0	10.1	235	520	4,079	0.005	67	1000
Research Cottrell	99.0			690	600			
Research Cottrell	98.5	9.15	235	809	250		73	52
				675	1,425		73	300
				625	670		72	114
Western	99.5			800	4,000			750
Western	99.5			700	470		73	350
Pollution Control Walther	99.0	8.91	310	655	1,428		76	250
Pollution Control Walther	99.0	8.98	250	721	1,274		76	125
Pollution Control Walther	99.3	8.58	260	705	163		75	30
Pollution Control Walther	99.3	8.78	295	775	440		75	100
Pollution Control Walther			288					
Pollution Control Walther	99.5	8.26	325	660	2,474		76	550
Pollution Control Walther	99.1	7.94	301	700	1,700		77	350
Pollution Control Walther	99.5	8.94	369	815	5,142		77	800 ea.
Pollution Control Walther	99.6	7.8	359	820	2,314		78	550
-	99.32	7.06	353	820	5,104		76	818 ea.
-	99.5	8.21	322	720	3,000		75	500 ea.
-	99.33	9.39	366	700	3,888		73	660 ea.

Section 6

Analysis of Factors Influencing ESP Performance

PARTICLE SIZE DISTRIBUTION

General Discussion

    The particulate matter suspended in industrial gas streams may be in the form of nearly perfect spheres, regular crystalline forms other than spheres, irregular or random shapes, or as agglomerates made up from combinations of these. It is possible to discuss particle size in terms of the volume, surface area, projected area, projected perimeter, linear dimensions, light scattering properties, or in terms of drag forces in a liquid or gas (mobility). Particle sizing work is frequently done on a statistical basis where large numbers of particles, rather than individuals, are sampled. For this reason the particles are normally assumed to be spherical. This convention also makes transformation from one basis to another more convenient.

    Experimental measurements of particle size normally cannot be made with a single instrument if the size range of interest extends over much more than a decimal order of magnitude. Presentations of size distributions covering broad ranges of sizes then must include data points which may have been obtained using different physical mechanisms. Normally the data points are converted by calculation to the same basis and put into tabular form or fitted with a histogram or smooth curve to represent the particle size distribution. Frequently used bases for particle size distributions are the relative number, volume, surface area, or mass of particles within a size range. The size range might be specified in terms of aerodynamic, Stokes, or equivalent Polystyrene Latex (PSL) bead diameter. There is no standard equation for statistical distributions which can be universally applied to describe the results given by experimental particle size measurements. However, the log-normal distribution function has been found to be a fair approximation for some sources of particulate and has several features which make it convenient to use. For industrial sources the best procedure is to plot the experimental points in a convenient format and to examine the distribution in different size ranges separately, rather than trying to characterize the entire distribution by two or three parameters. The ready availability of inexpensive programmable calculators which can be used to convert from one basis to another compensates greatly for the lack of an analytical expression for the size distribution.

Characterization Of Particle Size Distributions

    Figure 43 shows plots of generalized unimodal particle size

distributions which will be used to graphically illustrate the terms which are commonly used to characterize an aerosol.[50] Occasionally size distribution plots exhibit more than one peak. A size distribution with two peaks would be called bimodal. Such distributions can frequently be shown to be equivalent to the sum of two or more distributions of the types shown in Figure 43. If a distribution is symmetric or bell shaped when plotted along a linear abscissa, it is called a "normal" distribution (Figure 43-c). A distribution that is symmetric or bell shaped when plotted on a logarithmic abscissa is called "log-normal" (Figure 43-d).

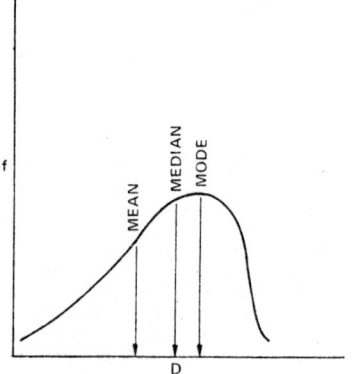

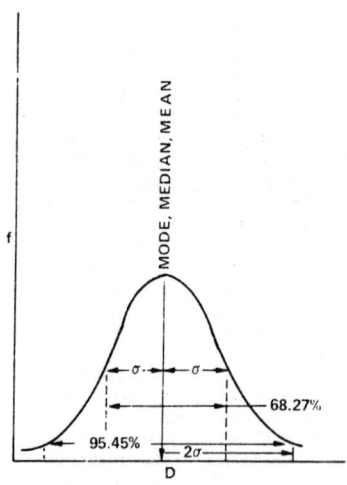

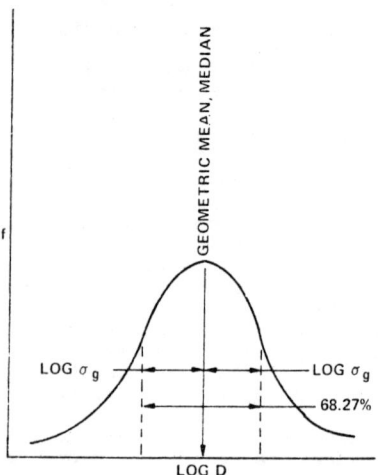

Figure 43. Examples of frequency or particle size distributions. D is the particle diameter.[50]

Interpretation of the frequency or relative frequency shown as f in Figure 43 is very subtle. One is tempted to interpret this as the amount of particulate matter of a given size. This interpretation is erroneous however and would require that an infinite number of particles be present. The most useful convention is to define f in such a way that the area bounded by the curve (f) and vertical lines intersecting the abscissa at any two diameters is equal to the amount of particulate matter in the size range indicated by the diameters selected. Then f is equal to the _relative_ amount of particulate matter in a narrow size _range_ about a given diameter.

The median divides the area under the frequency curve in half. For example, the mass median diameter (MMD) of a particle size distribution is the size at which 50% of the mass consists of particles of larger diameter, and 50% of the mass consists of particles having smaller diameters. Similar definitions apply for the number median diameter (NMD) and the surface median diameter (SMD).

The term "mean" is used to denote the arithmetic mean of the distribution. In a particle size distribution the mass mean diameter is the diameter of a particle which has the average mass for the entire particle distribution. Again, similar definitions hold for the surface and number mean diameters.

The mode represents the diameter which occurs most commonly in a particle size distribution. The mode is seldom used as a descriptive term in aerosol physics.

The geometric mean diameter is the diameter of a particle which has the logarithmic mean for the size distribution. This can be expressed mathematically as:

$$\log D_g = \frac{\log D_1 + \log D_2 + \ldots \log D_N}{N} \qquad (6\text{-a})$$

or as

$$D_g = \left(D_1 D_2 D_3 \ldots D_N \right)^{1/N} \qquad (6\text{-b})$$

The standard deviation (σ) and relative standard deviation (α) are measures of the dispersion (spread, or polydispersity) of a set of numbers. The relative standard deviation is the standard deviation of a distribution divided by the mean, where σ and the mean are calculated on the same basis; i.e., number, mass, or surface area. A monodisperse aerosol has a standard deviation and relative standard deviation of zero. For many purposes the standard deviation is preferred because it has the same dimensions (units) as the set of interest. In the case of a normal distribution, 68.27% of the events fall within one standard deviation of the mean, 95.45% within two standard deviations, and 99.73% within three standard deviations.

Table 6 summarizes nomenclature and formulae which are frequently used in practical aerosol research.[51] In actual practice most of the statistical analysis is done graphically.

Field measurements of particle size usually yield a set of discrete data points which must be manipulated or transformed to some extent before interpretation. The resultant particle size distribution may be shown as tables, histograms, or graphs.

Graphical presentations are the conventional and most convenient format and these can be of several forms.

TABLE 6. SUMMARY OF NOMENCLATURE USED TO DESCRIBE PARTICLE SIZE DISTRIBUTIONS[51]

Name	Symbol	Formula
Volume or Mass Mean Diameter	D_m	$D_m = \left(\dfrac{\sum_{j=1}^{N} n_j D_j^3}{N} \right)^{1/3}$
Surface Mean Diameter	D_s	$D_s = \left(\dfrac{\sum_{j=1}^{N} n_j D_j^2}{N} \right)^{1/2}$
Number Mean Diameter	D_n	$D_n = \dfrac{\sum_{j=1}^{N} n_j D_j}{N}$
Geometric Mean Diameter	D_g	$D_g = \left(\prod_{j=1}^{N} D_j \right)^{1/N}$
Surface-Volume Mean Diameter, or Sauter Diameter	D_{vs}	$D_{vs} = \dfrac{\sum_{j=1}^{N} D_j^3}{\sum_{j=1}^{N} D_j^2}$
Standard Deviation	σ	$\sigma = \left(\dfrac{\sum_{j=1}^{N} f_j (D_j - D_{m,n,s})^2}{N-1} \right)^{1/2}$
Relative Standard Deviation	α	$\alpha = \dfrac{\sigma}{D_{m,n,s}}$
Mass Median Diameter	MMD	Medians are most conveniently determined graphically. For many slightly skewed distributions Median = Mean + $\dfrac{1}{3}$(Mode-Mean).
Surface Median Diameter	SMD	
Number Median Diameter	NMD	

*j denotes a particular size interval, N is the total number of intervals, f_j is the relative, mass, surface area, or number of particles in the interval, and D_j is the diameter characteristic of the jth interval.

Cumulative mass size distributions are formed by summing all the mass containing particles less than a certain diameter and plotting this mass versus the diameter. The ordinate is specifically equal to $M(j) = \sum_{t=1}^{j} M_t$, where M_t is the amount of mass contained in the size interval between D_t and D_{t-1}. The abscissa would be equal to D_j. Cumulative plots can be made for surface area and number of particles per unit volume in the same manner. Examples of cumulative mass and number graphs are shown in Figures 44-b and 44-a, respectively, for the effluent from a coal-fired power boiler.[52] Although cumulative plots obscure some information, the median diameter and total mass per unit volume can be obtained readily from the curve. Because both the ordinate and abscissa extend over several orders of magnitude, logarithmic axes are normally used for both.

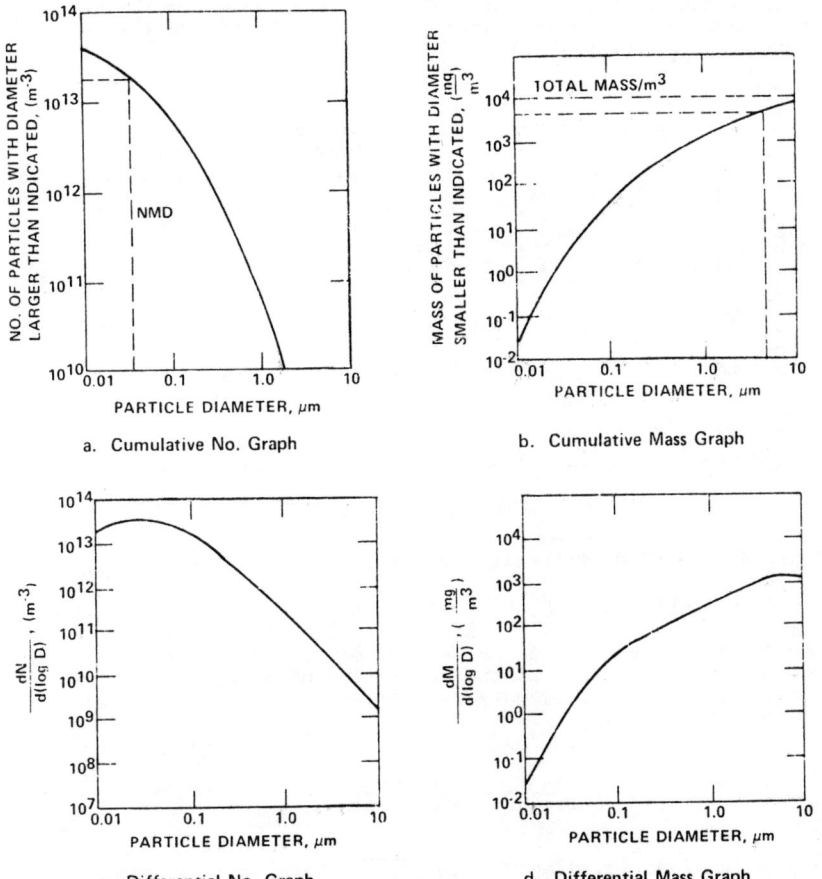

Figure 44. A single particle size distribution presented in four ways. The measurements were made in the effluence from a coal-fired power boiler.[52]

A second form of cumulative plot which is frequently used is the cumulative percent of mass, number, or surface area contained in particles having diameter smaller than a given size. In this case the ordinate would be, on a mass basis:

$$\text{Cumulative percent of mass less than size} = \frac{\sum_{t=1}^{j} M_t}{\sum_{t=1}^{N} M_t} \times 100\%. \quad (7)$$

The abscissa would be log D_j. Special log-probability paper is used for these graphs, and for log-normal distributions the data set would lie along a straight line. For such distributions the median diameter and geometric standard deviation can be easily obtained graphically. Figures 45-a and 45-b show cumulative percent graphs for the size distribution shown in Figure 44-a and a log-normal size distribution.[53]

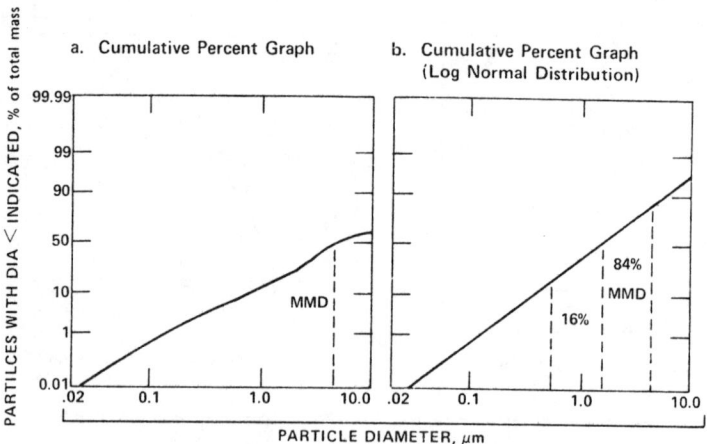

Figure 45. Size distributions plotted on log probability paper.[53]

Differential particle size distribution curves are obtained from cumulative plots by taking the average slope over a small size range as the ordinate and the geometric mean diameter of the range as the abscissa. If the cumulative plot were made on logarithmic paper, the frequency (slope) would be taking finite differences:

$$\frac{\Delta M}{\Delta (\log D)} = \frac{M_j - M_{j-1}}{\log D_j - \log D_{j-1}} \quad (8)$$

and the abscissa would be $D_g = \sqrt{D_j D_{j-1}}$ where the size range of interest is bounded by D_j and D_{j-1}. M_j and M_{j-1} correspond to the cumulative masses below these sizes. Differential number and surface area distributions can be obtained from cumulative graphs in precisely this same way. Differential graphs show visually the size range where the particles are concentrated with respect to the parameter of interest. The area under the curve in any size range is equal to the amount of mass (number, or surface area) consisting of particles in that range, and the total area under the curve corresponds to the entire mass (number, or surface area) of particulate matter in a unit volume. Again, because of the extent in particle size and the emphasis on the fine particle fraction, these plots are normally made on logarithmic scales. Figures 44-c and 44-d are examples of differential graphs of particle size distributions.

The particulate emissions from many industrial processes frequently follow the log-normal distribution law rather well. The particulate emissions from coal-fired boilers can often be approximated by a log-normal particle size distribution. For log-normal particle size distributions the geometric mean and median diameters coincide.

The normal distribution law is, on a mass basis:

$$f = \frac{dM}{dD} = \frac{1}{\sigma\sqrt{2\pi}} \exp\left[-\frac{(D-D_m)^2}{2\sigma^2}\right]. \tag{9}$$

The log-normal distribution law is derived from this equation by the transformation $D \rightarrow \log D$:

$$f = \frac{DM}{d(\log D)} = \frac{1}{\sqrt{2\pi} \log \sigma_g} \exp\left[-\frac{1}{2}\left(\frac{\log D - \log D_{gm}}{\log \sigma_g}\right)^2\right], \tag{10}$$

where σ_g, the geometric standard deviation, is obtained by using the transformation $D \rightarrow \log D$ in equation (10). This distribution is symmetric when plotted along a logarithmic abscissa and has the feature that 68.3% of the distribution lies within one geometric standard deviation of the geometric mean on such a plot. Mathematically, this implies that $\log \sigma_g = \log D_{84.14} - \log D_g$ or $\log D_g - \log D_{15.86}$ where $D_{84.14}$ is the diameter below which 84.14% of the distribution is found, etc. This can be simplified to yield:

$$\sigma_g \approx \frac{D_{84}}{D_g}, \tag{11}$$

$$\sigma_g \approx \frac{D_g}{D_{16}}, \text{ or} \tag{12}$$

$$\sigma_g \approx \left(\frac{D_{84}}{D_{16}}\right)^{\frac{1}{2}}. \tag{13}$$

When plotted on log-probability paper, the log-normal distribution is a straight line on any basis and is determined completely by the knowledge of D_g and σ_g. This is illustrated in Figure 45-b. Another important feature is the relatively simple relationships among log-normal distributions of different bases. If D_{gm}, D_{gs}, D_{gvs}, and D_{gN} are the geometric mean diameter of the mass, surface area, volume-surface, and number distribution, then:

$$\log D_{gs} = \log D_{gm} - 4.6 \log^2 \sigma_g , \qquad (14)$$

$$\log D_{gvs} = \log D_{gm} - 1.151 \log^2 \sigma_g , \text{ and} \qquad (15)$$

$$\log D_{gN} = \log D_{gm} - 6.9 \log^2 \sigma_g . \qquad (16)$$

The geometric standard deviation remains the same for all bases.

Particle size distributions may be expressed in terms of several different types of particle diameters. It is important to distinguish which type of particle diameter has been used in the construction of a particle size distribution since certain applications require that the particle size distribution be expressed in terms of a specific type of particle diameter.

If the density of a particle is known, the Stokes diameter (D_s) may be used to describe particle size. This is the diameter of a sphere having the same density which behaves aerodynamically as the particle of interest. For spherical particles, the Stokes number is equal to the actual dimensions of the particle.

The aerodynamic diameter (D_A) of a particle is the diameter of a sphere of unit density which has the same settling velocity in the gas as the particle of interest. The aerodynamic impaction diameter (D_{AI}) of a particle is an indication of the way that a particle behaves in an inertial impactor or in a control device where inertial impaction is the primary mechanism for collection.

If the particle Stokes diameter is known, the D_{AI} is equal to:

$$D_{AI} = D_s \sqrt{\rho C} , \qquad (17)$$

where ρ (gm/cm^3) is the particle density and C is the slip correction factor.

In optical particle sizing devices the intensity of light scattered by a particle at any given angle is dependent upon the particle size, shape, and index of refraction. It is impractical to measure each of these parameters and the theory for irregularly shaped particles is not well developed. Sizes based on light scattering by single particles are therefore usually estimated by comparison of the intensity of scattered light from the particle with the intensities due to a series of calibration spheres of very precisely known size. Most commonly these are polystyrene latex (PSL) spheres. Spinning disc and vibrating orifice aerosol generators can be used to generate monodisperse calibration aerosols of different physical properties. Because most manufacturers of optical particle sizing instruments use PSL spheres to calibrate their instruments, it is convenient to define an equivalent PSL diameter as the diameter of a PSL sphere which gives the same response with a particular optical instrument as the particle of interest.

Field Methods For Measuring Particle Size Distributions

General Considerations in Making Field Measurements--

An ideal particle size measurement device would be located <u>in situ</u> and give real time readout of particle size distributions and particle number concentration over the size range from 0.01 μm to 10 μm diameter. At the present time, however, partice size distribution measurements are made using several instruments which operate over limited size ranges and do not yield instantaneous data.

Particle sizing methods used in making field measurements may involve instruments which are operated in-stack, or out-of-stack where the samples are taken using probes. For in-stack sampling, the sample aerosol flow rate is usually adjusted to maintain near isokinetic sampling conditions in order to avoid concentration errors which result from under to oversampling large particles (dia. > 3μm) which have too high an inertia to follow the gas flow streams in the vicinity of the sampling nozzle. Since many particulate sizing devices have size fractionation points that are flow rate dependent, the necessity for isokinetic sampling in the case of large particles can result in undesirable compromises in obtaining data -- either in the number of points sampled or in the validity or precision of the data for large particles.

In general, particulate concentrations within a duct or flue are stratified to some degree with strong gradients often found for larger particles and in some cases for small particles. Such concentration gradients, which can be due to inertial effects, gravitational settling, passageway to passageway efficiency variations in the case of electrostatic precipitators, etc., require that multipoint (traverse) sampling be used.

Even the careful use of multipoint traverse techniques will not guarantee that representative data are obtained. The location of the sampling points during process changes or variations in precipitator operation can lead to significant scatter in the data. As an example, rapping losses in dry electrostatic precipitators tend to be confined to the lower portions of the gas streams, and radically different results may be obtained, depending on the magnitude of the rapping losses, and whether single point or traverse sampling is used. In addition, large variations in results from successive multipoint traverse tests can occur as a result of differences in the location of the sampling points when the precipitator plates are rapped. Similar effects will occur in other instances as a result of process variations and stratification due to settling, cyclonic flow, etc.

Choices of particulate measurement devices or methods for individual applications are dependent on the availability of suitable techniques which permit the required temporal and/or spatial resolution or integration. In certain instances the properties of the particulate are subject to large changes in not only size distribution and concentration, but also in chemical composition. Different methods or sampling devices are generally required to obtain data for long term process averages as opposed to the isolation of certain portions of the process in order to determine the cause of a particular type of emission.

Interferences exist which can affect most sampling methods.

Two commonly occurring problems are the condensation of vapor phase components from the gas stream and reactions of gas, liquid, or solid phase materials with various portions of the sampling systems. An example of the latter is the formation of sulfates due to appreciable (several milligram) quantities on several of the commonly used glass fiber filter media by reactions involving SO_x and trace constituents of the filter media. Sulfuric acid condensation in cascade impactors and in the probes used for extractive sampling is an example of the former.

If extractive sampling is used and the sample is conveyed through lengthy probes and transport lines, as is the case with several particle sizing methods, special attention must be given toward recognition, minimization, and compensation for losses by various mechanisms in the transport lines. The degree of such losses can be quite large for certain particle sizes.

In the following subsections, established field methods for measuring particle size distributions are briefly discussed. These are categorized according to the physical mechanism that is used to obtain the data: inertial (aerodynamic), optical, diffusional, or electrical. The purpose of the following subsections is to familiarize the reader with the various methods and instrumentation which are used to make field measurements of particle size distributions. The capabilities, limitations, advantages, and disadvantages of the various methods and instrumentation are presented. Detailed discussions of sampling procedures and data reduction techniques are given elsewhere[54,55] and will not be presented here.

Inertial (Aerodynamic) Methods--

Cascade impactors and cyclones are two types of inertial (aerodynamic) particle sizing instruments. These instruments employ the unique relationship between a particle's diameter and mobility in gas or air to collect and classify the particles by size. In order to avoid unnecessary complications in presenting data obtained with these instruments, particles of different shapes may be assigned aerodynamic diameters. Impactors and cyclones are well suited for industrial pollution studies because they are rugged and compact enough for in situ sampling.

Figure 46 is a schematic which illustrates the principle of particle collection which is common to all cascade impactors.[56] The sample aerosol is constrained to pass through a slit or circular hole to form a jet which is directed toward an impaction surface. Particles which have lower momentum will follow the air stream to lower stages where the jet velocities are progressively higher. For each stage there is a characteristic particle diameter which theoretically has a 50% probability of striking the collection surface. This particle diameter, or D_{50}, is called the effective cut size for that stage. Although single jets are shown in Figure 46 for illustrative purposes, the number of holes or jets on any one stage ranges from one to several hundred depending on the desired jet velocity and total volumetric flow rate. The number of jet stages in an impactor ranges from one to about twenty for various impactor geometries reported in the literature. Most commercially available impactors have between five and ten stages.

The particle collection efficiency of a particular impactor jet-plate combination is determined by the properties of the aerosol; such as the particle shape and density, the velocity of the air jet, and the viscosity of the gas; and by the design of the impactor stage, that is the shape of the jet, the diameter of the jet, and the jet-to-plate spacing.[57,58,59,60,61] There is also a slight dependence on the type of collection surface used (glass fiber, grease, metal, etc.).[62,63,64]

Most modern impactor designs are based on the semi-empirical theory of Ranz and Wong.[65] Although more sophisticated theories have been developed,[66,67,68] these are more difficult to apply. Since variations from ideal behavior in actual impactors dictate that they be calibrated experimentally, the theory of Ranz and Wong is generally satisfactory for the selection of jet diameters. Cohen and Montan,[57] Marple and Willeke,[58] and Newton, et al[60] have published papers that summarize the important results from theoretical and experimental studies to determine the most important factors in impactor performance:

(1) The jet Reynolds number should be between 100 and 3000.

(2) The jet velocity should be 10 times greater than the settling velocity of particles having the stage D_{50}.

(3) The jet velocity should be less than 110 m/sec.

(4) The jet diameter should not be smaller than can be attained by conventional machining technology.

(5) The ratio of the jet-plate spacing and the jet diameter or width (S/W) should lie between 1 and 3.

(6) The ratio of the jet throat length to the jet diameter (T/W) should be approximately equal to unity.

(7) The jet entries should be streamlined or countersunk.

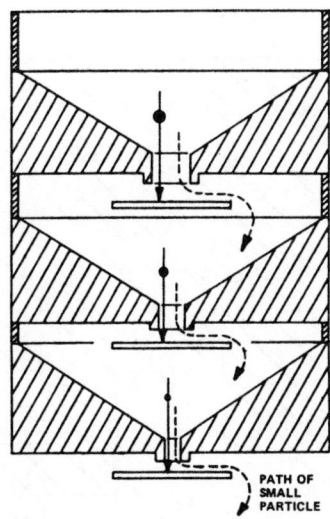

Figure 46. Schematic diagram, operation of cascade impactor.[56]

Smith and McCain[51] have observed that the jet velocity for optimum collection of dry particles may be as low as 10 m/sec, which places a more stringent criterion on impactor design and operation.

Figures 47 and 48 are charts that summarize the design criteria for cascade impactors.[51,58] It can be seen that it is almost impossible to achieve D_{50}'s of 0.2-0.3 μm without violating some of the recommended guidelines.

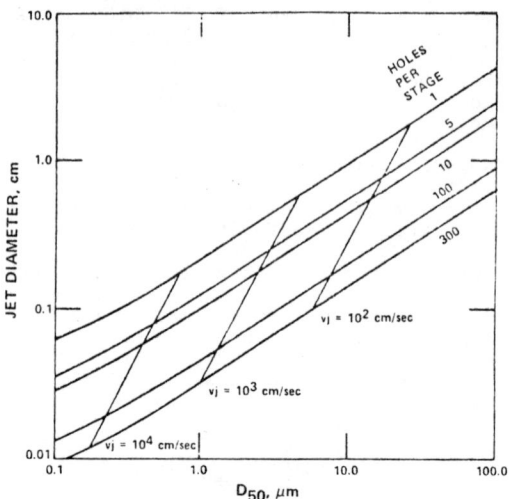

Figure 47. Approximate relationship among jet diameter, number of jets per stage, jet velocity, and stage cut point for circular jet impactors. From Smith and McCain.[51]

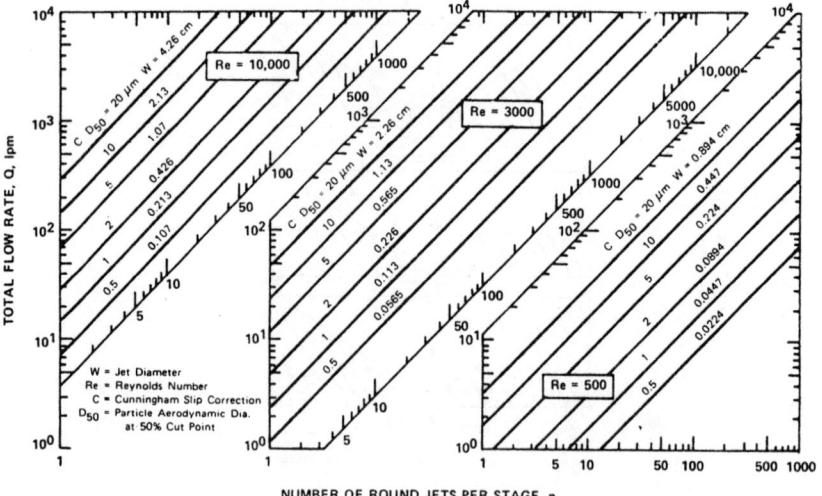

Figure 48. Design chart for round impactors. (D_{50} = aerodynamic diameter at 50% cut point.) After Marple.[58]

Table 7 lists six commercially available cascade impactors that are designed for in-stack use.[69] Table 47 in Appendix B shows some geometric and operating parameters for the commercial impactors.[69] Schematics of the commercial impactors are shown in Figure 49.[70]

TABLE 7. COMMERCIAL CASCADE IMPACTOR SAMPLING SYSTEMS[69]

Name	Nominal Flow Rate (cm³/sec)	Substrates	Manufacturer
Andersen Stack Sampler (Precollection Cyclone available)	236	Glass Fiber (available from manufacturer)	Andersen 2000, Inc. P.O. Box 20769 Atlanta, GA 30320
University of Washington Mark III Source Test Cascade Impactor (Precollection Cyclone available)	236	Stainless Steel Inserts, Glass Fiber, Grease	Pollution Control System Corp. 321 Evergreen Bldg. Renton, WA 98055
University of Washington Mark V	100	Stainless Steel Inserts, Glass Fiber, Grease	Pollution Control System Corp. 321 Evergreen Bldg. Renton, WA 98055
Brink Cascade Impactor (Precollection Cyclone available)	14.2	Glass Fiber, Aluminum, Grease	Monsanto EviroChem Systems, Inc. St. Louis, MO 63166
Sierra Source Cascade Impactor—Model 226 (Precollection Cyclone available)	118	Glass Fiber (available from manufacturer)	Sierra Instruments, Inc. P.O. Box 909 Village Square Carmel Valley, CA 93924
MRI Inertial Cascade Impactor	236	Stainless Steel, Aluminum, Mylar, Teflon. Optional: Gold, Silver, Nickel	Meteorology Research, Inc. Box 637 Altadena, CA 91001

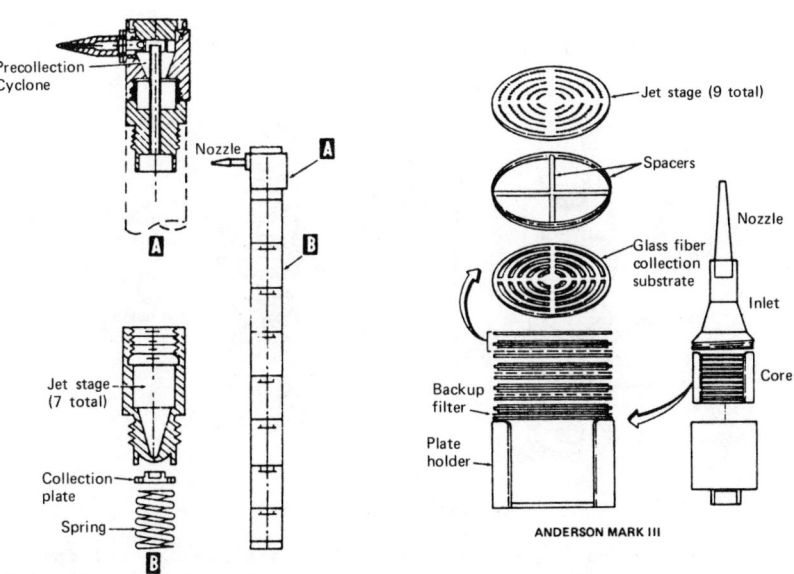

Figure 49. Schematics of five commercial cascade impactors.[70]
(Continued)

Figure 49. (Continued)

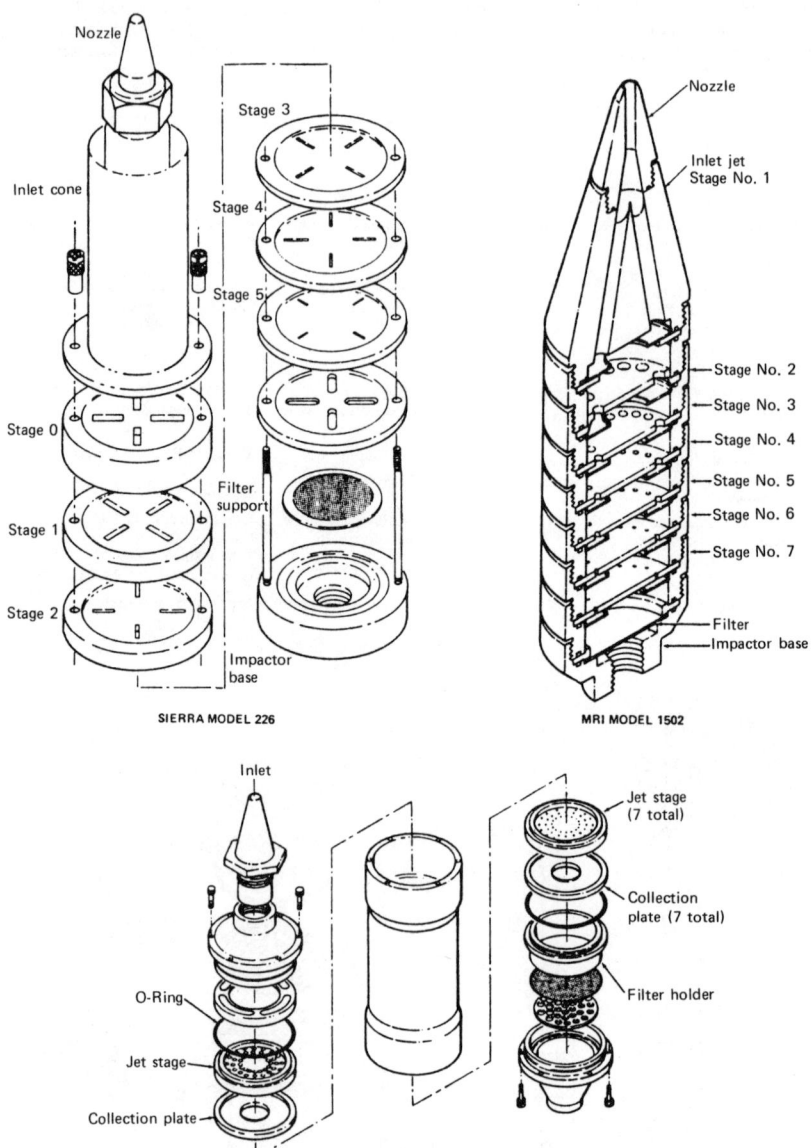

The impactors are all constructed of stainless steel for corrosion resistance. All of the impactors have round jets, except the Sierra Model 226, which is a radial slit design, and all have stages with multiple jets, except the Brink. It is customary to operate the impactors at a constant flow rate during a test so that the D_{50}'s will remain constant. The impactor flow rate is chosen, within a fairly narrow allowable range, to give a certain sampling velocity at the nozzle inlet. Streamlined nozzles of different diameters are provided to allow the sample to be taken at a velocity equal to that of the gas stream.

Since the impaction plates weigh a gram or more, and the typical mass collected on a plate during a test is on the order of 1-10 mg, it is often necessary to place a light weight collection substrate over the impaction plate to reduce the tare. These substrates are usually glass fiber filter material or greased aluminum foil. A second function of the substrates is to reduce particle bounce.

Cushing, et al have done extensive calibration studies of the commercial, in-stack, cascade impactors.[63] Figure 50 shows results from calibration of the Andersen Mark III impactor that are typical of the performance of the other types as well. Similar results have been reported by Mercer and Stafford,[68] Rao and Whitby,[62] and Calvert, et al[71] for impactors of different design. Notice that the calibration curve increases, as particle size increases, up to a maximum value that is less than 100%. The decrease in collection efficiency for large particles represents bounce and can introduce serious errors in the calculated particle size distribution.

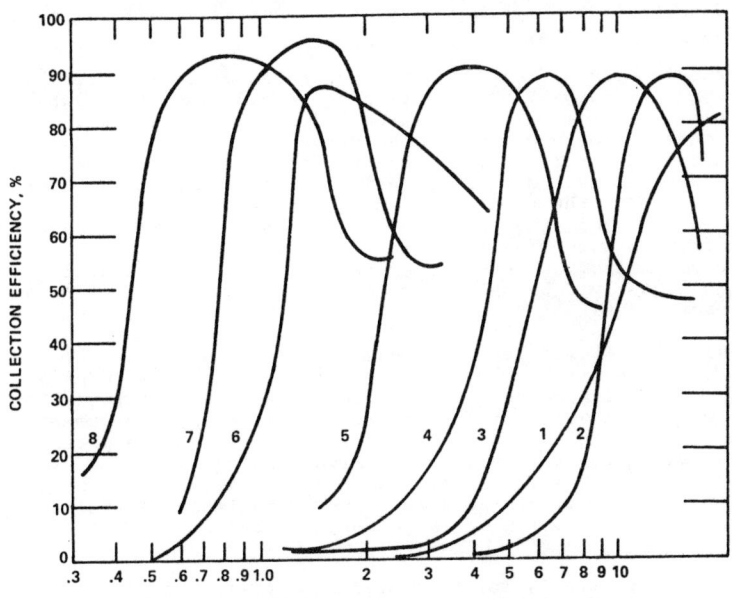

Figure 50. Calibration of an Anderson Mark III impactor. Collection efficiency vs. particle size for stages 1 through 8. After Cushing, et al.[63]

There has not been an extensive evaluation of cascade impactors under field conditions, although some preliminary work was reported by McCain, et al.[72] It is difficult to judge from existing data exactly how accurate impactors are, or how well the data taken by different groups or with different impactors will correlate. Problems that are known to exist in the application of impactors in the field are: substrate instability,[68,73] the presence of charge on the aerosol particles,[74] particle bounce,[62,68] and mechanical problems in the operation of the impactor systems.

It is usually impractical to use the same impactor at the inlet and outlet of an electrostatic precipitator when making fractional efficiency measurements because of the large difference in particulate loading. For example, if a sampling time of thirty minutes is adequate at the inlet, for the same impactor operating conditions and the same amount of sample collected, approximately 3000 minutes sampling time would be required at the outlet (a collection efficiency of 99% is assumed). Although impactor flow rates can be varied, they cannot be adjusted enough to compensate for this difference in particulate loading without creating other problems. Extremely high sampling rates result in particle bounce and in scouring of impacted particles from the lower stages of the impactor where the jet velocities become excessively high. Short sampling times may result in atypical samples being obtained as a result of momentary fluctuations in the particle concentration or size distribution within the duct. Normally, a low flow rate impactor is used at the inlet and a high flow rate impactor at the outlet. The impactors are then operated at their respective optimum flow rates, and the sampling times are dictated by the time required to collect weighable samples on each stage without overloading any single stage.

Particle size distribution measurements related to precipitator evaluation have largely been made using cascade impactors, which are effective in the size range from 0.3 to 20 μm diameter; although, in some cases, hybrid cyclone-impactor units or cyclones have also been used. The particle size distributions are normally calculated from the experimental data by relating the mass collected on the various stages to the theoretical or calibrated size cutpoints associated with the stage geometries. In the past, the reduction of data from an extensive field test has been excessively tedious and time consuming. However, computer programs are now available that significantly decrease the effort required to reduce and analyze impactor data.[75,76]

Figure 51 illustrates a typical <u>reverse flow cyclone</u>.[77] The aerosol sample enters the cyclone through a tangential inlet and forms a vortex flow pattern. Particles move outward toward the cyclone wall with velocities that are determined by the geometry and flow rate in the cyclone, and by their size. Large particles reach the walls and are collected. Figure 52 compares the calibration curve for a small cyclone with a typical impactor calibration curve.[78] The cyclone can be seen to perform almost as well as the impactor, and the problem of large particle bounce and reentrainment is absent.

An accurate theory for describing the operation of small cyclones has not yet been developed. Thus, cyclones used for particle sizing are presently designed and calibrated based on experience and experiment. As with impactors, cyclone performance may be conveniently expressed in terms of a characteristic D_{50}, which is the diameter of particles that are collected with 50% efficiency. In experiments with small cyclones, Chan and Lippmann[79] have observed that most cyclone performance data can be fitted by equations of the form

$$D_{50} = KQ^n , \qquad (18)$$

where:

    K = an empirical constant,

    Q = the sample flow rate, and

    n = an empirical constant.

Unfortunately, K and n are different for each cyclone geometry, and apparently are impossible to predict. In their study, Chan and Lippmann found K to vary from 6.17 to 4591, and n from -0.636 to -2.13. A similar study by Smith and Wilson[80] found K to vary from 44 to 14, and n from -0.63 to -1.11 for five small cyclones.

In addition to the flow rate dependence indicated in equation (18), cyclone D_{50}'s also are affected by temperature through the viscosity of the gas. Smith and Wilson found this dependence to be linear, but with a different slope for different cyclone dimensions and flow rates.

It is mandatory that the gas velocity and temperature through the cyclones be maintained at a constant setting while sampling, because the cyclone cut points are dependent upon the gas flow rate and temperature. This usually means that periods of non-isokinetic sampling may occur. Depending on the magnitude of the fluctuations in the velocity of the sampled stream, this may or may not introduce significant errors in the sizing process.

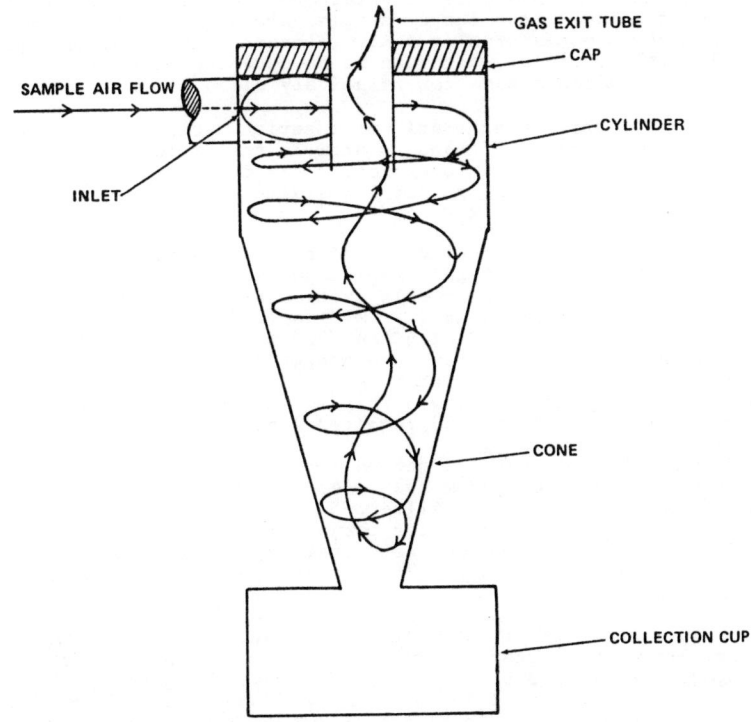

Figure 51. Hypothetical flow through typical reverse flow cyclone.[77]

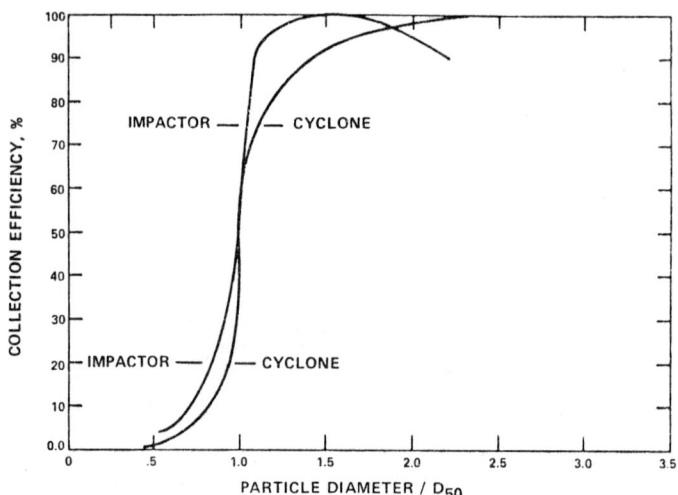

Figure 52. Comparison of cascade impactor stage with cyclone collection efficiency curve.[78]

A series of cyclones with progressively decreasing D_{50}'s can be used instead of impactors to obtain particle size distributions, with the advantages that larger samples are acquired and that particle bounce is not a problem. Also, longer sampling times are possible with cyclones, which can be an advantage for very dusty streams, or a disadvantage for relatively clean streams.

Figure 53 shows a schematic of a series cyclone system that was described by Rusanov[81] and is used in the Soviet Union for obtaining particle size information. This device is operated in-stack, but because of the rather large dimensions, requires a 20 cm port for entry.

Southern Research Institute, under EPA sponsorship, has designed and built a prototype three-stage series cyclone system for in-stack use.[82] A sketch of this system is shown in Figure 54. It is designed to operate at 472 cm^3/sec (1 ft^3/min). The D_{50}'s for these cyclones are 3.0, 1.6, and 0.6 micrometer aerodynamic at 21°C. A 47 mm Gelman filter holder, (Gelman Instrument Co., 600 South Wagner Road, Ann Arbor, MI 48106), is used as a back-up filter after the last cyclone. This series cyclone system was designed for in-stack use and requires a six inch sampling port.

Figure 55 illustrates a second generation EPA/Southern Research series cyclone system now under development which contains five cyclones and a back up filter.[80] It is a compact system and will fit through 4 inch diameter ports. The initial prototype was made of anodized aluminum with stainless steel connecting hardware. A second prototype, for in-stack evaluation, is made of titanium.

Figure 56 contains laboratory calibrations data for the five cyclone prototype system.[80] The D_{50}'s, at the test conditions, are 0.32, 0.6, 1.3, 2.6, and 7.5 µm. A continuing research program includes studies to investigate the dependence of the cyclone cut points upon the sample flow rate and temperature so that the behavior of the cyclones at stack conditions can be predicted more accurately.[80]

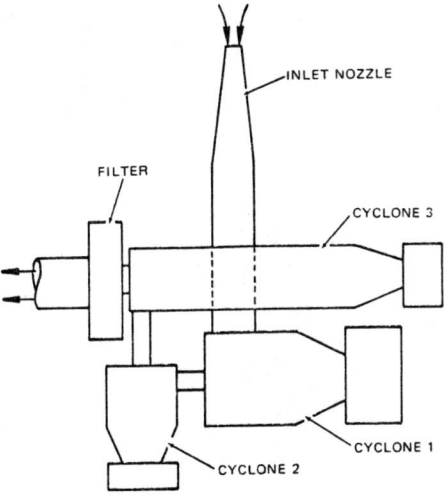

Figure 53. Series cyclone used in the U.S.S.R. for sizing flue gas aerosol particles. From Rusanov.[81]

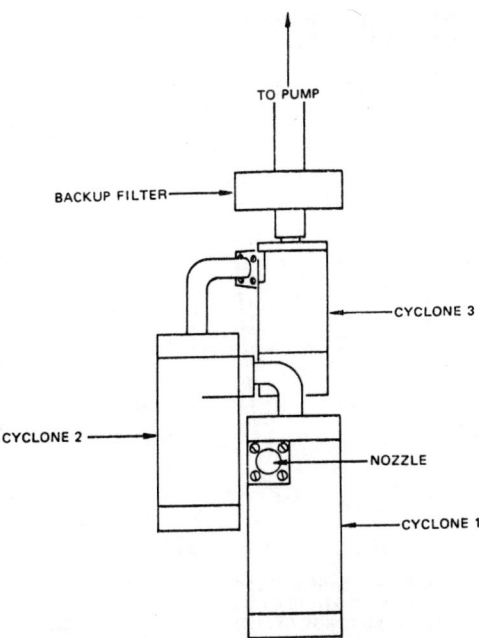

Figure 54. Schematic of the Southern Research Institute three series cyclone system.[82]

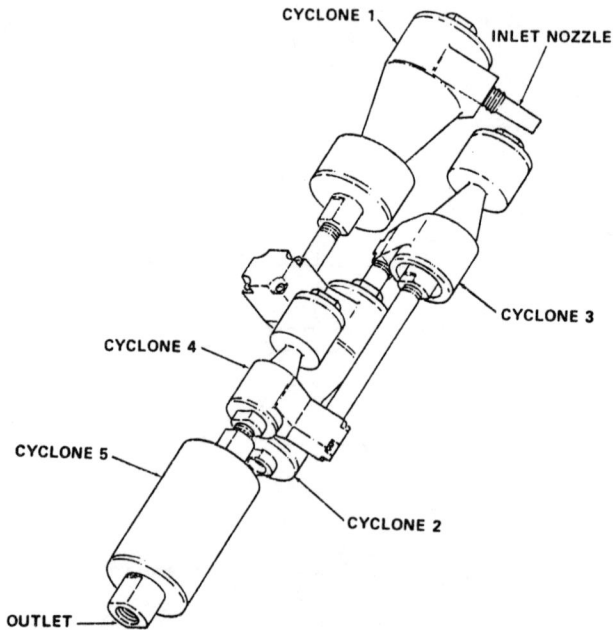

Figure 55. The EPA/Southern Research Institute five series cyclone system.[80]

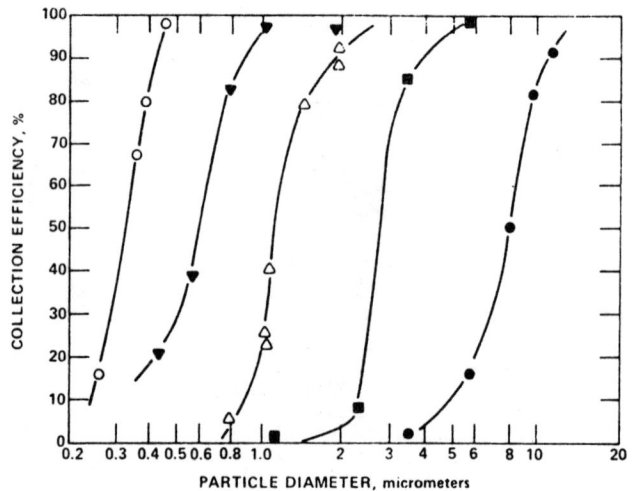

Figure 56. Laboratory calibration of the EPA/Southern Research Institute five series cyclone system. (Flow rate of 28.3 ℓ/min, temperature of 20°C, and particle density of 1 g/cm^3.)[80]

The Acurex-Aerotherm Source Assessment Sampling System (SASS) incorporates three cyclones and a back up filter.[83] Shown schematically in Figure 57,[84] the SASS is designed to be operated at a flow rate of 3065 cm^3/sec (6.5 ft^3/min) with nominal cyclone D_{50}'s of 10, 3, and 1 micrometer aerodynamic diameter at a gas temperature of 205°C. The cyclones, which are too large for in situ sampling, are heated in an oven to keep the air stream from the heated extraactive probe at stack temperature or above the dew point until the particulate is collected. Besides providing particle size distribution information, the cyclones collect gram quantities of dust (due to the high flow rate) for later chemical and biological analyses. The SASS train is available from Acurex-Aerotherm, Inc., 485 Clyde Avenue, Mountain View, California 94042.

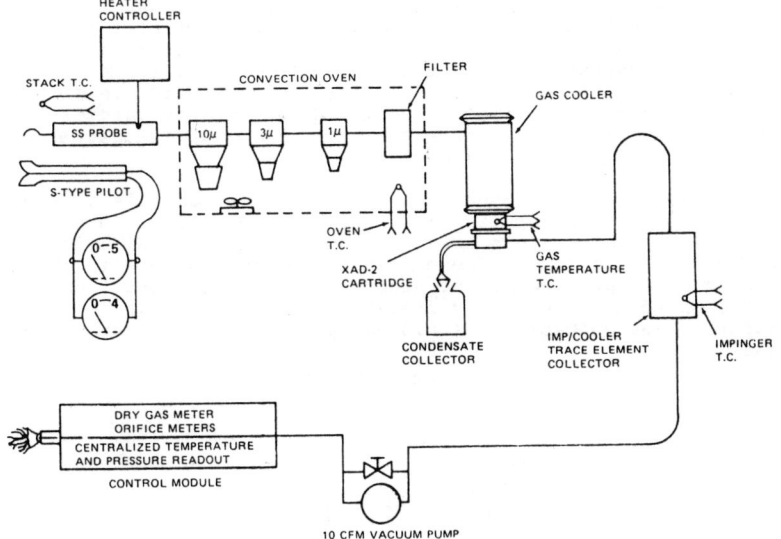

Figure 57. Schematic of the Acurex-Aerotherm source assessment sampling system (SASS).[84]

Small cyclone systems appear to be practical alternatives to cascade impactors as instruments for measuring particle size distributions in process streams. Cyclones offer several advantages:

Large, size-segregated samples are obtained.

There are no substrates to interfere with analyses.

They are convenient and reliable to operate.

They allow long sampling times under high mass loading conditions for a better process emission average.

They may be operated at a wide range of flow rates without particle bounce or reentrainment.

On the other hand, there are some negative aspects of cyclone systems which require further investigation:

Unduly long sampling times may be required to obtain large samples at relatively clean sources.

The existing theories do not accurately predict cyclones performance.

Cyclone systems are bulkier than impactors and may require larger ports for in-stack use.

Optical Methods--

Figure 58 is a schematic illustrating the principles of operation for optical particle counters.[85] A dilute aerosol stream intersects the focus of a light beam to form an optical "view volume". The photodetector is located so that no light reaches its sensitive cathode except that scattered by particles in the view volume. Each particle that scatters light with enough intensity will generate a current pulse at the photodetector, and the amplitude of the pulse can be related to the particle diameter. The rate at which the pulses occur is related to the particle concentration. Thus, optical particle counters yield real time information on particle size and concentration.

The simultaneous presence of more than one particle in the viewing volume is interpreted by the counter as a large single particle. To avoid errors arising from this effect, dilution to about 300 particles/cm^3 is generally necessary. Errors in counting rate also occur as a result of electronics deadtime and from statistical effects resulting from the presence of high concentrations of subcountable (D < 0.3 μm) particles in the sample gas stream.[86]

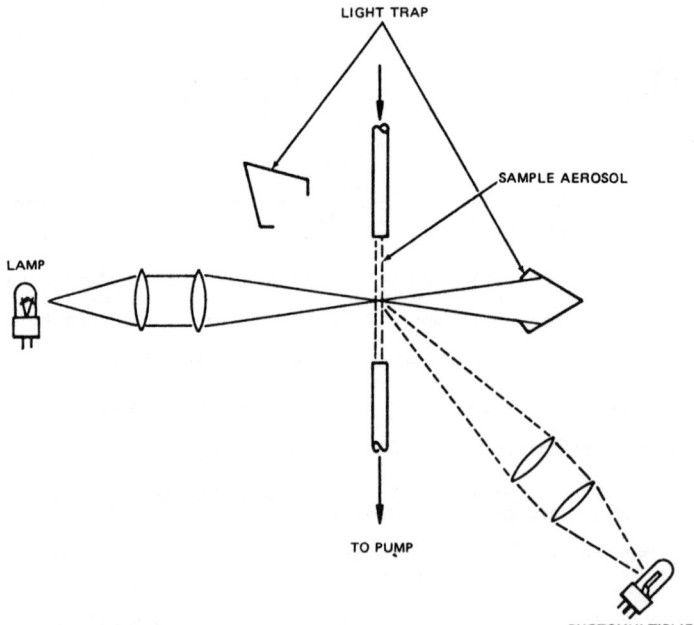

Figure 58. Schematic of an optical single particle counter.[85]

In an optical particle counter, the intensity of the scattered light, and amplitude of the resulting current pulse, depends on the viewing angle, particle refractive index, particle shape, and particle diameter. Different viewing angles and optical geometries are chosen to optimize some aspect of the counter performance. For example, the use of near forward scattering will minimize the dependence of the response on the particle refractive index, but with a severe loss of resolution near 1 μm diameter. The use of right angle scattering smooths out the response curve, but the intensity is more dependent on the particle refractive index. Figure 59 shows calibration data for near forward and right angle scattering particle counters.[87]

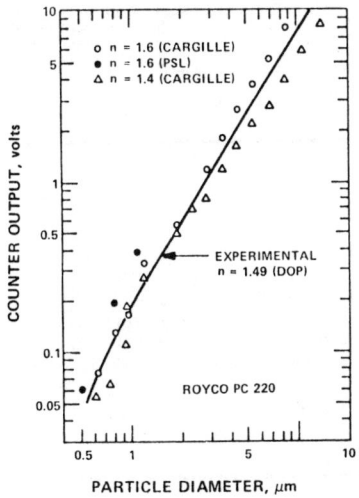
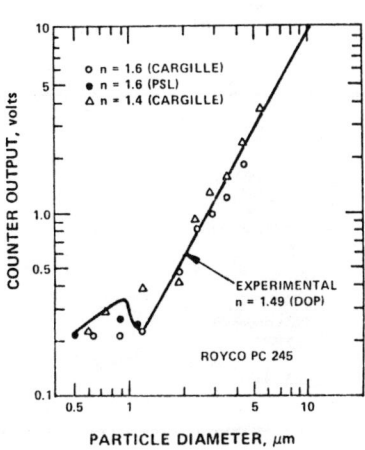

a. Right angle scattering. b. Near forward scattering.

Figure 59. Experimental calibration curves for two optical particle counters. After Willeke and Liu.[87]

Figure 60 illustrates some of the optical configurations that are found in commercial particle counters.[88] The pertinent geometric and operating constants of the counters are summarized in Table 8.[89]

The commercial optical counters that are available now were designed for laboratory work and have concentration limits of a few hundred particles per cubic centimeter. The lower size limit is nominally about 0.3 μm diameter. For use in studies of industrial aerosols, the gas sample must be extracted, cooled, and diluted; a procedure which requires great care to avoid introducing serious errors into calculations of the particle size distribution. The useful upper limit in particle size is limited by losses in the dilution system to about 2.0 μm diameter.[27] In addition, the particle diameter that is measured is not aerodynamic, and some assumptions must be made in order to compare optical with aerodynamic data. (It is possible to "calibrate" an optical counter, on a particulate source, to yield aerodynamic data. This is done by using special calibration impactors,[90] or settling chambers.[91]) Nevertheless, the ability to obtain real time information can sometimes be very important and the special problems in sampling with optical counters may be justified.

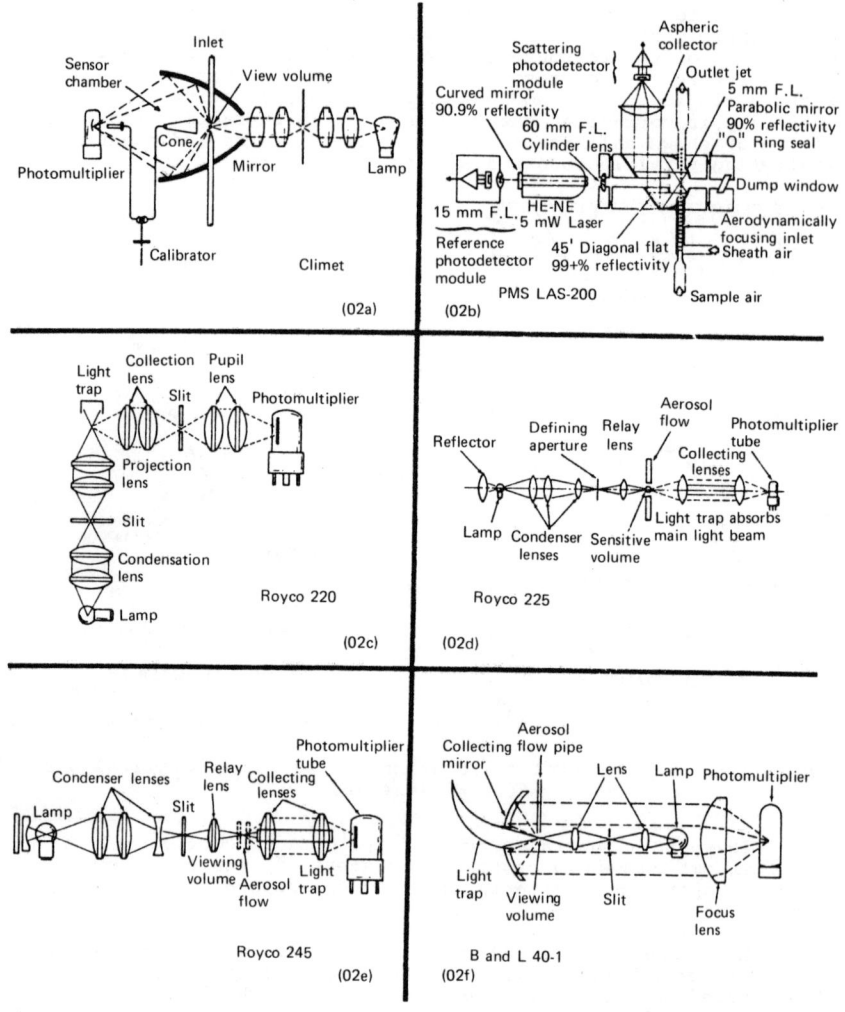

Figure 60. Optical configuration for six commercial particle counters.[88]

Diffusional and Condensation Nuclei Methods--

The classical technique for measuring the size distribution of submicron particles employs the relationship between particle diffusivity and diameter. In a diffusional sizing system, the test aerosol is drawn, under conditions of laminar flow, through a number of narrow, rectangular channels, a cluster of small bore tubes, or

TABLE 8. CHARACTERISTICS OF COMMERCIAL, OPTICAL, PARTICLE COUNTERS[8,9]

	Illuminating Cone Half Angle, γ	Light Trap Half Angle, α	Collecting Aperture Half Angle, β	Inclination Between Illuminating And Collecting Cone Axis, ψ	Viewing Volume	Sampling Rate
Bausch & Lomb Model 40-1 820 Linden Ave Rochester, NY 14625	13°	33°	53°	0°	0.5 mm³	170 cm³/min
Climet Models 201, 208 Climet Inst. Co. 1620 W. Colton Ave. Redlands, CA 92373	15	35	90	0	0.5	7,080
Climet Model 150	12	18	28	0	0.4	472
Royco Model 218 Royco Inst. 41 Jefferson Dr. Menlo Park, CA 94025	5	11	30	0	0.25	283
Royco Model 220	24	-	24	90	2.63	2,830
Royco Model 245	5	16	25	0	4.0	28,300
Royco Model 225	5	7	25	0	2.0	283 or 2,830
Tech Ecology Model 200 Tech Ecology, Inc. 645 N. Mary Ave. Sunnyvale, CA 94086	5	8	20	0	0.46	283
Tech Ecology Model 208	5	10	20	0	2.5	2,830
Particle Measurement Systems	0.5	35	120	0	0.003	120 or 1,200
*Model LAS-200 Particle Measuring Systems 1855 S. 57th Ct. Boulder, CO 80301						

*632.8 nm laser illum., all others are white light.

a series of small mesh screens (diffusion batteries). For a given particle diameter and diffusion battery geometry, it is possible to predict the rate at which particles are lost to the walls by diffusion, the rate being higher for smaller particles. The total number of particles penetrating the diffusion battery is measured under several test conditions where the main adjustable parameter is the aerosol retention time, and the particle size distribution is calculated by means of suitable mathematical deconvolution techniques. It is only necessary that the particle detector (usually a condensation nuclei counter) that is used at the inlet and outlet of the diffusion battery system responds to the total concentration, by number, of the particles in the size range of interest.

Figure 61-a shows a typical parallel channel diffusion battery, and Figure 61-b shows the aerosol penetration characteristics of this geometry at two flow rates.[92] The parallel plate geometry is convenient because of ease of fabrication and the availability of suitable materials, and also because sedimentation can be ignored if the slots are vertical, while additional information can be gained through settling, if the slots are horizontal.

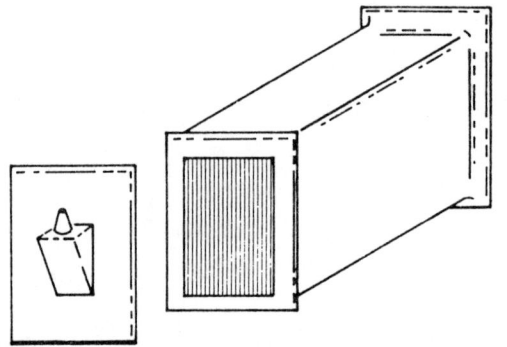

Figure 61-a. Parallel plate diffusion battery.[92]

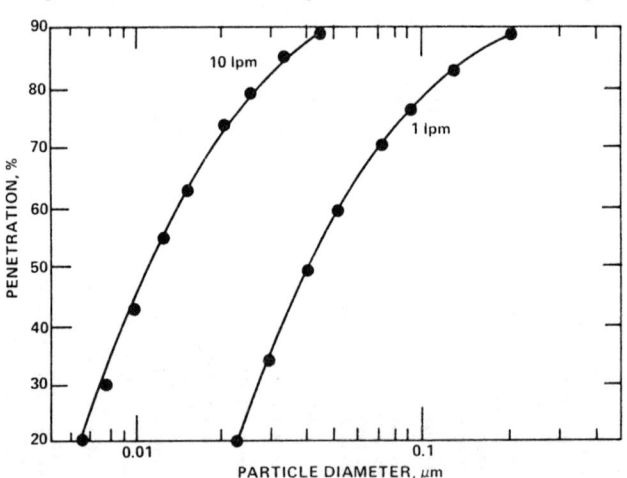

Figure 61-b. Parallel plate diffusion battery penetration curves for monodisperse aerosols (12 channels, 0.1 x 10 x 48 cm).[92]

Breslin et al[93] and Sinclair[94] report success with more compact, tube-type and screen-type arrangements in laboratory studies and a commercial version of Sinclair's geometry is available. (TSI Incorporated, 500 Cardigan Road, St. Paul, MN 55165). Although the screen-type diffusion battery must be calibrated empirically, it offers convenience in cleaning and operation, and compact size. Figure 62 shows Sinclair's geometry.[94] This battery is 21 cm long, approximately 4 cm in diameter, and weighs 0.9 kg, and contains 55 stainless steel screens of 635 mesh.

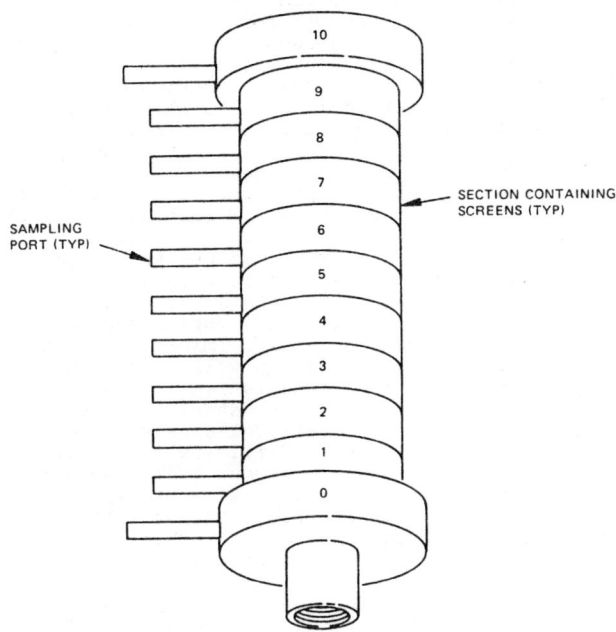

Figure 62. Screen-type diffusion battery. The battery is 21 cm long, 4 cm in diameter, and contains 55, 635 mesh stainless steel screens. After Sinclair.[94]

Diffusional measurements are less dependent upon the aerosol parameters than the other techniques discussed and perhaps are on a more firm basis from a theoretical standpoint.

Disadvantages of the diffusional technique are the bulk of the parallel plate diffusional batteries, although advanced technology may alleviate this problem; the long time required to measure a size distribution; and problems with sample conditioning when condensible vapors are present.

A practical limitation on the lower size limit for all methods used to determine ultrafine particle size distributions (diameters < 0.5 μm) is the loss of particles by diffusion in the sampling lines and instrumentation. These losses are excessive for particle diameters below about 0.01 μm where the samples are extracted from a duct and diluted to concentrations within the capability of the sensing devices.

Condensation nuclei (CN) counters function on the principle that particles act as nuclei for the condensation of water or other condensable vapors in a supersaturated environment. This process is used to detect and count particles with diameters in the 0.002 to 0.3 μm range (often referred to as condensation or Aitken nuclei). In condensation nuclei detectors, a sample is withdrawn from the gas stream, humidified, and brought to a supersaturated condition by reducing the pressure. In this supersaturated condition, condensation will be initiated on all particles larger than a certain critical size and will continue as long as the sample is supersaturated. This condensation process forms an homogeneous aerosol, predominantly composed of the condensed vapor containing one drop for each original particle whose size was greater than the critical size appropriate to the degree of supersaturation obtained; a greater degree of supersaturation is used to initiate growth on smaller particles. The number of particles that are formed is estimated from the light scattering properties of the final aerosol.

Because of the nature of this process, measurements of very high concentrations can be in error as a result of a lack of correspondence between particle concentration and scattering or attenuation of light. Additional errors can result from depletion of the vapor available for condensation. Certain condensation nuclei measuring techniques can also obtain information on the size distribution of the nuclei; that is, variations in the degree of supersaturation will provide size discrimination by changing the critical size for which condensation will occur. However, the critical size for initiating condensation is also affected by the volume fraction of water soluble material contained in the original aerosol particle, so the critical size will be uncertain unless the solubility of the aerosol particles is known.[95] At very high degrees of supersaturation (about 400%), solubility effects are only minor and essentially all particles in the original aerosol with diameters larger than 0.002 μm will initiate the condensation process. Figure 63, after Haberl, illustrates the condensation nuclei counter operating principle.[96]

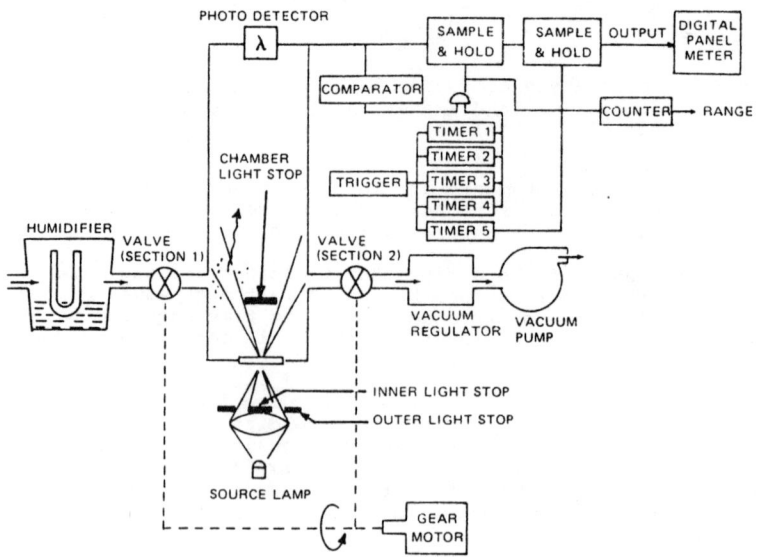

Figure 63. Diagram of a condensation nuclei counter. After Haberl and Fusco.[96]

Four models of CN counters are now available commercially. Two automatic, or motorized, types are the General Electric Model CNC-2 (General Electric-Ordnance Systems, Electronics Systems Division, Pittsfield, MA 01201) and the Environment-One Model Rich 100 (Environment-One Corporation, Schenectady, NY 12301). Small, manually operated, CN counters are also available from Gardner Associated, (Gardner Associates, Schenectady, NY 12301), and Environment-One.

The General Electric CN counter has mechanically actuated valves and is insensitive to moderate pressure variations at the inlet. The aerosol concentration is measured by the detection of scattered light from the test aerosol.

A disadvantage of the flow/valving arrangement in the General Electric counter is the intermittent (ℓ/sec) flow which introduces severe pressure pulsations into the sampling system. This problem has been minimized by the use of antipulsation devices consisting of a rubber diaphram[97] or two metal cylinders connected by a small orifice,[98] essentially pneumatic R-C networks.

The automatic Environment-One counter has some pneumatic valves. A pressure of more than 5 cm of water at the inlet can interrupt the operation. In the E-1, the aerosol concentration is measured by light extinction. The sampling rate of the E-1 counter can be adjusted from about 0.6 to 4.2 ℓ/min. Soderholm has reported a modification to the E-1 counter that replaces pneumatic valves with solenoidal ones.[99]

Fuchs[100] has reviewed diffusional sizing work up until 1956, while Sinclair,[101,97,102] Breslin, et al,[103] Twomey,[104] Sansone and Weyel,[105] and Ragland, et al,[90] have reported more recent work, both theoretical and experimental.

Figure 64 is a schematic diagram that illustrates an experimental setup for measuring particle size distributions by diffusional means, and Figure 65 shows penetration curves for four operating configurations.[106]

Because of the long retention time required for removal of particles by diffusion, measurements with diffusion batteries and CN counters are very time consuming. With the system described by Ragland, et al, for example, approximately two hours are required to measure a particle size distribution with diameters from 0.01 to 0.2 μm.[98] Obviously, this method is best applied to stable aerosol streams. It is possible that the new, smaller diffusion batteries will allow much shorter sampling times, but pulsations in flow may pose a serious problem for the low volume geometries.

Electrical Mobility Method--

An instrument that was developed for measuring laboratory and ambient aerosols over the 0.003 to 1 μm range of diameters, the electrical mobility analyzer, can also be applied to process streams with a suitable sample dilution and cooling interface.

Figure 66 illustrates the relationship between the diameter and electrical mobility of small aerosol particles.[107] If particles larger than those of minimum mobility are removed from the sample, the remaining particles exhibit a monotonically decreasing mobility with increasing diameter. Several aerosol spectrometers, or mobility analyzers, have been demonstrated that employ the

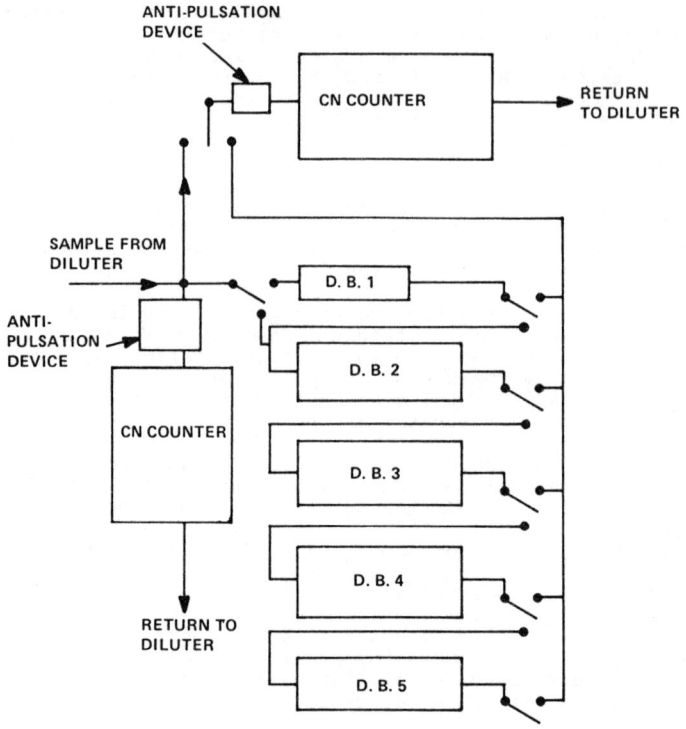

Figure 64. Diffusion battery and condensation nuclei counter layout for fine particle sizing.[106]

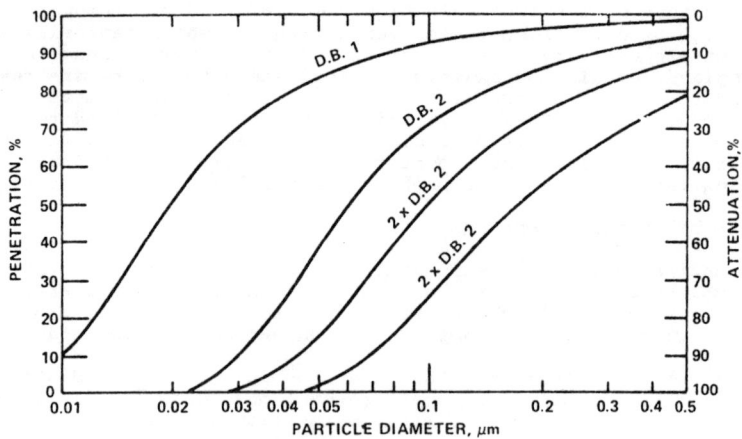

Figure 65. Theoretical parallel plate diffusion battery penetration curves.[106]

diameter-mobility relationship to classify particles according to their size.[108,109,110,111] Figure 67 illustrates the principle on which these devices operate.[112] Particles are charged under conditions of homogeneous electric field and ion concentration, and then passed into the spectrometer. Clean air flows down the length of the device and a transverse electric field is applied.

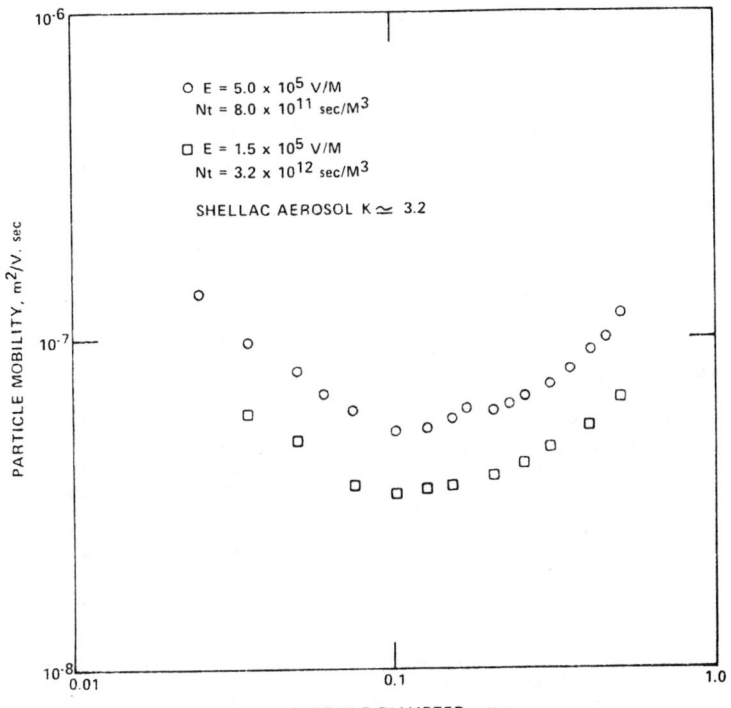

Figure 66. Particle mobility as a function of diameter for schellac aerosol particles charged in a positive ion field (after Cochet and Trillat[3]). K is the dielectric constant of the aerosol.[107]

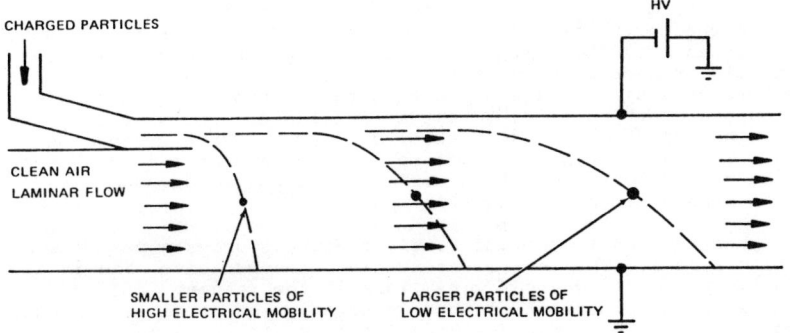

Figure 67. The electric mobility principle.[112]

From a knowledge of the system geometry and operating conditions, the mobility is derived for any position of deposition on the grounded electrode. The particle diameter is then readily calculated from a knowledge of the electric charge and mobility.

Difficulties with mobility analyzers are associated primarily with charging the particles (with a minimum of loss) to a known value and obtaining accurate analyses of the quantity of particles in each size range. The latter may be done gravimetrically,[108] optically,[109] or electrically.[110]

The concept described above has been used by Whitby, et al,[113,114] at the University of Minnesota, to develop a series of Electrical Aerosol Analyzers (EAA). A commercial version of the University of Minnesota devices is now marketed by TSI, Incorporated as the Model 3030 (Figure 68).[114] The EAA is designed to measure the size distribution of particles in the range from 0.0032 to 1.0 μm diameter. Since the concentration range for best operation is 1 to 1000 μg/m^3, dilution is required for most industrial gas aerosols.

The EAA is operated in the following manner. As a vacuum pump draws the aerosol through the analyzer (see Figure 68), a corona generated at a high voltage wire within the charging section gives the sample a positive electrical charge. The charged aerosol flows from the charger to the analyzer section as an annular cylinder of aerosol surrounding a cone of clean air. A metal rod, to which a variable, negative voltage can be applied, passes axially through the center of the analyzer tube. Particles smaller than a certain size (with highest electrical mobility) are drawn to the collecting rod when the voltage corresponding to that size is on the rod. Larger particles pass through the analyzer tube and are collected by a filter. The electrical charges on these particles drain off through an electrometer, giving a measure of current.

A step increase in rod voltage will cause particles of a larger size to be collected by the rod with a resulting decrease in electrometer current. This decrease in current is related to the additional number of particles being collected. A total of eleven voltage steps divide the 0.0032 to 1.0 micron size range of the instrument into ten equal logarithmic size intervals. Different size intervals can be programmed via an optional plug-in memory card.

The electrical aerosol analyzer can be operated either automatically or manually. In the automatic mode, the analyzer steps through the entire size range. For size and concentration monitoring over an extended period of time, the analyzer may be intermittently triggered by an external timer. The standard readout consists of a digital display within the control circuit module, although a chart recorder output is available. It is almost always advantageous to use a strip chart recorder to record the data. This allows the operator to identify a stable reading that may be superimposed on source variations and also gives a permanent record of the raw data.[115]

When the EAA is applied to fluctuating sources a peculiar problem arises. The instrument reading is cumulative, and it is impossible to tell whether variations in the reading reflect changes in the distribution or concentration of particles; hence, recordings that show rapid fluctuations in amplitude must be

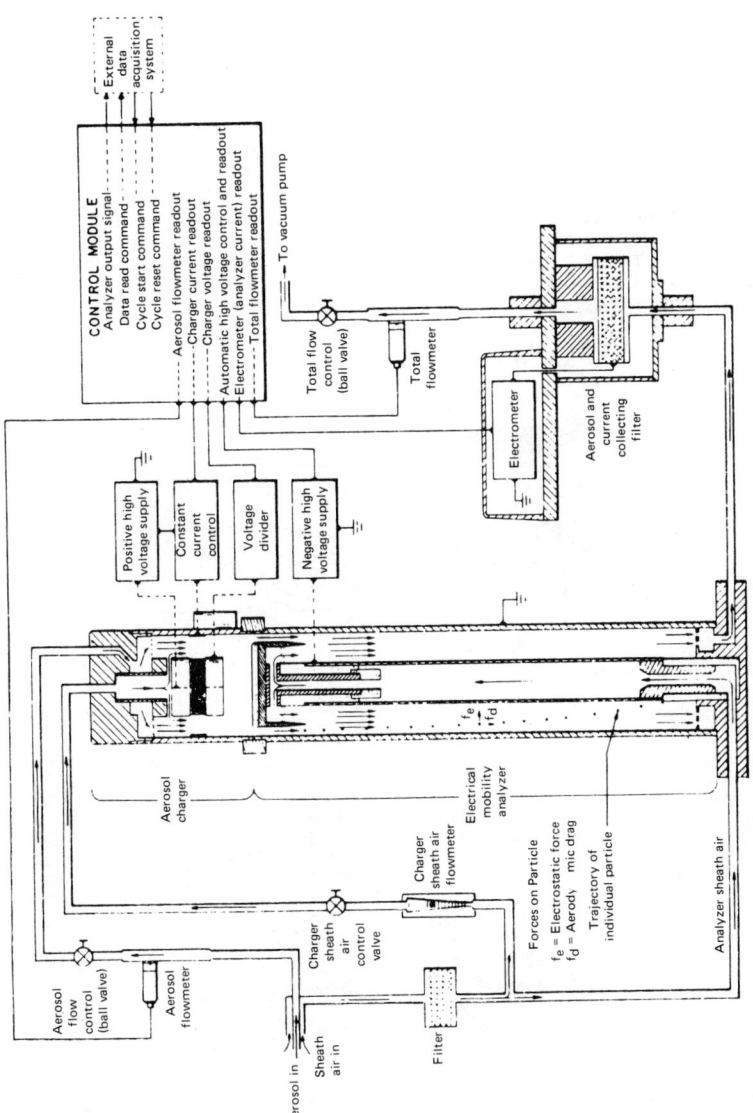

Figure 68. Flow schematic and electronic block diagram of the Electrical Aerosol Analyser.[114]

interpreted with great care. The lack of sensitivity can also be a problem at extremely clean sources.

The EAA requires only two minutes to perform a complete size distribution analysis, which generally makes it advantageous to use, especially on stable sources.

Other Specialized Particle Sizing Systems For Field Use

Respirable Particle Classifier (RPC) Impactor--

An in-stack sampling system, known as the **respirable particle classifier (RPC) impactor**, has been developed by Southern Research Institute to measure the particle emissions from stationary pollution sources in three size ranges.[116] The impactor, shown schematically in Figure 69, consists of a basic housing, a set of nozzles, a set of collection plates, and three sets of jet plates (two jet plates per set). The impactor body is anodized aluminum. The jet stages and collection stages are stainless steel.

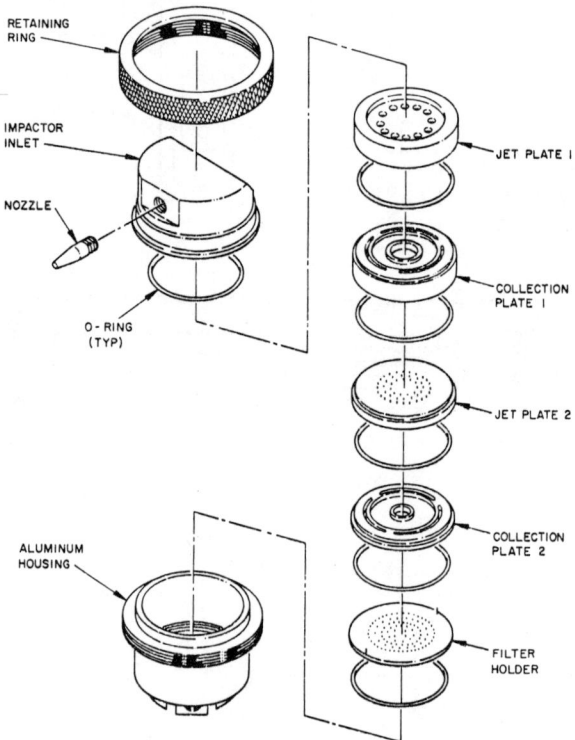

Figure 69. Respirable particle classifying (RPC) impactor.[116]

The impactor was tested on two coal-fired power plants. Concurrently with each test at Plant A a Brink Cascade Impactor was run to obtain a comparative size distribution. The results of the testing at Plant A, which occurred at the outlet of a precipitator collecting ash from a low sulfur Southeastern coal are shown in Figures 70-72. Concurrently with each test at Plant B an Andersen Stack Sampler was run with the new impactor to obtain a comparable size distribution. The results of the testing at Plant B, which occurred at the outlet of a precipitator collecting ash from a medium-high sulfur Southeastern coal, are shown in Figure 73.

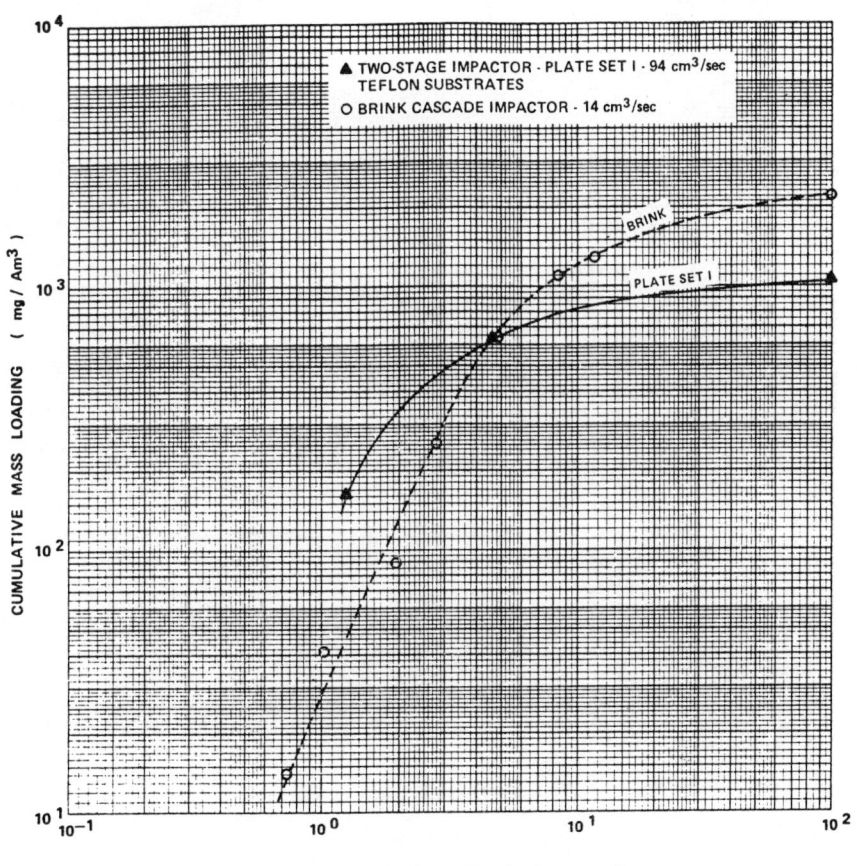

Figure 70. Cumulative mass loading versus particle diameter, March 11, 1975.[116]

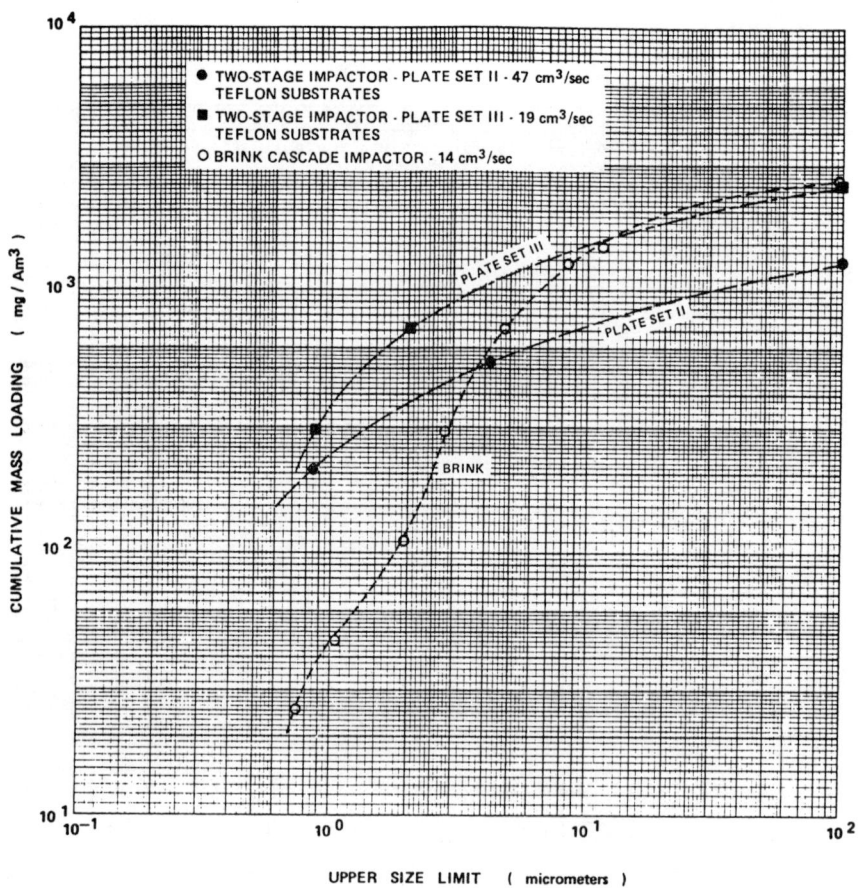

Figure 71. Cumulative mass loading versus particle diameter, March 12, 1975.[116]

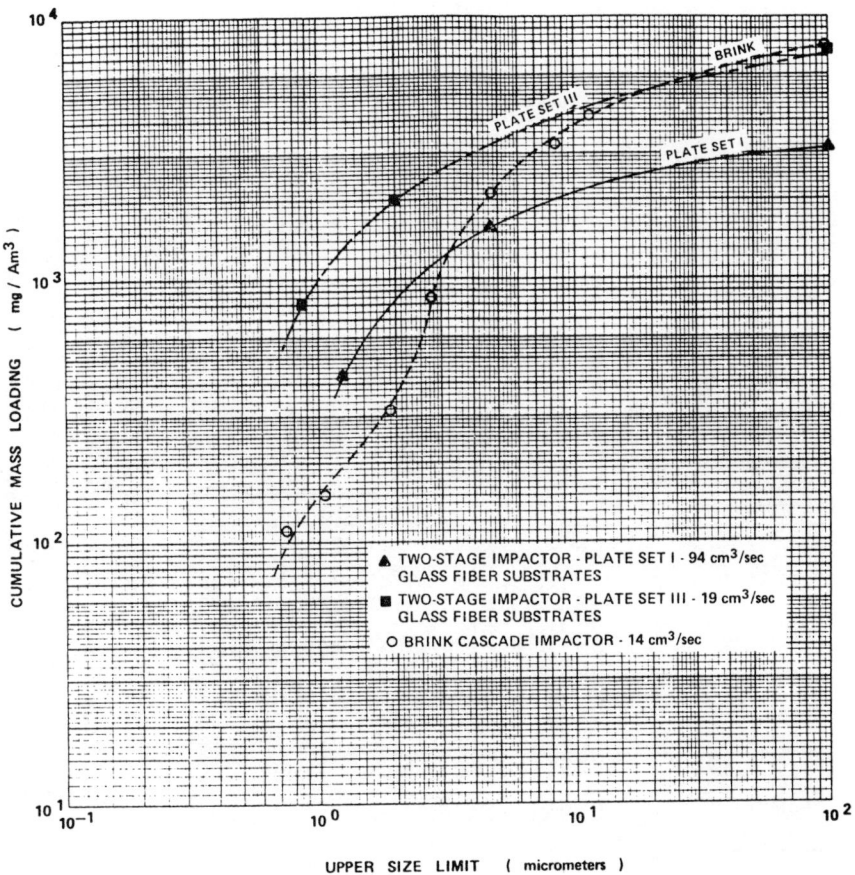

Figure 72. Cumulative mass loading versus particle diameter, March 13, 1975.[116]

100 Electrostatic Precipitator Manual

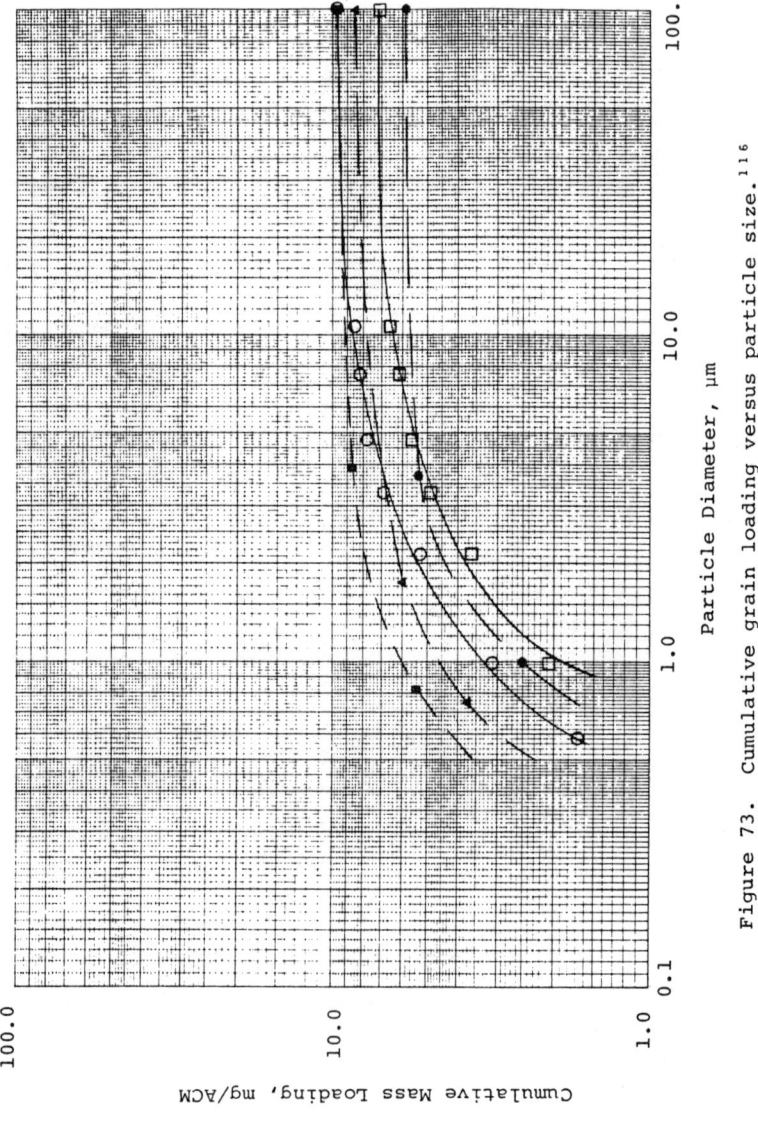

Figure 73. Cumulative grain loading versus particle size.[116]

o Andersen Stack Sampler – July 30, 1975 ● Two Stage Impactor – Jet Plate Set I – July 31, 1975
□ Andersen Stack Sampler – July 31, 1975 ■ Two Stage Impactor – Jet Plate Set II – July 31, 1975
▲ Two Stage Impactor – Jet Plate Set III – July 30, 1975

Large Particle Sizing System (LPSS)--

In order to more clearly define the mechanisms by which rapping losses occur in dry ESP's, time resolved data are required on the particulate concentrations and size distribution across typical portions of ESP exit planes. Conventional sampling methods generally require rather long integration times which are unsuited for examining 1 to 5 second transient events such as rapping puffs. Of the available measurement methods, only the optical single particle counters appear to offer the required combination of response time, dynamic range, and particle size resolution. Modified ambient air particle counters (Royco Model 225) are used as the measurement instrumentation in a large particle sizing system (LPSS) developed by Southern Research Institute.[117] The use of these counters requires extractive sampling and sample conditioning.

Due to instrumental limitations on the total concentration of aerosol particles in the sample gas stream arriving at the sensor, these particle counters require that the aerosol sample from the flue be diluted before measurement. Because of the very steep gradient in the size distribution on a number basis anticipated at the exit of a precipitator on a power boiler, the diluter was made as a size selective device which, under ideal conditions, dilutes small particles in the sample gas stream by fairly large factors while passing a relatively confined and undiluted stream of the lower concentration large particles directly to the particle sensor. Figure 74 illustrates the operational system for the particle dilution train.

The geometry of the diluter as shown in Figure 74 is such that the large particles, having high inertia, tend to pass directly from the inlet to the sample exit of the diluter while small particles (having relatively low inertia), are mixed to varying degrees, depending on their sizes with the dilution air thus producing a substantial reduction in concentration of small particles and condensable vapors in the sample stream while maintaining the concentration of the less numerous larger particles.

For the purpose of rapping studies, it is desirable to be able to investigate the concentrations during and between raps and monitor background fluctuations. For this purpose, five channel analog ratemeters were constructed as modifications to the particle counters to provide parallel monitoring of the instantaneous concentration of particles in five preselected size intervals. These analog ratemeters provide approximately a half second response time, thus permitting monitoring of concentration changes throughout a rapping puff.

The sampling probes are configured and installed in such a manner as to permit a vertical traverse to be made along the center line of one lane at the exit plane of the precipitator. Probe losses are minimized by installing the particle sampling train (probe, diluter, particle counter) underneath the outlet duct and extracting the aerosol sample through a vertical probe with a single 90° bend between the sampling point and the particle sensor. The probe and nozzle are constructed from a continuous length of 4 mm I.D. stainless steel tubing. As used in previous tests, the probe flow was laminar with a Reynolds number of 100. In this configuration the system could conceivably be calibrated to give absolute concentrations. However, at this time there is not enough data to warrant its use to detect more than relative concentration changes.

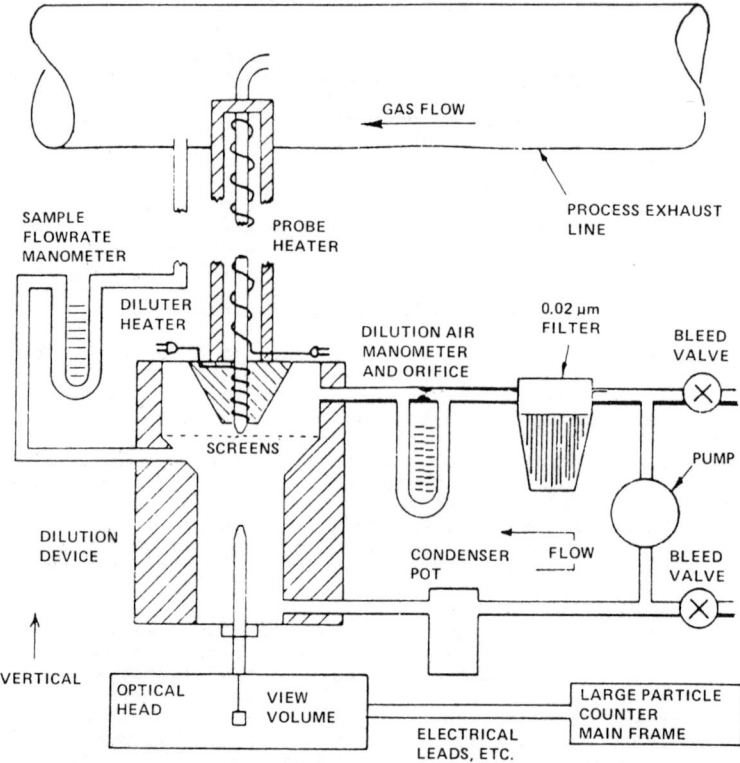

Figure 74. Large particle sizing system.[117]

For those circumstances in which it is not possible to sample from below the duct, a second sample extraction system is used. In this case the sample is removed at a high flow rate, 0.0019-0.0047 m^3/sec (4-10 cfm), through a large bore probe (4.06 cm diameter) and conveyed to a suitable location beside or on the top of the duct, at which point a secondary sample is extracted into the diluter and counter as illustrated in Figure 75.

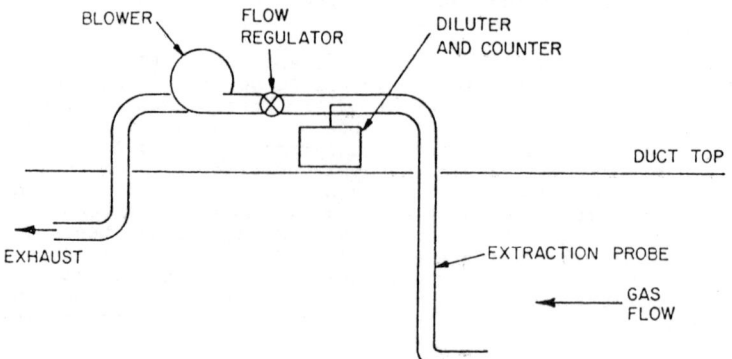

Figure 75. Extractive sampling system.[117]

This sampling method provides information on relative concentrations of particles of various sizes during and between puffs, but does not provide quantitative concentration data because of the uncertainties in the probe losses and in the degree to which the secondary sample represented the average concentration in the high flow rate probe. Automatic data acquisition can be accomplished as shown in the block diagram of the electronic package in Figure 76.

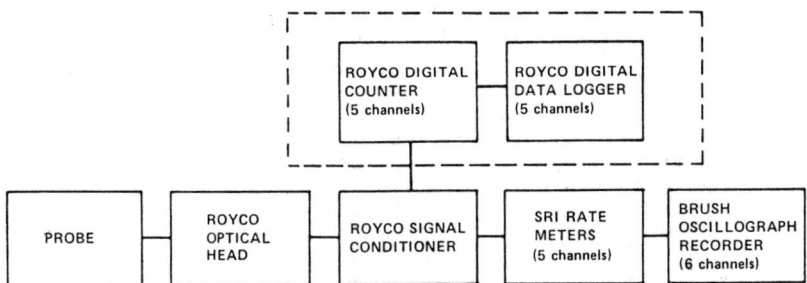

Figure 76. Block diagram of large particle sizing system.[117]

Laboratory Methods For Measuring Particle Size Distributions

Measurements of the size distribution of particles that have been collected in the field and transported to a laboratory must be interpreted with great caution, if not skepticism. It is difficult to collect representative samples in the first place, and it is almost impossible to reconstruct the original size distribution under laboratory conditions. For example, one can not distinguish from laboratory measurements, whether or not some of the particles existed in the process gas stream in agglomerates of smaller particles. Also, unwanted agglomerates can sometimes be formed in collecting and transporting particulate samples.

In spite of the limitations inherent in laboratory methods, they must be used in some instances to determine particle size and to segregate particles for analysis of their composition or other properties of interest. The following subsections contain discussions of some of the "standard" laboratory techniques used for particle size analysis of dust samples.

Sedimentation and Elutriation--

Elutriators and sedimentation devices separate particles that are dispersed in a fluid according to their settling velocities due to the acceleration of the earth's gravity.

Large particles in a quiescent aerosol will settle to the bottom region of the chamber more quickly than smaller particles that have smaller settling velocities. This principle is used in gravitational sedimentation and elutriation to obtain particle size distributions of polydisperse aerosols. In elutriation, the air is made to flow upward so that particles with settling velocities equal to or less than the air velocity will have a net velocity upward and particles which have settling velocities greater than the air velocity will move downward.

There are a number of commercial devices and methods having varying requirements of dust amounts and giving different ranges of size distributions, with a minimum size usually no smaller than two micrometers.[118,119] An important disadvantage is the inability of most sedimentation and elutriation devices to give good size resolution. Another disadvantage is the length of time (sometimes several hours) required to use some of the methods.

Popular methods of sedimentation include the pan balance, which weighs the amount of sediment falling on it from a suspension, and the pipette, which collects the particles in a small pipette at the base of a large chamber. Cahn's electronic microbalance, (Cahn Instrument Company, 7500 Jefferson St., Paramount, CA 90723), has an attachment that permits it to function as a settling chamber. Perhaps the most popular elutriator is the Roller particle size analyzer illustrated in Figure 77[119] (the Roller particle size analyzer is available from the American Standard Instrument Co., Inc., Silver Springs, MD).

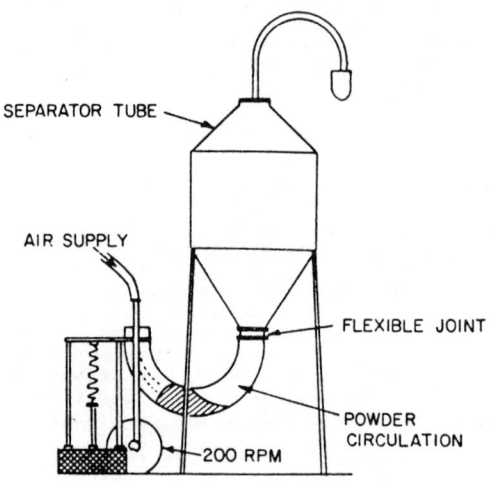

Figure 77. The Roller elutriator. After Allen.[119]

An instrument that measures the size distribution of particles in a liquid suspension is the X-ray Sedigraph, (Micromerities Instrument Corporation, 800 Goshen Springs Road, Norcross, GA 30071). The sample is continually stirred until the sampling period starts. The concentration of the particles is monitored by means of the extinction of a collimated x-ray beam. Upon sampling, the x-ray beam is moved upward mechanically to shorten the sampling time that is required. The particle-size distribution is plotted automatically. The reported range of sensitivity of the X-ray Sedigraph is from 0.1 to 100 µm.

Centrifuges--

Aerosol centrifuges provide a laboratory method of size-classifying particles according to their aerodynamic diameters. The advantage over elutriators is that the settling, or precipitation, process is speeded up by the large centrifugal acceleration. The sample dust is introduced in the device as an aerosol and enters a chamber which contains a centrifugal force field.

In one type of aerosol centrifuge, the larger particles overcome the viscous forces of the fluid and migrate to the wall of the chamber, while the smaller particles remain suspended. After the two size fractions are separated, one of them is reintroduced into the device and is fractionated further, using a different spin speed to give a slightly different centrifugal force. This is repeated as many times as desired to give an adequate size distribution. One of the more popular lab instruments using this technique is the Bahco microparticle classifier, which is illustrated in Figure 78,[120] and is available commercially from the Harry W. Dietert Company, Detroit 4, Michigan. The cutoff size can be varied from about two to fifty micrometers to give size distribution characterization of a 7 gm dust sample. A similar instrument is the B.C.U.R.A. (British Coal Utilization Research Association, Leatherhead, Surrey, U.K.) centrifugal elutriator which has a range of four to twenty-six micrometers.[121]

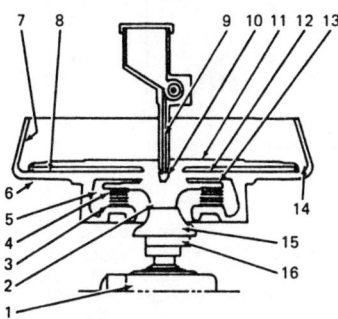

SCHEMATIC DIAGRAM

1. Electric Motor
2. Threaded Spindle
3. Symmetrical Disc
4. Sifting Chamber
5. Container
6. Housing
7. Top Edge
8. Radial Vanes
9. Feed Point
10. Feed Hole
11. Rotor
12. Rotary Duct
13. Feed Slot
14. Fan Wheel Outlet
15. Grading Member
16. Throttle

Figure 78. The Bahco microparticle classifier.[120]

In the second type of centrifuge, the device is run continuously, and the particle size distribution is determined from the position where the particles are deposited. Examples are a spiral centrifuge developed by Goetz, et al,[122,123,124] (Figure 79) and by Stöber and Flachsbart,[125] (Figure 80) that can classify polydisperse dust samples with particles from a few hundredths of a micron to approximately two micron in diameter. The conifuge, first built by Swayer and Walton[126] and modified several times since then,[127,128] is useful in the study of aerodynamic shape factor, but can also be used for the determination of size distributions especially for particles having aerodynamic diameters smaller than twenty-five micrometers (see Figure 81).[129] In continously operating centrifuges, the particles are generally deposited onto a foil strip, where their position yields a measure of their size, and their number is obtained by microscopy, radiation, or by weighing segments of the foil.

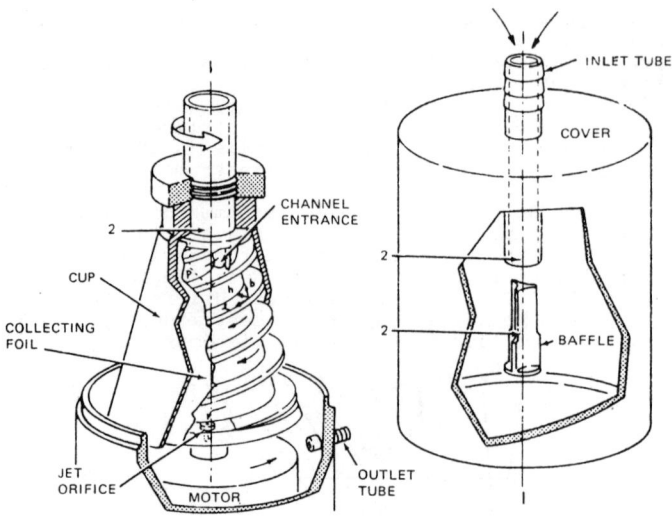

Figure 79. A cut-away sketch of the Goetz Aerosol Spectrometer spiral centrifuge. In assembled form the vertical axes (1) coincide and the horizontal arrows (2) coincide. After Gerber.[122,123,124]

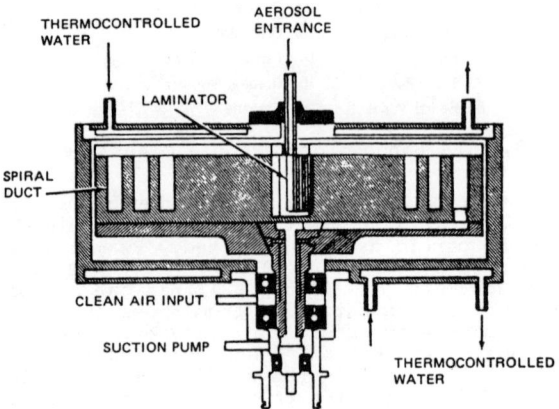

Figure 80. Cross-sectional sketch of the Stöber Centrifuge. After Stöber and Flachsbart.[125]

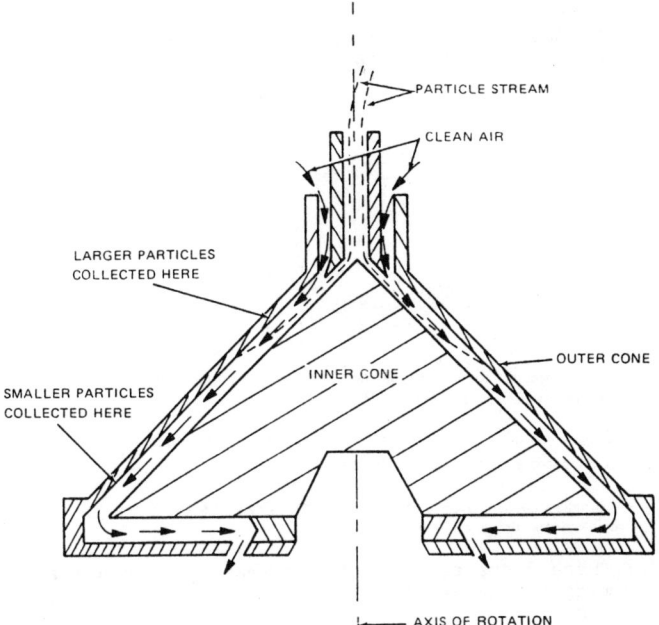

Figure 81. Cross-sectional sketch of a conifuge.[129]

Microscopy--

Microscopic analysis has long been regarded as the established, fundamental technique of counting and sizing particles that the human eye cannot comfortably see. Usually, the method involves one person, a microscope, and a slide prepared with a sample of the aerosol to be measured. A random selection of the particles would then be measured and counted, with notable characteristics of color, shape, transparency, or composition duly recorded. The most difficult task, especially since the advent of sophisticated computerized equipment has made counting and sizing easier, is the preparation of a slide which contains a representative sample of the aerosol.

It takes careful technique to obtain a slide sample which is not biased toward large or small particles, does not contain agglomerations which were not present in the stack, does not break up agglomerations which were present in the stack, is not too dense or too sparse, and has not been contaminated in the process of preparation. Different methods of slide preparation for optical and photographic microscopy are discussed by Cadle[118] and Allen.[119] A particularly good discussion of particle analysis through microscopy is given in Volume I of the McCrone Particle Atlas.[130] One main disadvantage of microscopic analysis is the type of diameter measured. Depending on the shape of the particles, several different types of diameter are used to characterize the size of the particle. Three commonly used types of diameter are shown in Figure 82 with their definitions.[131] However, for most control and standards work, the diameter of interest is the aerodynamic diameter, which is based on the particles' behavior in air. In these cases, the data from microscopic analysis is helpful only insofar as it can be related to the particular need of the experiment.

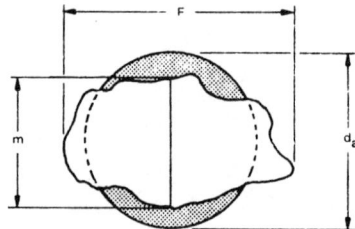

F - Feret's diameter, the distance between two tangents on opposite sides of the particle, parallel to a fixed direction.

M - Martin's diameter, the length of the line which bisects the image of the particle, parallel to a fixed direction.

d_a - Diameter of a circle having the same projected area as the particle in the plane of the surface on which it rests.

Figure 82. Three diameters used to estimate particle size in microscopic analyses.[131]

Particle sizes which can be easily studied on optical microscopes, range from about .2 to 100 micrometers. Electron microscopes have increased the size range of particles capable of being analyzed by microscopy down to 0.001 micrometers. Both scanning and transmission electron microscopes provide much information on surface features, agglomeration, size, composition and shape of particles in size ranges below that of optical microscopes.

Computerized scanning devices have increased the analyzing ability of present day microscopes and simplified counting and sizing.

Several commercial laboratories are equipped to provide physical and structural characterizations of dust samples quickly and fairly inexpensively.

Sieves--

Because of its relatively large lower particle size limit (50-75 micrometers), sieving has a limited use for characterizing most industrial sources today. However, for particles within its workable size range, sieving can be a very accurate technique, yielding adequate amounts of particles in each size range for thorough chemical analysis.

Sieving, one of the oldest ways of sizing particles geometrically, is the process by which a polydisperse powder is passed through a series of screens with progressively smaller openings until it is classified as desired. The lower size limit is set by the size of the openings of the smallest available screen, usually a woven wire cloth. Recently, micro-etched screens have become available. In the future, the lower size limit may be lowered by using membrane filters which can be made with smaller holes than woven, fine, wire cloth.

Sieves are available from several manufacturers in four standard size series: Tyler, U.S., British, and German. See Table 9 for a comparison of these series.[132] Tyler screens are manufactured by the W. S. Tyler Co., Cleveland, Ohio.

TABLE 9. COMPARISON TABLE OF COMMON SIEVE SERIES[132]

Tyler[a]		U.S.[b]		British Standard[c]		German DIN[d]		
Equiv. Mesh	Openings in mm.	Mesh No.	Openings in mm.	Mesh No.	Openings in mm.	DIN No.	Mesh per sq. cm.	Openings in mm.
3.5	5.613	3.5	5.66			1	1	6.000
4	4.699	4	4.76					
5	3.962	5	4.00					
6	3.327	6	3.36	5	3.353	2	4	3.000
7	2.794	7	2.83	6	2.812			
8	2.362	8	2.38	7	2.411	2.5	6.25	2.400
9	1.931	10	2.00	8	2.057	3	9	2.000
10	1.651	12	1.68	10	1.676	4	16	1.500
12	1.397	14	1.41	12	1.405			
14	1.168	16	1.19	14	1.204	5	25	1.200
16	0.991	18	1.00	16	1.003	6	36	1.020
20	0.833	20	0.84	18	0.853			
						8	64	0.750
24	0.701	25	0.71	22	0.699			
28	0.589	30	0.59	25	0.599	10	100	0.600
						11	121	0.540
32	0.495	35	0.50	30	0.500	12	144	0.490
35	0.417	40	0.42	36	0.422	14	196	0.430
42	0.351	45	0.35	44	0.353	16	256	0.385
48	0.295	50	0.297	52	0.295	20	400	0.300
60	0.208	60	0.250	60	0.251	24	576	0.250
65	0.208	70	0.210	72	0.211	30	900	0.200
80	0.175	80	0.177	85	0.178			
100	0.147	100	0.149	100	0.152	40	1600	0.150
115	0.124	120	0.125	120	0.124	50	2500	0.120
150	0.104	140	0.105	150	0.104	60	3600	0.102
170	0.088	170	0.088	170	0.089	70	4900	0.088
200	0.074	299	0.974	200	0.076	80	6400	0.075
250	0.061	230	0.062	240	0.066	100	10000	0.060
270	0.053	270	0.053	300	0.053			
325	0.043	325	0.044					
400	0.038	400	0.037					

[a] Tyler Standard Screen Scale Series.

[b] U.S. Sieve Series (Fine Series), National Bureau of Standards LC-584 and ASTME-11.

[c] British Standard Sieve Series, British Standards Institution, London BS-410:1943.

[d] German Standard Sieve Series, German Standard Specification DIN 1171.

Other methods of size classification using sieving principles are currently being studied and improved. Wet sieving is useful for material originally suspended in a liquid or which forms aggregates when dry-sieved. Air-jet sieving, where the particles are "shaken" by a jet of air directed upward through a portion of the sieve, has been found to be quicker and more reproducible

than hand or machine sieving, although smaller amounts of powder
(5 to 10 g) are generally used. Felvation[133] (using sieves in
conjunction with elutriation) and "sonic sifting"[134] (oscillation
of the air column in which the particles are suspended in a set
of sieves) are similar techniques that employ this principle.

Coulter Counter--

Figure 83 illustrates the principle by which Coulter counters
(Coulter Electronics, Inc., 590 West 20th Street, Hialeah, FL 33010)
operate.[135] Particles suspended in an electrolyte are forced
through a small aperture in which an electric current has been
established. The particles passing through the aperture displace
the electrolyte, and if the conductivity of the particle is different from the electrolyte, an electrical pulse of amplitude
proportional to the particle-electrolyte interface volume will
be seen. A special pulse height analyzer is provided to convert
the electronic data into a size distribution. A bibliography of
publications related to the operation of the Coulter counter has
been compiled by the manufacturer and is available on request.

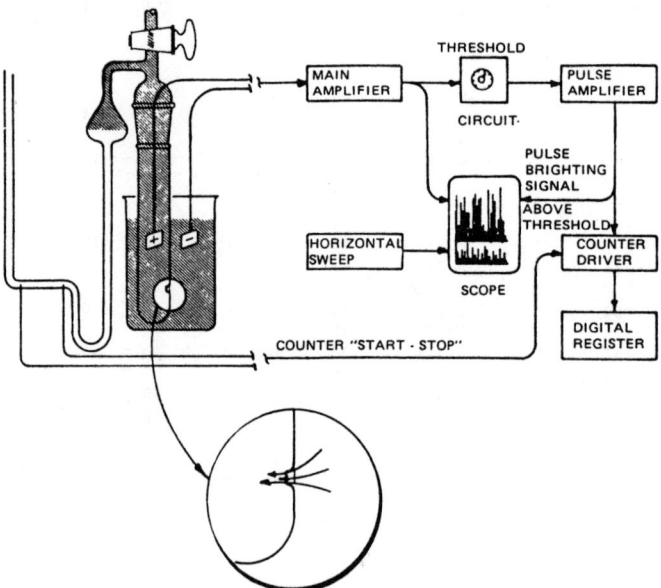

Figure 83. Operating principle of the Coulter Counter. Courtesy
of Coulter Electronics.[135]

Effect Of Particle Size Distribution On ESP Performance

The distribution of the various particle sizes entering a
given precipitator can have a significant effect on the maximum
overall mass collection efficiency that can be obtained. This is
due to particles of different diameters having different effective
migration velocities and collection efficiencies in a precipitator.
Figure 84 shows some typical data for effective migration velocity
and collection efficiency as a function of particle diameter.[136]

The data were obtained by making measurements with impactors at the inlet and outlet of a full-scale precipitator collecting fly ash particles and having a specific collection area of 55.7 m²/(m³/sec) and an average current density of 20 nA/cm². In general, there is a minimum in the collection efficiency versus particle diameter curve somewhere in the range between 0.3 and 0.9 μm. From the type of relationship shown in Figure 84, it is evident that different inlet size distributions will produce different overall mass collection efficiencies provided other operating variables do not change significantly.

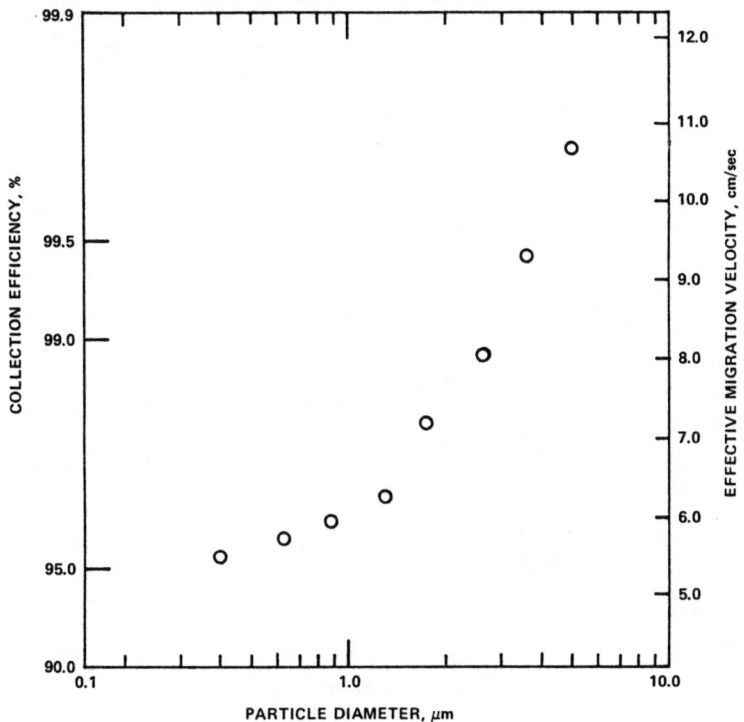

Figure 84. Typical data for effective migration velocity and collection efficiency as a function of particle diameter.[136]

Figure 85 shows the theoretically calculated effect of inlet particle size distribution on overall mass collection efficiency.[137] Although the particle size distribution will influence to some extent the voltage-current characteristics of precipitators collecting fly ash particles, the curves were generated by assuming the voltage-current characteristics remain constant in order to obtain trends. In the calculations, the specific collection area and current density were held fixed at 25 m²/(m³/sec) and 26 nA/cm², respectively. It is clear that both the mass median diameter and geometric standard deviation have a strong effect on overall mass collection efficiency. The overall mass collection efficiency increases with increasing MMD and decreasing σ_p.

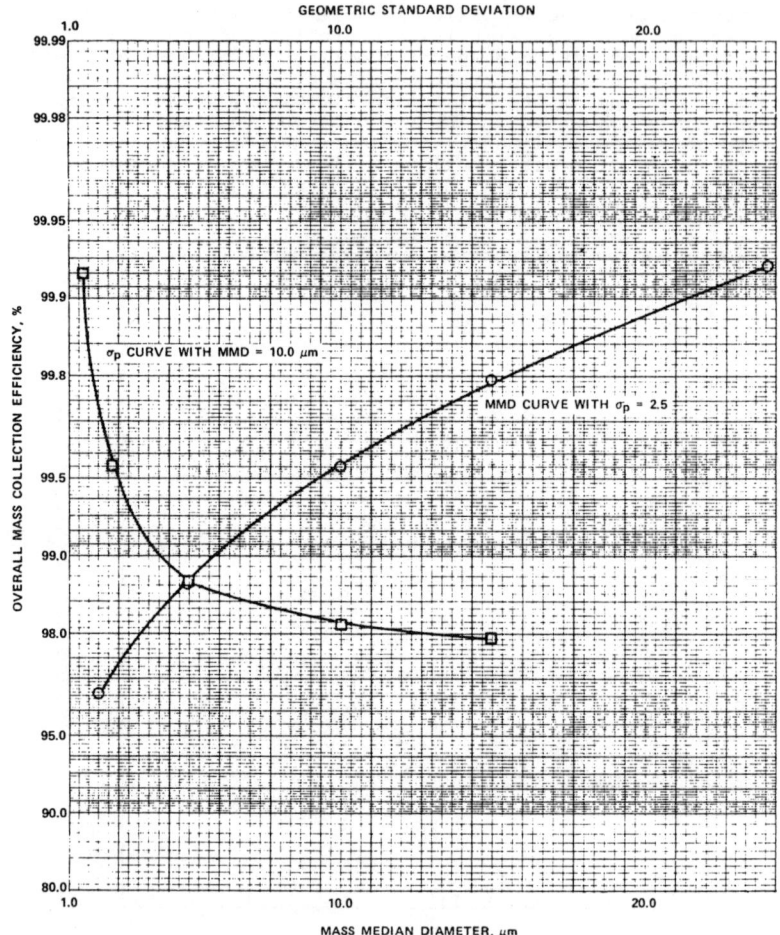

Figure 85. Effect of particle size distribution on overall mass collection efficiency.[137]

The above considerations point out the importance of considering the effect of variations in particle size distribution on overall mass collection efficiency for a given specific collection area and set of electrical operating conditions. Any program to evaluate the performance of a precipitator should include measurements of the particle size distribution at the inlet and outlet of the precipitator. In designing a new precipitator, particle size distribution measurements on a gas stream which is similar to the one to be treated should be considered. A precipitator should be designed with the capability of meeting emissions standards with a somewhat less favorable particle size distribution than that currently existing or that anticipated in order to provide a margin of safety. This is necessary because changes in the process which produces the emissions may result in a less favorable particle size distribution.

As mentioned earlier, the particle size distribution also influences the voltage-current characteristics of precipitators collecting fly ash particles. In addition, the particle size distribution effects the opacity of the effluent from the precipitator. These topics will be discussed in later sections.

Measured Size Distributions From Various Installations

Plant Number One--

Particle size distributions were obtained at the inlet and outlet to a cold-side electrostatic precipitator collecting ash from low sulfur Western coals. The precipitator, which is preceded by a mechanical collector, consists of six fields. The first and second fields each have 5,351 m^2 (57,600 ft^2) of collecting area, while the third through the sixth fields have 6,688.8 m^2 (72,000 ft^2) of collecting area, for a total of 37,457.3 m^2 (403,200 ft^2). This gives a specific collection area of 99.2 m^2/(m^3/sec) (504 ft^2/1000 cfm) for the design volume of 377.6 m^3/sec (800,000 acfm). The precipitator has twelve-inch plate spacing and operates at approximately 149°C (300°F).

The determination of the cumulative inlet particle size distribution between 0.25 μm and 10.0 μm, shown in Figure 86, was performed using two modified Brink cascade impactors (seven stages, precollector cyclone, and back up filter). Outlet particle size distributions were measured using Andersen stack samplers. Rapping and nonrapping outlet size distributions on a cumulative basis are shown in Figures 87 and 88. Figure 89 shows the rap and no-rap data for the ultra fine system and the rap and no-rap impactor derived efficiencies. The estimated no-rap efficiencies were based on the data from the large-particle, real-time system and are subject to large uncertainties because of poor counting statistics for the larger particles, coupled with the limited time span over which the data were taken. However, it is obvious that very high collection efficiencies are achieved in the particle diameter range from 0.05 to 20.0 μm.

Plant Number Two--

Particle size distributions were obtained at the inlet and outlet to a cold-side electrostatic precipitator collecting ash from high sulfur Eastern coals. The precipitator consists of three fields and is divided into two collectors, A and B. The test program was performed on Collector A, the collecting area of which is 7,374.4 m^2 (79,380 ft^2). This gives a specific collection area of 34.475 m^2/(m^3/sec) (175 ft^2/1000 cfm) for the design volume flow of 213.82 m^3/sec (453,000 acfm). The precipitator has 27.94-cm (11-inch) spacing and operates at approximately 149°C (300°F).

Cumulative mass loadings for two groups of inlet runs are given in Figures 90 and 91. Outlet cumulative mass loadings for Outlet Group 1 (reduced load, normal precipitator operation), Outlet Group 2 (normal operation), and Outlet Group 5 (one-half current density) are given in Figures 92, 93, and 94. The corresponding fractional efficiency curves are presented in Figures 95 through 97.

A comparison of Figure 96 with 97 clearly shows the detrimental effect of reduced current on collection efficiency.

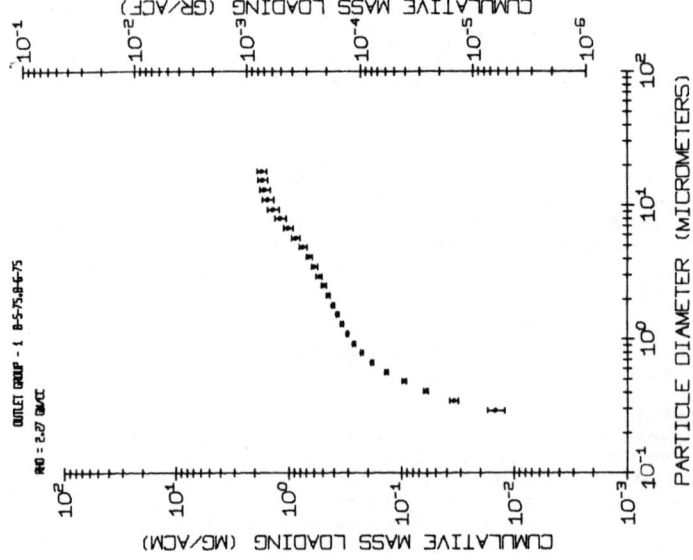

Figure 87. Plant 1 cumulative outlet distribution, rappers on, for a cold-side electrostatic precipitator collecting ash from a low sulfur Western coal.

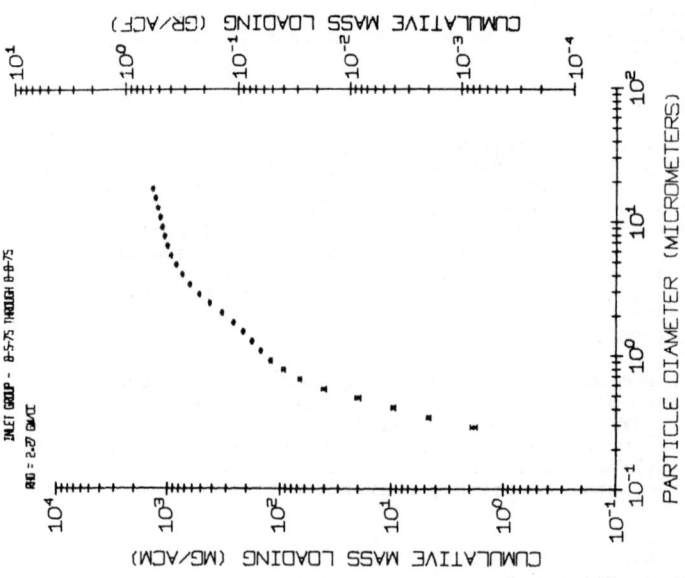

Figure 86. Plant 1 cumulative inlet distribution between 0.25 μm and 10.0 μm particle diameter for a cold-side electrostatic precipitator collecting ash from a low sulfur Western coal.

Analysis of Factors Influencing Performance 115

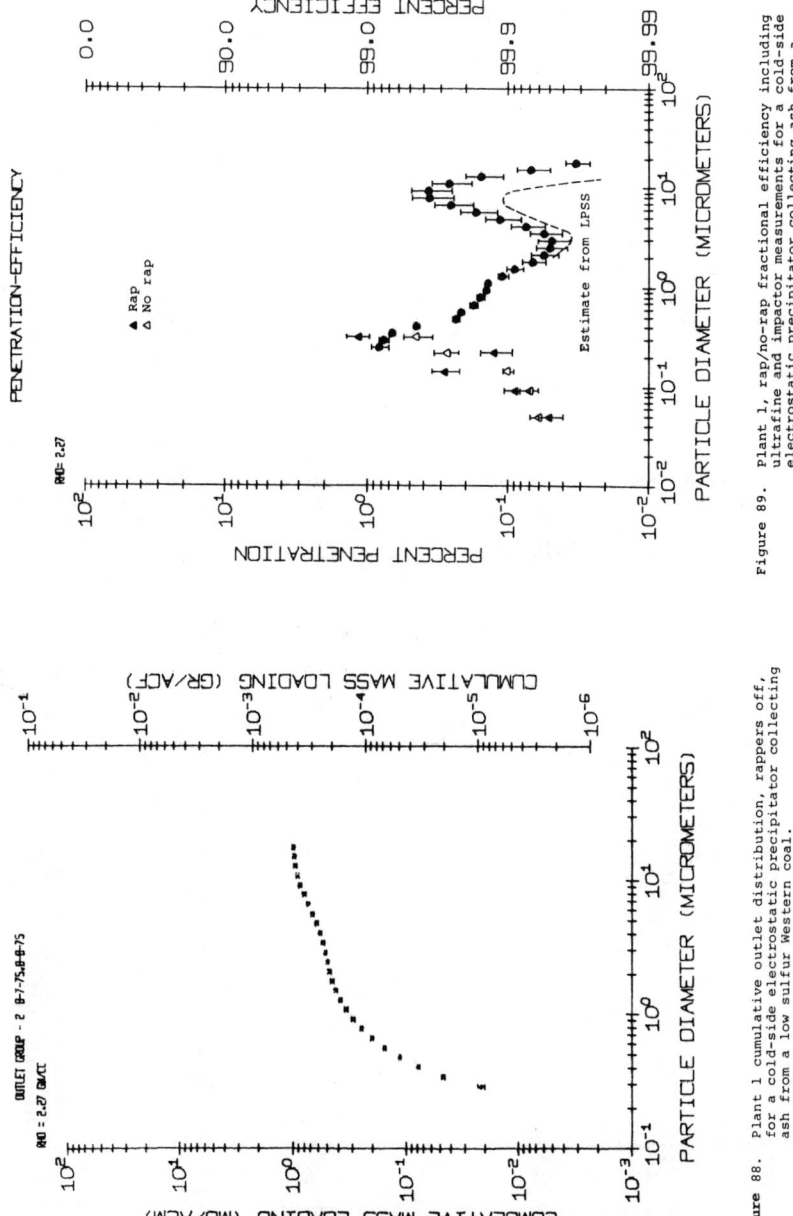

Figure 88. Plant 1 cumulative outlet distribution, rappers off, for a cold-side electrostatic precipitator collecting ash from a low sulfur Western coal.

Figure 89. Plant 1, rap/no-rap fractional efficiency including ultrafine and impactor measurements for a cold-side electrostatic precipitator collecting ash from a low sulfur Western coal.

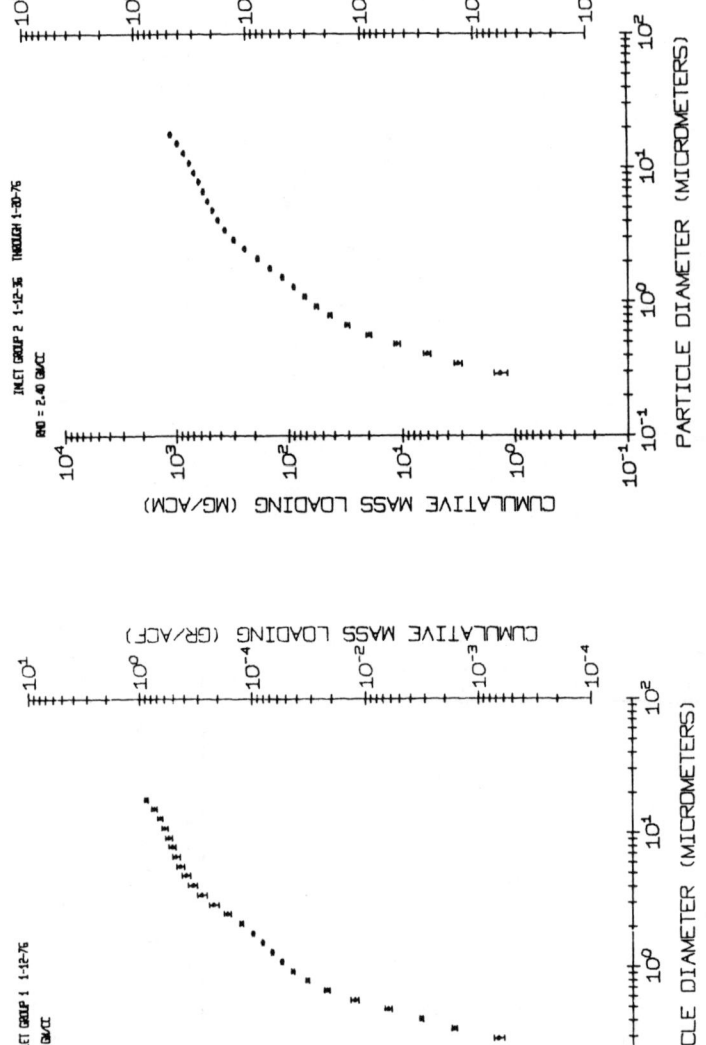

Figure 90. Plant 2 inlet cumulative size distribution for a cold-side electrostatic precipitator collecting ash from a high sulfur Eastern coal.

Figure 91. Plant 2 average inlet cumulative size distribution for a cold-side electrostatic precipitator collecting ash from a high sulfur Eastern coal.

Analysis of Factors Influencing Performance   117

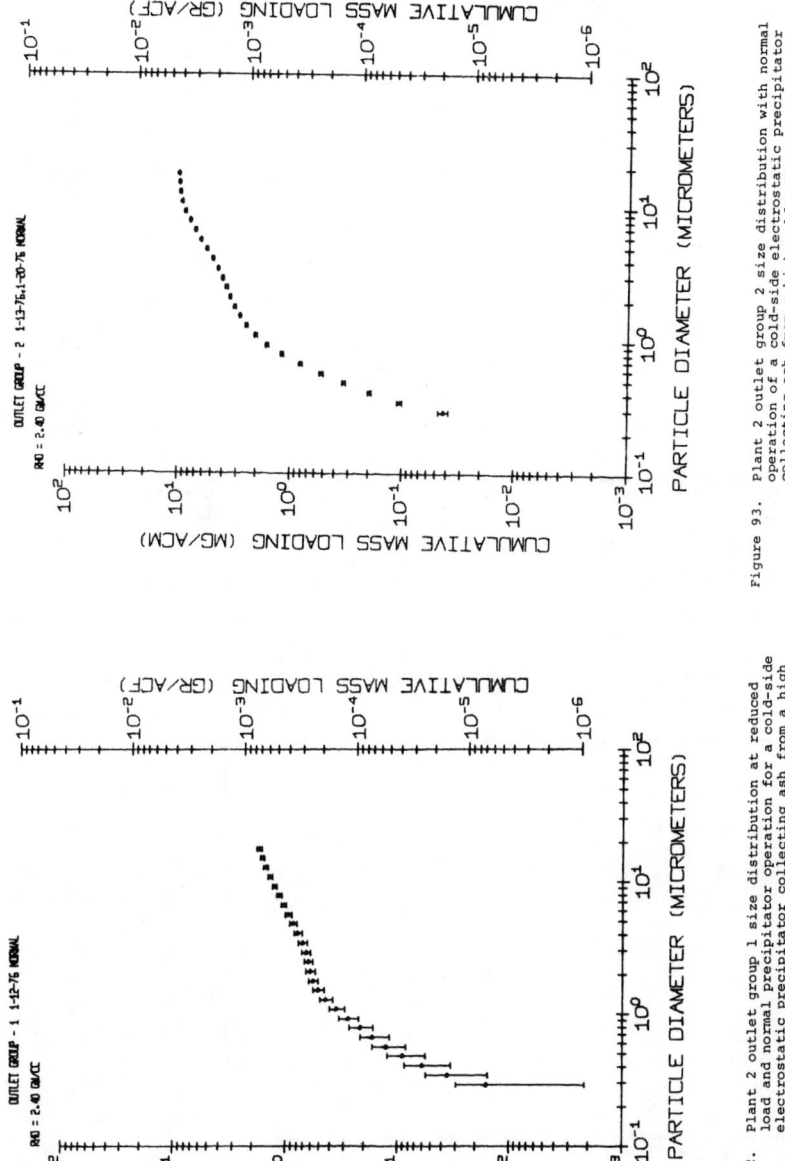

Figure 92. Plant 2 outlet group 1 size distribution at reduced load and normal precipitator operation for a cold-side electrostatic precipitator collecting ash from a high sulfur Eastern coal.

Figure 93. Plant 2 outlet group 2 size distribution with normal operation of a cold-side electrostatic precipitator collecting ash from a high sulfur Eastern coal.

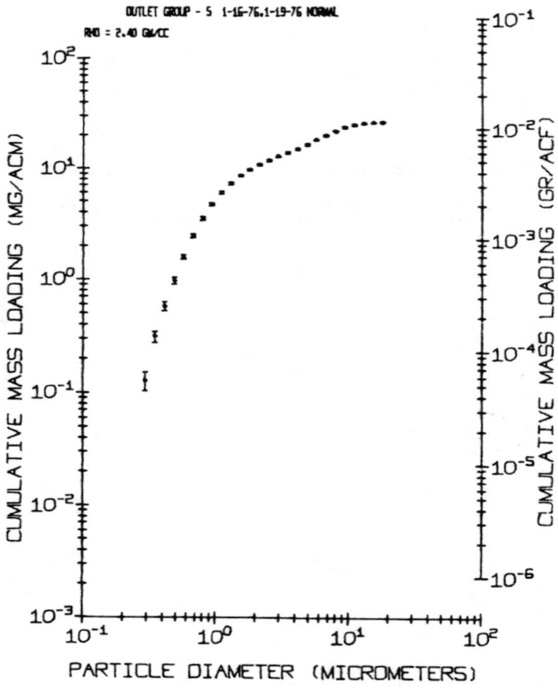

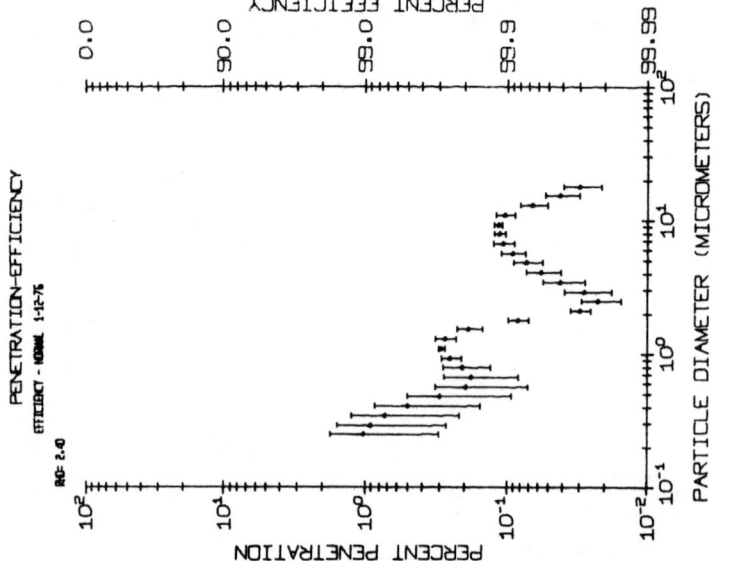

Figure 94. Plant 2 outlet group 5 size distribution for a cold-side precipitator operating at one-half current density collecting ash from a high sulfur Eastern coal.

Figure 95. Plant 2 fractional efficiency, outlet group 1 for reduced load and normal operation of a cold-side electrostatic precipitator collecting ash from a high sulfur Eastern coal.

Analysis of Factors Influencing Performance 119

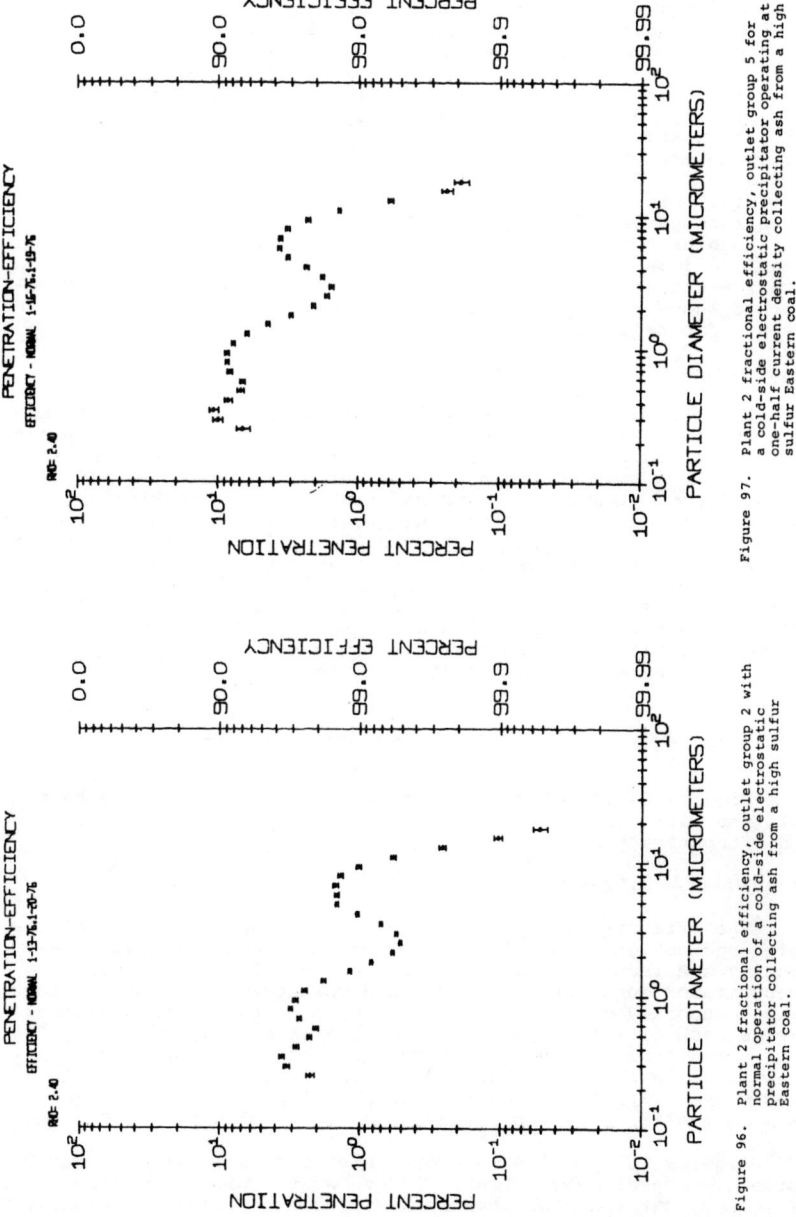

Figure 96. Plant 2 fractional efficiency, outlet group 2 with normal operation of a cold-side electrostatic precipitator collecting ash from a high sulfur Eastern coal.

Figure 97. Plant 2 fractional efficiency, outlet group 5 for a cold-side electrostatic precipitator operating at one-half current density collecting ash from a high sulfur Eastern coal.

Plant Number Three--

   The size distributions from this plant were obtained at the inlet and outlet of a cold-side electrostatic precipitator collecting ash from high sulfur Eastern coals. A mechanical collector precedes the precipitator which consists of four fields in the direction of gas flow and is divided into collectors A and B. The test program was conducted on collector B, the total collecting area of which is 5900.64 m^2 (63,516 ft^2), giving a specific collection area of 43.48 $m^2/(m^3/sec)$ (220.9 $ft^2/1000$ cfm) for the design volume of 135.70 m^3/sec (287,500 acfm). The precipitator has 25.4 cm (12 inch) plate spacings and operates at approximately 160°C (320°F). Inlet and outlet cumulative size distributions are given in Figures 98 and 99, and Figure 100 shows the fractional efficiencies for normal operation. Figure 101 contains the fractional efficiency data for normal operating conditions obtained from the ultra fine system and the impactors. Reasonable agreement is shown between the ultra fine system and the impactors in the overlap region.

Plant Number Four--

   The size distributions from plant number four were obtained from the inlet and outlet of a hot-side electrostatic precipitator collecting ash from a low sulfur Eastern coal. The precipitator consists of A and B casings each of which has two inlet and two outlet ducts. Tests were conducted on casing B (consisting of Chambers B1 and B2). Casing B has four fields in series, each of which has a collecting area of 3912.95 m^2 (42,120 ft^2). Although the precipitator was designed to have an SCA of 53.15 $m^2/(m^3/sec)$ (270 $ft^2/1000$ acfm) for a total volume flow of 590 m^3/sec (1,250,000 acfm), the gas flow for the two chambers tested was about 430,000 acfm, which resulted in an SCA of approximately 390 $ft^2/1000$ acfm. The collecting electrodes have nine inch spacing and the precipitator operates at approximately 343°C (650°F).

   Figures 102 and 103 present inlet and outlet size distributions resulting from impactor measurements made on Duct B1 and B2 (casing B).

   Figure 104 illustrates the fractional efficiencies obtained with the ultra fine sizing system and impactors for duct B1 with and without rapping.

Plant Number Five--

   The size distributions from this plant were obtained at the inlet and outlet of a cold-side electrostatic precipitator collecting ash from low sulfur Western coals. The electrostatic precipitator consists of six divided chambers, the test program being conducted on Chamber 5. Each chamber has five electrical fields each of which has a collection area of 3518.96 m^2 (37,879 ft^2). The precipitator has 25 cm (9.75 in) plate spacing, operates at 88 to 120°C (190 to 250°F), and is designed to handle 1100 m^3/sec (2,330,000 acfm). The actual specific collection area on the tested chamber was approximately 590 $ft^2/1000$ acfm.

   Figures 105 and 106 show the inlet and outlet size distributions, respectively. Figure 107 shows the ultrafine fractional efficiency data and the impactor derived fractional efficiencies under normal conditions.

Analysis of Factors Influencing Performance 121

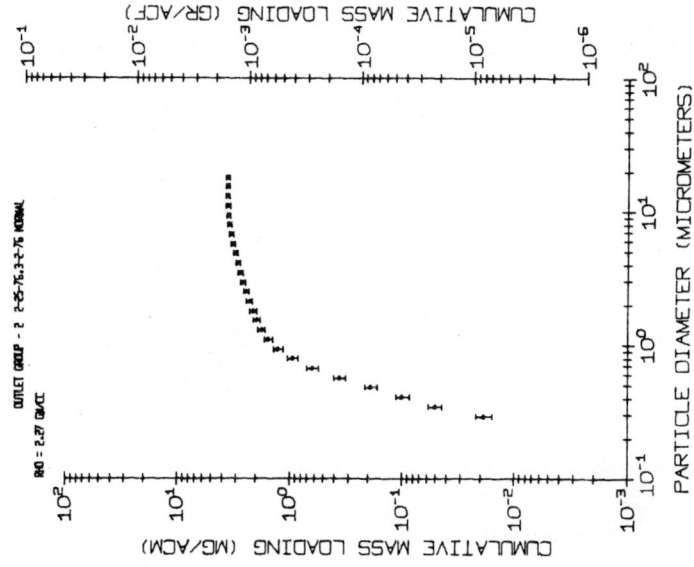

Figure 98. Plant 3 inlet cumulative size distribution for a cold-side electrostatic precipitator collecting ash from a high sulfur Eastern coal.

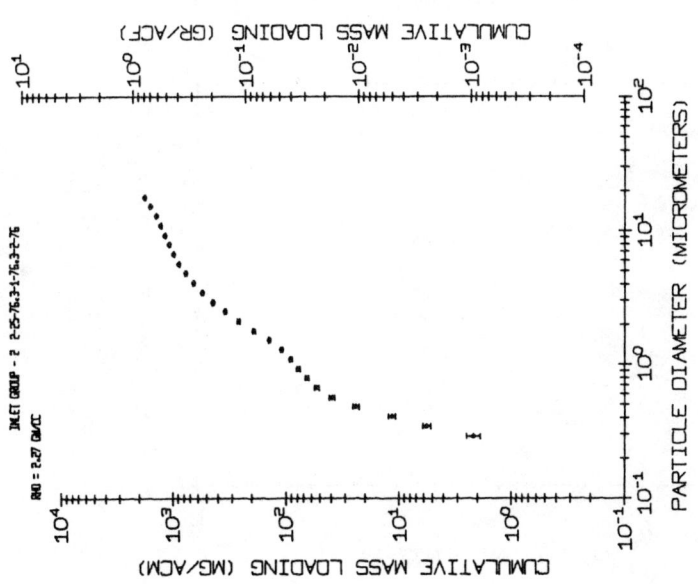

Figure 99. Plant 3 outlet cumulative size distribution for a cold-side electrostatic precipitator collecting ash from a high sulfur Eastern coal.

122 Electrostatic Precipitator Manual

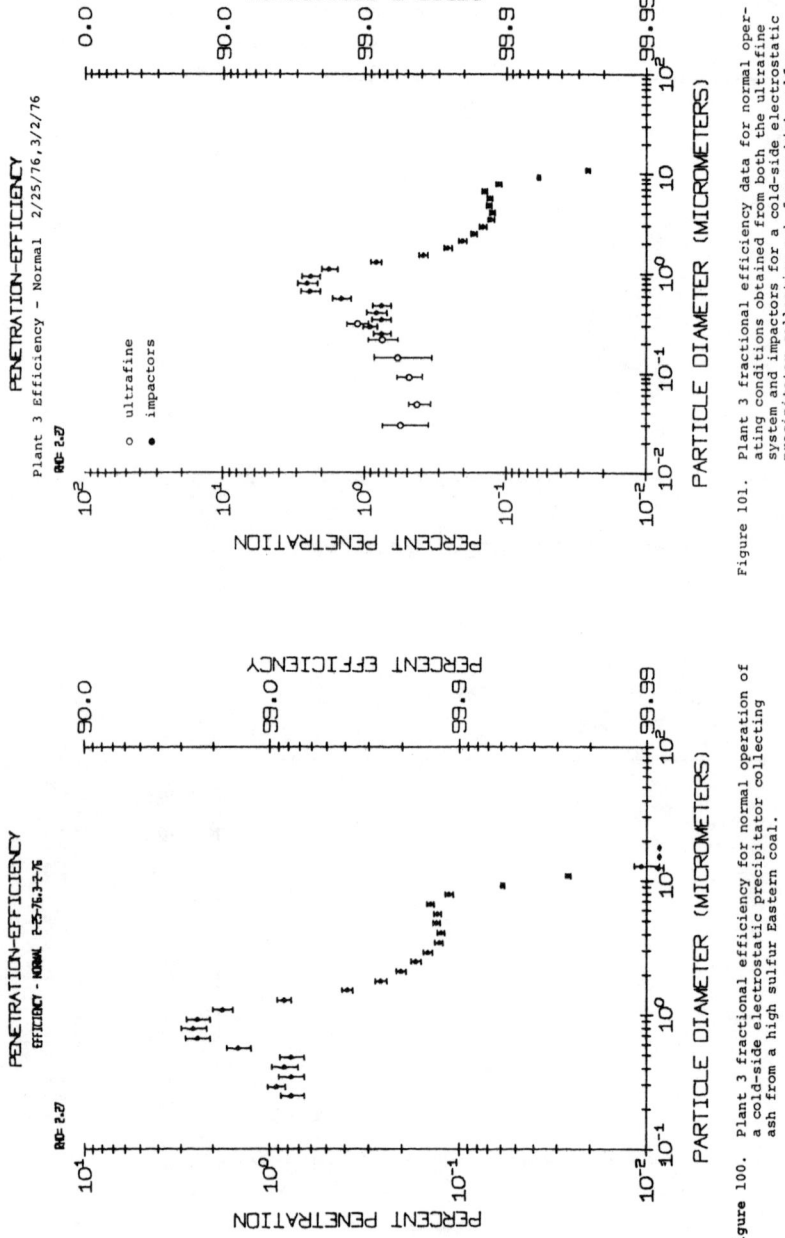

Figure 100. Plant 3 fractional efficiency for normal operation of a cold-side electrostatic precipitator collecting ash from a high sulfur Eastern coal.

Figure 101. Plant 3 fractional efficiency data for normal operating conditions obtained from both the ultrafine system and impactors for a cold-side electrostatic precipitator collecting ash from a high sulfur Eastern coal.

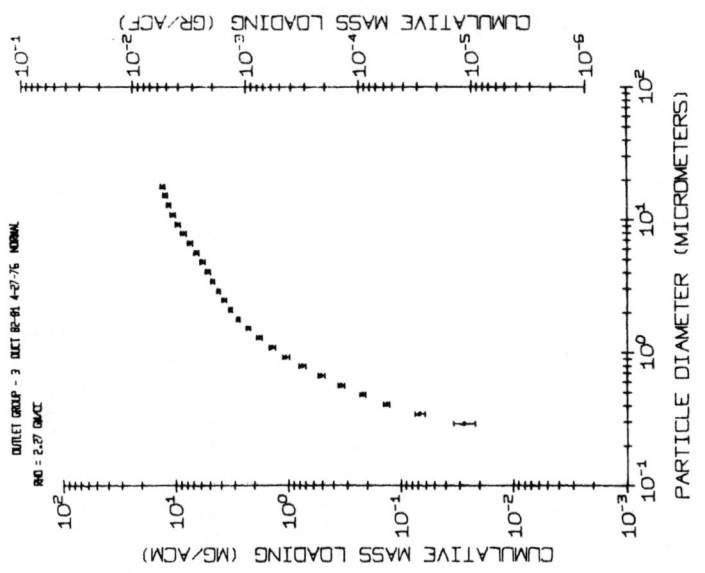

Figure 102. Plant 4 inlet cumulative size distribution resulting from impactor measurements made on ducts B1 and B2 of a hot-side precipitator collecting ash from a low sulfur Eastern coal.

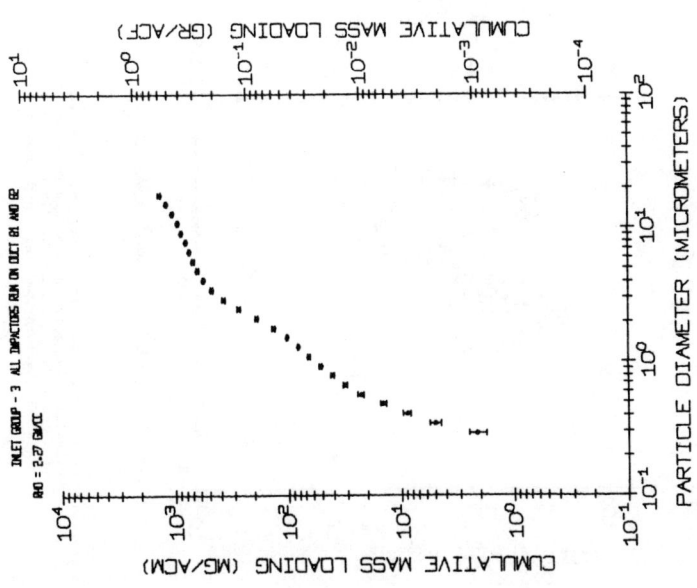

Figure 103. Plant 4 outlet cumulative size distribution resulting from impactor measurements made on ducts B1 and B2 of a hot-side precipitator collecting ash from a low sulfur Eastern coal.

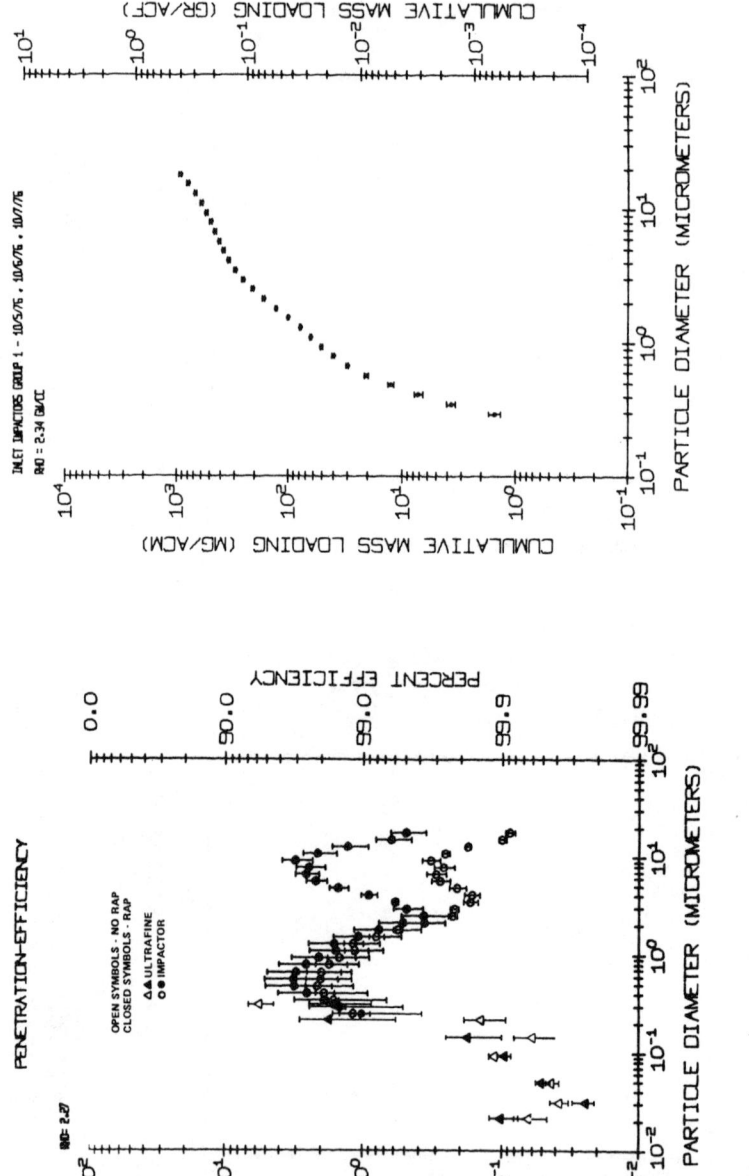

Figure 104. Plant 4 fractional efficiency data obtained with the ultrafine sizing system and impactors for duct B1 of a hot-side precipitator collecting ash from a low sulfur Eastern coal, with and without rapping.

Figure 105. Plant 5 inlet cumulative size distribution resulting from impactor measurements on a cold-side precipitator collecting ash from a low sulfur Western coal.

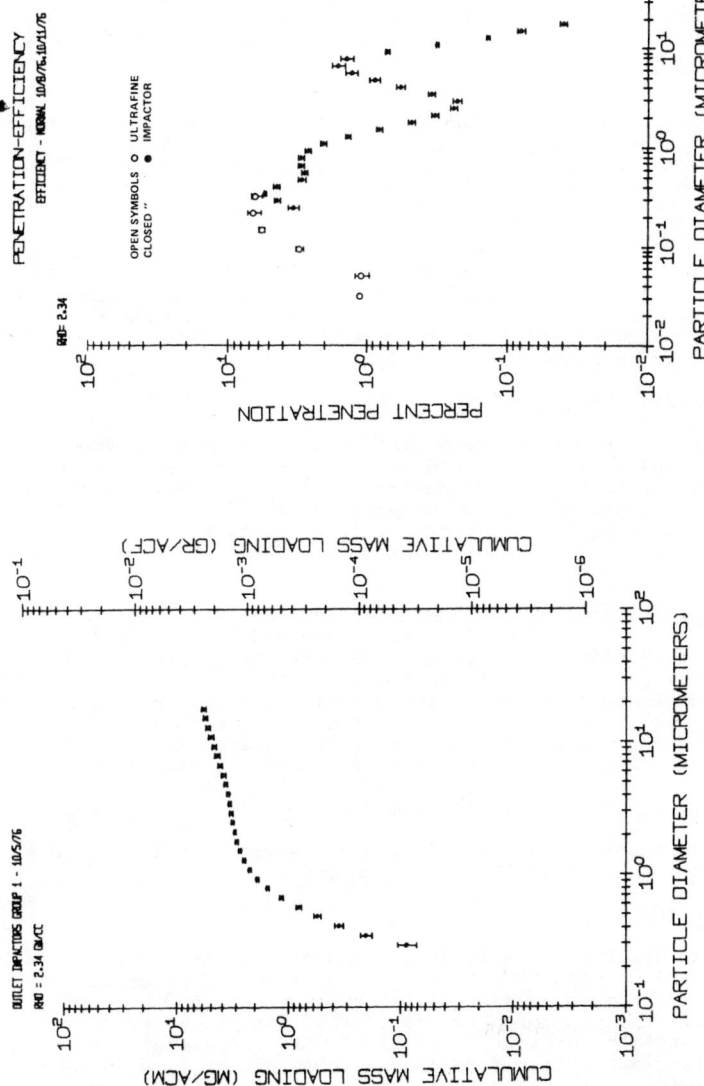

Figure 106. Plant 5 outlet cumulative size distribution resulting from impactor measurements on a cold-side precipitator collecting ash from a low sulfur Western coal.

Figure 107. Plant 5 fractional efficiency data obtained with the ultrafine sizing system and impactors under normal conditions on a cold-side precipitator collecting ash from a low sulfur Western coal.

126   Electrostatic Precipitator Manual

Plant Number Six--

Particle size distributions were obtained at the inlet and outlet to a hot-side electrostatic precipitator collecting ash from a low sulfur Western coal. This hot-side precipitator operates at approximately 360°C (680°F). The precipitator consists of two separate collectors, each of which has eight isolatable chambers, the test program being conducted on the number eight chamber. There are in each chamber six electrical fields in the direction of gas flow, and each field has a total collecting area of 1170.54 m^2 (12,600 ft^2). The complete precipitator installation was designed to handle 1859.68 m^3/sec (3,940,000 acfm) at 350°C which results in a design specific collection area of 60.43 m^2/(m^3/sec) (307 ft^2/1000 cfm).

The inlet and outlet size distributions are shown in Figures 108 and 109, respectively. Figure 110 shows the ultrafine and impactor fractional efficiencies for normal conditions.

Plant Number Seven--

The size distributions from plant number seven were obtained at the inlet to a cold-side precipitator collecting ash from high sulfur coals. The in situ particle size distribution measurements were conducted at the inlets to both the A and B sides of the precipitator using modified Brink cascade impactors and the results are shown in Figures 111 and 112.

Plant Number Eight--

The size distributions from plant number eight were obtained at the inlet and outlet to a cold-side electrostatic precipitator collecting ash from low sulfur Western coals. Figure 113 shows a graph of the fractional collection efficiencies for the small particle fraction using Brink impactor instrumentation.

Plant Number Nine--

Using Brink impactors at the inlet and Andersen impactors at the outlet, particle size measurements were made at the inlet and outlet of a cold-side electrostatic precipitator collecting ash from medium sulfur (1.0-1.5%) Southeastern coals. The precipitator consists of collectors A and B each of which has two collectors in series. Tests were conducted on A side only which has a collecting area of 28,877 m^2 (311,000 ft^2). There are twelve electrical sections in the direction of gas flow. Gas flow at full load ($\sim$700 MW) is about 520 m^3/sec at 149°C, giving a specific collecting area of 55 m^2/(m^3/sec) or 283 ft^2/1000 cfm.

Inlet and outlet size distributions and fractional efficiency data are shown in Figures 114, 115, and 116.

Plant Number Ten--

The size distributions for plant number ten were obtained at the inlet and outlet of a hot-side electrostatic precipitator collecting ash from low-medium (1.0%) sulfur Western coal. The electrostatic precipitator consists of four individual precipitators of two sections consisting of 13,582 m^2 of plate area collecting particulate matter from a gas stream with a flow rate of about 1.33 x 10^4 m^3/min at a temperature of 371°C at full load (357 MW). However, tests were conducted at a load with volume flow rates on the order of 9628 m^3/min which corresponds to a specific collection area of 310 ft^2/1000 acfm.

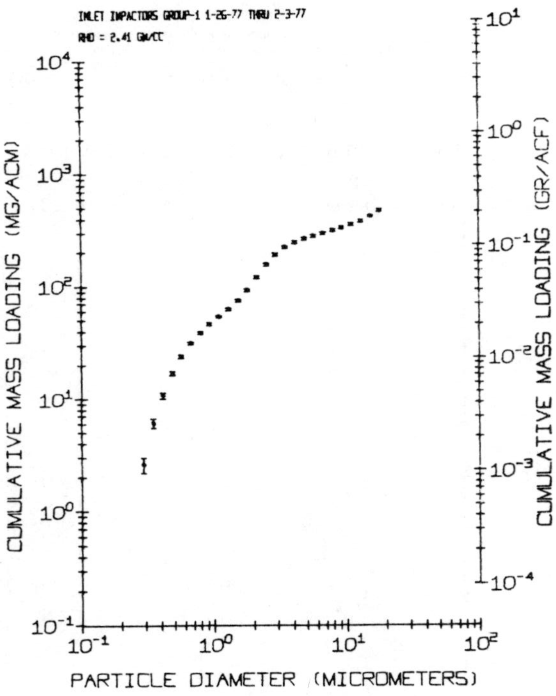

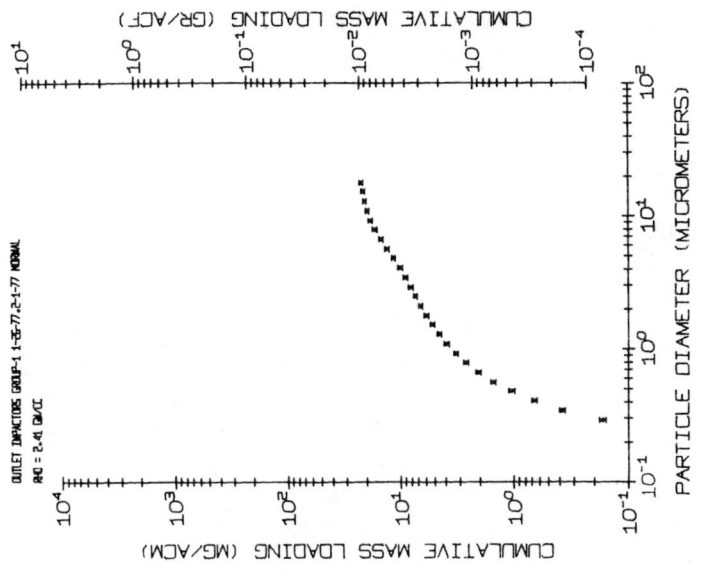

Figure 108. Plant 6 inlet cumulative size distribution resulting from impactor measurement on a hot-side precipitator collecting ash from a low sulfur Western coal.

Figure 109. Plant 6 outlet cumulative size distribution resulting from impactor measurements on a hot-side precipitator collecting ash from a low sulfur Western coal.

128 Electrostatic Precipitator Manual

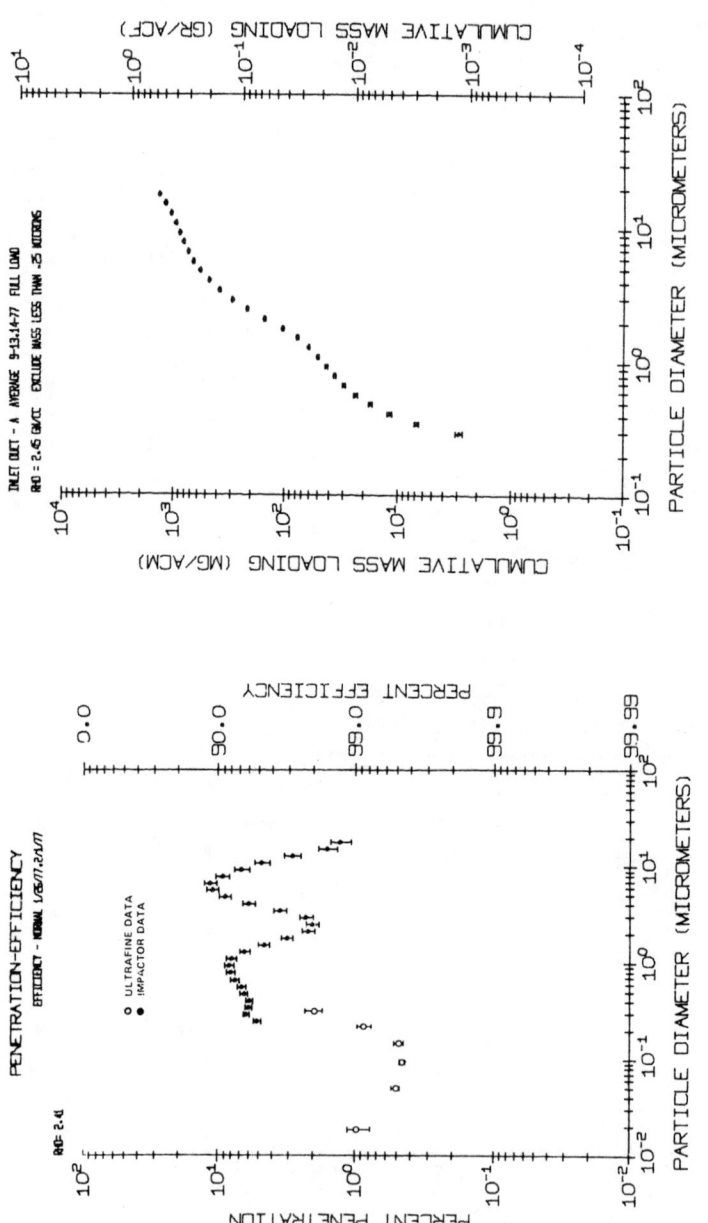

Figure 110. Plant 6 fractional efficiency data obtained with the ultrafine sizing system and impactors under normal conditions on a hot-side precipitator collecting ash from a low sulfur Western coal.

Figure 111. Plant 7 inlet cumulative size distribution resulting from impactor measurements on a cold-side precipitator collecting ash from a high sulfur coal.

Analysis of Factors Influencing Performance   129

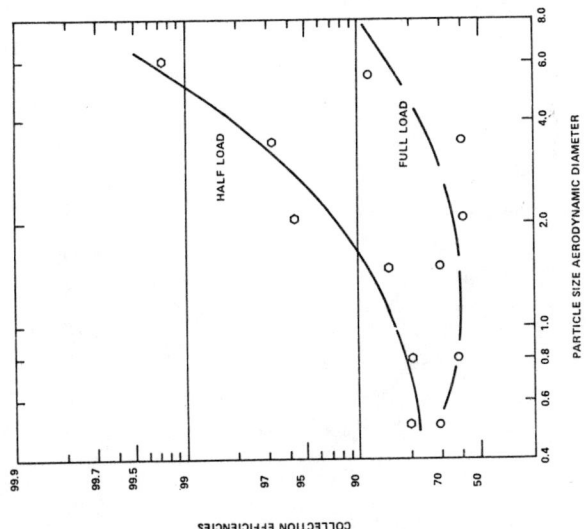

Figure 113. Plant 8 fractional collection efficiencies for small particle fraction obtained with Brink impactors on a cold-side precipitator collecting ash from a low sulfur Western coal.

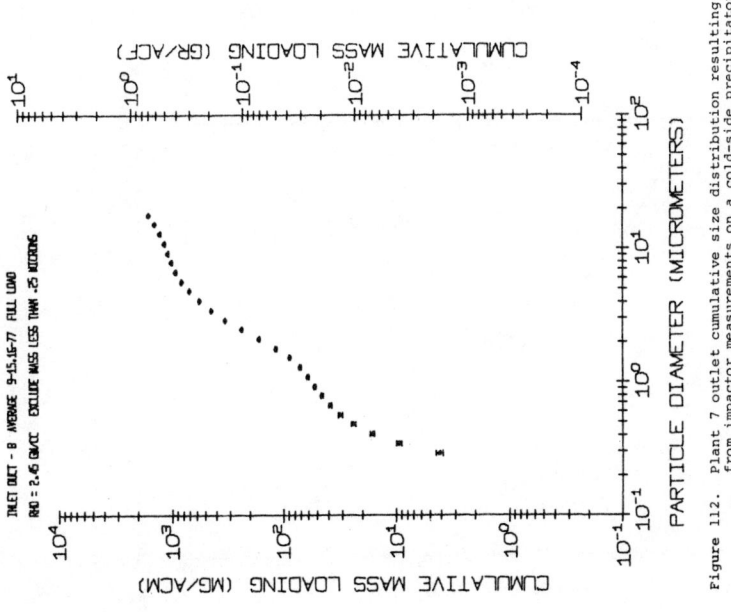

Figure 112. Plant 7 outlet cumulative size distribution resulting from impactor measurements on a cold-side precipitator collecting ash from a high sulfur coal.

130   Electrostatic Precipitator Manual

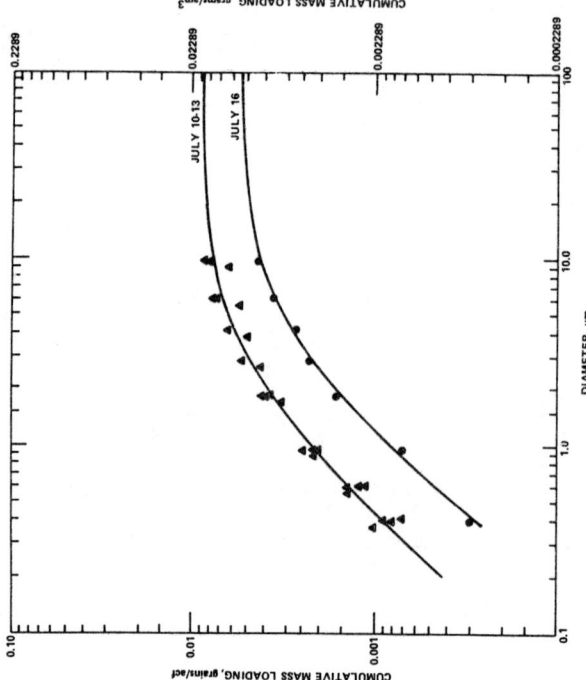

Figure 115. Plant 9 outlet cumulative size distribution obtained from an Andersen impactor on a cold-side precipitator collecting ash from a medium sulfur (1.0-1.5%) Southeastern coal.

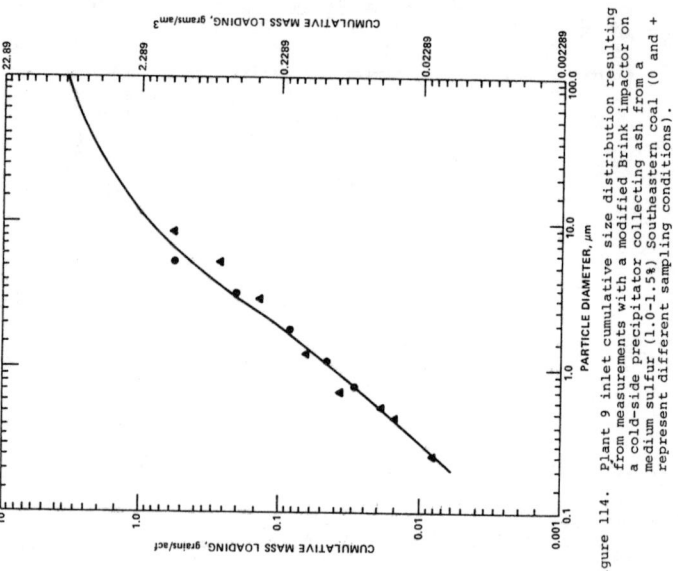

Figure 114. Plant 9 inlet cumulative size distribution resulting from measurements with a modified Brink impactor on a cold-side precipitator collecting ash from a medium sulfur (1.0-1.5%) Southeastern coal (0 and + represent different sampling conditions).

Analysis of Factors Influencing Performance   131

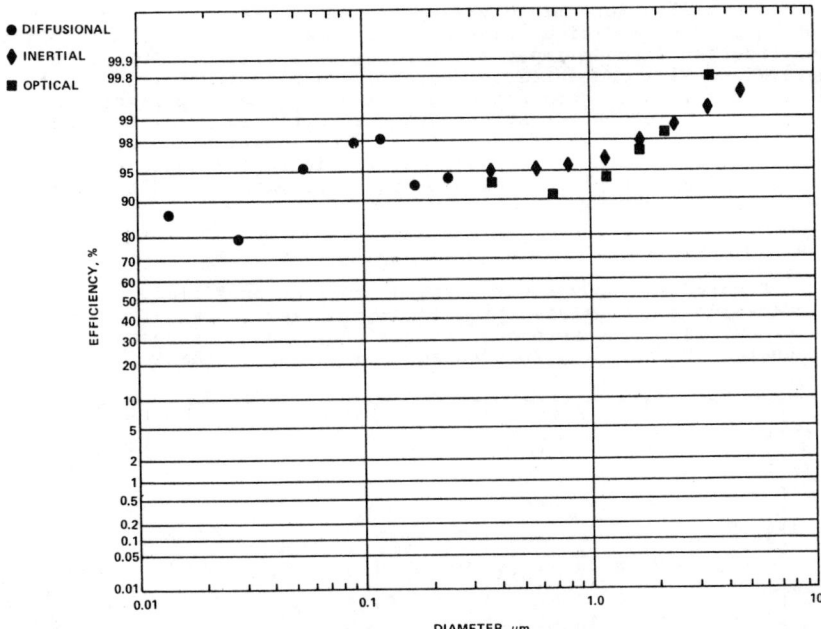

Figure 116. Plant 9 fractional efficiency measurements on a cold-side precipitator collecting ash from a medium sulfur (1.0-1.5%) Southeastern coal.

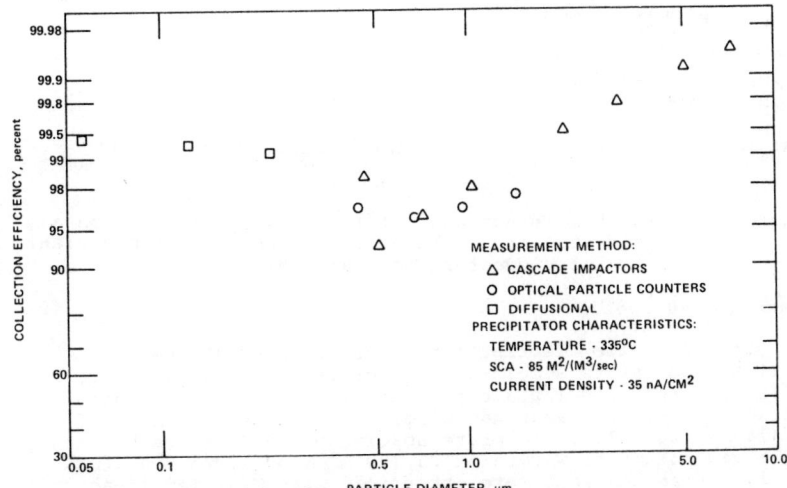

Figure 117. Plant 10 fractional efficiency measurements on a hot-side precipitator collecting ash from a low-medium sulfur (1.0%) Western coal.

Inertially determined size/mass concentration data were obtained using modified Brink Cascade impactors for inlet sampling and Andersen Cascade impactors for outlet sampling. Optically determined size/concentration data over a size range from about 0.3 to 2.0 μm were obtained using Climet and Royco particle counters. Size/concentration data were obtained by diffusional methods using diffusion batteries and condensation nuclei counters simultaneously with the optical data. Figure 117 shows the fractional collection efficiencies of the precipitator and the measurement methods used.

Plant Number Eleven--

Figure 118 shows the fractional collection efficiencies of a cold-side electrostatic precipitator collecting ash from a plant burning Midwestern coal and refuse. The measurements were conducted at three load/percentage coal-refuse combinations.

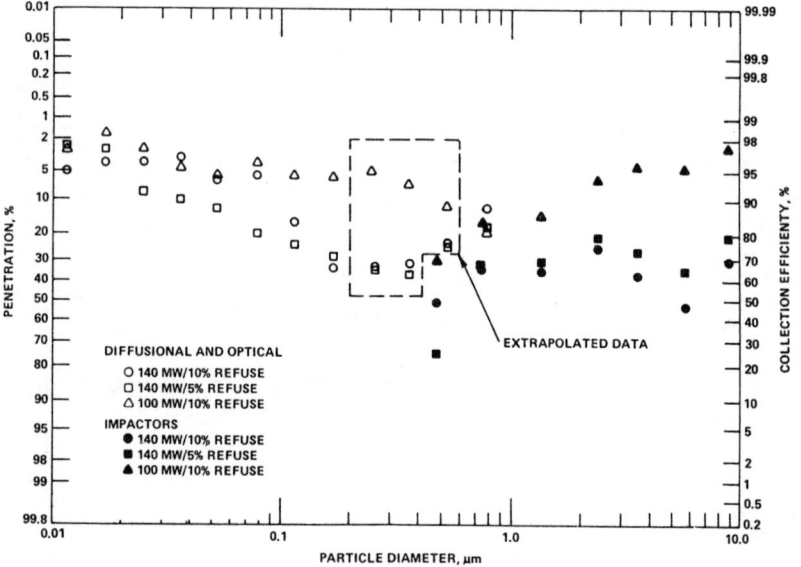

Figure 118. Plant 11 fractional efficiency measurements on a cold-side precipitator collecting ash from a plant burning Midwestern coal and refuse.

Plant Number Twelve--

Particle size distributions were obtained at the inlet and outlet to a cold-side electrostatic precipitator, collecting ash from a plant burning high sulfur ($\sim$2%) Eastern coals. The precipitator has a collection electrode area of 19,414 m^2 (208,980 ft^2), plate spacing of 25.4 cm (ten inches), and a gas volume flow rate of 28,700 m^3/min (1,025,000 cfm). The particle size analyses were determined with modified impactor type devices and the results are shown in Figures 119 and 120. The inlet particle size distribution is unusually large for a power plant. The modified impactor devices were equipped with cyclone collectors which remove the coarsest size fraction prior to introduction into the impactor.

Analysis of Factors Influencing Performance 133

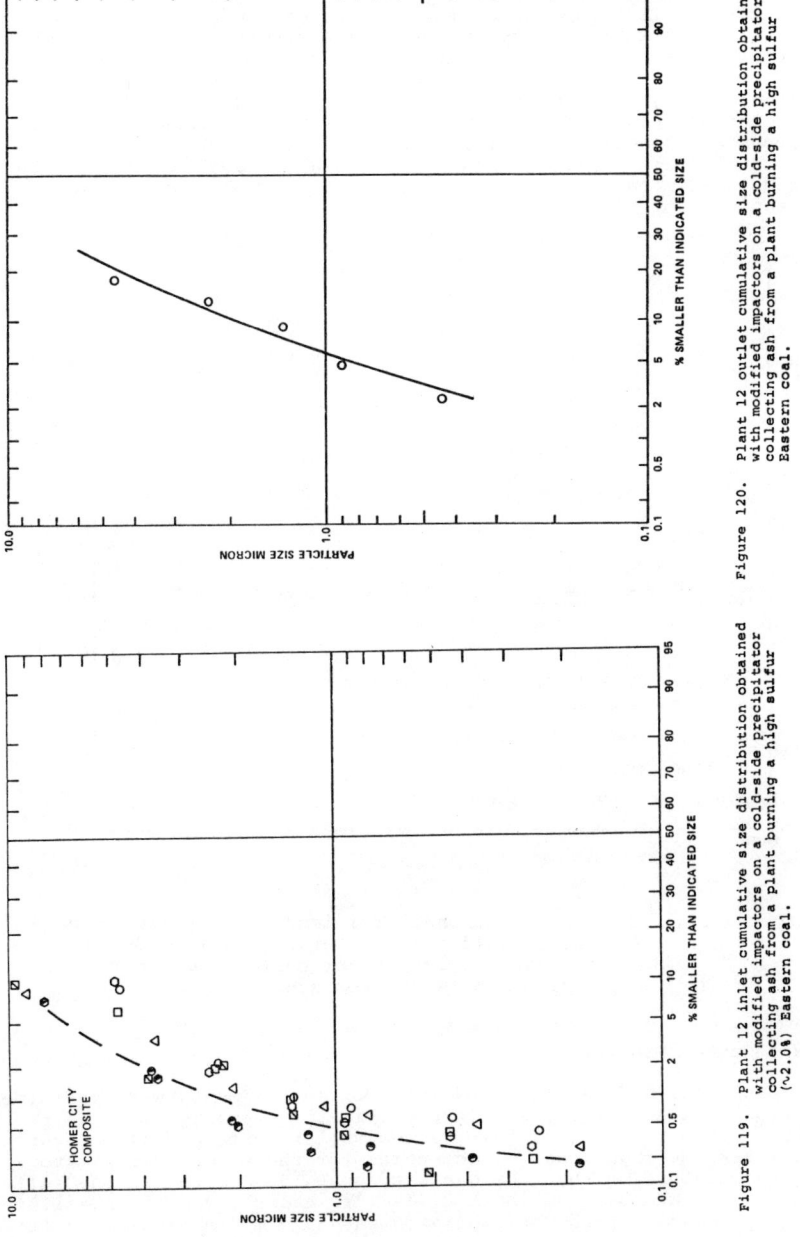

Figure 119. Plant 12 inlet cumulative size distribution obtained with modified impactors on a cold-side precipitator collecting ash from a plant burning a high sulfur (~2.0%) Eastern coal.

Figure 120. Plant 12 outlet cumulative size distribution obtained with modified impactors on a cold-side precipitator collecting ash from a plant burning a high sulfur Eastern coal.

Plant Number Thirteen--

The performance of a high efficiency cold-side electrostatic precipitator located in the Midwest was measured with special emphasis on the efficiency of the precipitator as a function of particle size over the range from 0.01 μm to 5 μm. The Midwestern coal burned by the power plant was high in sulfur (∼3.6%) content. Particle size measurements were performed using cascade impactors, a Climet optical particle counter, and diffusion batteries with CN counters to obtain particle size distributions. Figure 121 shows the fractional efficiencies calculated from the optical and diffusional data. Impactor data are not shown because of likely contamination.

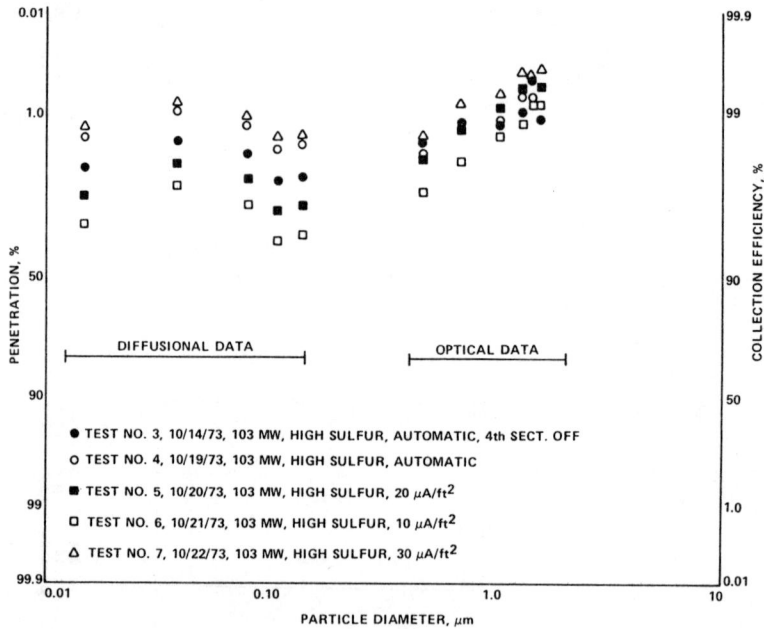

Figure 121. Plant 13 fractional efficiency data measured by optical and diffusional methods on a cold-side electrostatic precipitator collecting ash from a high sulfur (3.6%) Midwestern coal.

Plant Number Fourteen--

Optical, diffusional, and impactor measurements were performed on a pilot-scale electrostatic precipitator treating flue gas resulting from the combustion of a low sulfur Western coal. Figure 122 gives the fractional efficiencies for the pilot precipitator. The temperature of the flue gas was about 160°C (320°F), the sulfur content of the coal was about 0.47% (dry basis), and the specific collecting area of the precipitator was 66.9 $m^2/(m^3/sec)$ (340 ft^2/1000 cfm).

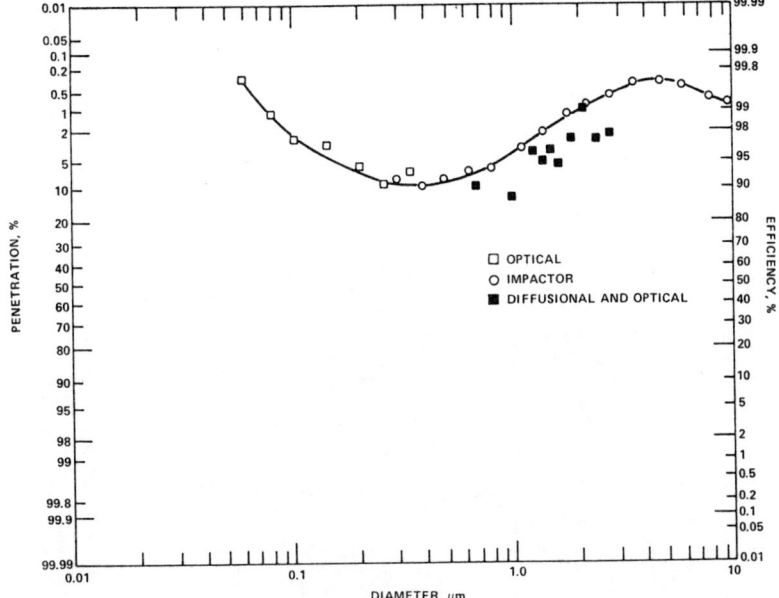

Figure 122. Plant 14 fractional efficiency data obtained by using optical, diffusional, and impactor measurements performed on a pilot-scale precipitator collecting ash from a low sulfur Western coal.

Plant Number Fifteen--

The size distributions shown in Figures 123 - 128 were obtained at the inlet and outlet to a pilot scale electrostatic precipitator collecting ash resulting from the combustion of a low sulfur Western coal. Inlet particle sizing was performed using two six-stage Brink impactors with precollector cyclones and back up filters. The outlet particle sizing was accomplished with an eight-stage Andersen impactor with back up filter. In the case of the Brink impactor, foil substrates were coated with silicone vacuum grease and baked prior to use if the temperature of the flue gas was less than 204°C (400°F). Otherwise, ungreased aluminum foil substrates were used. Glass fiber filter substrates were used in the Andersen impactor.

The inlet size distribution curves are shown in Figure 123, and the outlet size distributions are shown in Figures 124 through 128.

Summary Of Inlet Particle Size Distributions

Inlet particle size distributions from most of the plants previously discussed have been organized into various areas of interest. Figure 129 shows the inlet size distributions of those plants whose electrostatic precipitators were preceded by a mechanical collector. Figure 130 shows inlet size distributions of ash collected from hot-side electrostatic precipitator installations. Figure 131 gives the inlet size distributions from cold-side electrostatic precipitators collecting ash produced from both high and low sulfur coals.

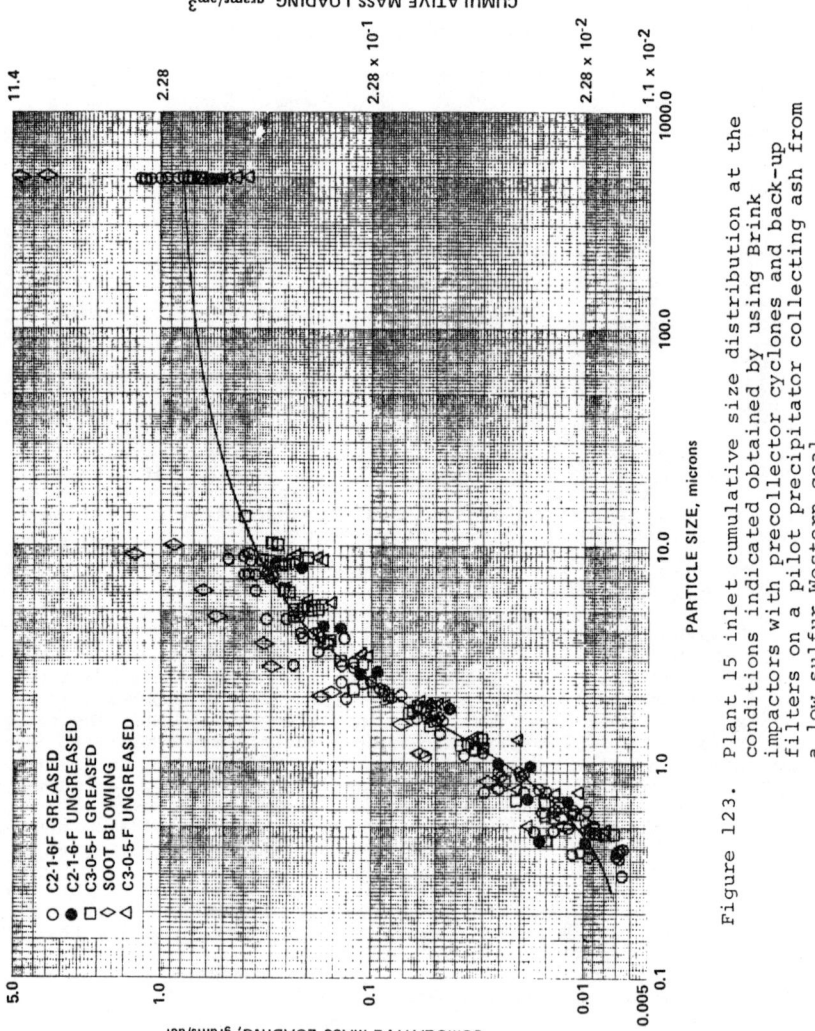

Figure 123. Plant 15 inlet cumulative size distribution at the conditions indicated obtained by using Brink impactors with precollector cyclones and back-up filters on a pilot precipitator collecting ash from a low sulfur Western coal.

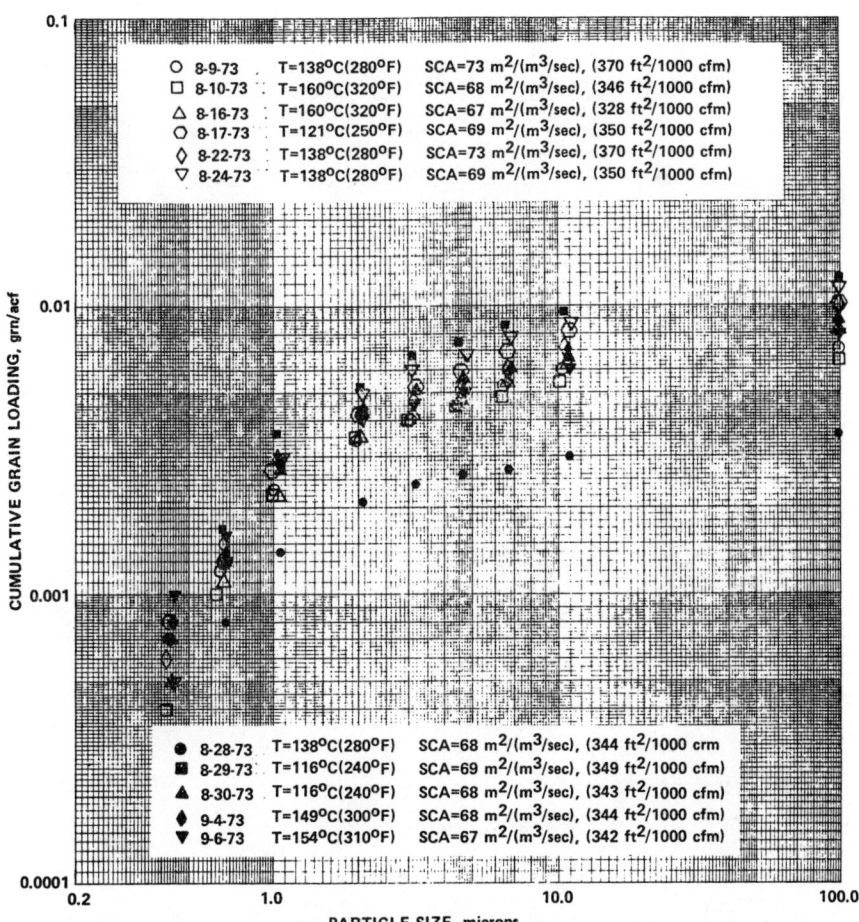

Figure 124. Plant 15 outlet cumulative particle size distribution at the conditions indicated obtained by using an Andersen impactor with a back-up filter on a pilot precipitator collecting ash from a low sulfur Western coal.

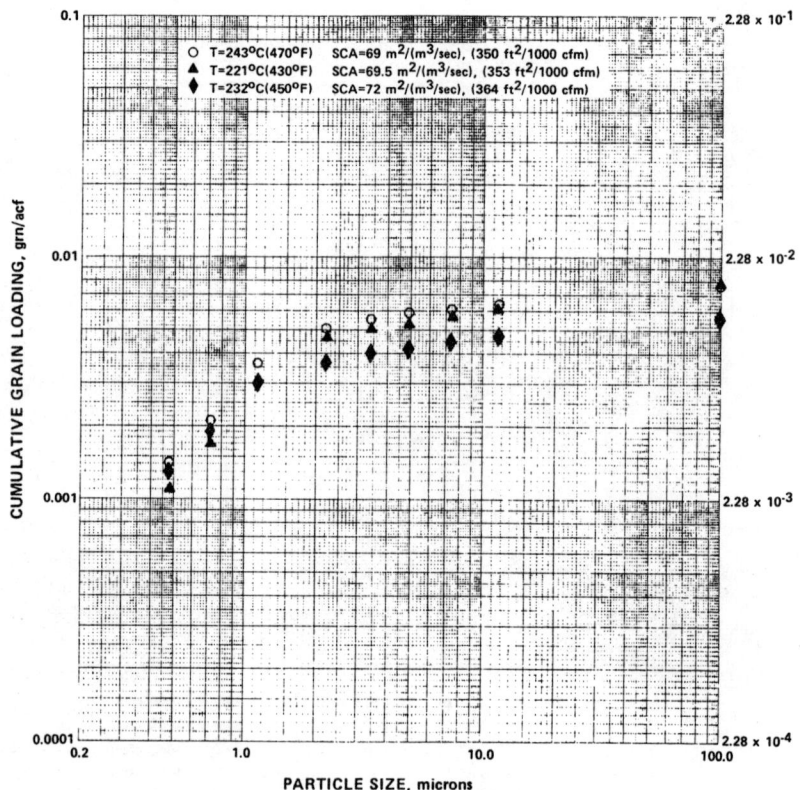

Figure 125. Plant 15 outlet cumulative particle size distribution at the conditions indicated obtained by using an Andersen impactor with a back-up filter on a pilot precipitator collecting ash from a low sulfur Western coal.

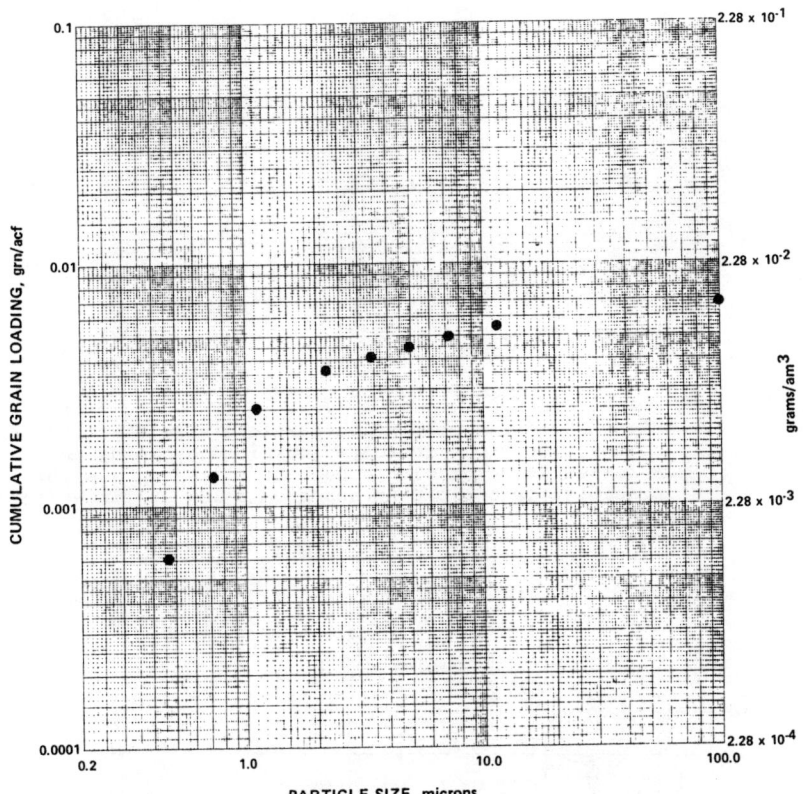

Figure 126. Plant 15 outlet cumulative particle size distribution at the conditions indicated obtained by using an Andersen impactor with a back-up filter on a pilot precipitator collecting ash from a low sulfur Western coal.

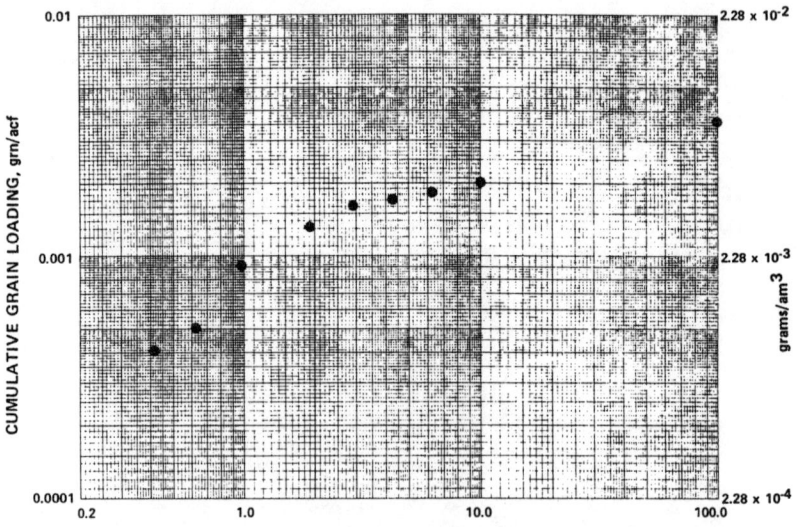

Figure 127. Plant 15 outlet cumulative particle size distribution at the conditions indicated obtained by using an Andersen impactor with a back-up filter on a pilot precipitator collecting ash from a low sulfur Western coal.

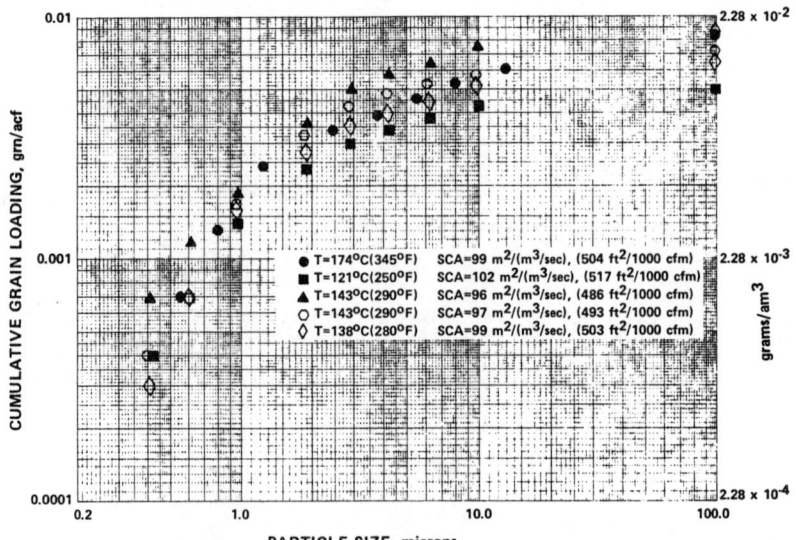

Figure 128. Plant 15 outlet cumulative particle size distribution at the conditions indicated obtained by using an Andersen impactor with a back-up filter on a pilot precipitator collecting ash from a low sulfur Western coal.

Analysis of Factors Influencing Performance   141

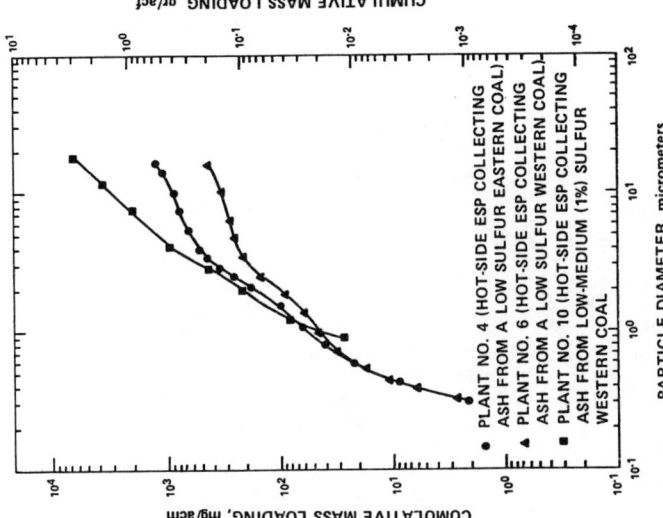

Figure 130. Inlet size distributions of hot-side ESP installations.

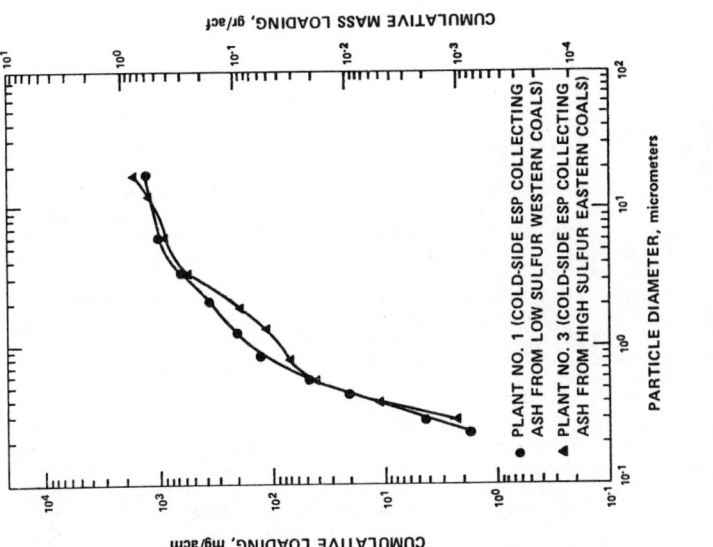

Figure 129. Inlet size distributions of cold-side ESPs preceded by mechanical collectors.

142   Electrostatic Precipitator Manual

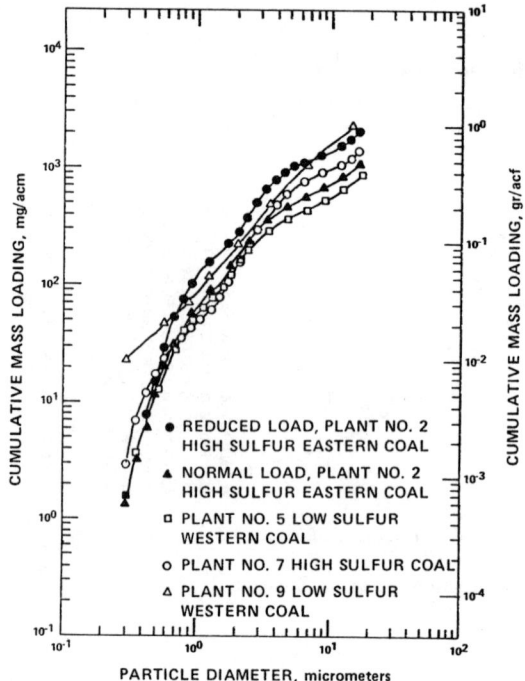

Figure 131. Inlet size distributions of cold-side ESPs collecting ashes from high sulfur and low sulfur coals.

SPECIFIC COLLECTION AREA

The specific collection area (SCA) which is defined as the ratio of the total collection area to the total gas volume flow rate is an important parameter that influences the performance of a precipitator. The SCA can be changed by changing either the collection plate area or the gas volume flow rate or both. In effect, changes in SCA result in changes in the treatment time experienced by the particles. Thus, increasing the SCA of a precipitator increases the collection efficiency. In designing a precipitator, the total gas volume flow rate will be known so that the SCA is determined by the choice of total collection plate area. In existing precipitators, the total collection plate area is fixed but the SCA can change due to changes in the gas volume flow rate.

The SCA provides the most flexible variable in designing a precipitator. Although the SCA has economic and practical limitations, it has no physical limitations and can be increased indefinitely. Even though a curve of collection efficiency versus SCA will level off for the larger values of SCA due to the exponential nature of the collection mechanism, greater efficiency can always be obtained from increased SCA.

Figure 132 shows experimental fractional efficiency data

obtained from a laboratory precipitator collecting dioctyl phthalate (DOP) droplets under essentially idealized conditions at two different SCAs at two different current densities.[138] In these experiments, all variables could be kept essentially constant except the SCA which was changed by changing the gas velocity. The fractional efficiency data were obtained by making particle size distribution measurements with a Brink impactor at the inlet and outlet of the precipitator. For a given current density, the experimental data show the increase in particle collection efficiency with increased SCA.

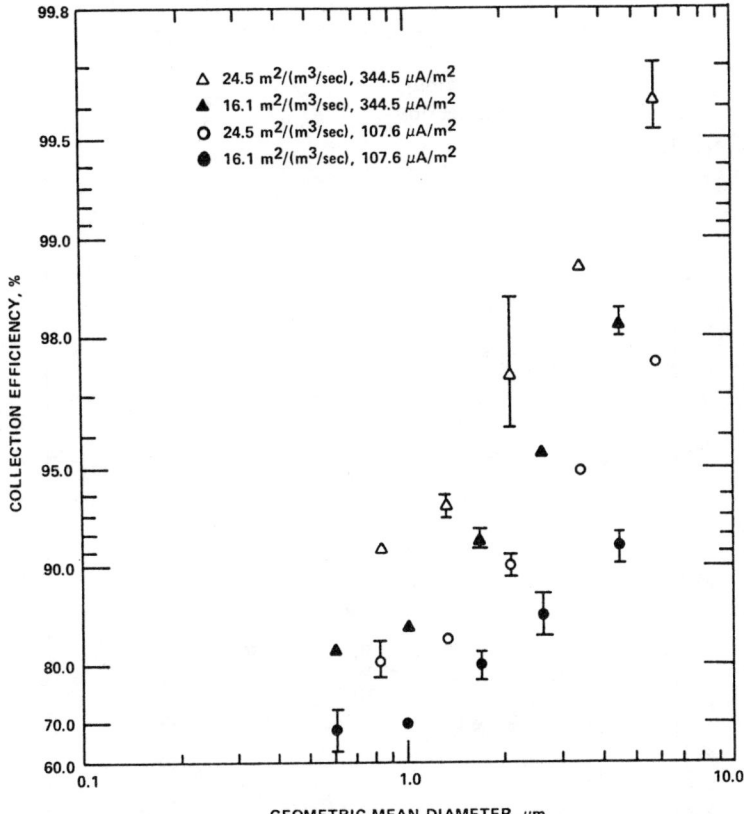

Figure 132. Experimental fraction efficiency data obtained from a laboratory precipitator collecting dioctylphthalate (DOP) droplets under essentially idealized conditions at two different SCAs at two different current densities.[138]

Figure 133 shows experimental data on the effects of SCA on overall mass collection efficiency. The data were obtained from pilot plant studies on the flue gas from a coal fired boiler. Test velocities through the precipitator were varied from 1.13 to 2.53 m/sec. The precipitator had two electrical sections in the direction of gas flow. The inlet section was maintained at approximately 41.7 nA/cm² while the outlet section was maintained at approximately 69.5 nA/cm². Although attempts were made to hold flue gas temperatures and boiler operating conditions identical for each test, inlet gas temperatures ranged from 146 to 174°C and the inlet mass loading ranged from 0.011 to 0.018 kg/DNCM. The data represent periods during which both the discharge and collection electrodes were rapped. Although effects due to changes in gas temperature, inlet mass loading, and particle reentrainment, and nonideal conditions influence the data to some extent, the data show the definite increase in overall mass collection efficiency with increased SCA.

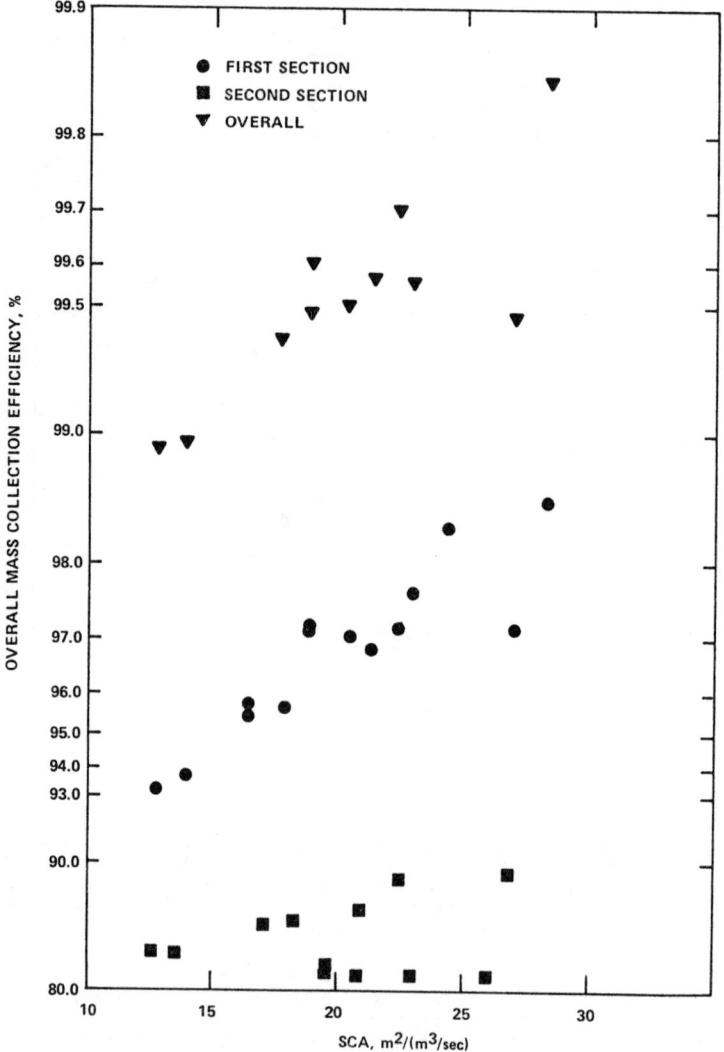

Figure 133. Effects of SCA on overall mass collection efficiency.

Data showing the effect of SCA on the overall mass collection efficiency of a full-scale precipitator collecting fly ash particles are given in Figure 134. The precipitator had a collection electrode area of 7,698 m², three electrical sections in the direction of gas flow, thirty-six gas passages, and a plate height of 8.9 m. The SCA was varied by changing the boiler load. The temperature and resistivity of the ash ranged from 180 to 200°C and 0.4 to 1.0 x 10^{12} ohm-cm, respectively. As with the pilot plant data discussed previously, the effect of SCA can not be completely isolated since all other variables can not be held rigidly constant when the SCA is changed. However, the data again show the definite increase in overall mass collection efficiency with increased SCA.

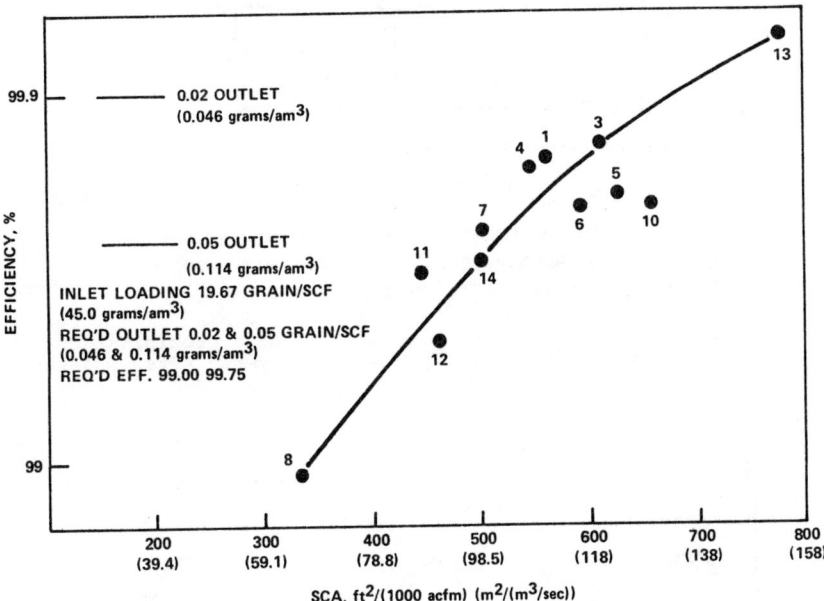

Figure 134. Measured efficiency as a function of specific collection area.

VOLTAGE-CURRENT CHARACTERISTICS

Electrical Circuitry For A Precipitator

The electrical equivalent circuit of a precipitator is shown in Figure 135.[139] The voltage normally applied to a precipitator is either half-wave or full-wave rectified 60 Hertz ac. Neglecting for a moment the effects of C_D and R_D, the capacitor, C_p, charges on the increasing portion of the voltage waveform and discharges on the decreasing portion. The current from the discharging capacitance flows through the resistance R_G tending to maintain the peak voltage applied. There is an exponential decay of this voltage dependent on the time constant of the $R_G C_p$ circuit. This time constant is given by:[140]

$$T = R_G C_p \quad , \tag{19}$$

where T is the time in seconds for the voltage waveform to decrease to approximately 37% of its peak value after the voltage is removed. The current, I, will flow in the return leg of the circuit only during the charging of the capacitor. During the remainder of the cycle, the current supplied to R_G is the discharge current from C_p. These relationships are shown in Figure 136. In this example T is assumed to be greater than 8 milliseconds or 1/2 cycle of the line voltage.

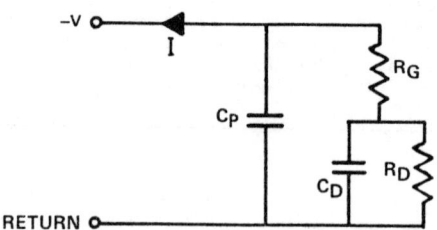

V = VOLTAGE APPLIED ACROSS ELECTRODES IN VOLTS
I = TOTAL CONVENTIONAL CURRENT FLOW IN AMPERES
C_P = EQUIVALENT CAPACITANCE OF THE ELECTRODE SYSTEM IN FARADS
R_G = EQUIVALENT RESISTANCE OF THE INTER-ELECTRODE REGION IN OHMS
C_D = EFFECTIVE CAPACITANCE OF THE DUST LAYER IN FARADS
R_D = EFFECTIVE RESISTANCE OF THE DUST LAYER IN OHMS

Figure 135. Electrical equivalent circuit of a precipitator electrode system with a dust layer. After Oglesby and Nichols.[139]

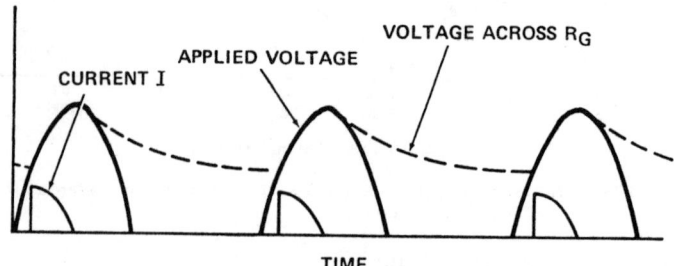

Figure 136. Voltage-current relationship in an ideal capacitor/resistor parallel combination.

Normally the effective impedance presented by the parallel combination of C_D and R_D is negligible compared to the impedance of R_G. Thus, the time domain response of the precipitator is determined by the combination of C_p and R_G. However, this is not true when the dust layer is in a breakdown condition and possibly exhibiting back corona. The breakdown may effectively short out the dust layer and a portion of R_G thereby reducing the time constant, T, and increasing the current, I. This change in time constant may be monitored on an oscilloscope presentation of the voltage waveform and used to support evidence that breakdown of the dust layer is occurring.

The voltages and currents in a precipitator are most often measured by the installed power set instrumentation as root-mean-square (rms) or effective values. The capacitances and resistances vary slowly with time so that the equivalent circuit of a precipitator in normal operation can be approximated as a pure resistance across the terminals of a DC source. The voltage-current relationship is simply $V = RI$ where R is the effective value of the resistance in ohms, V is in rms volts, and I is in rms amperes. An actual precipitator departs from ohmic behavior in that R is a non-linear function of the current. The graphical presentation of precipitator voltage versus secondary current is not the straight line generated with an ohmic resistance, but generally curved and referred to as a V-I curve.

Measurement Of Voltage-Current Characteristics[141]

Many precipitator control rooms have panel meters for each transformer/rectifier (T/R) set which display the primary and secondary voltages and currents and the sparking rate. The secondary voltage-current characteristics are needed in order to analyze the electrical operation of a precipitator. Thus, panel meters for measuring both secondary voltages and currents should be provided. If a precipitator is not equipped with panel meters for measuring secondary voltages, or if calibrations of existing meters are desired, temporary voltage divider networks and accurate voltmeters can be installed on the precipitator side of the rectifier networks as shown at point number 1 in Figure 137 to obtain secondary voltage measurements. In practice, the voltage dividers are inserted in parallel across the high voltage bus sections of the precipitator. Typically, the resistor R_2 has a value of about 1×10^9 ohms and R_1 has a value of about 12×10^3 ohms. Because of the voltage drop across R_2, this resistor should be well insulated electrically.

If it is necessary to measure the secondary current, a voltmeter can be placed across resistor R_3 in the Surge Arrester network in the return leg of the secondary circuit. The resistor R_3 is typically 50 ohms or less. The entire precipitator secondary current passes through this resistance. The voltage developed across R_3 is proportional to the current. Some manufacturers utilize a meter calibrated to read current based on the detection of this voltage. Other manufacturers may place a current meter with very low internal impedance across R_3 and allow all the precipitator current to pass through the meter. In this case, the resistor R_3 is in the circuit to prevent isolating the power set if the meter is removed from the circuit. Point number 2 in Figure 137 shows the relation of these components to the remainder of the system.

In order to calibrate the secondary current meter it must first be determined whether the meter is a voltage or current sensing device. If this cannot be determined from the precipitator operation and maintenance manual, a test must be made. If it is a voltage detecting type current meter, a volt meter placed across the resistor will read within a few percent of the same voltage whether the T/R set current meter is attached or not. If the measured voltage is low with the T/R set meter in the circuit, the T/R set meter is a current sensing device. Calibrating a voltage sensing meter requires accurately measuring the resistance of the resistor, out of the circuit, and recording the voltages for various currents. Then, Ohm's law is applied to obtain the true currents. Comparison of the true currents with the meter readings yields a calibration curve for the meter. If the power set has a current sensing meter, a calibrated current meter of appropriate capacity is inserted in series with the meter to be calibrated. Measurement of various currents with the two meters and comparison of the readings yield a calibration curve for the uncalibrated meter.

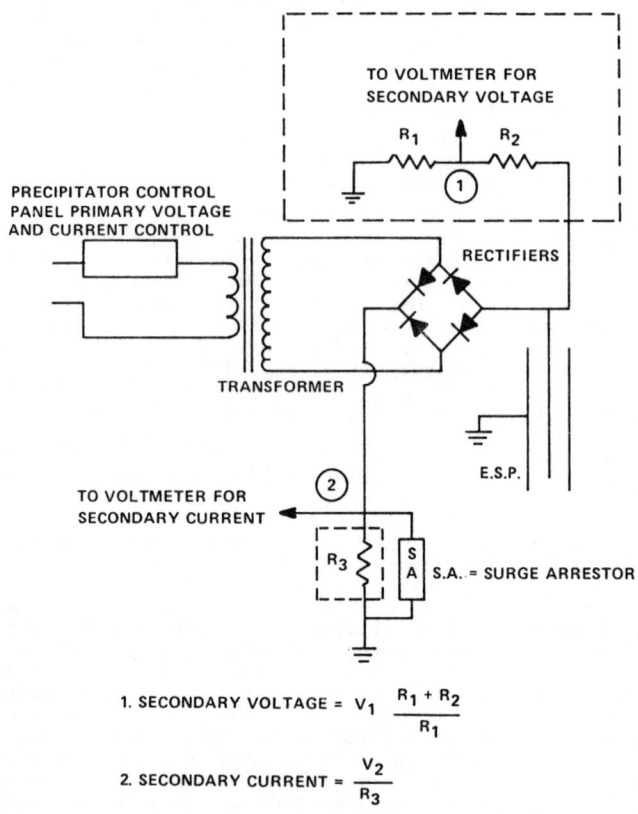

1. SECONDARY VOLTAGE = $V_1 \dfrac{R_1 + R_2}{R_1}$

2. SECONDARY CURRENT = $\dfrac{V_2}{R_3}$

Figure 137. Voltage divider network for measuring precipitator secondary voltages and currents.

Figure 138 is a facsimile of a data sheet used to collect data from which voltage-current relationships may be plotted. In the general heading, information is recorded which will identify the test, the power supply (T/R Set), the plate area fed by the power set, and the determined calibration factors for the voltage and current. Data is taken as the manual set control is gradually increased until some current flow is detected. This is recorded as the corona starting voltage. Subsequent points are taken by increasing the control for some increment of current and recording the meter readings at that point. Readings are taken until some limiting factor is reached. This factor is recorded on the right-hand side of the data sheet and is usually excessive sparking or a current or voltage limitation of the power set.

POWER SET
VOLTAGE-CURRENT CURVE DATA SHEET

DATE/TIME _____ T/R SET NO. _____ COLLECTING AREA _____
VOLTAGE DIV. MULT. _____
T/R SET DCMA CORRECTION _____

PRIMARY VOLTS	PRIMARY AMPS	DCKV T/R SET METER	DCMA T/R SET METER	SPARK RATE	DCMA CORR.	DC VOLTS VOLTAGE DIV.	DCKV CORR.	μA/ft²	NA/cm²	TERMINAL POINT DETERMINED BY: (CIRCLE ONE)
										1. SPARKING
										2. SEC. CURRENT LIMIT
										3. SEC. VOLTAGE LIMIT
										4. OTHER _____
										COMMENTS:

Figure 138. Sample V-I curve data sheet.

The columns as shown in Figure 138 usually completed for each point include those labeled PRIMARY VOLTS, PRIMARY AMPS, DCKV T/R SET METER, DCMA T/R SET METER, SPARK RATE, and DC VOLTS VOLTAGE DIV. At a later time the DCMA correction factor is applied to the T/R set meter reading and the DCMA CORR. column is completed.* The DCKV CORR. column is completed by multiplying the DC VOLTS VOLTAGE DIV. column by the voltage divider multiplier. The last two columns are completed by dividing the DCMA CORR. by the appropriate collecting area in square feet or square centimeters and applying a multiplicative factor of 10^{-3}. A plot is then made on linear graph paper of the DCKV CORR. vs $\mu A/ft^2$ or nA/cm^2 depending on the experimental requirement. A typical voltage-current curve is shown in Figure 139.[142]

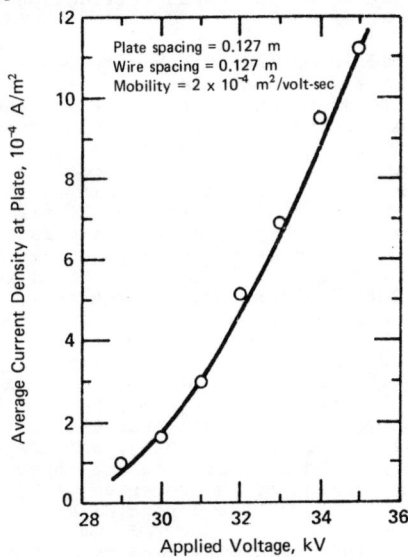

Figure 139. Typical voltage-current curve derived experimentally in a laboratory wire-duct precipitator. After McDonald.[142]

___

*On a dual half-wave installation where the voltage is measured on one independent HV bus but the current is the sum of both sections, the secondary current must also be multiplied by the ratio of the plate area of the section under test to the total plate area in order to approximate the secondary current in that power supply leg.

Voltage is plotted linearly along the horizontal axis and current density linearly along the vertical axis. Current density at the collection plate is used rather than total current supplied to give a basis for comparison. This curve was obtained with 2.67 mm diameter wires in a laboratory scale precipitator.

Effect Of Electrode Geometry

Geometrical factors which affect the electrical characteristics of a wire-plate precipitator include the plate-to-plate spacing, wire-to-wire spacing, wire radius, plate area per power set, and roughness of the wire. Each of these factors contributes its own distinctive effect on the electrical characteristics.

The plate-to-plate and wire-to-wire spacings affect the spatial distribution of the current density, electric field, and space charge density. For the same applied voltage, wire radius, and wire-to-wire spacing, the effect of increasing the plate-to-plate spacing is one of distributing the ionic current, originating from the region near the wire, and the potential difference over increasing surface areas. This leads to lower and less rapidly varying values of current density, electric field intensity, and space charge density in the region outside the corona sheath. For the same applied voltage, wire radius, and plate-to-plate spacing, the effect of decreasing the wire-to-wire spacing is to increase the uniformity of the current density and electric field distributions. It should be noted, however, that there is an optimum wire-to-wire spacing which will yield a maximum current, and reduction of the wire-to-wire spacing below this value will lead to reduced currents due to an increased interaction of the electric fields near the wires.

Increasing the radius of the corona wire leads to higher corona starting voltages and lower electric field intensities at the surface of the wire at corona onset. For a given applied voltage, above the corona starting voltage, the corona current will decrease as the wire radius is increased. For the same average current density at the plate, the space charge density near the wire decreases as the corona wire radius is increased. The average current density at the plate is maintained because of the higher applied voltages which are necessary to produce ionization as the corona wire radius is increased. The higher applied voltages result in higher values of electric field intensity outside the region of ionization and, consequently, faster migration of the ions towards the plate.

Figures 140 through 145 show _theoretically calculated_ trends caused by changes in wire radius, plate-to-plate spacing, and wire-to-wire spacing.[4] The symbols r_w, S_x, S_y, and b in the figures represent the wire radius, one-half the plate-to-plate spacing, one-half the wire-to-wire spacing, and the charge carrier mobility, respectively. The curves in all of these figures were obtained by taking the values of the relative gas density (δ) and the roughness factor (f) of the corona wires to be unity.

Figures 140, 141, and 142 demonstrate the effects of wire size, plate-to-plate spacing, and wire-to-wire spacing on voltage-current characteristics. Figure 140 shows that an increase in wire size leads to a higher starting voltage and lower currents for the same applied voltages. An increase in wire size shifts the voltage-current curve to the right but does not substantially alter the shape of the curve. Figure 141 demonstrates that increasing the plate-to-plate spacing has only a slight effect on raising the starting voltage but leads to a large drop in current for the same applied voltage at voltages above corona start. An increase in plate-to-plate spacing rotates the voltage-current curve to the right (produces a decrease in the slope of the curve). Figure 142 shows that the wire-to-wire spacing has little effect on voltage-current characteristics over a wide range of values. Increasing the wire-to-wire spacing generally shifts the voltage-current curve to the left although the curves for different wire-to-wire spacings may intersect one another.

Figures 143, 144, and 145 illustrate how the average electric field at the plate E_p and the average electric field between the electrodes E_a vary as a function of the average current density at the plate for different wire sizes, plate-to-plate spacings, and wire-to-wire spacings. Figure 143 shows that for the same average current density at the plate the average electric field at the plate increases slowly with increasing wire size. Figure 144 demonstrates that for the same average current density at the plate the average electric field at the plate increases rapidly with increasing plate-to-plate spacing. Figure 145 indicates that the wire-to-wire spacing has only a small effect on the average electric field at the plate for any given average current density at the plate.

In most practical applications, the geometries will differ to some extent from a true wire-plate design. For example, discharge electrodes may have a design other than round wire, discharge electrodes may be supported in a rigid frame, and the collection electrodes may contain protrusions such as baffles and flanges. However, the general trends discussed above for wire-plate geometries will be evidenced in the geometries utilized commercially.

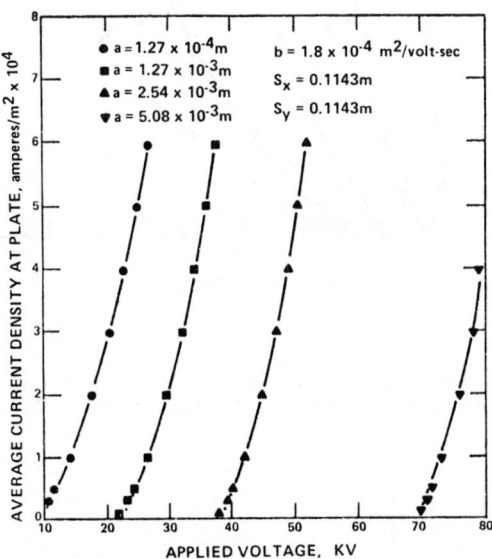

Figure 140. Theoretical curves showing the effect of wire size on voltage-current characteristics.

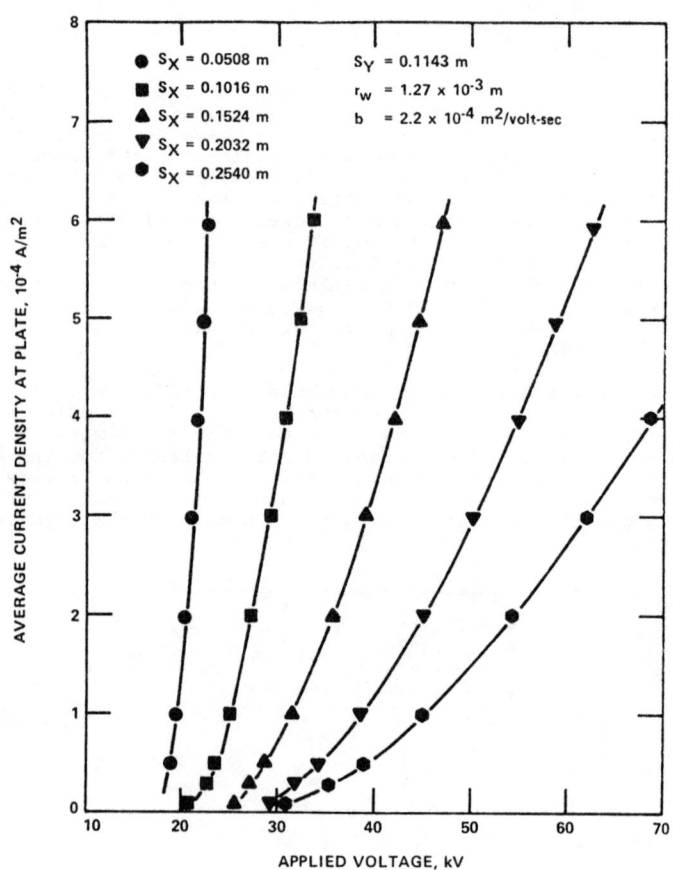

Figure 141. Theoretical curves showing the effect of plate-to-plate spacing on voltage-current characteristics.

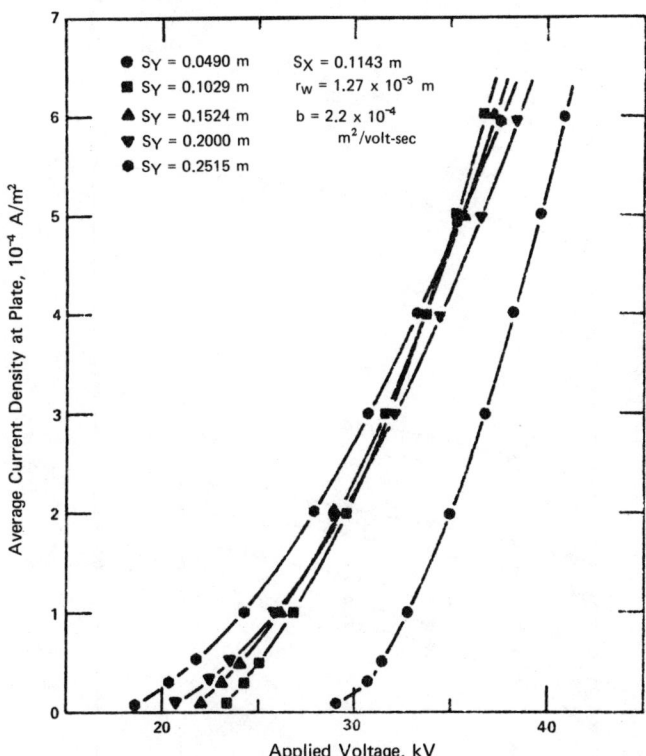

Figure 142. Theoretical curves showing the effect of wire-to-wire spacing on voltage-current characteristics.

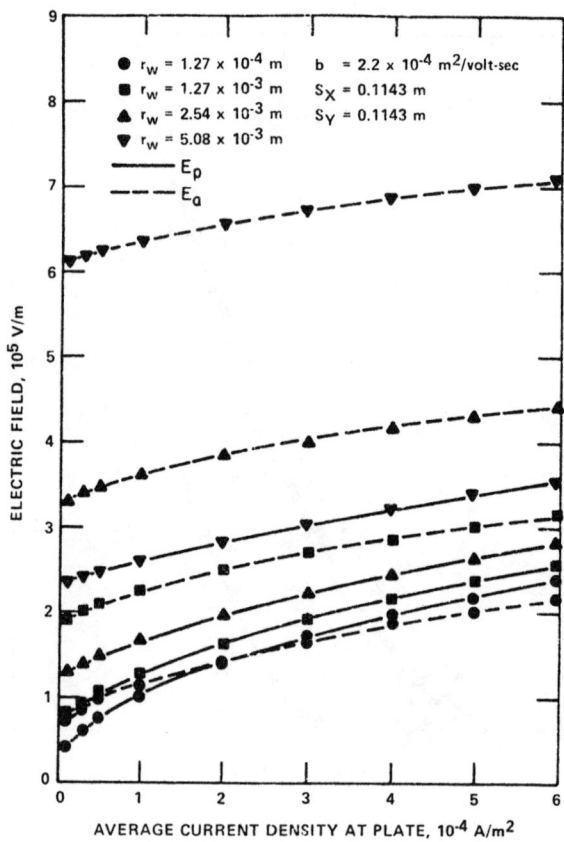

Figure 143. Theoretical curves showing the effect of wire size on the electric field and current density.

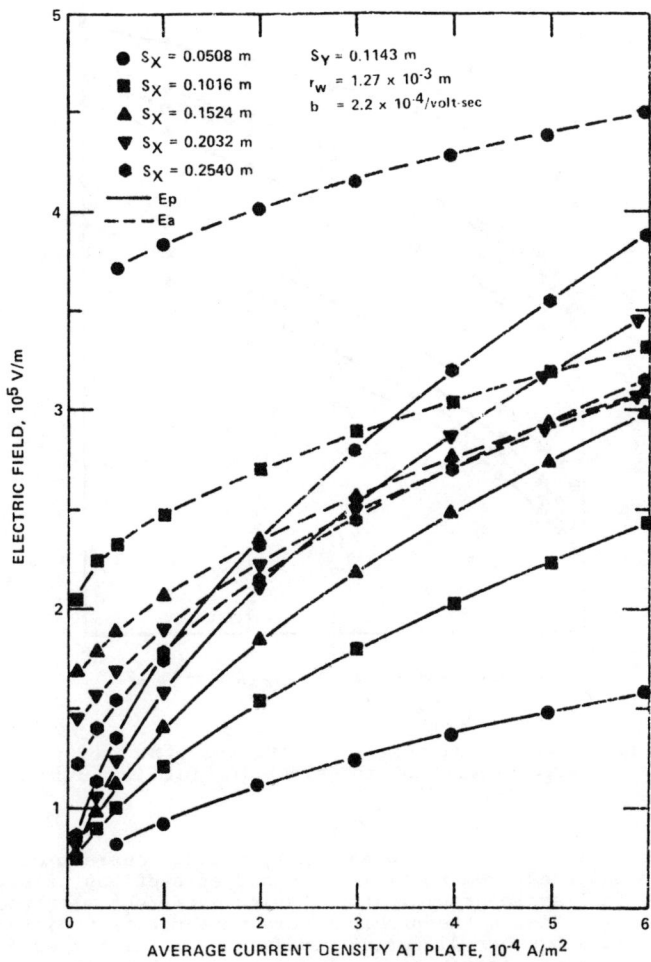

Figure 144. Theoretical curves showing the effect of plate-to-plate spacing on the electric field and current density.

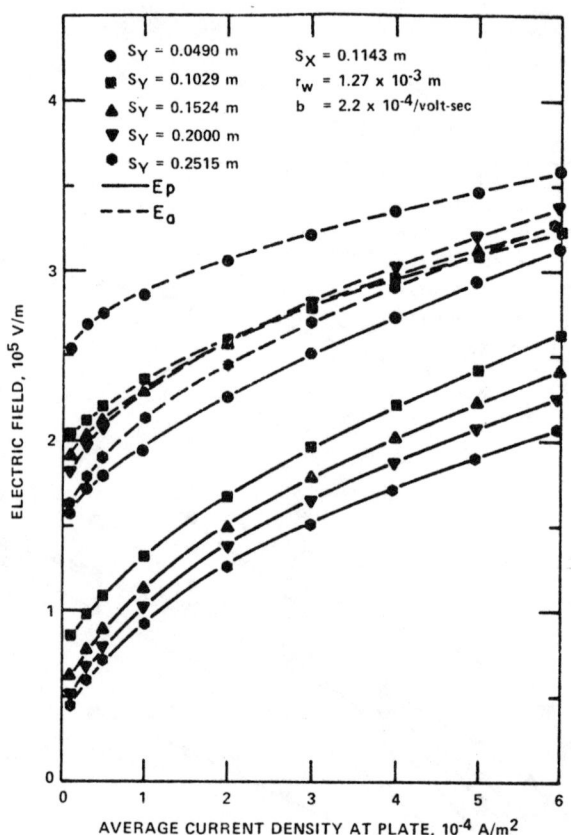

Figure 145. Theoretical curves showing the effect of wire-to-wire spacing on the electric field and current density.

Figures 146 and 147 show secondary voltage-current characteristics obtained from the inlet and outlet sections of several cold-side full-scale precipitators having different electrode geometries. Although the precipitators were treating different types of fly ash under differing conditions of inlet mass loading and particle size distribution, temperature, ash resistivity, gas velocity, etc., the effects of geometry are still evidenced in the voltage-current curves. Larger plate-to-plate spacings tend to rotate the voltage-current curve to the right and tend to lead to higher applied voltages for a given current density. Larger effective discharge electrode diameters tend to shift the entire voltage-current curve to the right. In practice, these effects may be obscured to some extent by the surface properties of the discharge electrodes and the presence of particles on the discharge and collection electrodes. These effects will be discussed later.

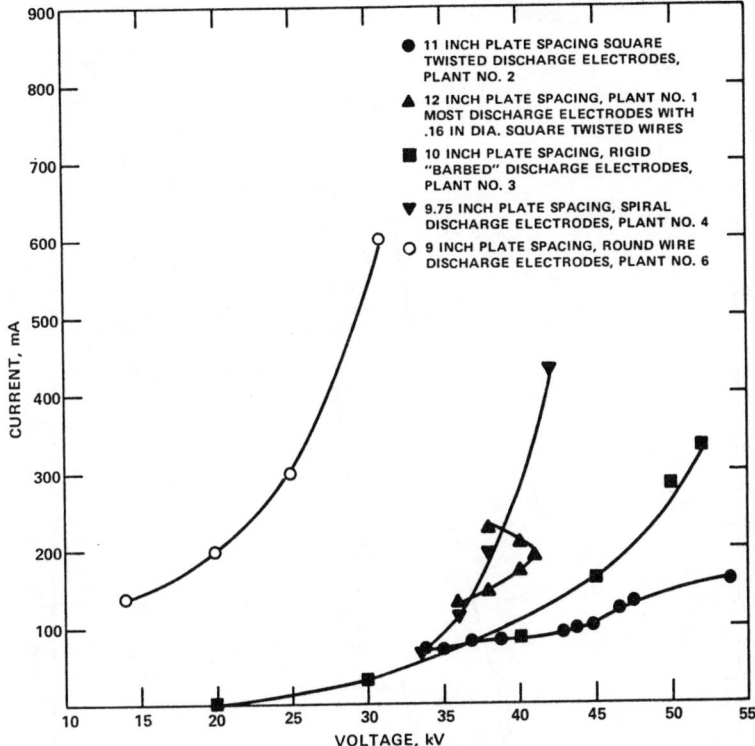

Figure 146. Secondary voltage-current curves obtained from the inlet sections of several cold-side full-scale precipitators having different electrode geometries.

The collection plate area per power set is another geometrical factor of importance in determining the electrical characteristics of a precipitator because it affects the sparkover voltage. The optimum spark rate for n wires will be the same as for one wire since a spark in any of the n wires causes the voltage to collapse momentarily on all wires. Therefore, the optimum operating voltage for n wires will be lower than for one wire. If the optimum operating voltage at the optimum spark rate for one corona wire is V_1 (kV), then the optimum operating voltage V_n (kV) at the same optimum spark rate for n identical corona wires is[143]

$$V_n = V_1 - \frac{1}{b} \log_e n ,\qquad (20)$$

where b is an empirical constant with a value on the order of one.

Equation (20) can be related to plate area by substituting the quantity (total plate area)/(plate area per corona wire) for n. The relationship of the optimum operating voltage to the number of corona wires is shown in Figure 148 for $V_1 = 50$ kV and various values of b. In practice, once the plate height and wire-to-wire

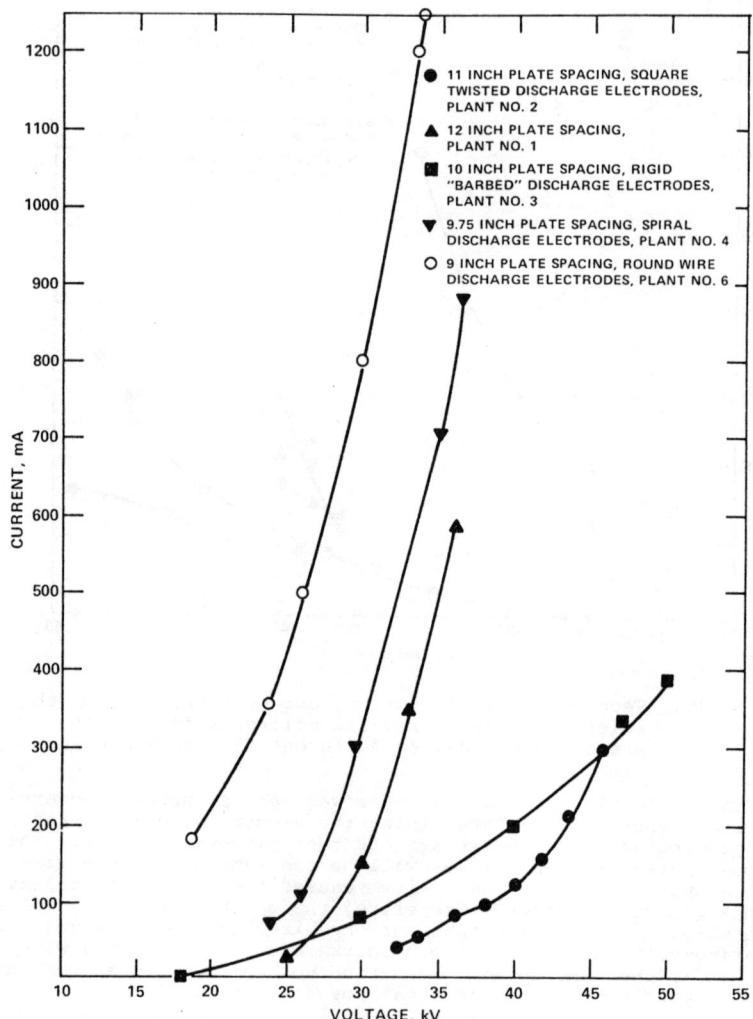

Figure 147. Secondary voltage-current curves obtained from the outlet sections of several cold-side full-scale precipitators having different electrode geometries.

spacing are established, the plate area per corona wire becomes
fixed. Thus, the number of wires per power set is determined by
the plate area per power set. For best precipitator performance,
the plate area serviced by a single power set should be made small
enough to avoid a large reduction in the optimum operating voltage.
This normally leads to a precipitator design with a high degree of
electrical sectionalization. A high degree of electrical sectionalization is also beneficial for two other reasons: (1) the outage
of electrical sections produces less degradation in precipitator
performance and (2) particle reentrainment due to sparking and
rapping is less severe.

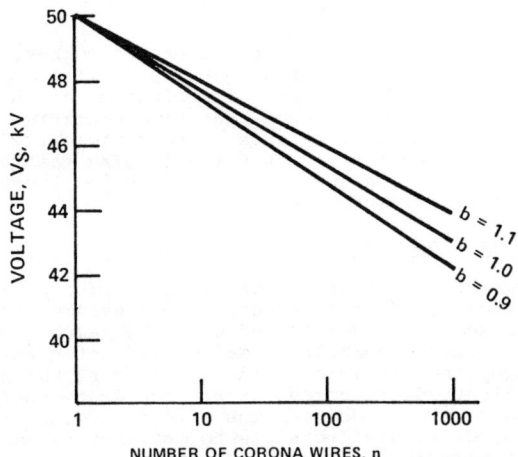

Figure 148. Sparking voltage as a function of number of corona wires.

The roughness (or surface condition) of the corona wires is
another geometrical factor which influences the electrical characteristics of a precipitator. A roughness factor f is used to
designate the degree of roughness of round corona wires. The
roughness of the wires affects the electrical characteristics
by influencing the corona starting voltage, the electric field
intensity at the surface of the wires at corona onset, and the
space charge densities near the wires. Values of the roughness
factor normally lie in the range 0.5-1.0,[144] and a 0.1 change
in value will result in considerably different electrical characteristics.

For clean, smooth wires used in laboratory experiments in
air, the roughness factor can be taken as unity with good results.[144,145] If the surface of a wire is specked with dirt,
rough, or scratched, the roughness factor will be less than unity
and difficult to determine quantitatively. These types of imper-
fection on the surface of the wire give rise to local regions
which have a smaller radius of curvature than the wire. Higher
electric field intensities will exist where these imperfections
are located and corona discharge will occur at reduced voltages
at these locations before spreading to the entire surface of the
wire at higher voltages. This results in a nonuniform current
along the length of the wire. The effect of these imperfections
is to decrease the corona starting voltage, electric field intensity at the surface of the wire at corona onset, and, for
any given applied voltage, to increase the space charge density

near the wire and the average current density at the plate. In the practical observation of the effect of wire roughness, the voltage-current characteristic will shift to the left with decreasing values of roughness factor in a manner which is similar to decreasing the wire radius.

In industrial applications, the wires can accumulate ash to the extent that they are completely covered. In these cases, there is an effect of increasing the wire radius which is different from accounting for the imperfections on the surface of a wire. Thus, a new radius for the discharge electrode is established and imperfections will exist on the surface defined by this radius.

Since the roughness factor depends on the number, type, extent, and radii of the imperfections on the surface of a wire, the possibility of a non-empirical determination of this parameter is quite remote. This means that representative values of the roughness factor must be determined empirically by making measurements of voltage-current curves and using the roughness factor as an adjustable parameter in the existing theories in order to fit the experimental data.

Effect Due To Gas Properties

The voltage-current characteristics of a precipitator are affected significantly by the temperature, pressure, and composition of the gaseous conduction medium. The temperature and pressure of the gaseous conduction medium influence the corona starting voltage, the electric field intensity at the surface of the discharge electrode at corona onset, the space charge density near the discharge electrode, and the effective mobility of the molecular ions. Certain effects due to temperature and pressure can be analyzed through changes in the gas density (δ):

$$\delta = \delta_0 \frac{T_0}{T} \cdot \frac{p}{p_0} , \qquad (21)$$

where

δ_0 = gas density at T_0 and p_0 (kg/m^3),

T_0 = standard temperature (273°K),

T = actual temperature of the gas (°K),

p_0 = standard atmospheric pressure (760 mm Hg), and

p = actual pressure of the gas (mm Hg).

The parameter δ decreases with increasing temperature and decreasing pressure. As δ decreases, the corona starting voltage, the electric field intensity at the surface of the discharge electrode at the corona onset, and the sparkover voltage all decrease. These effects can be explained by examining the influence of δ on the space charge density near the discharge electrode. As δ decreases, the effective mobility of the ions increases due to a reduced number of collisions with neutral molecules. For a given applied voltage, this leads to a decrease in space charge density near the discharge electrode and an increase in the average current density at the collection electrode. The decrease in space charge density near the discharge electrode results in the attainment of a given discharge current at a lower value of electric field intensity at the surface of the discharge electrode. Thus, in order to maintain a given average current density at the plate as δ decreases, the applied voltage must be lowered so that the lower electric field intensities which result will move the ions away from the region near the discharge electrode at a slower rate.

Figures 149, 150, and 151 contain experimental data showing the effects of temperature and pressure on voltage-current characteristics and sparkover voltage.[146]

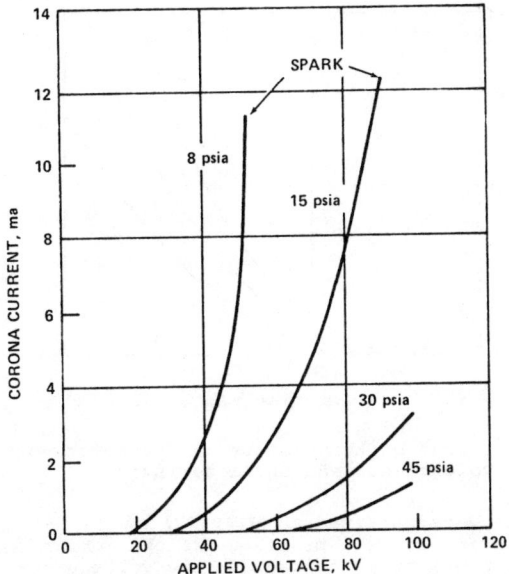

Figure 149. Effect of air pressure on sparkover voltage and voltage-current characteristics.[146]

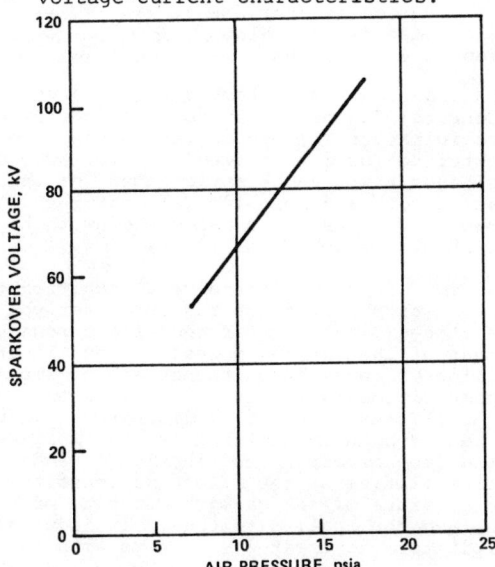

Figure 150. Effect of air pressure on sparkover voltage.[146]

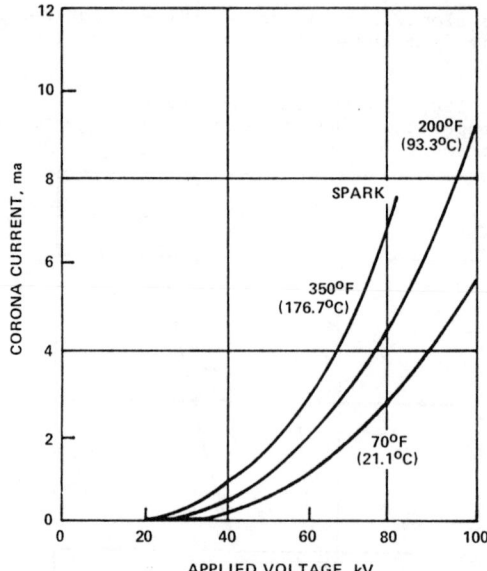

Figure 151. Effect of temperature on sparkover voltage and voltage-current characteristics.[146]

These data were obtained in wire-cylinder geometries for negative corona in air. In practice, if the temperature is increased or the pressure is decreased, the voltage-current curve will shift to the left and will acquire steeper slopes. The shift is due to the decrease in corona starting voltage, and the steeper slopes are due to an increase in the effective ion mobility. The data in these figures also demonstrate how the sparkover voltage decreases as δ decreases.

Figures 152 and 153 show voltage-current curves obtained from outlet electrical fields in several full-scale, cold-side and hot-side precipitators. These curves approximate those that would be obtained for a particle-free gas. The curves show the range of voltages that can be anticipated for the essentially clean flue gases at cold and hot-side temperatures. The lower voltages and smaller voltage range associated with high temperature operation should be noted.

The composition of the gas can have a significant effect on the voltage-current characteristics and sparkover voltage which are obtained in a precipitator. For negative corona discharge, the concentrations of the various molecular constituents and the electron affinities of these constituents are of importance. Different gas compositions will result in different effective charge carrier mobilities in a corona discharge. In general, the current is carried by both molecular ions and free electrons. The extent of the free electron contribution depends on the electron-trapping capabilities of the molecular constituents, the temperature and pressure of the gas, the spacing of the collection electrodes, and the applied voltage. In industrial application of precipitators to treat gas streams emanating from the combustion of coal, the contribution of free electrons to the total current is normally not considered to be significant.

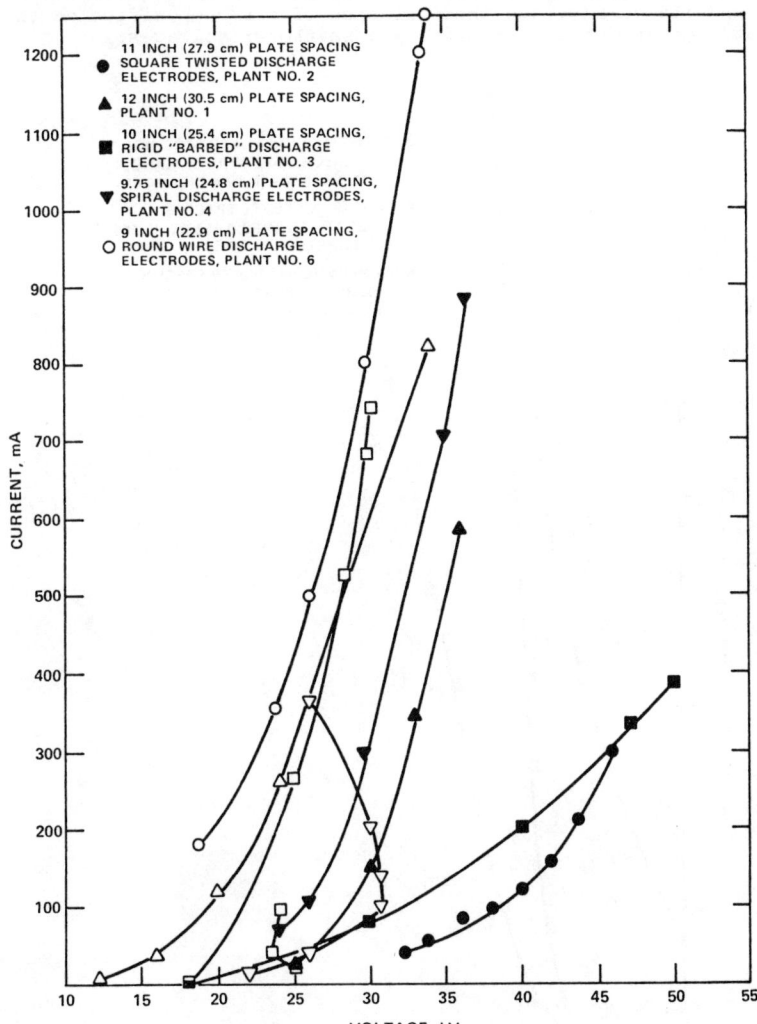

Figure 152. Voltage-current curves obtained from outlet electrical fields in several cold-side electrostatic precipitators.

The flue gas resulting from the combustion of coal and entering a precipitator contains the electron-trapping gases O_2, CO_2, H_2O, SO_2, SO_3, and NO_x in approximate concentrations of 2.0-8.0%, 11.0-16.0%, 5.0-14.0%, 150-3000 ppm, 0.0-30.0 ppm, and 200-800 ppm, respectively. The order of importance of the constituent gases with respect to electron-trapping capabilities is SO_2, O_2, H_2O, and CO_2. Minimum amounts of these gases required to produce a significant effect on the electrical conditions are:

SO_2, 0.5-1.0%; O_2, 2.0-3.0%; and H_2O, about 5.0%. The effect of CO_2 can normally be ignored due to the presence of the other electron-trap gases. The experimental data[147] in Figures 154, 155, and 156 demonstrate the influence of gas composition on the voltage-current characteristics and sparkover voltages.

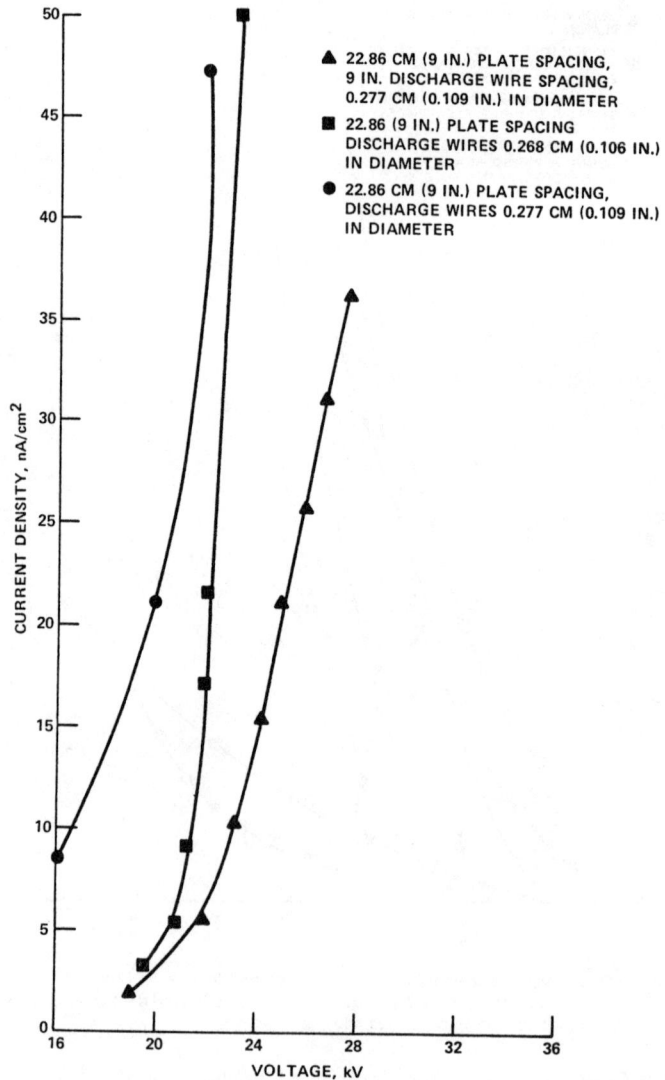

Figure 153. Voltage-current curves obtained from outlet electrical fields in several hot-side electrostatic precipitators.

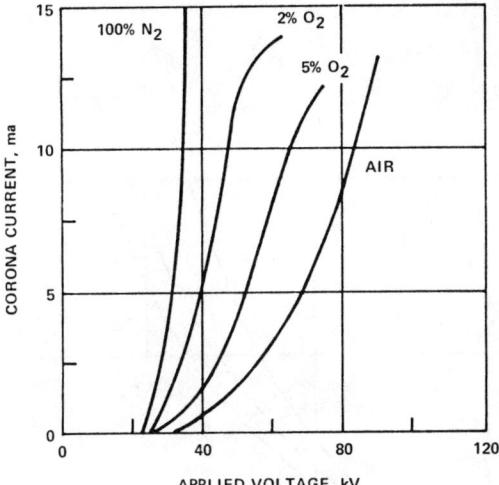

Figure 154. Influence of gas composition on the voltage-current characteristics.[147]

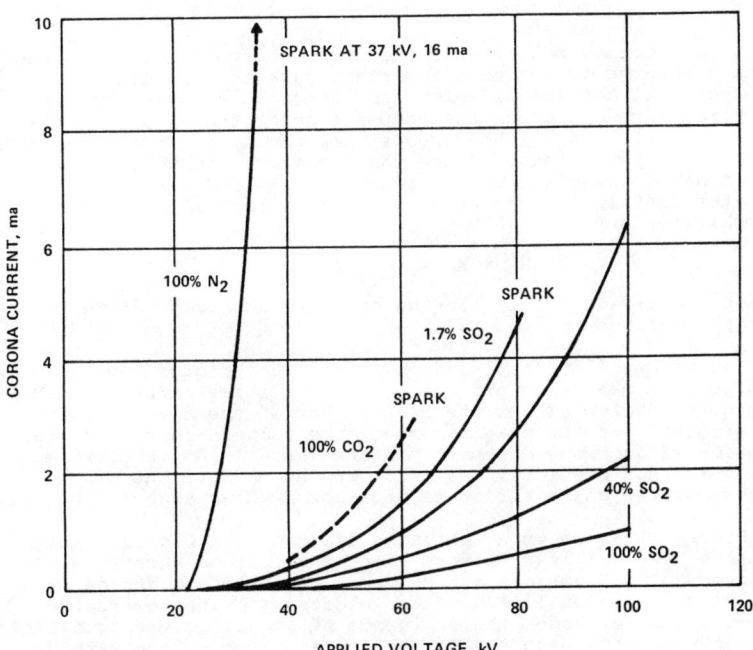

Figure 155. Influence of gas composition on the voltage-current characteristics and sparkover voltages.[147]

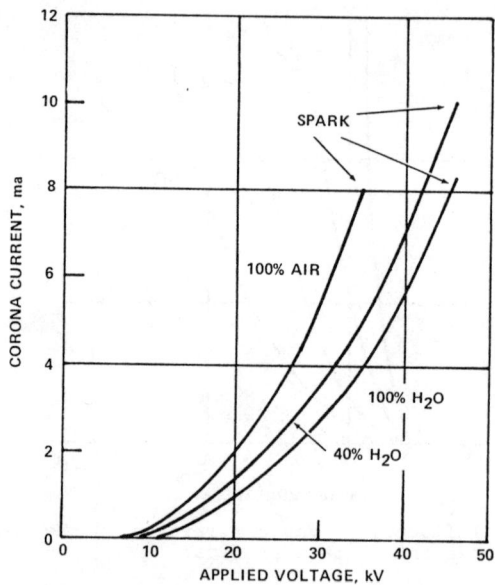

Figure 156. Influence of gas composition on the voltage-current characteristics and sparkover voltages.[147]

The effective mobility of the ionic charge carriers in the corona discharge is the most important parameter in determining the electrical conditions which can be established in the gas. This parameter depends on the composition of the gas, the relative concentrations of the gaseous components, and the temperature and pressure of the gas. Since the effective ion mobility (K) is a function of temperature and pressure, measured values of this parameter are usually reported in terms of the reduced effective ion mobility (K_0):

$$K_0 = K \frac{p}{760} \cdot \frac{273}{T + 273}, \qquad (22)$$

where p (Torr) and T (°C) are the pressure and temperature at which the measurement was made.

Laboratory[148,149,150] and in situ[151] techniques have been developed for measuring effective ion mobilities. The laboratory techniques involve either the measurement of the time of flight of the ions[148] or the measurement of the voltage-current characteristics of a corona discharge in the gas.[150] An in situ technique which has been utilized involves the measurement of the voltage-current characteristics of a corona discharge in the gas.[151]

Figure 157-a shows a schematic diagram of a mobility tube which has been utilized to make time of flight measurements of ion mobilities.[148] Electrons are released from a photocathode by a pulse of ultraviolet light. These electrons drift toward the collector (anode) under the influence of a uniform electric field. The electrons attach to neutral molecules close to the cathode. The negative ions then drift toward the collector. The grids are normally transparent so that if a voltage pulse is applied to the grids some of the ions or electrons in the vicinity of the grid are absorbed and the average collector current is decreased. By varying the delay time of the grid voltage pulse with respect to

the light pulse, a waveform of the ion current as a function of
delay time is obtained. A typical ion-current waveform is shown
in Figure 157-b. An ion-current waveform is obtained for each
grid so that the drift velocity can be obtained from the difference
of ion transit times. Then, the ratio of the drift velocity to
the electric field strength yields the ion mobility.

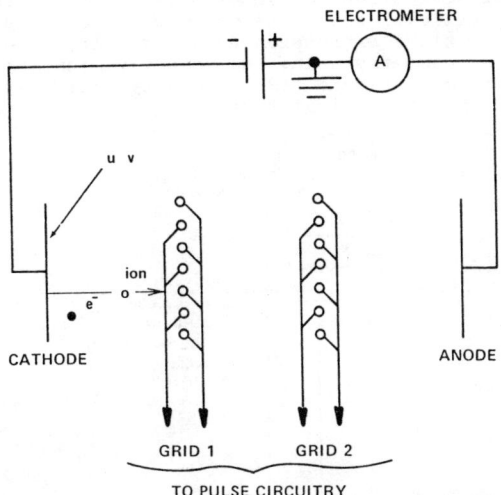

Figure 157-a. Schematic diagram of mobility tube.[148]

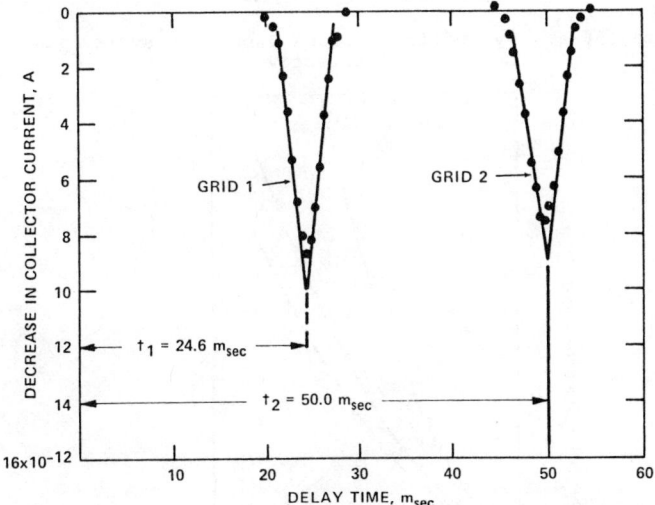

Figure 157-b. Ion-current waveform obtained for $E/N = 3.1 \times 10^{-18}$
V·cm^2, $N = 8.0 \times 10^{18}$ cm^{-3}, and $T = 300°K$. The
waveform obtained at the second is smaller in peak
height and broader than that obtained at the first
grid because of diffusion effects. The loss of ions
to the grids under these conditions was negligible.[148]

Figure 158-a shows a schematic diagram of a laboratory apparatus which has been utilized to determine effective ion mobilities from the measurement of the voltage-current characteristics of a corona discharge in the gas.[150] A simulated flue gas composition flows through a wire-cylinder corona discharge system and is maintained at the desired temperature. In this technique, a voltage-current curve is measured for corona discharge in the particular gas. The measured voltage-current curve is fit to an analytical expression relating voltage and current for wire-cylinder geometry using the effective ion mobility as an adjustable parameter. Figure 158-b shows typical voltage-current data along with the theoretical fits.

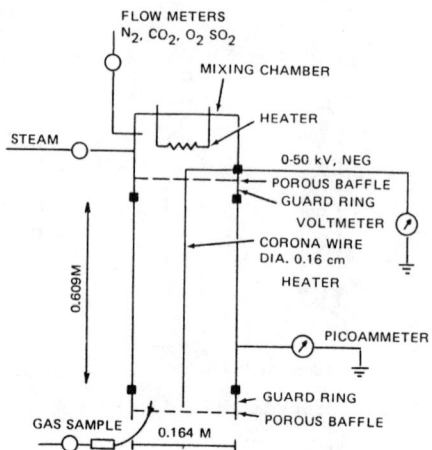

Figure 158-a. Cylindrical corona discharge system for determining effective mobility.[150]

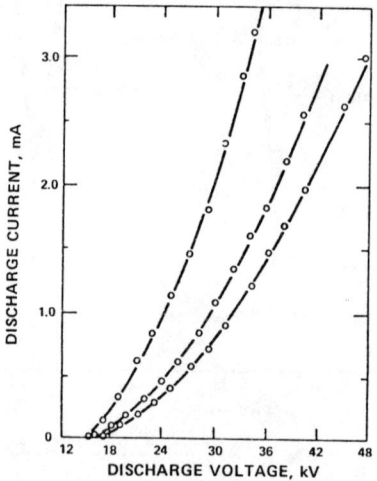

Figure 158-b. Negative corona voltage-current characteristics for simulated flue gas with H_2O volume concentration of (1) 0.6%, (2) 8.4%, and (3) 17.8%. Solid line theory, circles data.[150]

Figure 159 shows a schematic diagram of an "ion mobility probe" which has been utilized to make in situ measurements of effective ion mobilities.[151] The probe, which is made of stainless steel, can be inserted through a standard 10.16 cm (4 in.) test port into a flue gas environment at temperatures up to 400°C. The flue gas is filtered and pulled through a wire-cylinder corona discharge system. A voltage-current curve is measured for corona discharge in the flue gas. The data obtained are analyzed in the same manner as the laboratory technique discussed previously. The in situ technique has the advantage of utilizing the true flue gas composition in a particular application. Table 10 gives some measured values of effective ion mobility for various gas compositions. Several of the values are for gas compositions which are very similar to those obtained from the combustion of coal.

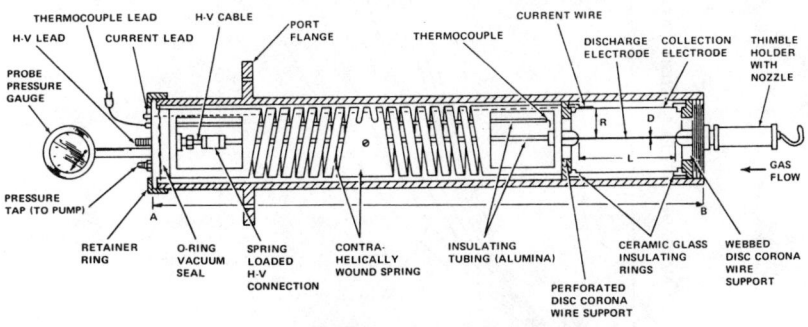

R - COLLECTION ELECTRODE CYLINDER RADIUS = 4.32 cm
L - EFFECTIVE DISCHAGE ELECTRODE LENGTH = 22.86 cm
D - DISCHARGE ELECTRODE DIAMETER = 88.9 mm
A-B - TOTAL PROBE LENGTH = 1.22 m

Figure 159. Schematic diagram of an in situ "ion mobility probe".

TABLE 10. REDUCED EFFECTIVE NEGATIVE ION MOBILITIES FOR VARIOUS GAS COMPOSITIONS

| Gas Composition (vol %) | | | | | Reduced Effective Ion Mobility |
N_2	CO_2	O_2	SO_2	H_2O	(cm^2/V-sec)
—	—	—	—	100.0	0.67±0.17*
—	—	100.0	—	—	2.46±0.06**
—	100.0	—	—	—	1.08±0.03**
—	—	—	100.0	—	0.35***
............(Laboratory Air)............					1.03†
............(Laboratory Air)............					1.26-1.96††
79.4	14.7	4.6	0.2	0.6	5.39†††
73.5	13.6	4.2	0.2	8.4	2.93†††
65.9	12.2	3.8	0.2	17.8	2.23†††
71.0	11.2	3.7	0.0	14.0	2.35†††
75.7	11.6	3.2	0.0	9.4	3.02†††
75.1	11.5	3.2	0.1	9.9	2.74†††
78.5	10.9	3.6	0.0	7.0	3.36†††
78.3	19.8	3.6	0.1	7.0	2.67†††
77.9	10.8	3.6	0.3	7.0	2.70†††
77.6	10.7	3.7	0.7	7.0	2.43†††

*J.J. Lowke and J.A. Rees, *Australian J. Phys. 16*, 447 (1963).
**E.W. McDaniel and H.R. Crane, *Rev. Sci. Instru. 28*, 684 (1959).
***E.W. McDaniel and M.R.C. McDowell, *Phys. Rev. 114*, 1028 (1959).
†B.Y.H. Liu, K.T. Whitby, and H.H.S. Yu, *J. Appl. Phys. 38*, 1592 (1967).
††J. Bricard, M. Cabane, G. Modelaine, and D. Vigla, *Aerosols and Atmospheric Chemistry.* Edited by G.M. Hidy, New York, NY, 27 (1972).
†††H.W. Spencer, III, "Experimental Determination of the Effective Ion Mobility of Simulated Flue Gas." In Proceedings of 1975 IEEE-IAS Conference, September 28, 1975, Atlanta, GA.

Figures 160 and 161 contain theoretical projections showing the effect of effective ion mobility on the voltage-current and electric field current density relationships in a wire-plate geometry.[4]

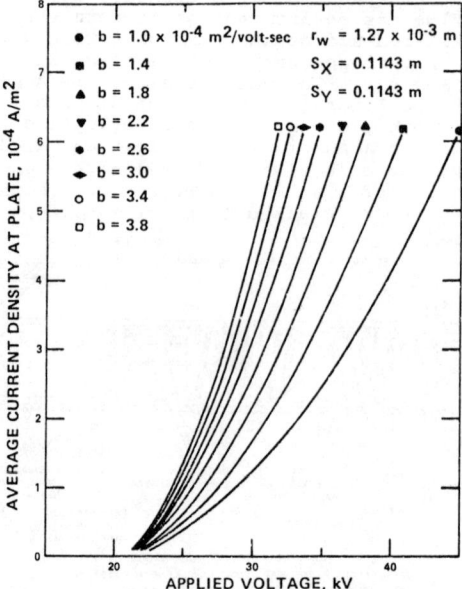

Figure 160. Theoretical curves showing the effect of effective mobility on voltage-current characteristics.[4]

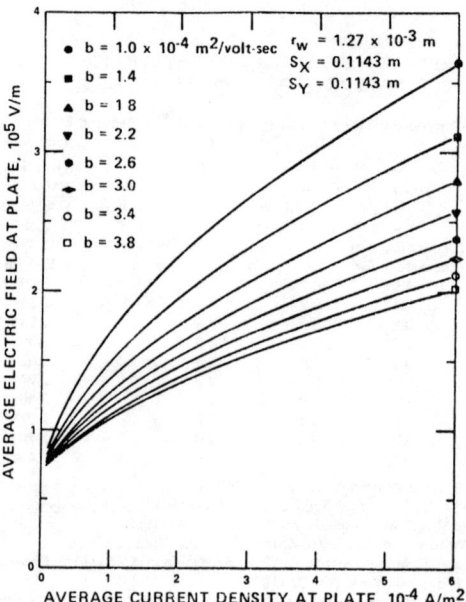

Figure 161. Theoretical curves showing the effect of effective mobility on the electric field and current density.[4]

Figure 162 shows theoretically predicted collection efficiency versus reduced effective ion mobility curves for several particle sizes contained in a typical inlet particle size distribution found in the combustion of coal and for a mass loading of 9.16×10^{-3} kg/m^3 (4.0 grains/acf).

These curves indicate that the effective ion mobility can have a significant effect on particle collection efficiencies. For example, a reduced effective ion mobility of 2.2×10^{-4} m^2/v·s leads to a collection efficiency of 81.8% for a 0.55 µm particle, whereas a value of 3.5×10^{-4} m^2/v·s yields 77.8%.[138]

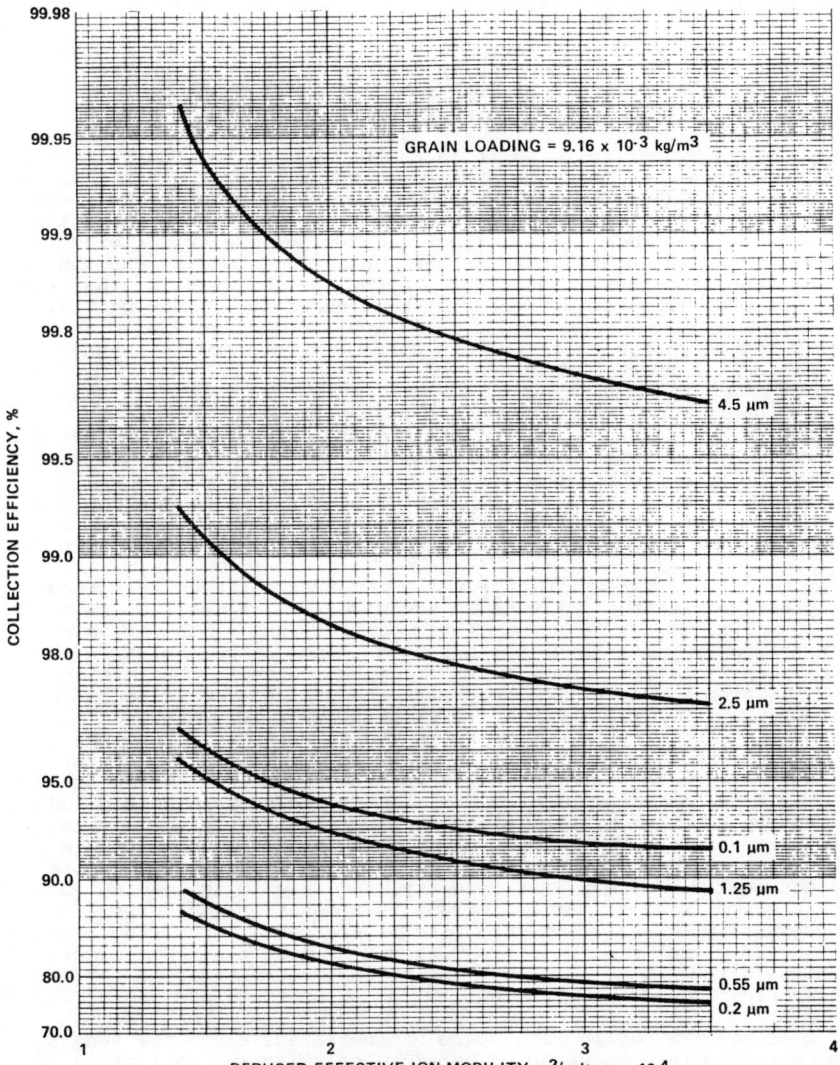

Figure 162. Collection efficiency as a function of reduced effective ion mobility for several particle sizes.

Effects Due To Particles

Particles affect the voltage-current characteristics due to their presence in the gas stream and due to their accumulation on the discharge and collection electrodes. An analysis of measured secondary voltage-current characteristics is essential for determining how the operating electrical conditions are affecting precipitator performance. A correct analysis of measured secondary voltage-current characteristics depends on an understanding of the possible effects due to particles. In many cases, the analysis of voltage-current characteristics is complicated due to competing effects caused by the particles.

The particles in the gas stream become charged due to collisions with the ions created in the corona discharge. The charged particles are much less mobile than the ions and move relatively slowly toward the collection electrode. A particulate charge distribution (particulate space charge) which has an electric field associated with it is established in the interelectrode space. At points near the surface of the discharge electrode, the electric field due to the charged particles is opposite in direction to the electric field produced by the surface charge on the discharge electrode. This effect reduces the electric field strength which is effective in the ionization process near the discharge electrode. Thus, for a given applied voltage, the rate of ion production or current will be reduced due to the particles. The lowering in current for a given applied voltage due to the presence of particles in the gas stream is referred to as the "particulate space charge effect".

In practice, the particulate space charge effect can be detected by examining the voltage-current curves from successive electrical sections in the direction of gas flow. In progressing from the inlet to the outlet, the voltage-current curves will shift to the left. This shift to the left is due to a reduction in the particulate space charge effect as charged particles are removed from the gas stream along the length of the precipitator.

Figure 163 shows voltage-current curves obtained from three successive electrical fields in the direction of gas flow in a full-scale, cold-side precipitator collecting fly ash particles. This figure demonstrates the expected effect of particles on the voltage-current curves. The shift of the curve to the left in moving from the inlet to the outlet and the reduced voltages in the outlet electrical field are of particular importance in analyzing the effects of particles and in determining whether or not the electrical fields are behaving properly. In certain cases, the voltage-current curve for the second electrical field may lie to the right of that for the first electrical field and then the behavior of the following electrical fields is similar to that shown in Figure 163. This is again due to particulate space charge and depends on several factors including allowable electrical conditions, gas velocity, and inlet mass loading and particle size distribution. Thus, this behavior would not necessarily indicate abnormal precipitator behavior. Also, for high efficiency precipitators with six or more electrical fields in the direction of gas flow, the voltage-current curves for the electrical fields near the outlet should approach one another.

Since the particulate space charge effect along the length of a precipitator depends on particle charging and collection, changes in the gas volume flow through a precipitator will result in changes in the voltage-current characteristics. In precipitators installed on coal-fired boilers, the gas volume flow can vary due to changes in the power generation load. Figures 164 through 167 show the effect of gas volume flow on the voltage-current characteristics as estimated by using a mathematical model.[152] The parameters used in the calculations are typical of full-scale, cold-side precipitators. There are four, nine foot long electrical sections in the direction of gas flow. The electrode geometry consists of plate-to-plate and wire-to-wire spacings of 22.86 cm and a wire radius of 0.138 cm. The inlet particle size distribution (MMD = 25 μm, σ_p = 2.8) is representative of fly ash particles. Although changes in load will also result in changes in other parameters such as temperature, resistivity, gas composition, etc. which have their own influences on the voltage-current characteristics, it has been assumed that these parameters remain constant. Figures 164, 165, and 166 show the effect of particles in the different electrical sections for high, medium, and low gas flow rates, respectively. Figure 167 compares the voltage-current characteristics of the first and last electrical sections at each gas flow rate.

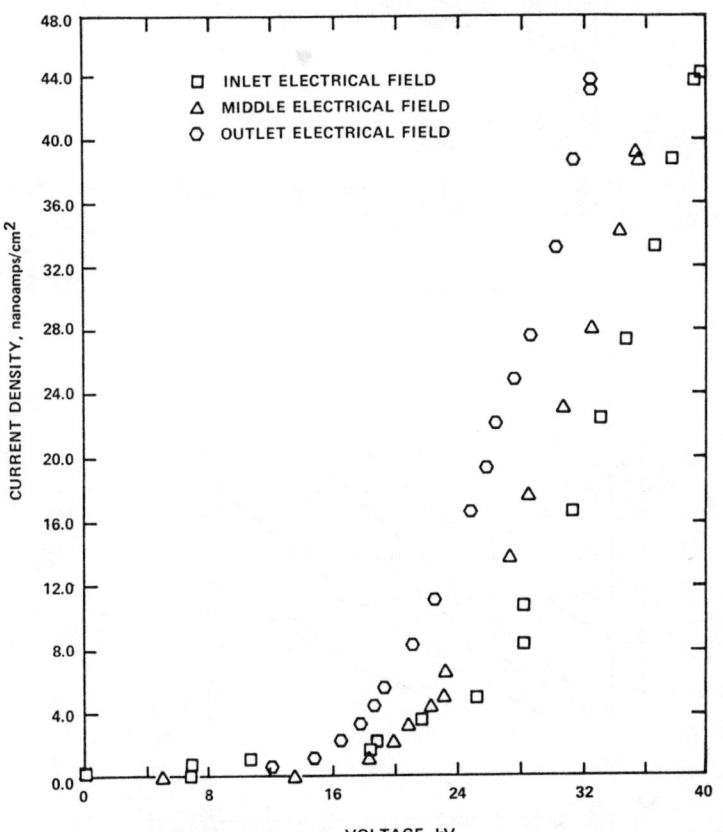

Figure 163. Secondary voltage-current curves demonstrating the particulate space charge effect in a full-scale, cold-side precipitator collecting fly ash.

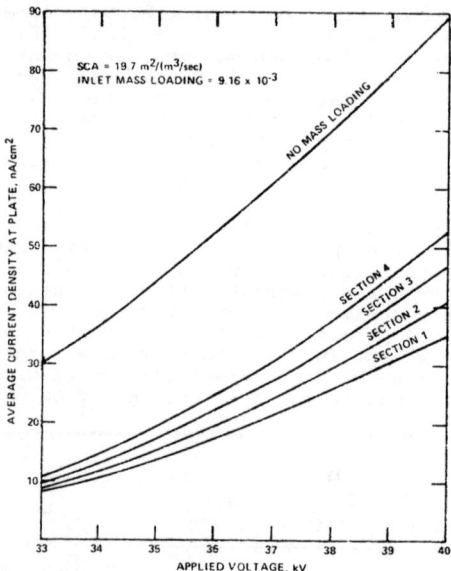

Figure 164. Theoretical voltage-current curves for a specific collection area of 19.7 m²/(m³/sec).[152]

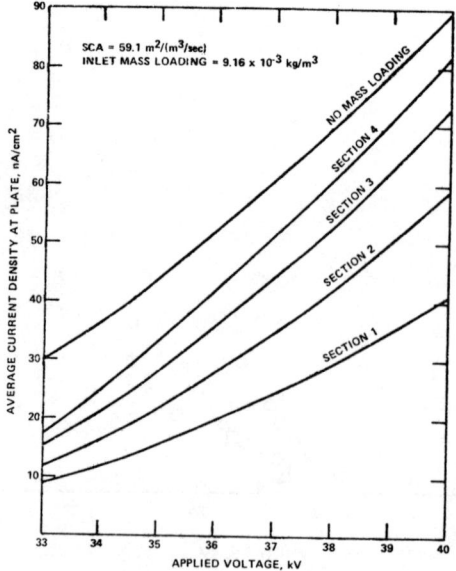

Figure 165. Theoretical voltage-current curves for a specific collection area of 59.1 m²/(m³/sec).[152]

Analysis of Factors Influencing Performance   175

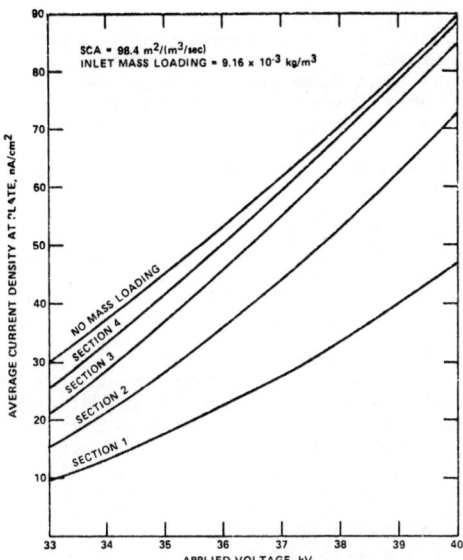

Figure 166. Theoretical voltage-current curves for a specific collection area of $98.4 \, m^2/(m^3/sec)$.[152]

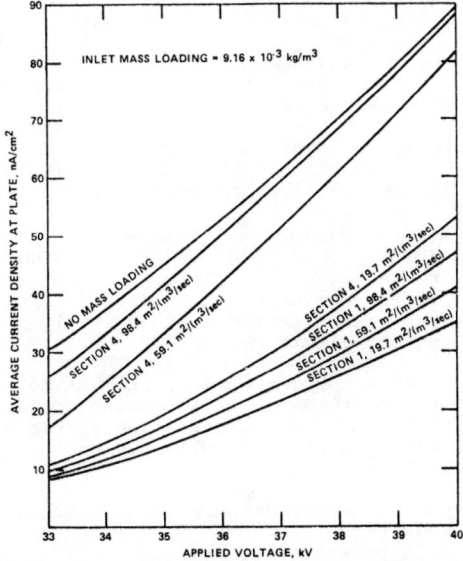

Figure 167. Comparison of theoretical voltage-current curves for different specific collection areas.[152]

The collection of particles on the discharge and collection electrodes may influence the voltage-current characteristics in several ways. The effect of particulate collection on the discharge electrodes has been discussed earlier in terms of geometrical factors. Effects due to particulate collection on the collection electrodes may result because of dielectric breakdown of the collected layer, an effective reduction in the discharge-to-collection electrode spacing, or a nonnegligible voltage drop across the collected layer.

The most significant factor affecting the operation of a precipitator collecting fly ash particles is the dielectric breakdown properties of the collected particulate layer. Precipitators collecting particulate layers with resistivities greater than 10^{11} ohm-cm are limited to low voltages and currents due to dielectric breakdown of the particulate layer. As discussed earlier, back corona can exist well in advance of sparking conditions for collected particulate layers with resistivities greater than 10^{11} ohm-cm. A precipitator should be operated at voltages which do not produce back corona or excessive sparking in order to avoid detrimental effects to precipitator performance. A detailed discussion concerning the resistivity of collected fly ash will be given later.

The thickness of the collected particulate layer is manifested in the voltage-current characteristics. There is a voltage drop across the layer that is given by equation (3). This voltage drop depends on the average current density in the layer and the resistivity and thickness of the layer. This voltage drop is not effective in producing current in the corona discharge. For low resistivity (<10^{10} ohm-cm) fly ash and layers of larger thickness (>1 cm), the effect of the layer might be to shift the voltage-current curve to the left. This can result due to the combination of a negligible voltage drop across the layer for current densities preceding sparkover and an effective reduction in discharge-to-collection electrode spacing. However, normally, the combination of resistivity and thickness of the layer is such that the effect of the voltage drop across the layer dominates the effect on the voltage-current characteristic. In this case, the voltage-current curve is shifted to the right due to the effect of the layer. When a particulate layer is present, the additional voltage which is necessary to produce a given clean collection electrode current density is given approximately by equation (3).

Figure 168 contains experimental data showing the normal effect that a fly ash layer on the collection electrode has on the voltage-current characteristics.[153] The condition illustrated is for dc voltage-current curves with and without a one centimeter thick dust layer with a resistivity of 1 x 10^{11} ohm-centimeter on a typical second electrical field.

In analyzing voltage-current curves, it must be kept in mind that the effect of the particulate layer on the collection electrode is in addition to the particulate space charge effect in the gas and the effect of particles collected on the discharge electrodes.

Thus, the various possible effects discussed in this section may compensate to some extent for one another and obscure the individual effects. All possible effects and their different combinations should be considered in order to obtain the best possible explanation of measured voltage-current characteristics.

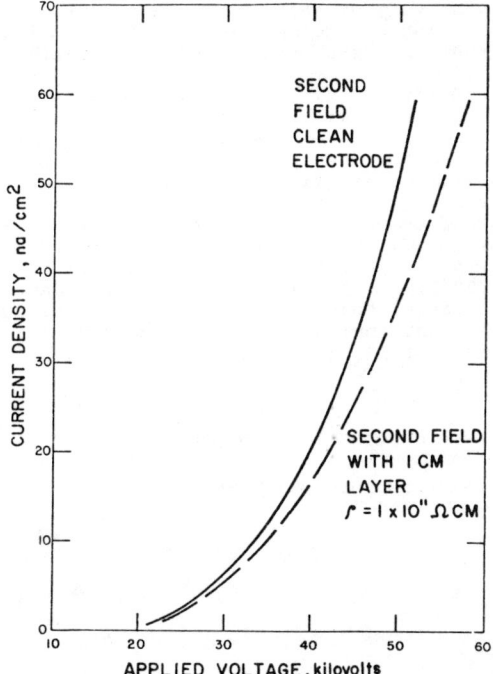

Figure 168. Voltage vs. current characteristic for second field clean electrode and 1 cm layer of 1×10^{11} ohm-cm dust.

Effects Due To Chemical Conditioning Agents

The addition of certain chemical conditioning agents into the gas stream prior to treatment by a precipitator may result in improved electrical operating conditions. It is also possible that the use of certain chemical conditioning agents will result in worse electrical operating conditions. Depending on the type of chemical conditioning agent and the appropriate method of utilization, the agent may be added to the gas stream in either the vapor, liquid, or solid phase.

Chemical conditioning agents may affect the voltage-current characteristics by (1) modifying the resistivity of the collected particulate layer, (2) changing the composition of ionic charge carriers in the gas, (3) introducing a space charge effect, or (4) changing the adhesive and cohesive properties of the collected material. Chemical conditioning agents have been used primarily as a means of lowering the resistivity of the collected fly ash. In principle, the added agents come into intimate contact with the particles in the gas stream and/or in the collected layer and produce a larger number of charge carriers or more mobile charge carriers in the collected layer than in the unconditioned environment. The availability of a larger number of charge carriers or more mobile charge carriers to transfer the current through the layer results in a reduction in the electrical resistivity of the layer. The possibilities also exist that the addition of certain chemical conditioning agents will have no effect on the value of the electrical resistivity or that the value will be increased as a consequence of binding charge carriers which would have been free to carry current in the unconditioned environment.

Figures 169 and 170 contain data from two different cold-side industrial precipitators showing the effect on the voltage-current characteristics of adding SO_3 in the vapor state to the gas stream at a location prior to the precipitator.[154,155] In both cases, the electrical conditions are significantly improved due to a reduction in the resistivity of the collected ash layer resulting from the injection of the SO_3. For the data in Figure 169, the addition of 25 ppm of SO_3 lowered the value of ash resistivity from approximately 6×10^{12} ohm-cm at 118°C (245°F) to 4×10^{10} ohm-cm at 143°C (290°F). Although the conditioned data were obtained two months later than the unconditioned data due to difficulties with the conditioning system, the data definitely show the pronounced effect of SO_3 injection.

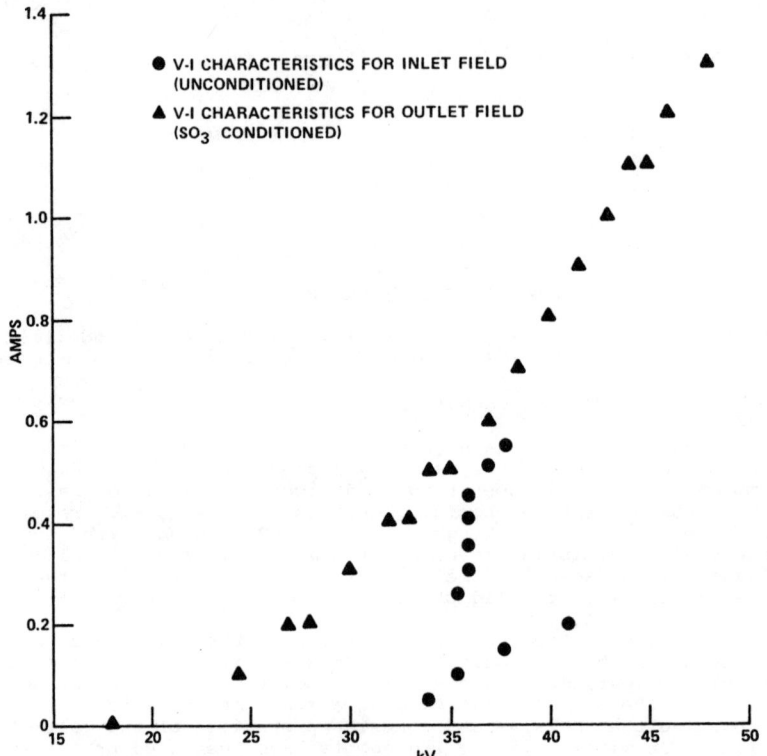

Figure 169. Effect on the voltage-current characteristics of adding SO_3 in the vapor state to the gas stream at a location prior to the precipitator.

For the data in Figure 170, the addition of 8 ppm of SO_3 lowered the value of ash resistivity from approximately 6×10^{11} ohm-cm to 6×10^{10} ohm-cm at 165°C (330°F). At this installation, the unconditioned and conditioned data were obtained within the same week. The reduction in resistivity at this installation is less dramatic than that at the other installation just discussed due to the higher temperature of the gas. At the higher temperature, the SO_3 was not near the dew point. This results in less surface adsorption on the particles than would occur at lower gas temperatures. However, the pronounced effect of SO_3 on ash resistivity is still evidenced.

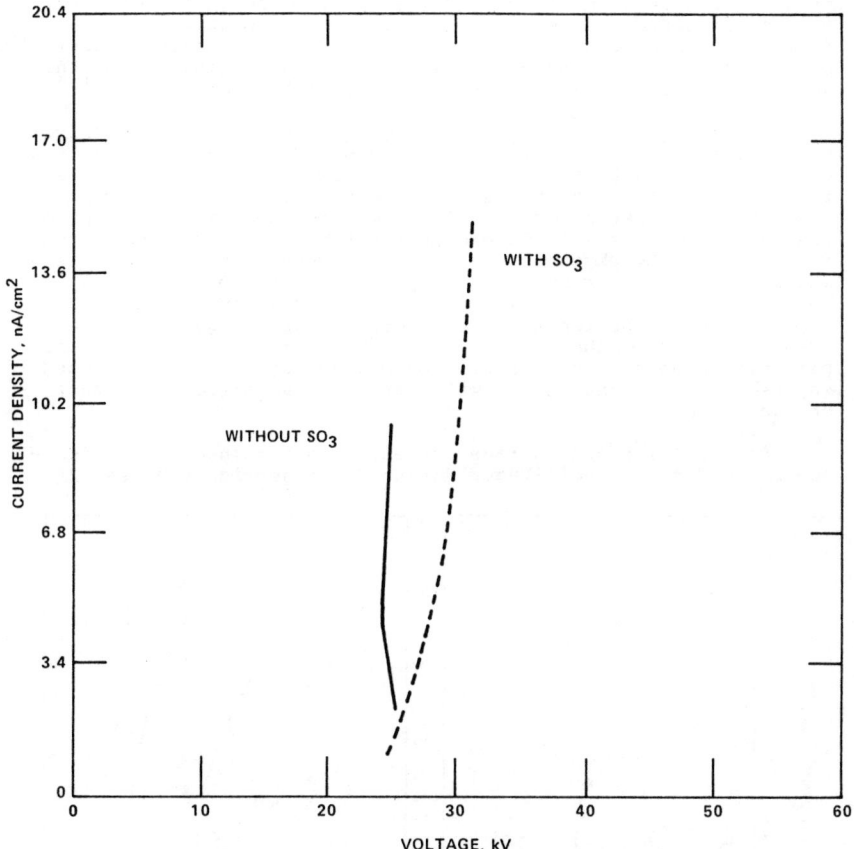

Figure 170.   Effect on the voltage-current characteristics of adding SO_3 in the vapor state to the gas stream at a location prior to the precipitator.

Normally, the SO_3 is injected in concentrations of 25 ppm by volume or less. When SO_3 is injected into the gas stream, the precipitator will respond with rapid improvement in the voltage-current characteristics. The voltage-current characteristics will then continue to improve further until an equilibrium condition is reached where the resistivity no longer changes.

Although not as general in application to fly ash as SO_3, other chemical conditioning agents have been found to improve the electrical conditions by reducing the resistivity of certain types of fly ash.[156,157,158] Studies are presently in progress that may determine which chemical conditioning agents are effective in reducing the resistivity of fly ashes of various compositions.[159,160] At the present, however, the use of conditioning agents other than SO_3 is based on trial and error and past experience where different agents are tried until one is found that produces the desired effect. There is evidence[161,162,163] that sodium compounds can be introduced into the gas stream in the form of solid particles as a means of reducing fly ash resistivity. Studies are now in progress to determine the feasibility of this approach in full-scale, cold-side industrial precipitators.[155]

The addition of conditioning agents may also have an effect on the electrical properties of the gas. This could result in an effect on the voltage-current characteristics that is separate from the effect of the resistivity of the collected material. The nature and extent of the effect of the conditioning agent on the gas properties will depend on the type of agent, the concentration, and the physical state of the agent. If the agent is added to the gas stream in the vapor state, the molecules may become ionized in the corona discharge. This would change the composition of the ionic current carriers in the gas. If the agent is added to the gas stream in the form of liquid or solid particles, these particles will be charged by the ions produced in the corona discharge and will introduce a particulate space charge effect.

Figure 171 shows voltage-current data obtained from a full-scale, cold-side precipitator during SO_3 injection studies.[164]

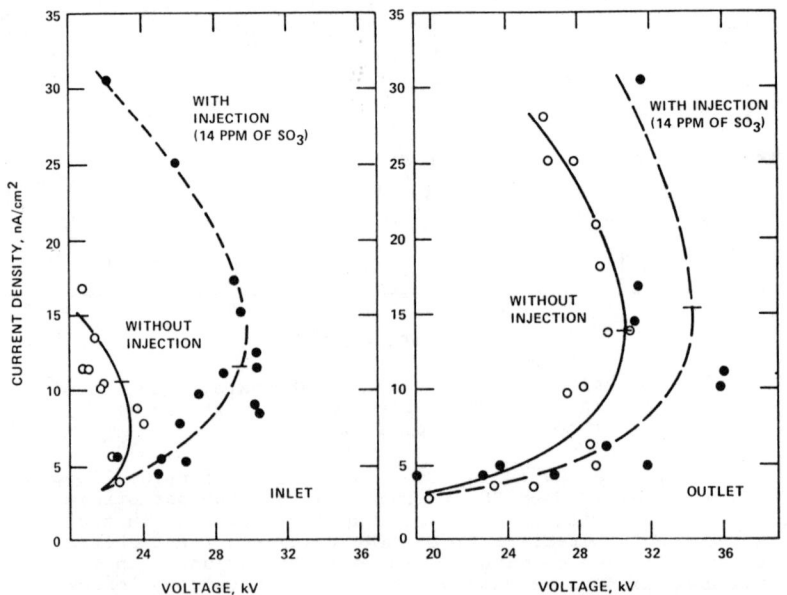

Figure 171. Current density vs. voltage for a full-scale, cold-side precipitator without and with SO_3 conditioning.

These data show that the injection of SO_3 increased voltages and currents and decreased the spark rate. The segments of the curves with positive slopes portray the data obtained with no sparking or very light sparking. The segments of the curves with negative slopes in regions of high current density represent the experimental results with moderate to heavy sparking. The short horizontal lines intersecting each curve indicate the average values of current density observed with the power supplies under automatic control. An interesting feature of the data in Figure 171 is the indication that the injection of sulfur trioxide permitted both higher current densities and higher voltages to be reached without the occurrence of excessive sparking. Shifts in the voltage curves to the right along the voltage axis at least suggest the possibility of a space-charge effect resulting from the introduction of less mobile charge carriers in the gas stream.

One possibility is that the added concentration of sulfur trioxide assumed most of the ionic space charge and the new ions thus introduced carried current with a lower mobility than the normally occurring ions produced from oxygen, water vapor, and sulfur dioxide. An alternative possibility is that part of the added sulfur trioxide was condensed as a fine mist of sulfuric acid and then, electrically charged, caused a very pronounced shift in charge carriers from gaseous ions to relatively immobile acid particles.

Figures 172 and 173 show voltage-current data obtained from a full-scale, cold-side precipitator during NH_3 injection studies with low and high sulfur coals.[164] Figure 174 shows data from a different precipitator indicating the almost instantaneous response of the electrical conditions to the injection of NH_3.[164] Here, much higher voltages could be achieved for the same current density. This type of response normally does not occur in those cases when the conditioning agent primarily affects the ash layer. Figure 175 shows voltage-current data obtained from the precipitator.[164] In these studies, the significant shift to the right of the voltage-current characteristics was attributed to a particulate space charge effect caused by ammonium sulfate particles formed due to the chemical reaction of SO_3 and NH_3.

The addition of conditioning agents can also affect the adhesive and cohesive properties of the collected fly ash. Adhesive forces refer to those between the ash layer and the collection electrode while cohesive forces refer to those between the particles in the ash layer. Certain types of conditioning agents may interact with certain types of fly ash to form a collected layer which is more favorable for precipitator operation due to modification of the adhesive or cohesive properties of the ash. In order to avoid excessive particle reentrainment, it is desirable that the collected layer adhere favorably to the collection electrode and that the particles in the layer bind together so that large agglomerates are reentrained during a rap.

The degradation in performance of a precipitator that results from rapping reentrainment can be reduced if the deposit of fly ash on the electrodes can be made more cohesive. Dalmon and Tidy[165] recognized that sulfuric acid vapor may have this effect as the principal mode of conditioning for an ash that is not high in resistivity. Other investigators have recognized that added ammonia may also have this effect in addition to space-charge enhancement.[166]

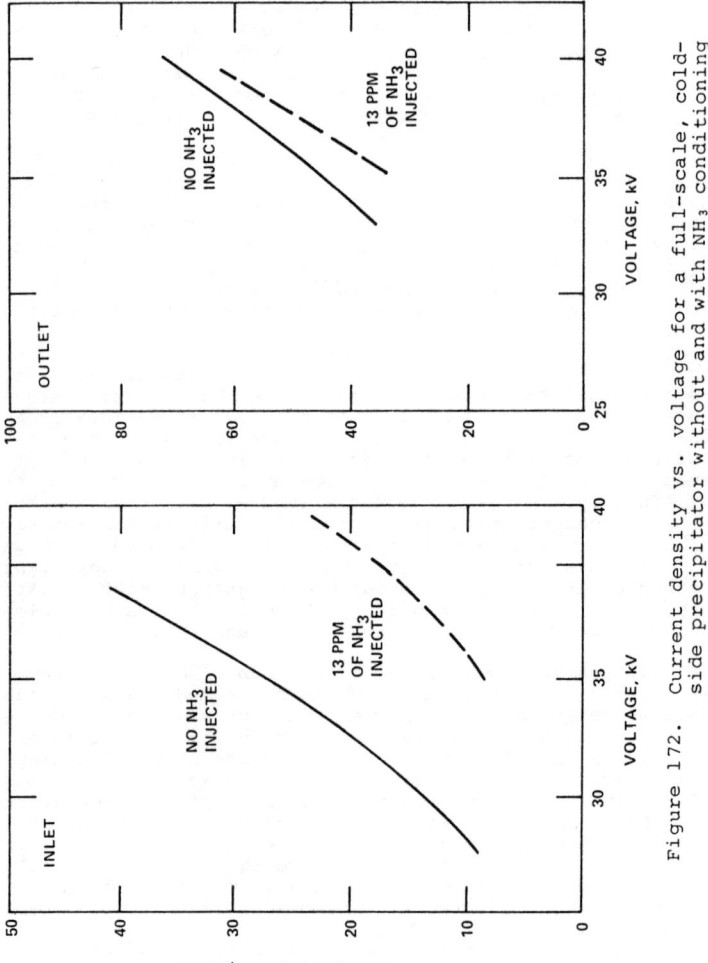

Figure 172. Current density vs. voltage for a full-scale, cold-side precipitator without and with NH$_3$ conditioning low sulfur coal.

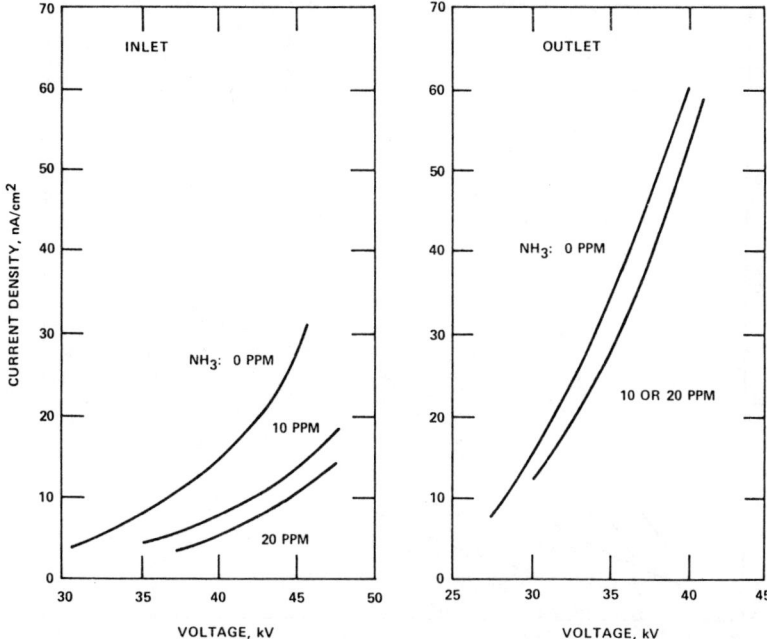

Figure 173. Current density vs. voltage for a full-scale, cold-side precipitator without and with NH_3 conditioning high sulfur coal.

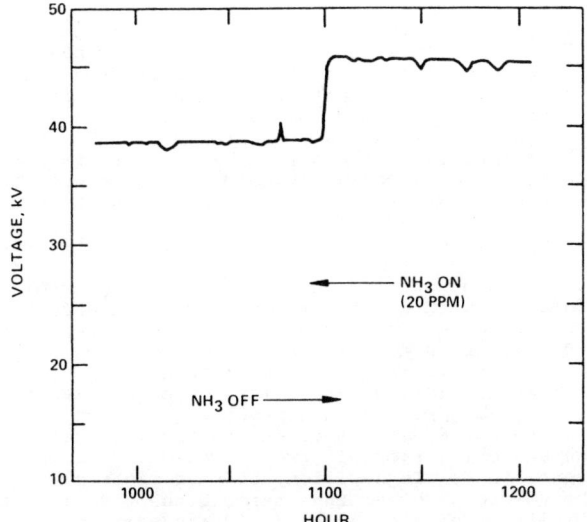

Figure 174. Rapidity of the effect of ammonia injection on the voltage supplied to the inlet electrical field of a full-scale, cold-side precipitator (high-sulfur coal).

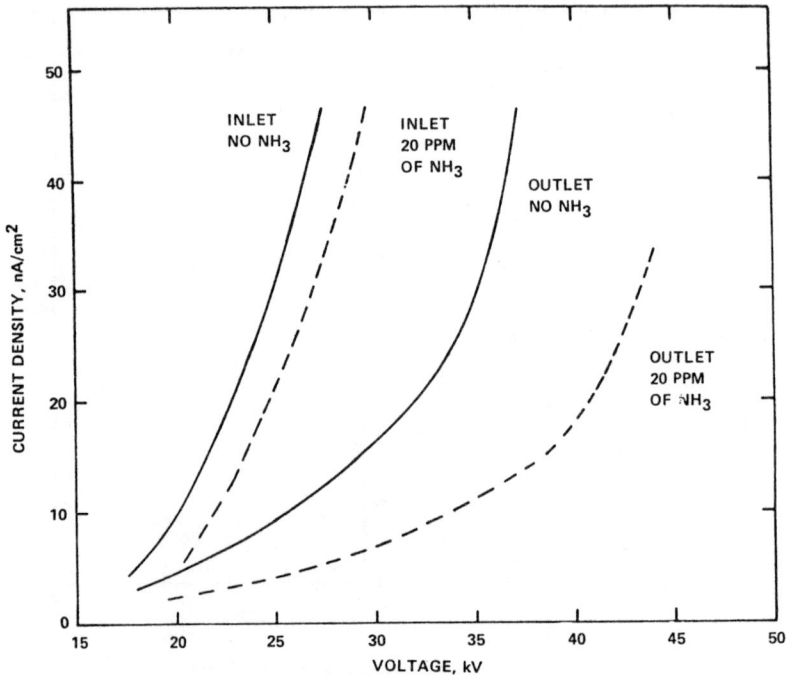

Figure 175. Current density vs. voltage for a full-scale, cold-side precipitator without and with NH_3 conditioning (high sulfur coal).

Figure 176 is the reproduction of an obscurometer chart that shows the effect of ammonia in reducing reentrainment.[166] Before the gas was added, the optical instrument registered a series of spikes that were coincidental with rapping puffs. After the gas was added, a gradual suppression of these spikes occurred. Once again, after injection of the gas was stopped, the spikes gradually returned. It is important to observe that the effect of ammonia on rapping puffs was only gradually observed, as expected from the requirement of a change in cohesive forces on the ash residing on the electrodes, whereas the effect of ammonia on space charge--a gas-phase property--was earlier shown to be rapidly detected.

It should also be pointed out that certain types of conditioning agents may interact with certain types of fly ash under certain conditions to form a very sticky or cement-like material on the discharge and collection electrodes. This situation has occurred at some installations where conditioning agents were used. If this occurs, the existing rapping forces may not be sufficient to remove the collected fly ash from the discharge and collection electrodes. Eventually, in this type of situation, the buildup of material on the discharge and collection electrodes would result in very poor electrical conditions. In addition, hoppers might plug up and current paths to ground other than through the gas might be created. The effect of a given conditioning agent on the stickiness of the fly ash layer should be examined in the laboratory or with a pilot unit before injecting the agent into an industrial precipitator.

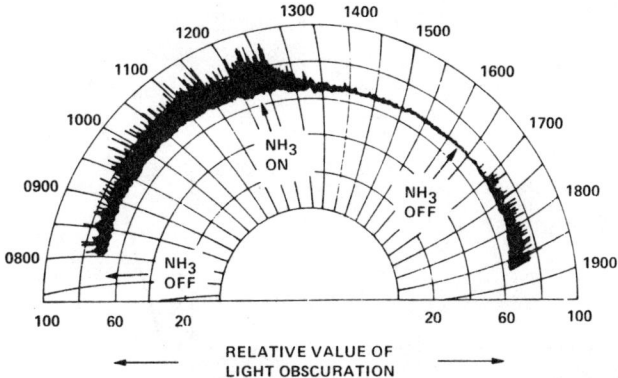

Figure 176. Reduction of rapping reentrainment by ammonia.

Effect Of Voltage-Current Characteristics On Precipitator Performance

The voltage-current characteristics that can be obtained prior to sparkover or back corona are indicative of how effective a precipitator will be in removing particles from the gas stream. Ideally, the voltage-current curves should extend over a wide range of applied voltage between corona start and the maximum useful current so that a stable operating point can be chosen and should extend to high useful values of applied voltage and current. Low values of operating applied voltage and/or current will result in reduced performance and will require the use of a larger precipitator in order to recover the performance loss.

Figure 177 shows voltage-current curves obtained from the three electrical sections of a laboratory precipitator when particles are present in the gas stream.[167] The carrier gas was ambient air. The particulate source was an atomizer which produces an aerosol of dioctyl phthalate (DOP) containing many different particle diameters. Although the precipitator is divided into four baffled and independent electrical sections, the last two sections were connected together. The experimental data were obtained with a plate-to-plate spacing of 25.4 cm, wire-to-wire spacing of 12.7 cm, wire radius of 0.1191 cm, and gas velocity of 0.976 m/sec. These parameters are also characteristic of full-scale precipitators.

Figure 178 shows the theoretically calculated effect of applied voltage and current density for the experimental conditions described above.[167] The curves in Figure 177 were used to determine various applied voltages and current densities. The overall mass collection efficiency was calculated for 2, 5, 10, 20, 25.8, 35, and 45 nA/cm^2. The inlet mass loading and particle size distribution and the gas temperature and pressure were also measured and used in the calculations. In making the calculations, it was assumed that the normalized standard deviation of the gas velocity distribution (σg) and the gas bypassage of electrified regions (S) were negligible ($\sigma g = 0$, $S = 0$). Calculations were made with no rapping reentrainment (the actual case) and with rapping reentrainment by simulating what would occur if the collected material was fly ash.

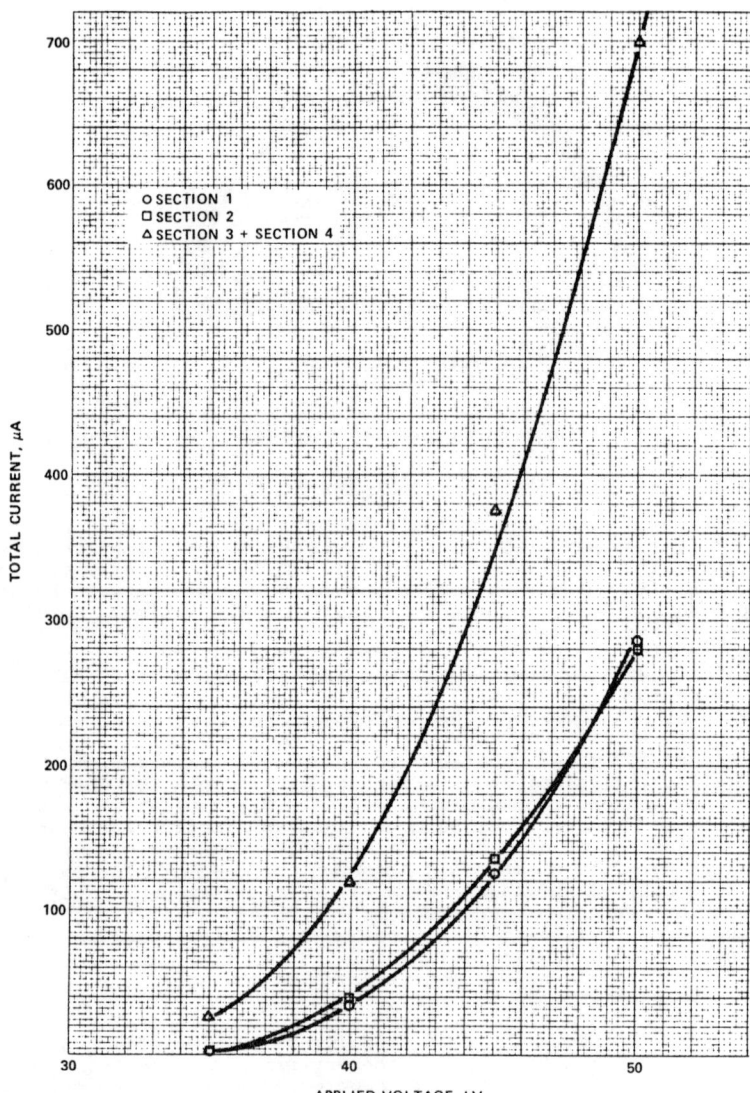

Figure 177. Experimental voltage-current curves from a wire-plate laboratory precipitator.[167]

The curves in Figure 178 show that if the precipitator is restricted to operate at low values of current density due to sparking or back corona, then significantly reduced overall mass collection efficiencies will result. Thus, in designing a precipitator, the effect of possible changes in the allowable current density due to changes in the gas or particulate properties must be taken into account. If reduced current densities are a possibility, then the possible reduction in collection efficiency must be compensated for by an excess in specific collection area (or, more appropriately, collection plate area). The curves in Figure 178 also show that more mass will exit the precipitator due to rapping reentrainment for the lower values of current density than the higher values. At 2 nA/cm², the model predicts that approximately 3.1% of the mass entering the precipitator will exit due to rapping reentrainment. This is a consequence of more mass reaching the outlet sections for the lower current density. Thus, when considering the effect of reductions in current density on precipitator performance, one must take into account not only fundamental reductions in collection efficiency due to lowered migration velocities but also possible increased reductions due to increased rapping reentrainment.

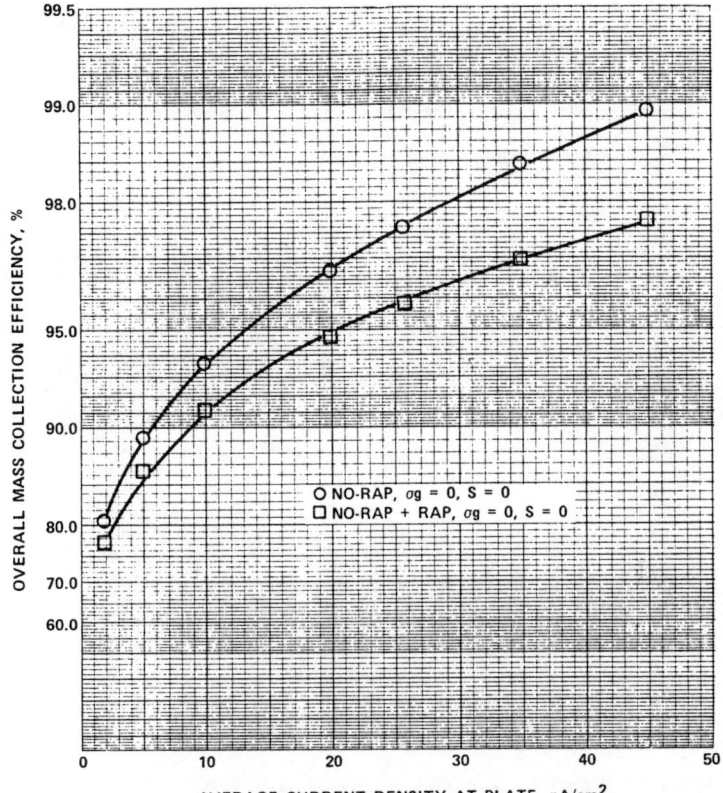

Figure 178. Theoretically calculated effect of current density on overall mass collection efficiency.[167]

As a further consideration, the increase in precipitator performance that can be achieved by increasing the applied voltage and current depends on where the change takes place on the voltage-current curve. In fly ash applications where high current densities can be achieved, the voltage-current curves become very steep for higher applied voltages with large increases in current resulting from small increases in applied voltage. For precipitators operating with voltages and currents on the steep portion of the voltage-current curve, increasing the current will not have a very pronounced effect on improving precipitator performance because only small changes in applied voltage will be realized and the applied voltage plays the dominant role in limiting particle charge and controlling the electric field.

The experimental data in Figure 179 show the effect of applied voltage and current density on the collection efficiency and migration velocity of different particle diameters.[138] The data were obtained from the laboratory precipitator just discussed with all parameters being the same except the gas velocity which is increased to 1.49 m/sec. Since these data are essentially free of nonideal effects, they clearly demonstrate the significance of the maximum allowable applied voltage and current density on precipitator performance.

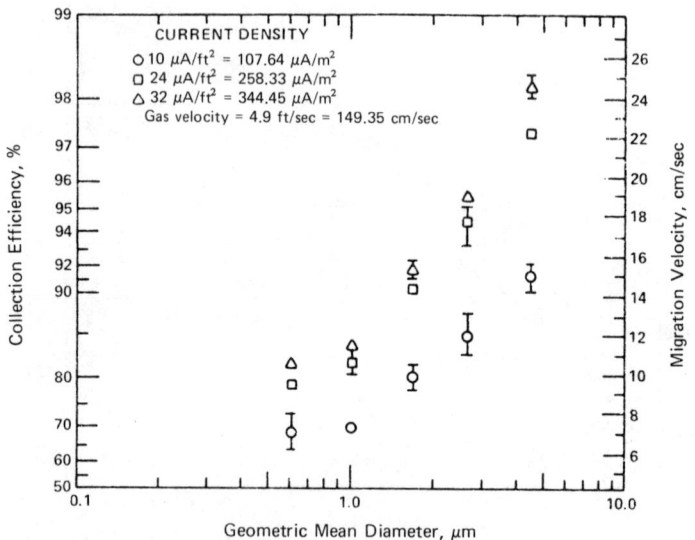

Figure 179. Experimental fractional efficiencies and migration velocities for negative corona with a wire of radius 0.119 cm and gas velocity of 1.49 m/sec.[138]

Figure 180 shows measured overall mass collection efficiencies obtained from full-scale, cold-side precipitators collecting fly ash plotted as a function of specific collection area for various average current densities. Although certain factors which will affect overall mass collection efficiency such as inlet mass loading and particle size distribution, geometry, gas composition, applied voltage, ash resistivity, gas velocity distribution, extent of gas sneakage, and particle reentrainment characteristics are most likely different for the various precipitators, the data definitely show the trend of increased performance with increased current density.

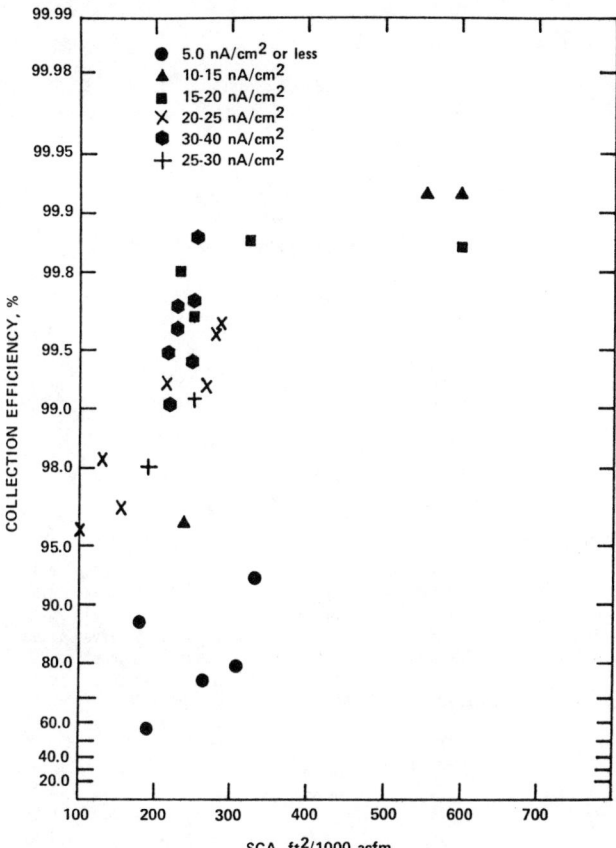

Figure 180. Measured overall mass collection efficiencies obtained from full-scale, cold-side precipitators collecting flyash plotted as a function of specific collection area for various average current densities.

Measured Secondary Voltage-Current Data From Full-Scale Precipitators Collecting Fly Ash

Since the secondary voltage-current data from the individual power supplies in an electrostatic precipitator provide valuable information for use in (1) troubleshooting and diagnosing precipitator problems, (2) theoretically predicting the performance of a precipitator, and (3) interpreting the influence of the ash resistivity and particulate space charge suppression of the corona current on performance, data from several representative precipitators will be presented and discussed. These data should acquaint one with the various practical situations that can be encountered during field measurements on a full-scale precipitator. In examining the voltage-current curves, one should be systematically looking for the effects described previously. Chemical analyses of coal, fly ash, and gas samples are given in Tables 11 - 13 for those plants where the data were available. These data are needed for proper interpretation of the voltage-current curves.

TABLE 11

AS RECEIVED, PROXIMATE CHEMICAL ANALYSES OF COAL SAMPLES FROM COLD-SIDE UNITS

Plant #	Moisture	Volatile Matter	Fixed Carbon	Ash	Sulfur	Btu/lb
1	13.94	37.78	43.07	5.21	0.41	10,557
2	2.04	39.05	47.91	11.00	3.28	12,421
3	3.35	34.84	45.27	16.54	3.09	11,399
4	17.22	28.67	40.96	13.15	0.51	9,316
5	-	-	-	-	1.40	-
6	Midwestern coal - chemical analyses not available					
7	8.80	-	-	12.30	0.77	10,211

AS RECEIVED, PROXIMATE CHEMICAL ANALYSES OF COAL SAMPLES FROM HOT-SIDE UNITS

Plant #	Moisture	Volatile Matter	Fixed Carbon	Ash	Sulfur	Btu/lb
8	2.97	29.42	48.96	18.67	0.93	11,613
9	8.26	38.90	43.56	9.28	0.45	11,006
10	4.3	-	-	24.5	1.02	9,800

TABLE 12

CHEMICAL ANALYSES OF ASH SAMPLES FROM COLD-SIDE AND HOT-SIDE UNITS

COLD-SIDE UNITS

Plant #	Li_2O	Na_2O	K_2O	MgO	CaO	Fe_2O_3	Al_2O_3	SiO_2	TiO_2	P_2O_5	SO_3	LOI	Soluble SO_4
1*	0.02	0.26	1.72	3.61	8.71	5.49	24.64	50.55	1.22	0.50	0.75	0.61	0.73
2*	0.02	0.55	2.49	0.95	5.64	24.38	18.30	45.08	1.31	0.30	1.86	3.97	
**	0.02	0.54	2.49	0.95	4.73	22.72	18.52	45.69	1.45	0.30	2.77	5.72	
3***	0.03	0.67	2.12	1.00	4.95	13.13	21.76	50.23	1.96	0.78	2.29	10.92	1.56
4**	0.02	1.38	0.54	1.1	5.8	6.1	13.2	70.8	0.87	0.05	0.5	1.0	
5	0.08	0.42	2.4	1.1	1.8	9.0	28.2	49.4	2.2	0.59	0.35	5.2	1.10
6	Midwestern coal - chemical analyses not available												
7*	0.02	0.46	2.4	2.6	11.8	6.0	19.3	55.7	0.97	0.57	0.62		

HOT-SIDE UNITS

Plant #	Li_2O	Na_2O	K_2O	MgO	CaO	Fe_2O_3	Al_2O_3	SiO_2	TiO_2	P_2O_5	SO_3	LOI	Soluble SO_4
8	0.04	0.40	3.1	1.3	1.2	6.7	28.8	55.2	2.4	0.24	0.59	4.0	0.29
9*	0.014	1.78	1.2	1.7	7.4	5.0	23.9	56.4	2.1	0.49	0.40	0.10	
10	0.02	1.40	1.19	0.97	4.91	3.71	26.96	57.25	1.05	0.16	0.41	0.61	

*Hopper ash sample
**Ash obtained from high volume sampler
***Isokinetically collected ash sample

TABLE 13

GAS ANALYSES FROM COLD-SIDE AND HOT-SIDE UNITS

COLD-SIDE UNITS

Plant No.	Temp, °C	Volume, % CO_2	O_2	H_2O	SO_2, ppm	SO_3, ppm
1	138	13.5	4.5	7.1	276	<0.5
	149	13.0	7.0	8.4	223	<0.5
2	152	15.0	4.0	~8.0	3081	11.9
3	154	12.5	5.5	~8.0	2521	6.4
4	~105	13.6	6.2	7.9	490	<0.5
5	~155	-	3.1	10.5	1433	3.8
6	Not available					
7	115	13.8	5.2	8.2	520	9.9
	99	11.9	7.4	7.8	-	-

HOT-SIDE UNITS

Plant No.	Temp, °C	CO_2	O_2	H_2O	SO_2, ppm	SO_3, ppm
8	332	15.3	5.0	8.5	788	2.3
9	350	15.2	3.3	9.1	370	<0.5
10	Not available					

Measured Cold-Side Curves--

Plant 1 - Cold-side ESPs collecting ash from low sulfur Western coal--The electrostatic precipitator installed on Unit 1 of Plant 1 consists of six fields (Figure 181). The first and second fields each have 5,351 m² (57,600 ft²) of collecting area while the third through the sixth fields have 6,688.8 m² (72,000 ft²) of collecting area, for a total of 37,457.3 m² (403,200 ft²). This gives a specific collection area of 99.2 m²/(m³/sec)(504 ft²/1000 cfm) for the design volume of 377.6 m³/sec (800,000 acfm). Each field has two double half wave transformer rectifiers. The arrangement of the TR sets is shown in Figure 181. The precipitator has 12" plate spacing and operates at approximately 149°C (300°F). Flue gas is supplied and withdrawn through two inlet and two outlet ducts, and a mechanical collector precedes the precipitator. The precipitator employs a drop hammer type of rapping system in which two plates are rapped simultaneously. The first two fields are rapped six times per hour, the third and fourth fields are rapped three times per hour and the fifth and sixth once per hour.

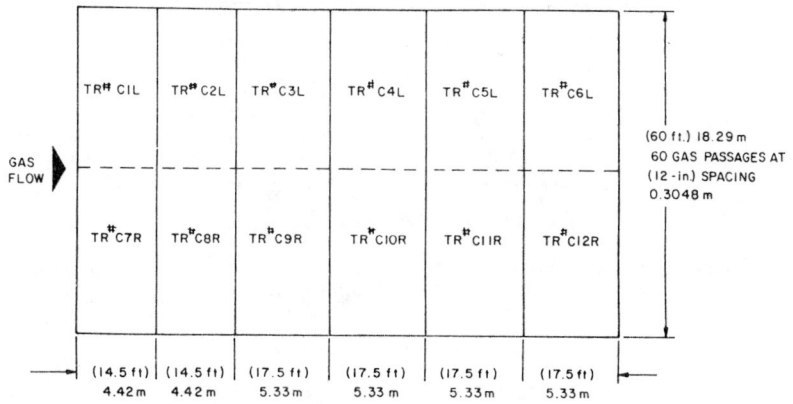

Figure 181. Precipitator layout for Plant 1, Unit 1.

Voltage-current density curves from transformer-rectifier sets for a cold-side precipitator at Plant 1 are shown in Figures 182 and 183, and average operating conditions are given in Table 14. A comparison of the breakdown point in the V-j characteristics and the operating points under automatic control indicates that the power supplies were operating at the point at which maximum voltage was obtained. The V-j characteristics also indicate that the automatic control of the C1L power set was not operating properly. The breakdown point in the V-j characteristics of the C12R power set was abnormally low compared to the other outlet power sets. This indicates a problem with either the power set or the precipitator internals.

The low current densities and shapes of the voltage-current curves for all the electrical fields indicate that the resistivity of the ash layer was a limiting factor. Since the inlet mass loading was fairly large and contained a relatively fine particle size distribution, the particulate space charge effect was greatest in the third electrical field instead of the first or second. The low current densities resulted in relatively long residence times in order to fully charge the fine particles.

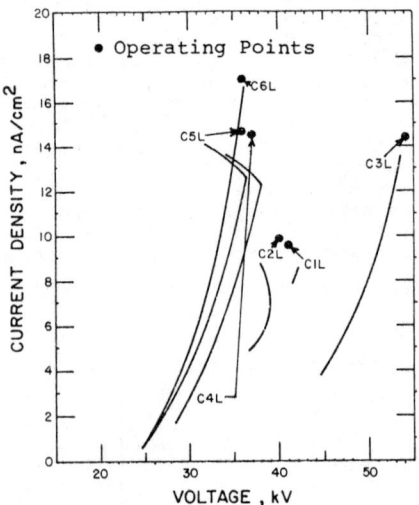

Figure 182. Voltage vs. current density for left or north side of Unit 1 precipitator of Plant 1.

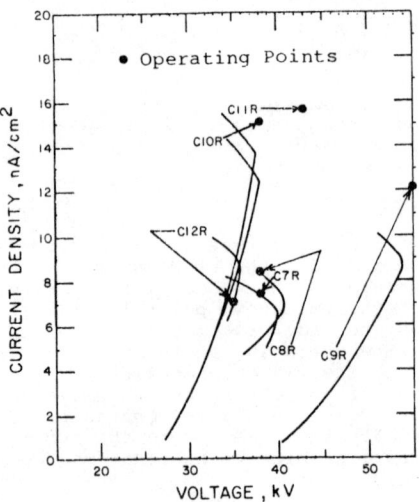

Figure 183. Voltage vs. current density for right or south side of Unit 1 precipitator of Plant 1.

TABLE 14. AVERAGE ELECTRICAL READINGS, PLANT 1

Transformer Rectifier Set	Current mA	Current Density nA/cm²	Voltage kV	Plate Area m²
C7R	196	7.33	38	2,675
C1L	256	9.57	41	2,675
C7R and C1L	452	8.45	39.5	5,350
C8R	224	8.37	38	2,675
C2L	262	9.79	40	2,675
C8R and C2L	486	9.08	39	5,350
C9R	405	12.10	55	3,344
C3L	476	14.20	54	3,344
C9R and C3L	881	13.15	54.5	6,688
C10R	500	15.00	38	3,344
C4L	482	14.40	37	3,344
C10R and C4L	982	14.70	37.5	6,688
C11R	524	15.70	43	3,344
C5L	486	14.50	36	3,344
C11R and C5L	1,010	15.10	39.5	6,688
C12R	234	7.00	35	3,344
C6L	570	17.00	36	3,344
C12R and C6L	824	12.00	35.5	6,688

Plant 2 - Cold-side ESPs collecting ash from high sulfur Eastern coal--The electrostatic precipitator installed on Unit 4 of Plant 2 consists of three fields in the direction of gas flow as Figure 184 illustrates. The precipitator is physically divided into two collectors (A & B). The test program conducted at Plant 2 was performed on the "A" side of the #4 precipitator. The total collecting area for the "A" side is 7,374.4 m² (79,380 ft²), 2458.13 m² (26,460 ft²) per field. This gives a specific collection area of 34.475 m²/(m³/sec) (175 ft²/1000 cfm) for the design volume flow of 213.82 m³/sec (453,000 acfm) per collector. Each collector has three double half-wave transformer rectifiers, one per field. The precipitator has 27.94 cm (11 in.) plate spacing and operates at approximately 149°C (300°F). The precipitator employs a drop hammer type of rapping system in which two plates are rapped simultaneously with each hammer. The first field is rapped ten times per hour, the second field is rapped six times per hour, and the third field is rapped one time per hour. The emitting electrodes are square twisted wires with an approximate diameter of .419 cm (.165 in.) and are 10.0 m (32' 9 3/4") long. There are 12 wires per lane per field for a total of 1512 wires. The discharge electrodes are held in a rigid frame, and each frame holds 4 wires.

The average daily operating voltages and currents during the testing period are given in Table 15. Figure 185 contains the secondary voltage and current relationships obtained on the 3A TR set with a voltage divided resistor assembly attached. Figures 186 and 187 contain the V-I relationships for TRs 1A and 2A. These voltage-current relationships along with resistivity data of around 1.0×10^{10} Ω-cm indicate that the electrical operating conditions at this installation are not limited by dust resistivity. The nature of the voltage-current curve for the inlet set (1A) suggests that a combination of space charge effects and dust accumulation

on the electrodes cause the sparking at the relatively low values
of current density at which the inlet set operates.

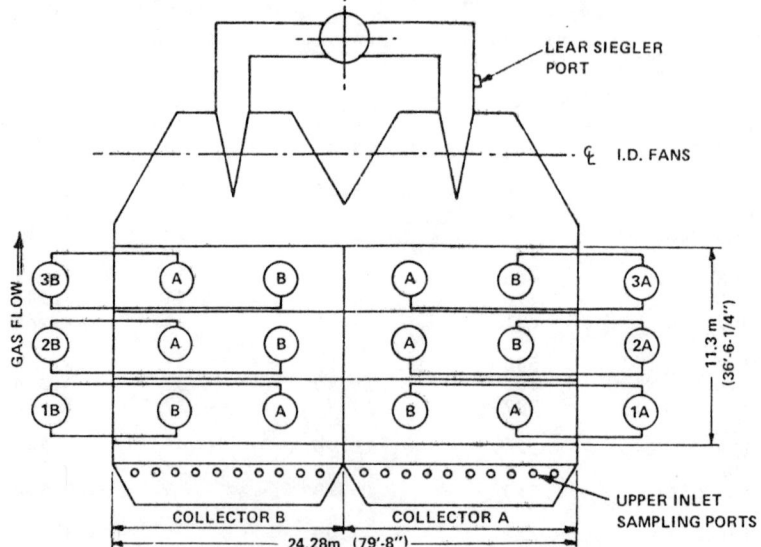

Figure 184. Precipitator layout for Plant 2, Unit 4.

TABLE 15. AVERAGE ELECTRICAL OPERATING CONDITIONS
DURING SAMPLING PERIODS

| Day | TR | Primary | | Secondary | | | | Current Density |
| | | | | Bushing #A | | Bushing #B | | |
		Amps	Volts	Amps	KV	Amps	KV	nA/cm^2
1	1A	37.5	360.	.155	44.8	.145	41.5	12.2
	2A	82.5	377.5	.23	42.5	.21	42.6	18.
	3A	97.	390.	.3	42.2	.295	42.4	24.3
2	1A	29.7	312.5	.133	41.1	.14	40.1	11.1
	2A	78.2	370.8	.223	42.4	.2	41.8	17.3
	3A	92.5	385.8	.283	42.3	.275	42.6	22.8
3	1A	32.8	326.	.116	41.9	.138	43.	10.4
	2A	82.	401.	.23	47.0	.218	45.8	18.3
	3A	94.	401.	.3	44.3	.3	44.7	24.5
4	1A	32.8	331.3	.14	41.8	.148	42.4	11.8
	2A	80.	389.2	.222	45.2	.204	44.2	17.4
	3A	93.8	400.1	.297	44.2	.294	45.8	24.1
5	1A	--	238.	.05	31.7	.08	31.1	5.3
	2A	37.4	338.	.1	42.9	.094	41.2	7.9
	3A	42.6	350.	.14	41.1	.137	41.7	11.3
8	1A	--	243.	.05	32.9	.064	32.3	4.7
	2A	35.4	344.	.10	43.5	.092	44.3	7.8
	3A	41.2	350.	.14	42.	.136	41.5	11.3
9	1A	31.5	331.3	.14	42.5	.15	42.	11.8
	2A	76.5	383.3	.21	44.5	.198	43.7	16.7
	3A	88.8	398.3	.283	44.6	.268	44.6	22.5

Analysis of Factors Influencing Performance   195

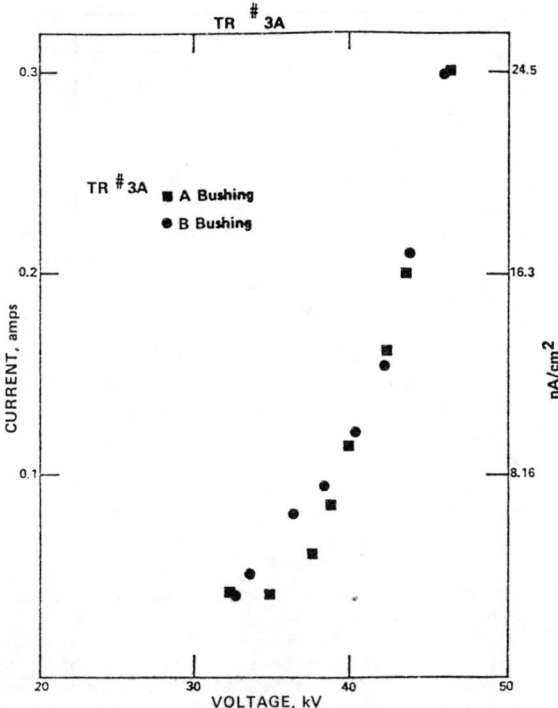

Figure 185. Voltage vs. current relationship for transformer rectifier #3A, Plant 2.

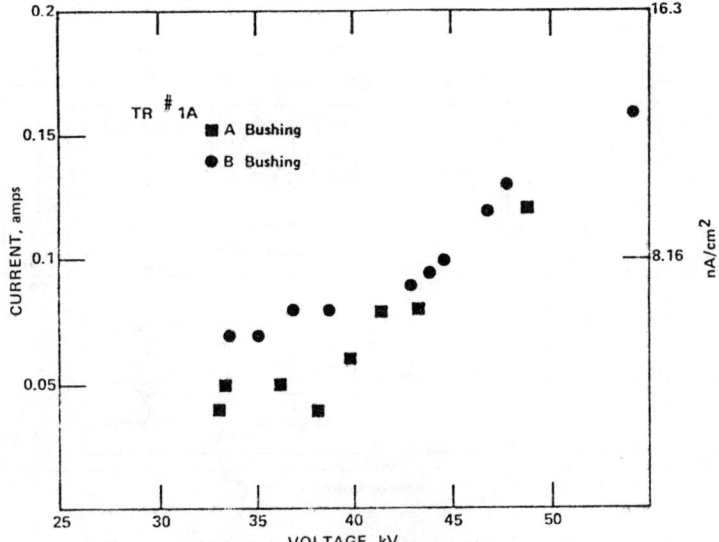

Figure 186. Voltage vs. current for transformer rectifier #1A, Plant 2.

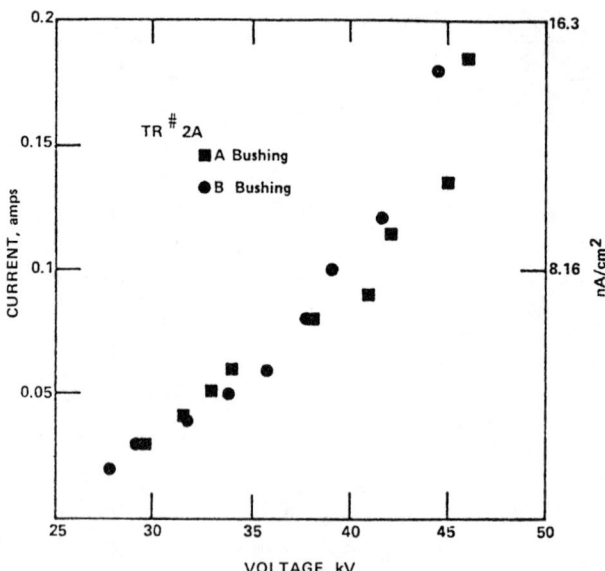

Figure 187. Voltage vs. current for transformer rectifier #2A, Plant 2.

Plant 3 - Cold-side ESPs collecting ash from high sulfur Eastern coal--A mechanical collector, which was reported to have been reworked when the precipitator was installed, precedes the electrostatic collector at Plant 3, Unit 5. The precipitator consists of four fields in the direction of gas flow (Figure 188) and is physically divided into two collectors (A and B).

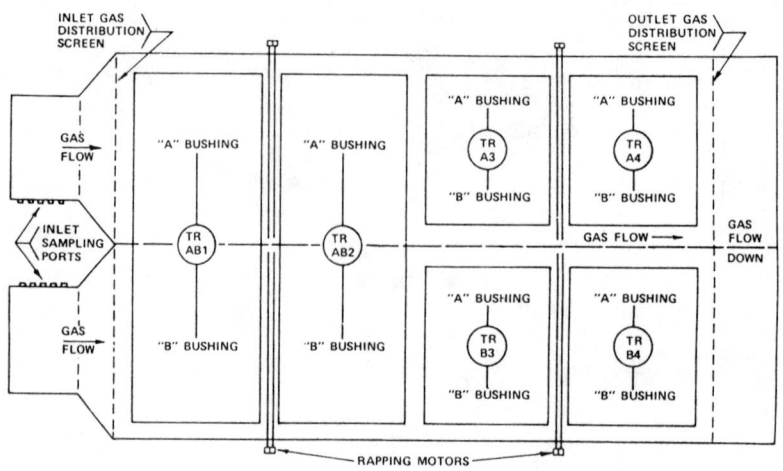

Figure 188. Plant 3, Unit 5 precipitator layout.

The test program conducted at Plant 3 was conducted on the "B" side of the #5 precipitator. The total collecting area for the "B" side is 5900.64 m² (63,516 ft²) with 1475.16 m² (15,878 ft²) per field.

This gives a specific collection area of 43.48 m²/(m³/sec) (220.9 ft²/1000 cfm) for the design volume of 135.70 m³/sec (287,500 acfm) per collector. The precipitator has six full-wave transformer rectifiers; each transformer rectifier has an "A" and "B" bushing, as seen in Figure 188.

The precipitator has 25.4 cm (10 in.) plate spacings and operates at approximately 160°C (320°F). A tumbling hammer type of rapping system is employed in which each collecting plate is rapped with a hammer. The first two fields are rapped every six minutes, whereas half of the third and fourth fields are rapped every six minutes. The emitting electrodes are rigid "barbed" electrodes which are 0.502 m (1'7 3/4") apart in the direction of gas flow.

Table 16 contains the average electrical conditions, and Figures 189 through 192 illustrate the secondary voltage-current relationships at Plant 3. These curves indicate excellent electrical operating conditions.

TABLE 16. AVERAGES OF HOURLY ELECTRICAL READINGS
PLANT 3, "B" SIDE OF PRECIPITATOR 5

Day	TR#	KV	MA	nA/cm²
1	5AB1	53.0	339	23.1
	5AB2	53.0	504	34.3
	5B3	51.0	708	48.2
	5B4	47.5	679	46.2
5	5AB1	51.6	198	13.5
	5AB2	52.0	347	23.6
	5B3	47.9	466	31.7
	5B4	48.5	675	45.9
6	5AB1	44.6	292	19.9
	5AB2	51.7	402	27.3
	5B3	48.9	529	36.0
	5B4	48.5	675	45.9

198 Electrostatic Precipitator Manual

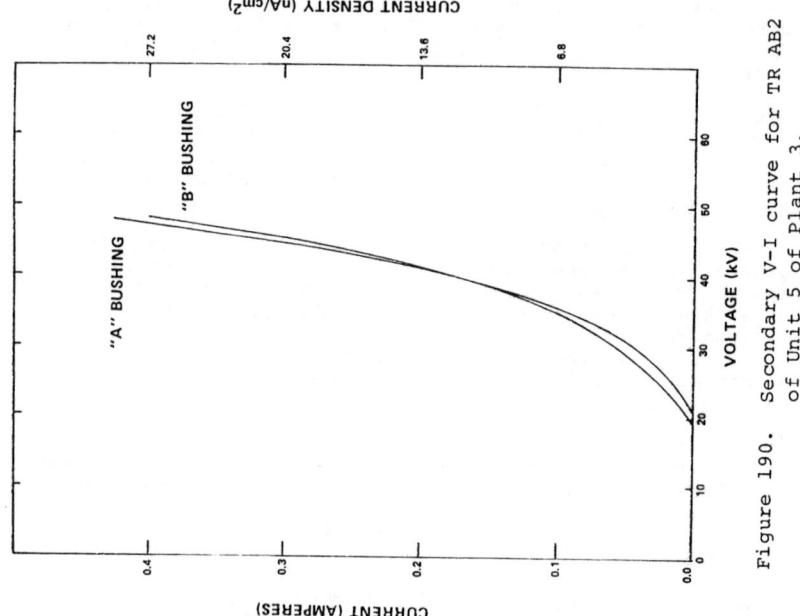

Figure 190. Secondary V-I curve for TR AB2 of Unit 5 of Plant 3.

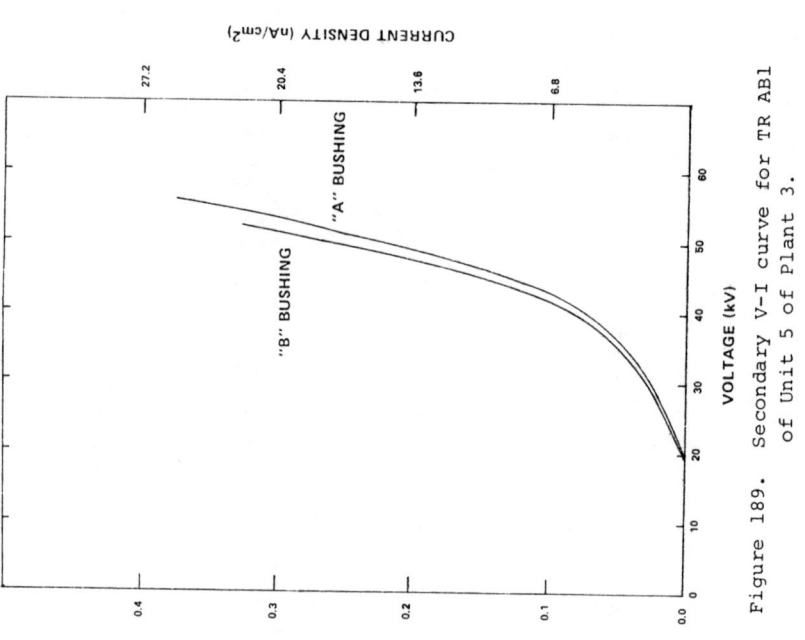

Figure 189. Secondary V-I curve for TR AB1 of Unit 5 of Plant 3.

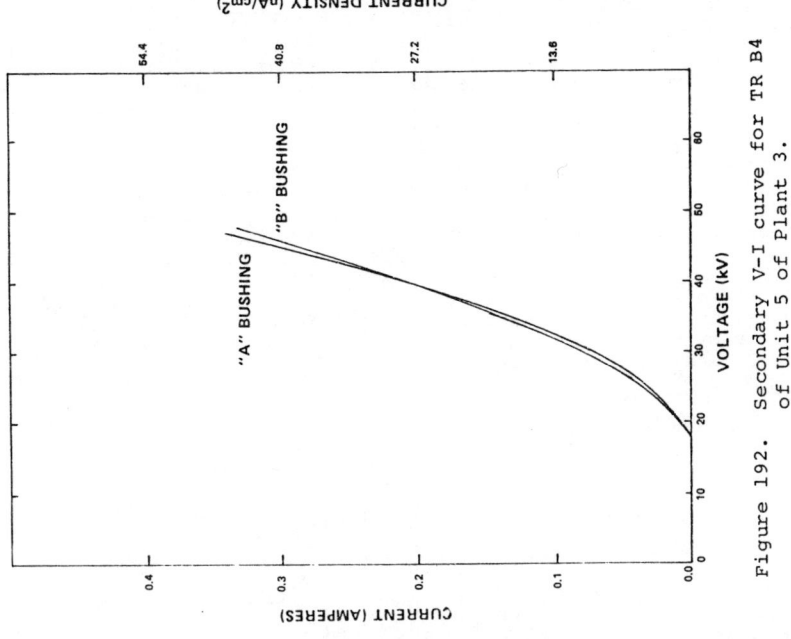

Figure 191. Secondary V-I curve for TR B3 of Unit 5 of Plant 3.

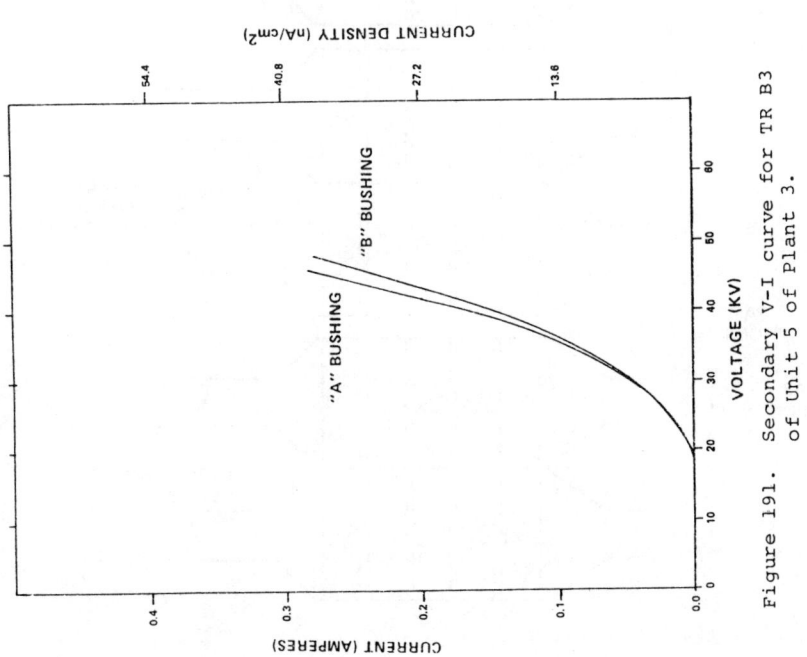

Figure 192. Secondary V-I curve for TR B4 of Unit 5 of Plant 3.

200   Electrostatic Precipitator Manual

Plant 4 - Cold-side ESPs collecting ash from low sulfur Western coals--The electrostatic precipitator installed on Unit 1 of Plant 4 consists of six physically divided chambers (Figure 193). The test program was conducted on the #5 chamber of the Unit 1 precipitator. Each chamber of the precipitator has 44 lanes and five electrical fields in the direction of gas flow. Each electrical field is 3.2 m (10.5 ft) long and has a total collection area of 3518.96 m^2 (37,879 ft^2). The precipitator has 25 cm (9.75 in) plate spacing, and spiral discharge electrodes with a radius of 1.24 mm (0.49 in). Tumbling hammers are used to rap both the collecting plates and high voltage discharge frames. The precipitator operates at 88 to 120°C (190 to 250°F) and was designed to handle 1100 m^3/sec (2,330,000 acfm) at 121°C (250°F), which results in a design specific collection area of 95.97 m^2/(m^3/sec)(487.6 ft^2/1000 acfm). However, the actual SCA measured on the tested chamber was approximately 590 ft^2/1000 acfm. Rapping frequencies in the direction of gas flow are 10, 5, 5, 2, and 1 per hour, respectively.

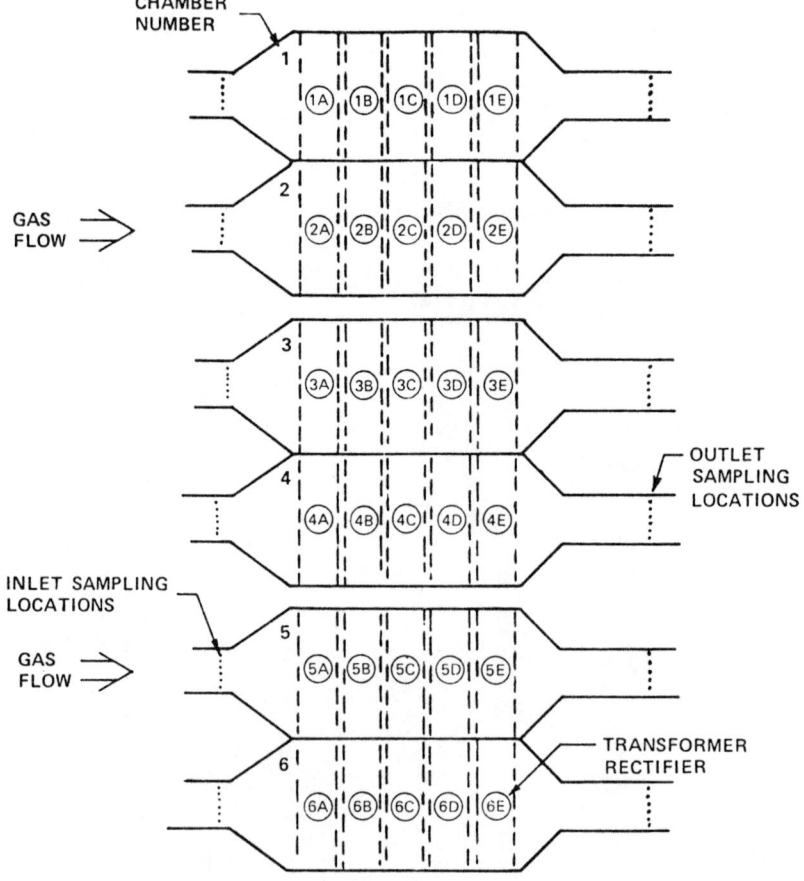

Figure 193. Plant 4, Unit 1 precipitator layout.

Table 17 contains average secondary voltage and current readings, and Figure 194 contains secondary voltage and current curves obtained at Plant 4. The location of the operating points for sets B, C, D, and E with respect to the V-I curves suggests that a significant portion of the secondary current is being consumed by sparking or back corona.

TABLE 17. OPERATING SECONDARY VOLTAGES AND CURRENTS DAILY AVERAGES UNIT 1, CHAMBER 5

Day	TR#	KV[1]	A	C.D. $\mu A/ft^2$	$\mu A/m^2$
1	5A	38.9	.238	6.28	67.6
	5B	36.5	.624	16.47	177.3
	5C	33.0	.910	24.02	258.6
	5D	37.8	.590	15.58	167.7
	5E	35.1	.860	22.70	244.4
2	5A	39.6	.245	6.47	69.6
	5B	36.5	.800	21.12	227.3
	5C	32.9	.850	22.44	241.6
	5D	37.3	.785	20.72	223.
	5E	35.0	.862	22.76	245.
3	5A	39.6	.260	6.86	73.8
	5B	35.9	.724	19.11	205.7
	5C	33.0	.722	19.06	205.2
	5D	37.0	.846	22.33	240.4
	5E	34.6	.864	22.81	245.5
4	5A	39.6	.322	8.50	91.5
	5B	35.9	.794	20.96	225.6
	5C	32.2	.959	25.32	272.6
	5D	37.4	.754	19.91	214.3
	5E	34.6	.894	23.60	254.
7	5A	39.5	.217	5.73	61.7
	5B	34.5	.887	23.42	252.1
	5C	31.6	.870	22.97	247.3
	5D	36.4	.653	17.24	185.6
	5E	34.1	.827	21.83	235.

[1] Corrected meter readings.

Plant 5 - Cold-side ESP collecting ash from medium sulfur Southeastern coal--Figure 195 illustrates the gas flow and precipitator arrangement. Some of the electrical sets were not operating on the B side precipitator, apparently due to broken corona wires; therefore, tests were conducted on the A side only. Each precipitator consists of two collectors in series, each of which has 144 gas passages, with 0.229 m plate-to-plate spacing (9 in.), 9.14 m high plates (30 ft), and 5.45 m in length (18 ft). Thus, each precipitator consists of 144 gas passages 9.14 m high (30 ft), 10.97 m long (36 ft), for a total collecting area of 28877 m^2 (311,000 ft^2) per precipitator. The precipitators each have twelve electrical sections arranged in series with the gas flow, such that the individual sections power 1/12 of the plate area and 1/12 of the length. Gas flow at full load ($\sim$700 MW) for each precipitator is about 520 m^3/sec (1.1 x 10^6 cfm) at 300°F. The specific collecting area at these conditions would be 55 m^2/(m^3/sec) or 283 ft^2/1000 cfm.

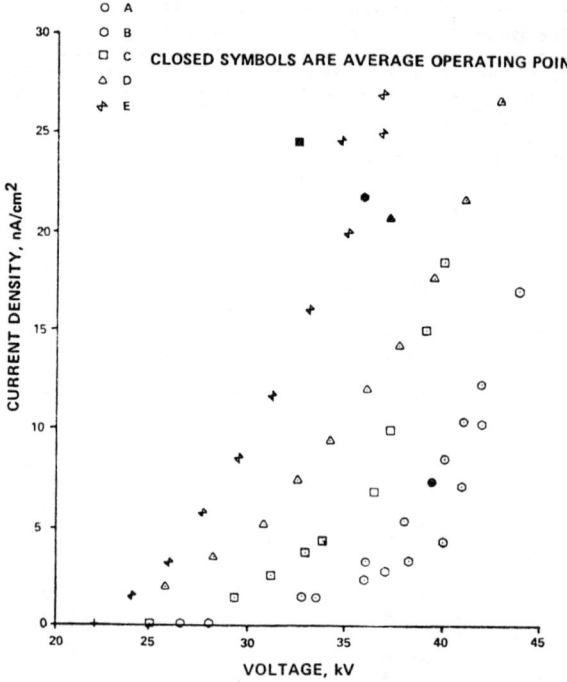

Figure 194. Secondary current-voltage relationship, Plant 4, Unit 1, Chamber 5.

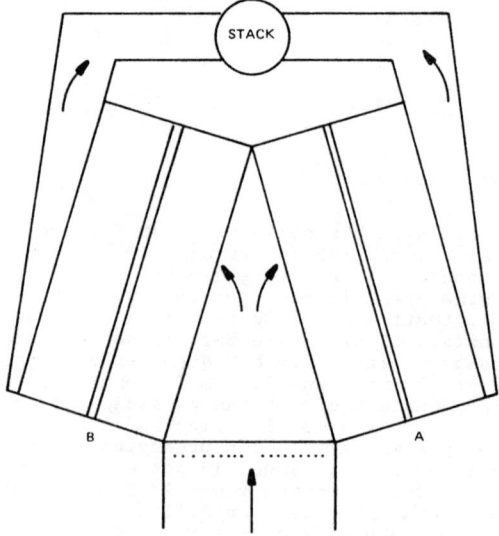

Figure 195. Precipitator layout at Plant 5, Unit 10.

Figure 196 shows the voltage-current relationship obtained for power sets in the front and rear section of Precipitator B. These data were taken from the "B" side in order that measurements in progress on the "A" side would not be disrupted. The difference shown between the two power supplies may be caused, in part, by space charge suppression of corona current caused by the higher dust loading experienced by set 10BF, and, in part, by differences in electrode alignment. Neither set shows any indication of back corona. Although operating current density is limited to an average of around 20 nA/cm^2 and the power supplies exhibit an increased sparking tendency as the sulfur content of the coal drops, the operation of this unit is not seriously imparied by high resistivity. A dust of excessively high resistivity often results in the occurrence of back corona at a lower voltage than the sparkover voltage, and would be indicated by drastically reduced precipitator performance and by the shape of the voltage-current relationships for the power supplies.

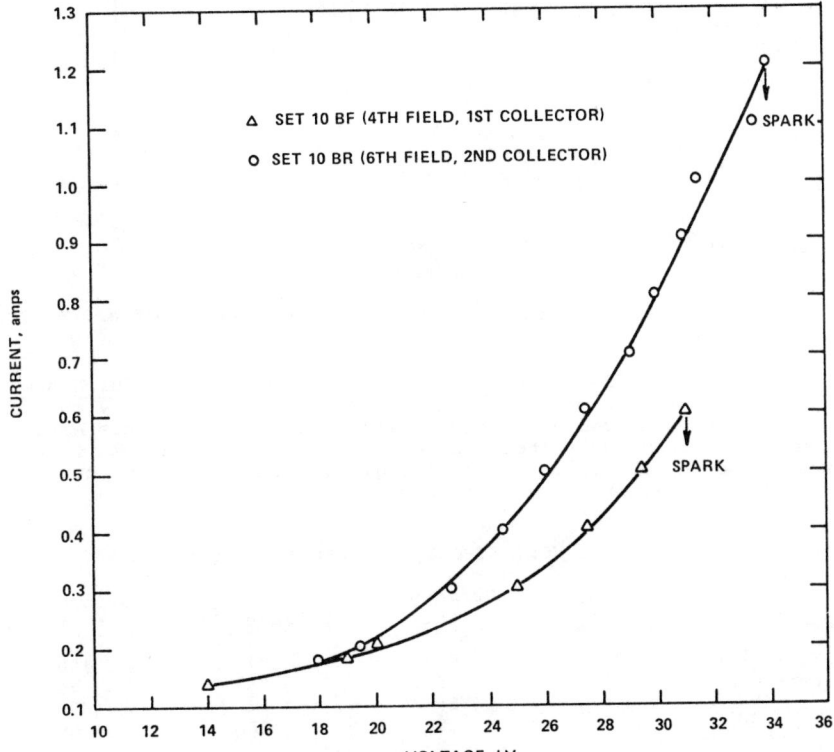

Figure 196. Voltage-current relationships obtained on precipitator "B", Plant 5, Unit 10.

Plant 6 - Cold-side ESP collecting ash from Midwestern coal--
Secondary voltage-current curves are shown for the inlet and outlet fields for this Midwestern power plant in Figure 197. The resistivity of the ash during the test series was between 10^{11} and 10^{12} Ω-cm.

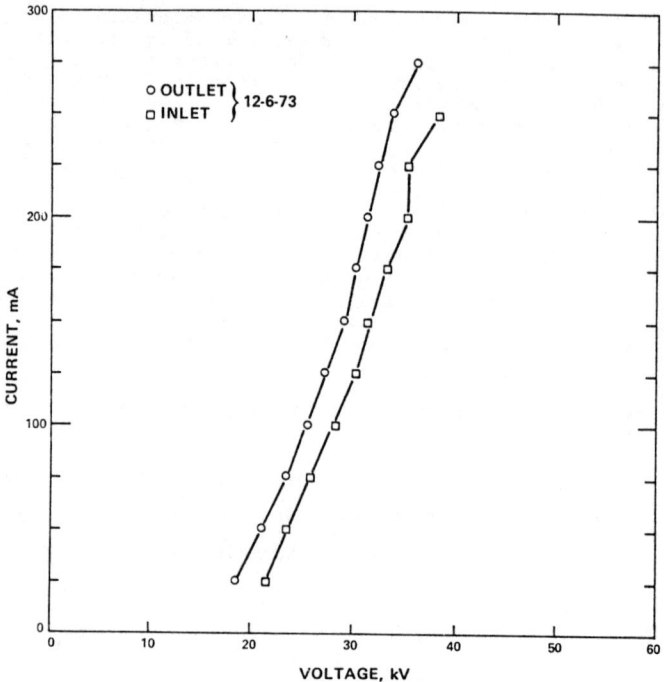

Figure 197. Secondary voltage vs. current curves from Plant 6.

Plant 7 - Cold-side ESP collecting ash from low sulfur Western coal--The operating points of the ESP primary and secondary current meters were monitored routinely during a field test at this Western plant. The results of these current and voltage measurements are summarized in Table 18. Figure 198 gives a representative curve of the voltage-current characteristics.

TABLE 18. VOLTAGE CURRENT OPERATING DATA

Day and Time		Primary Voltage	Primary Current-A	Secondary Voltage (kV)	Secondary Current (Ma)
1	10:10 A	150	90	27	400
	11:35 A	110	15	25	50
	2:00 P	110	15	21	50
	5:00 P	110	15	30	50
2	9:30 A	130	40	30	150
	11:30 A	142	47	30	150
	4:30 P	145	47	29	150
3	9:00 A	140	40	32	150
	11:30 A	141	42	31	150
	1:45 P	160	43	40	150
	11:45 P	170	43	41	150
4	4:30 P	170	44	40	150

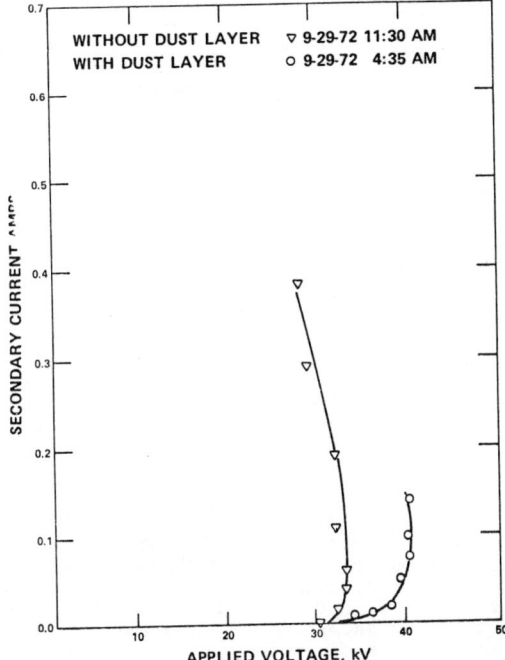

Figure 198. Voltage-current characteristics of Section 1B inlet, Plant 7.

Measured Hot-Side Curves--

Plant 8 - Hot-side ESP collecting ash from low sulfur Eastern coal--The electrostatic precipitator installed on Unit 3 of Plant 8 is a hot electrostatic precipitator which operates at approximately 343°C (650°F). The precipitator consists of two separate casings, A & B, each of which has two inlet and two outlet ducts. Tests at Plant 8 were conducted on the "B" side of the #3 precipitator, or one half of the unit. The "B" side precipitator has four fields in series; each field has a total collection area of 3912.95 m^2 (42,120 ft^2) and is powered by one transformer-rectifier. Figure 199 illustrates the hot side precipitator layout. This unit is a retrofit which was installed in series with an existing cold side precipitator. The collecting electrodes have 22.9 cm (9 in.) spacing, are 9.14 m (30 ft) high and are 2.74 m (9 ft) deep per field. The collecting plates are rapped by solenoid activated drop hammers. Each drop hammer is activated at least once every two minutes. The emitting electrodes have 22.9 cm (9 in.) spacing, both parallel and perpendicular to the gas flow and are .277 cm (.109 in.) in diameter. The emitting electrodes are vibrated twice every hour with electric vibrators. Although the precipitator was designed to have an SCA of 53.15 m^2/(m^3/sec)(270 ft^2/1000 acfm) for a total volume flow of 590 m^3/sec (1,250,000 acfm), the gas flow for the two chambers tested was about 430,000 acfm, which resulted in an SCA of approximately 390 ft^2/1000 acfm.

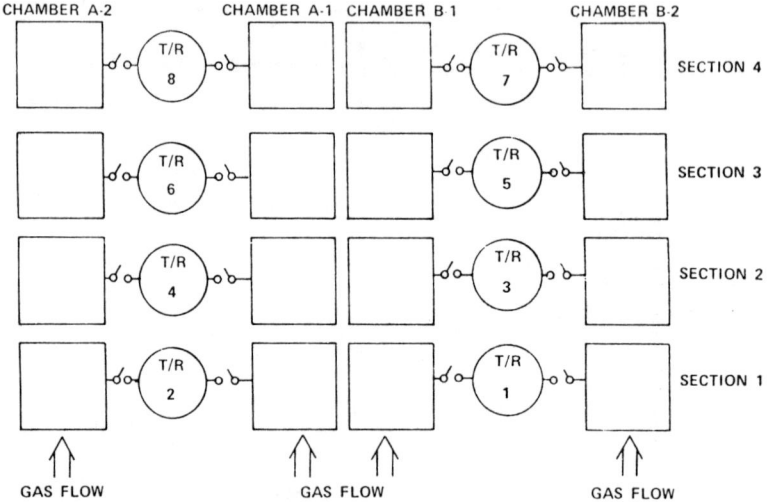

Figure 199. Plant 8, Unit 3 precipitator configuration.

Table 19 gives the average electrical condition data for the hot-side precipitator power supplies tested at Plant 8. Voltage-current curves for the indicated power supplies are shown in Figure 200. These data indicate good electrical operating conditions for a hot-side precipitator, and show the expected decrease in voltage from inlet to outlet for a given current due to decreasing particulate space charge.

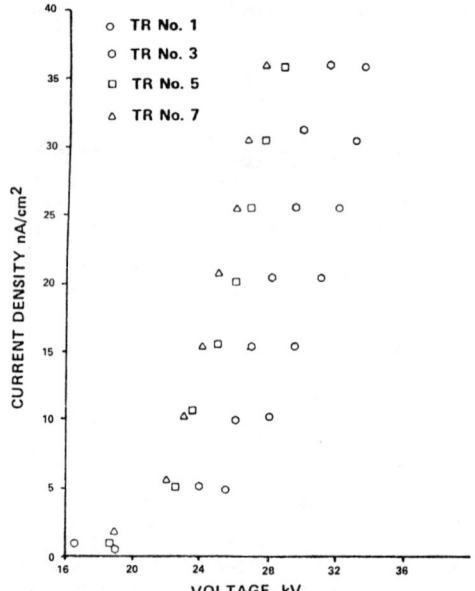

Figure 200. V-I curves for Unit 3, Plant 8.

TABLE 19. AVERAGE ELECTRICAL OPERATING CONDITIONS (PLANT 8)

Day	TR Set	Primary Voltage Volts	Primary Current Amps	Secondary Voltage kV	Secondary Current Densities nA/cm^2	Spark Rates Sparks/min
1	1	269.14	207.43	35.88	31.2	27.86
	3	265.00	240.43	33.95	42.7	4.71
	5	226.29	233.29	29.13	36.3	4.43
	7	265.86	235.57	27.50	37.3	4.86
2	1	269.63	199.50	35.00	30.7	28.75
	3	268.13	241.00	34.30	37.6	2.
	5	226.00	237.88	29.00	37.3	2.
	7	232.38	234.88	27.50	37.1	5.
3	1	272.00	200.57	36.17	30.7	27.43
	3	272.00	210.71	34.83	37.3	5.40
	5	226.14	237.00	29.10	37.1	2.29
	7	230.71	232.71	27.27	36.8	4.86
4	1	272.71	201.86	35.14	30.9	24.43
	3	272.86	239.71	35.40	37.3	10.00
	5	227.43	236.86	29.40	37.1	5.14
	7	230.29	232.57	27.27	36.8	5.
5	1	274.14	212.43	35.67	31.2	24.57
	3	228.14	234.29	35.00	37.1	12.14
	5	228.14	237.14	29.43	37.3	4.00
	7	230.29	233.14	28.60	37.1	4.43
6	1	275.00	196.00	35.50	30.2	35.14
	3	259.50	214.70	33.65	31.7	26.80
	5	230.00	239.14	29.50	37.8	5.14
	7	232.71	236.14	27.25	37.6	4.57
Avg. All Tests	1	272.10	202.97	35.56	30.82	28.03
	3	260.94	230.14	34.52	37.28	10.18
	5	227.36	236.89	29.26	37.15	3.83
	7	237.04	234.17	27.57	37.12	4.79

Plant 9 - Hot-side ESP collecting ash from low sulfur Western coal--The electrostatic precipitator installed on Unit 3 of Plant 9 is a hot precipitator which operates at approximate 360°C (680°F). This precipitator consists of two separate collectors, each of which has 8 isolatable chambers. The test program was conducted on the #8 chamber of the upper precipitator on Unit 3. Each precipitator chamber has thirty-five 22.9 cm (9 in.) lanes or gas passages and six electrical fields in the direction of gas flow. Each field is 1.83 m (6 ft) deep and has a total collecting area of 1170.54 m^2 (12,600 ft^2). The discharge electrodes have a diameter of 0.268 cm (.1055 in.) and are powered in each field by a full-wave transformer-rectifier which also powers a field in the adjacent chamber. Figures 201 and 202 illustrate the duct work and chamber arrangement. The complete precipitator installation was designed to handle 1859.68 m^3/sec (3,940,000 acfm) at 350°C (662°F), which results in a design specific collection area of 60.43 m^2/(m^3/sec)(307 ft^2/1000 cfm). Both collecting and discharge electrodes are rapped with solenoid activated magnetic impulse rappers. The original rapper program installed with the precipitator had a program time of 90 minutes. During that 90 minutes, all rappers (both plate and discharge) in the first and second fields were activated nine times, all rappers in the third and fourth fields were activated five times, and all rappers in the fifth and sixth fields were activated four times.

During the test program on this unit, two different rapping programs, each of which separated the wire and plate rappers, were examined. The first program tested had rapping frequencies in the direction of gas flow of 8, 8, 3, 3, 1, 1 per 73 minute period. Wire rappers were activated eight times in the 73 minute period. The second program examined had rapping frequencies in the direction of gas flow of 3, 3, 1, 1, 1, 1 per 22 minute period. Wire rappers were activated once in the 22 minute period. Although the different rapping programs probably affected the voltage-current curves to some extent, the data did not show a significant difference in the curves.

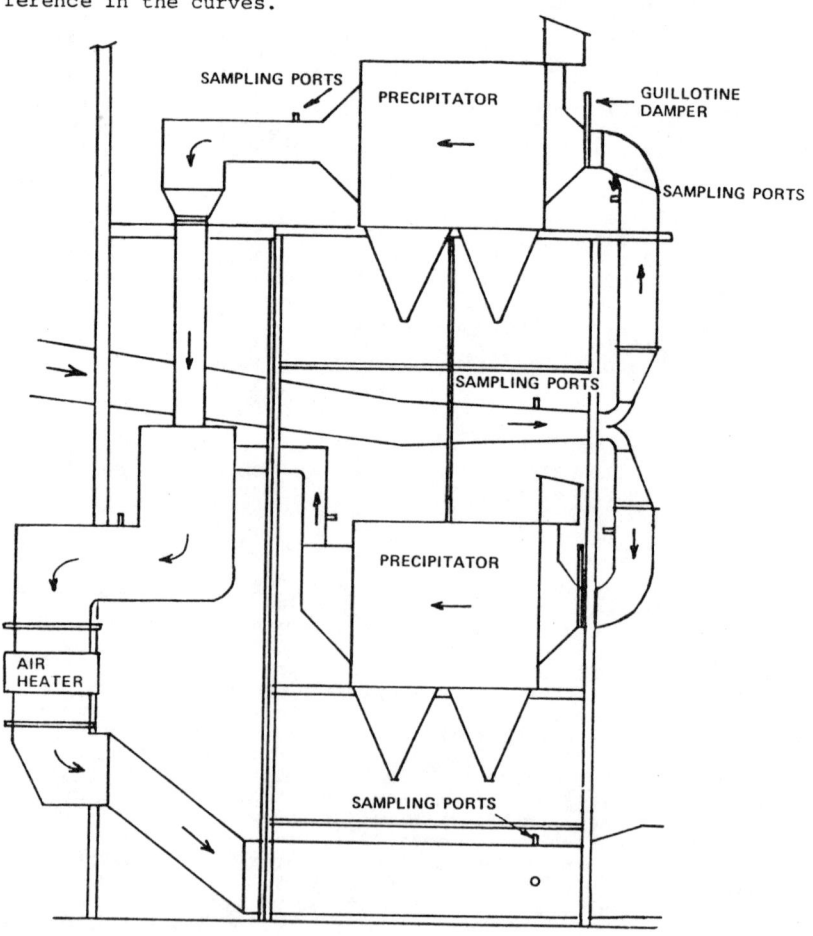

Figure 201. Ductwork arrangement for Plant 9, Unit 3.

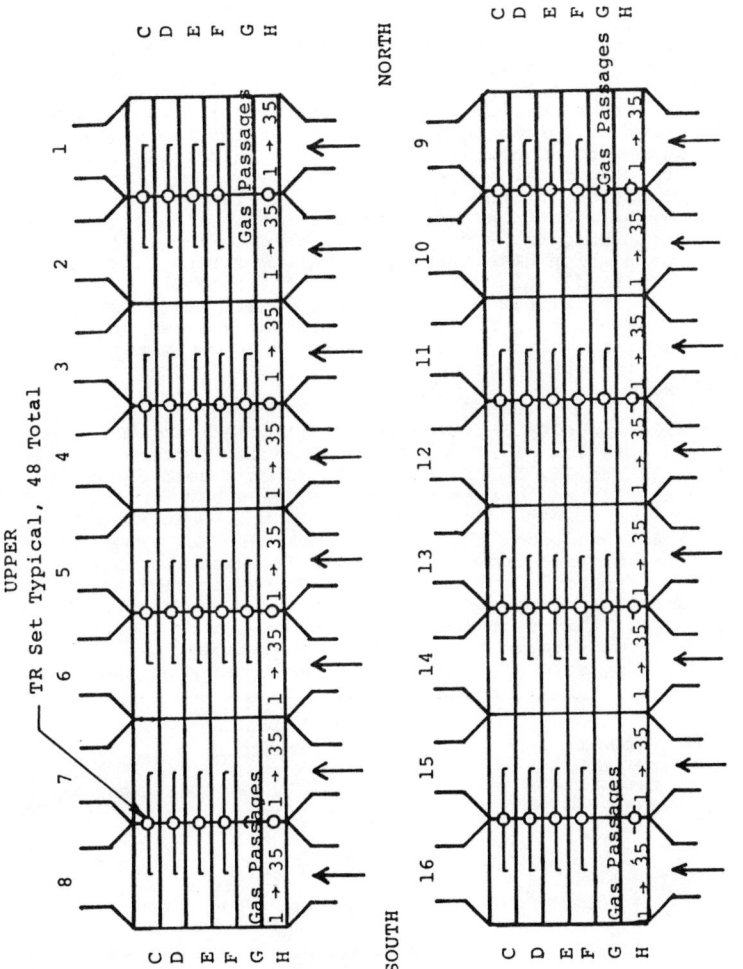

Figure 202. Chamber arrangement for Plant 9, Unit 3.

Figures 203 and 204 show the full load secondary voltage-current relationships for the inlet (H) and outlet (C) fields for two of the chambers (7 and 8) of a hot precipitator at Plant 9. Also shown are the average operating points for Plant 9 and similar secondary V-I curves and operating points for Plant 8. Note that for full load conditions the inlet V-I curves for the Plant 8 and Plant 9 units are similar in shape, but Plant 8 achieves higher current densities prior to sparkover. The outlet V-I curves, however, are significantly different in shape, with Plant 8 achieving substantially higher operating voltages. The design of the electrodes in the two precipitators is similar, and the operating temperatures differ by only -12°C (10°F). These observed differences in electrical operating parameters result in significant differences in theoretical prediction of collection efficiency for the two units. Laboratory measurements of resistivity indicate that resistivity and breakdown strength of the dust under laboratory conditions do not offer an explanation for the low voltages at Plant 9. It is possible, however, that the effective dielectric strength of the dust under field conditions may be lower than in the laboratory. Two other causes which have been hypothesized for the low voltages are: (1) unexpectedly high values of effective mobility for the flue gas due to the effect of reduced gas density, and (2) electrode geometry problems. Figure 205 gives V-I curves for all fields of two of the chambers of the Plant 9 precipitator. The effects of particulate space charge in progressing from inlet to outlet are evident.

Figures 203 and 204 indicate the dramatic effect on the power supply characteristics of reducing the operating temperature (outlet values) from 329 to 252°C (625 to 485°F) as unit load is dropped to 400 MW. Both inlet and outlet sets became severely spark-rate limited, and the operating points under automatic control were much lower under half load conditions than they were at 800 MW. The collection efficiency dropped from 99.26 to 92.17% (mass train data), even though the specific collecting area of the precipitator was doubled as gas flow decreased. The electrical operating characteristics suggest that dust resistivity increased to the point that breakdown was occurring in the deposited dust layer, and that the resulting sparking severely limited the performance of the unit.

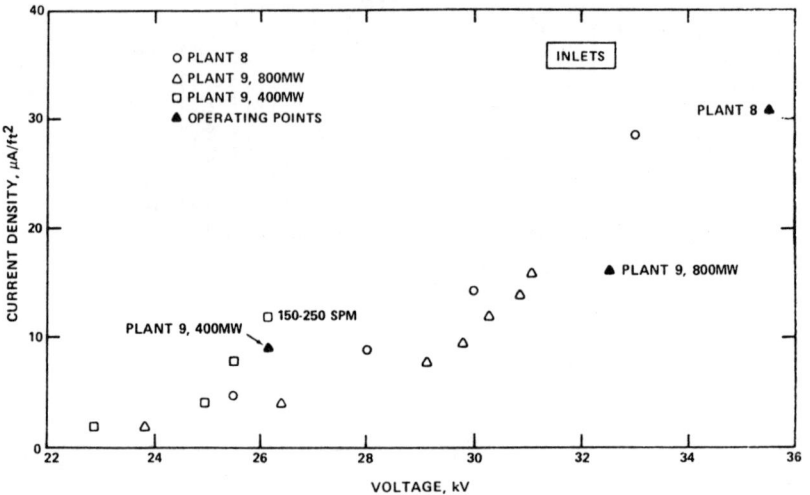

Figure 203. Inlet voltage current curves for Plants 9 and 8.

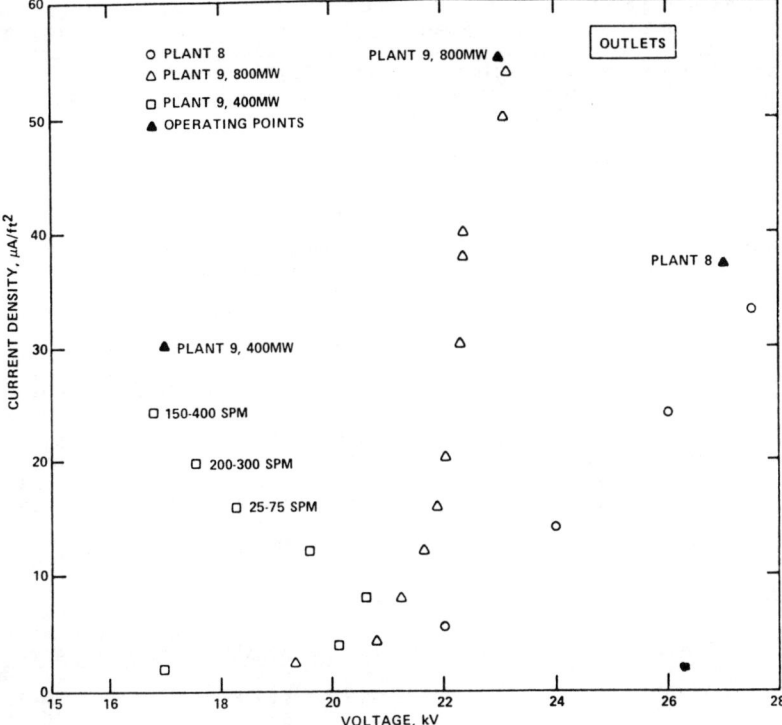

Figure 204. Outlet voltage current curves for Plants 9 and 8.

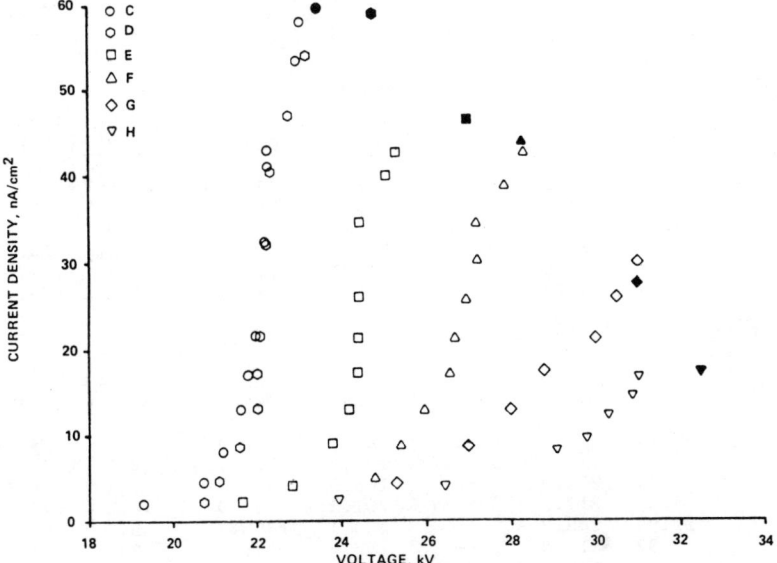

Figure 205. Voltage current curve, Unit 3, Chambers 7 and 8, Plant 9 (solid symbols are operating points).

Although the V-I curves show that better collection efficiency could be obtained if the power supplies were adjusted to operate at lower sparking rates, the available data clearly indicate that serious degradation in collection efficiency may result in this unit as load and temperature are reduced unless proper control is provided for the TR sets.

Table 20 contains the averages of panel meter readings from the test series. Note that the voltages for all fields were substantially reduced during the half load condition.

TABLE 20. AVERAGES OF HOURLY METER READINGS CHAMBER 7 AND 8 (kV VALUES ARE FROM VOLTAGE DIVIDER DATA)

Day 1

Field	DCKV	ACV	Spark	ACA	DCMA	Current Density nA/cm²	(μA/ft²)
H	31.77	215.0	50-200	95.9	390.7	16.7	(15.5)
G	28.38	212.1	50-150	110.0	518.9	22.2	(20.6)
F	27.20	220.0	25	185.0	1017.1	43.6	(40.4)
E	24.75	212.9	--	197.5	1096.1	47.0	(43.5)
D	22.75	204.3	--	251.4	1332.1	57.1	(52.9)
C	22.32	205.0	--	252.4	1387.1	59.4	(55.0)

Day 2

Field	DCKV	ACV	Spark	ACA	DCMA	Current Density nA/cm²	(μA/ft²)
H	31.81	215.0	100-200	107.9	413.6	17.7	(16.4)
G	29.47	223.9	65	136.8	658.6	28.2	(26.1)
F	27.69	228.9	--	188.0	1023.6	43.8	(40.6)
E	24.78	214.6	--	195.7	1085.4	46.6	(43.1)
D	22.62	205.0	--	250.5	1335.0	57.2	(53.0)
C	22.28	205.0	--	251.9	1383.2	59.3	(54.9)

Day 3

Field	DCKV	ACV	Spark	ACA	DCMA	Current Density nA/cm²	(μA/ft²)
H	25.52	169.0	50-150	59.0	234.0	10.0	(9.3)
G	23.86	139.0	50-200	25.0	110.0	4.8	(4.4)
F	23.44	152.0	25-100	44.0	135.0	5.8	(5.4)
E	18.99	159.0	50-250	120.0	568.0	24.3	(22.5)
D	19.12	150.0	25-200	105.0	440.0	18.9	(17.5)
C	16.93	163.0	25-100	155.0	750.0	32.2	(29.8)

Day 4

Field	DCKV	ACV	Spark	ACA	DCMA	Current Density nA/cm²	(μA/ft²)
H	31.42	212.0	50-100	88.3	360.0	15.4	(14.3)
G	29.57	217.8	50-150	107.8	526.1	22.6	(20.9)
F	27.67	223.9	60	168.3	921.7	39.5	(36.6)
E	25.08	217.2	--	201.1	1100.0	47.2	(43.7)
D	22.93	205.2	15	253.1	1305.6	55.9	(51.8)
C	23.02	211.0	--	253.2	1376.1	59.0	(54.6)

(CONT'D)

TABLE 20. (CONT'D)

Day 6

Field	DCKV	ACV	Spark	ACA	DCMA	Current Density nA/cm^2	(μA/ft^2)
H	31.45	214.6	50-150	90.4	374.6	16.1	(14.9)
G	29.61	221.1	25-100	119.6	602.7	25.8	(23.9)
F	27.72	225.0	20	189.1	1007.7	43.2	(40.0)
E	24.87	213.2	--	200.7	1102.7	47.3	(43.8)
D	22.44	204.3	--	253.6	1350.9	57.9	(53.6)
C	22.85	209.7	--	248.3	1351.4	57.9	(53.6)

Day 7

Field	DCKV	ACV	Spark	ACA	DCMA	Current Density nA/cm^2	(μA/ft^2)
H	29.75	194.4	50-200	55.1	227.7	9.7	(9.0)
G	28.20	203.9	50-150	100.7	477.1	20.4	(18.9)
F	26.22	214.7	25-100	185.9	998.2	42.8	(39.6)
E	23.70	210.1	10	198.2	1088.8	46.7	(43.2)
D	22.18	199.9	10	243.0	1291.5	55.3	(51.2)
C	22.33	207.2	--	252.9	1387.1	59.4	(55.0)

Day 8

Field	DCKV	ACV	Spark	ACA	DCMA	Current Density nA/cm^2	(μA/ft^2)
H	30.49	200.7	25-150	67.3	284.7	12.2	(11.3)
G	28.53	198.3	25-150	78.0	378.7	16.2	(15.0)
F	26.72	213.8	25-100	167.7	877.0	37.6	(34.8)
E	24.36	214.5	25	195.1	1088.0	46.7	(43.2)
D	22.50	200.1	--	248.1	1322.0	56.7	(52.5)
C	21.31	199.3	--	248.1	1348.3	57.8	(53.5)

Day 9

Field	DCKV	ACV	Spark	ACA	DCMA	Current Density nA/cm^2	(μA/ft^2)
H	30.45	192.4	50-200	43.1	198.2	8.5	(7.9)
G	28.38	182.3	50-150	45.8	206.8	8.9	(8.2)
F	27.00	201.8	25-100	142.9	701.4	30.0	(27.8)
E	25.06	214.6	50	194.8	1040.7	44.6	(41.3)
D	22.31	196.9	--	221.4	1129.6	48.4	(44.8)
C	22.20	201.6	--	245.3	1324.6	56.8	(52.6)

Day 10

Field	DCKV	ACV	Spark	ACA	DCMA	Current Density nA/cm^2	(μA/ft^2)
H	30.33	190.0	50-150	46.7	193.3	8.3	(7.7)
G	28.31	180.0	50-150	35.0	165.0	7.0	(6.5)
F	26.48	187.3	50	78.3	340.0	14.6	(13.5)
E	24.40	211.7	25	196.7	1060.0	45.5	(42.1)
D	22.04	190.0	--	188.0	906.7	38.9	(36.0)
C	21.18	198.3	--	239.0	1275.0	54.6	(50.6)

214 Electrostatic Precipitator Manual

Plant 10 - Hot-side ESP collecting ash from a Western power plant burning low sulfur coal--The layout of the precipitator and pertinent information are shown in Figure 206. The precipitator power supply secondary voltage and current measurements are given in Table 21. The current density was consistently lower on the left side inlet (section A), possibly due in part to some electrode misalignment for this field. Figure 207 shows typical secondary voltage-current curves obtained from this unit.

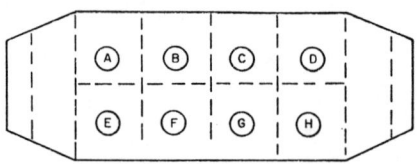

HOT-SIDE PLANT 10

INLET DUCT 7.6m x 1.60m(24'-11"x 5'-3")

OUTLET DUCT 8.84m x 1.60m(29'x 5'-3")

TOTAL AREA   146,160 FT^2 = 13,579 m^2

PLATE SPACING 9" = 23 cm

CORONA WIRE DIA. 0.1055 - .27 cm

4 PPTRS TOTAL INSTALLED

AREA FOR EACH POWER SUPPLY 1.7x10^3 m^2 = 18,270 FT^2

GUAR. 99.5 @ 350 MW

TEST 1.   VOL FLOW - 24 11/12' x 5.25' 2431 ft/min = 318,005 = 8600 m^3/min

SCA = 146,160   318,005 = 459 ft^2/kcfm = 1.5 m^2 -sec/m^3

CURRENT DENSITY = 42 μA/ft^2 = 45 nA/cm^2

AVG. EFF. 7 TESTS 3.085 INLET, .0228 OUTLET = 99.26

Figure 206.   Precipitator information and layout for the hot-side Plant 10 collector.

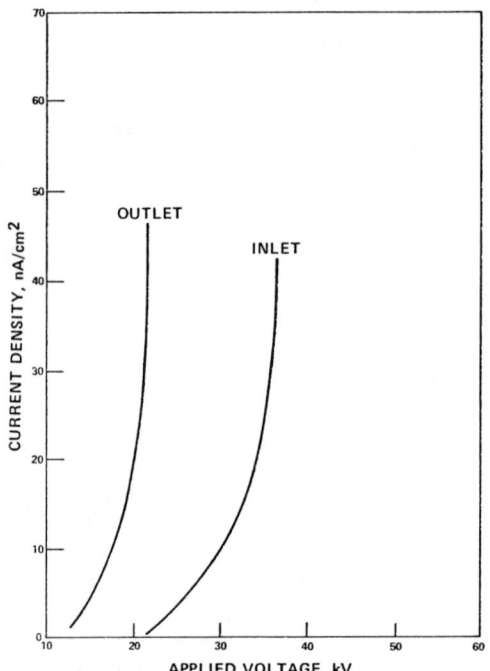

Figure 207.   Typical secondary voltage-current curves obtained from a hot-side ESP collecting ash from a Western power plant burning low sulfur coal.

TABLE 21. HOT-SIDE PLANT 10 SECONDARY VOLTAGE-CURRENT READINGS

	A		B		C		D	
	Voltage (kV)	Current (ma)	Voltage (kV)	Current (ma)	Voltage (kV)	Current (ma)	Voltage (kV)	Current (ma)
April 29, 1974	36.3	485	29.2	700	22.0	675	20.0	755
April 30, 1974	37.5	523	31.9	650	24.3	686	21.0	737
May 1, 1974	37.4	634	32.2	684	25.4	702	22.8	706
May 2, 1974	36.3	550	32.6	635	24.9	688	20.5	708
May 3, 1974	36.2	550	31.4	680	24.2	702	21.8	722
Average	36.8	548.4	31.5	670.0	24.2	690.6	21.2	725.6
Average Current Density, nA/cm²		32		39.4		40		42.7

	E		F		G		H	
	Voltage (kV)	Current (ma)	Voltage (kV)	Current (ma)	Voltage (kV)	Current (ma)	Voltage (kV)	Current (ma)
April 29, 1974	21.5	900	19.0	925	16.3	995	15.5	922.5
April 30, 1974	20.4	936	17.4	936	16.4	989	16.3	927
May 1, 1974	21.0	940	19.0	936	17.0	994	17.0	932
May 2, 1974	23.0	915	20.0	940	17.7	948	16.7	910
May 3, 1974	22.0	920	19.2	910	16.7	994	16.8	920
Average	21.6	922.2	18.9	929.4	16.8	984.0	16.5	922.3
Average Current Density, nA/cm²		54.2		54.7		57.9		54.3

Note: Each power set is connected to 1.7×10^7 cm² (18270 ft²)

RESISTIVITY OF COLLECTED FLY ASH

Effect Of Ash Resistivity On Precipitator Performance

In many instances, the useful operating current density in a precipitator is limited by the resistivity of the collected particulate layer. If the resistivity of the collected particulate layer is sufficiently high, electrical breakdown of the layer will occur at a value of current density which in most cases is undesirably low. Depending on the value of the applied voltage, the breakdown of the collected particulate layer will result in either a condition of sparking or the formation of stable back corona from points on the particulate layer. Excessive sparking and back corona are detrimental to precipitator performance and should be avoided.

Figure 208 shows an experimentally determined relationship between maximum allowable current density and resistivity.[168] It points out the severe drop in maximum allowable current density as the resistivity increases over the range $0.5 - 5.0 \times 10^{11}$ ohm-cm. Ash resistivities of 2×10^{10} ohm-cm or less generally allow extremely good electrical conditions to exist in a full-scale precipitator. Ash resistivities of 1×10^{12} ohm-cm or greater will cause back corona to ensue at relatively low applied voltages and will make it difficult to characterize precipitator operation.

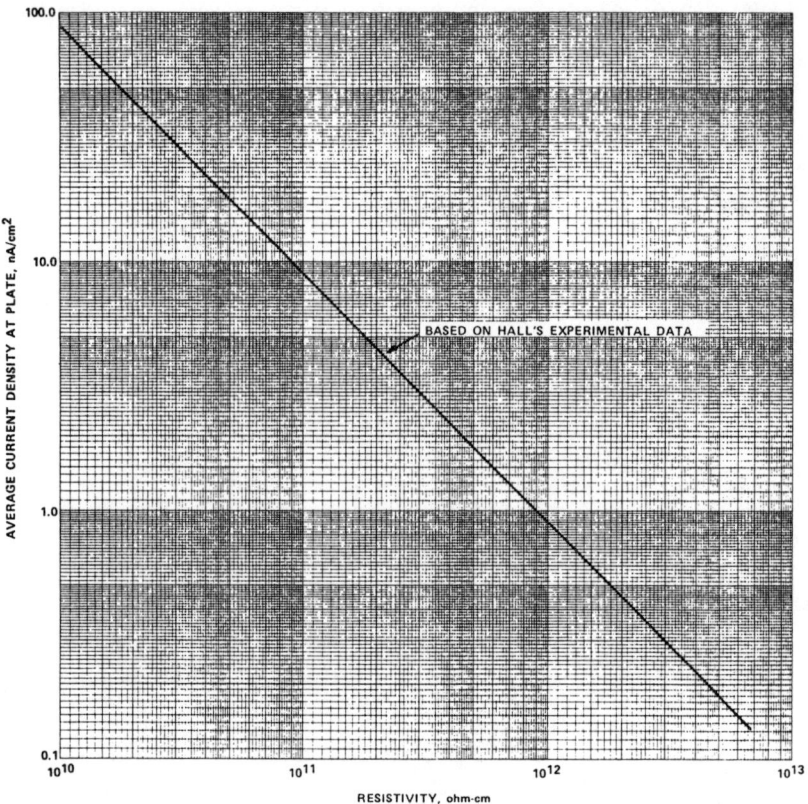

Figure 208. Experimentally determined effect of resistivity on allowable current density in a precipitator.[168]

Figure 209 shows the theoretically calculated effect of resistivity on overall mass collection efficiency for a particular cold-side, full-scale precipitator collecting fly ash particles. Measurements of inlet mass loading and particle size distribution, voltage-current characteristics, and gas velocity, volume flow, temperature, and pressure were used in the calculations. The operating applied voltage and current density for a given value of resistivity were determined by using the measured voltage-current characteristics and the data in Figure 208.

The curve in Figure 209 demonstrates how sensitive precipitator performance is to the resistivity of the collected ash. For this particular situation, the calculations project that an increase in resistivity from 10^{10} to 5×10^{11} ohm-cm will result in a decrease in overall mass collection efficiency from 98.1 to 81%. This example points out why a knowledge of the resistivity of the collected ash layer is crucial in designing a precipitator. The problem is made even more difficult since the resistivity can change significantly with changes in the composition, moisture content, and temperature of the flue gas. In addition, changes in resistivity due to changes in the coal producing the emissions must also be considered. Thus, in designing a precipitator, proper allowance must be made to account for possible values of resistivity that are larger than that anticipated.

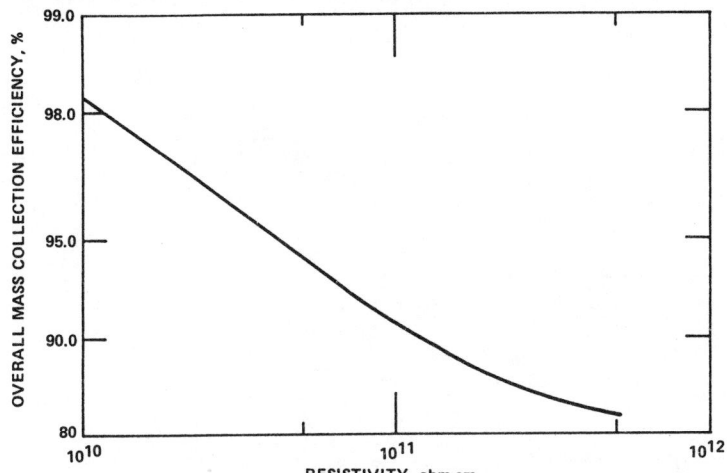

Figure 209. Effect of resistivity on overall mass collection efficiency.

Figure 210 shows measured overall mass collection efficiencies as a function of specific collection area for different cold-side, full-scale precipitators collecting fly ashes with various values of measured resistivity. Although the data from the different precipitators cannot be compared on the same basis due to differences in operating conditions and mechanical features, the data definitely show that precipitator performance decreases with increasing ash resistivity.

The resistivity of the collected ash layer also influences the electrostatic force which may hold the entire layer on the collecting surface or which may tend to pull the ash off from that surface.[169] The electrostatic force depends on the charge on the surface of the ash layer. The expression for this force can be derived by writing expressions for the voltage gradient in the gas and for that in the ash and using the principle of virtual work to find the force. The resulting equation for the force is[169]

$$F = (\varepsilon_0/2) \left[E^2 - \left(\frac{J\rho\varepsilon_1}{\varepsilon_0}\right)^2 \right], \qquad (23)$$

where

F = force per unit area (a positive force tends to pull the ash off the collecting electrode) [nt/m^2],

ε_0 = permittivity of free space (coul/V·m),

ε_1 = permittivity of the ash (coul/V·m),

ρ = resistivity of the ash (ohm-cm), and

E = potential gradient in the gas adjacent to the ash surface (V/m)

When $E = J\rho\varepsilon_1/\varepsilon_0$, the charge on the ash surface changes sign, and the force reverses its direction of action. Thus, depending on the values of E, J, ρ, and $\varepsilon_1/\varepsilon_0$, the force may act either to hold the ash to or to pull the ash from the collection electrode.

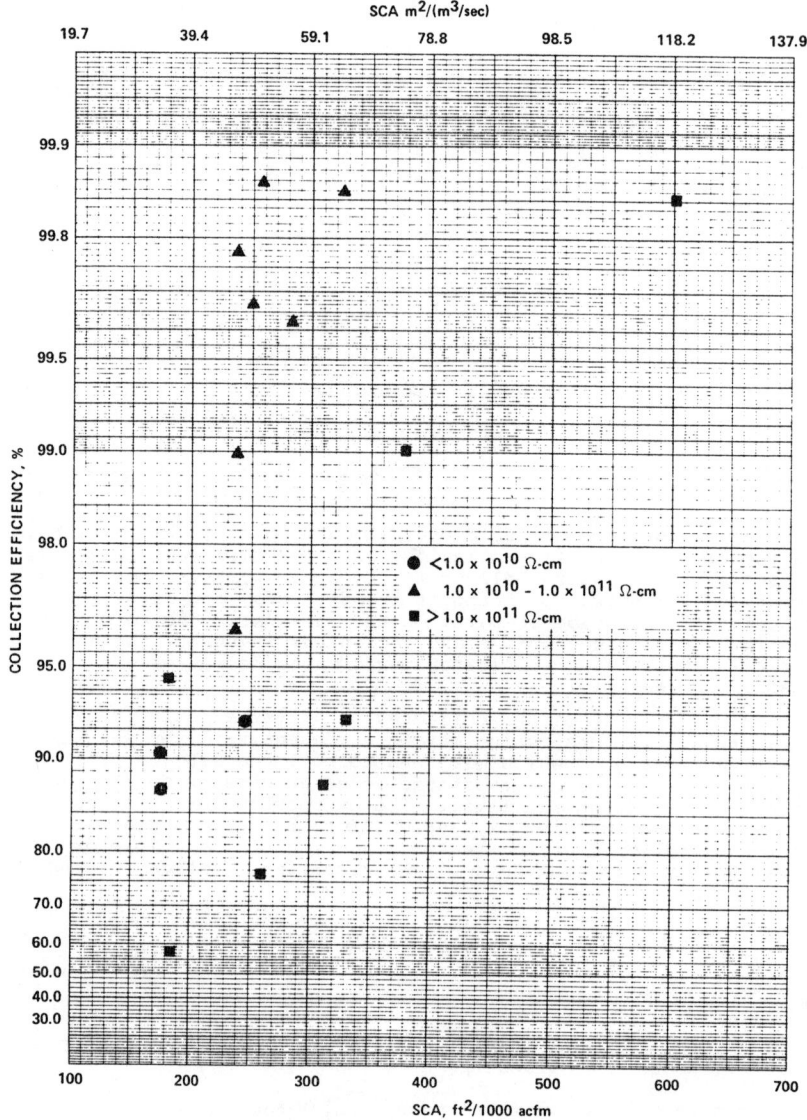

Figure 210. Measured overall mass collection efficiencies as a function of specific collection area for cold-side, full-scale precipitators collecting fly ashes of various values of measured resistivity.

Figure 211 shows the electrostatic force on the dust layer as a function of current density for several values of resistivity as predicted from equation (23). In obtaining the curves, it has been assumed that a dielectric constant ($\varepsilon_1/\varepsilon_0$) of 4 and a value of $E = 2.5$ kV/cm are typical for full-scale precipitators collecting fly ash particles. For the different values of resistivity, the curves were determined up to the maximum allowable current densities given in Figure 208.

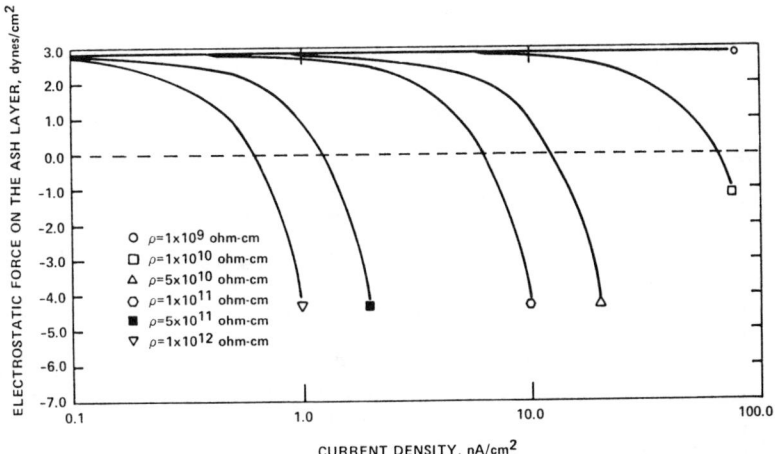

Figure 211. Electrostatic force on the dust layer as a function of current density for several values of resistivity.

As can be seen from Figure 211, ashes with resistivities between 10^9 and 10^{10} ohm-cm may be difficult to collect due to their tendency to come off the collection electrode. This situation results in excessive particle reentrainment, especially if high gas velocities exist in the precipitator. It is also interesting to note that for all values of resistivity there is a lower range of current densities in which the electrostatic force will be such as to pull the ash layer off the collection electrode. Thus, the precipitator must be operated near the maximum allowable current density for ashes with resistivities greater than 10^{10} ohm-cm in order to ensure that the electrostatic force will tend to hold the ash layer to the collection electrode.

Measured Voltage-Current Curves Demonstrating Back Corona

If the resistivity of the collected ash layer is high (greater than 10^{11} ohm-cm), back corona may occur at low applied voltages. The presence and onset of back corona can usually be detected from the measured, secondary voltage-current curves. Figure 212 shows voltage-current curves which demonstrate the behavior resulting from the occurrence of back corona.[154] The data are from a full-scale, cold-side precipitator collecting fly ash with a measured resistivity of approximately 6×10^{12} ohm-cm. With this high a value of resistivity, it would be anticipated that back corona would occur at low voltages without the presence of excessive sparking.

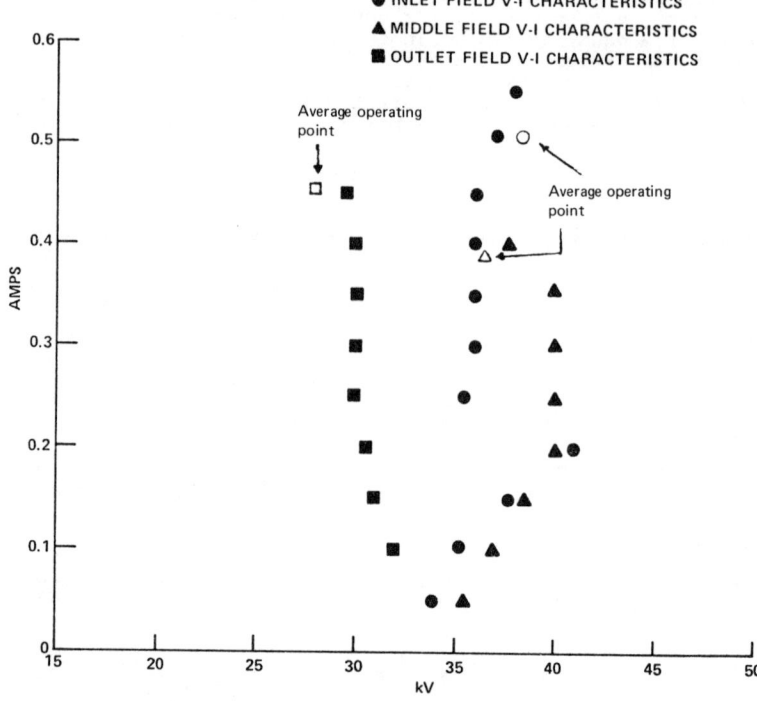

Figure 212. Voltage-current curves which demonstrate the behavior resulting from the occurrence of back corona.

At some point on a voltage-current curve demonstrating back corona, the applied voltage necessary to produce an increased current will drop below that which was previously needed to produce a lower current. This results in the slope of the curve changing from positive to negative. Practically speaking, the curve starts to bend to the left at some value of applied voltage. This is referred to as the "knee" of the curve. The inception of back corona is assumed to occur at an applied voltage which is just a little greater than that at the "knee". Once back corona is initiated, the collected ash layer breaks down electrically and discharges positive ions into the gas stream. This results in the measurement of a large negative current. The breakdown of the layer sustains itself at reduced voltages so that reduction of the applied voltage still results in increased current. Also, once back corona is initiated, the applied voltage may have to be completely turned off before the breakdown of the layer will cease.

In measuring voltage-current curves where back corona may occur, it is the practice of some investigators to record a curve by going upward in voltage and to record a second curve by going downward in voltage. If the two curves are essentially the same, then back corona does not exist or is not a serious problem. However, if the downward curve is shifted significantly to the left, then extensive back corona exists. This shift to the left of the downward curve is referred to as a "hysteresis effect".

From the curves and operating points shown in Figure 212, it is obvious that this precipitator was operating in back corona. This was also evidenced in very low measured mass collection efficiencies which were inconsistent with the relatively high measured operating current densities. This precipitator probably would have performed better if it had been operated nearer the "knee" of the voltage-current curves for the inlet and outlet electrical fields. Although the measured currents would be significantly reduced, the extreme ill effects of bipolar charging would be avoided.

It should be noted that the particulate space charge effect is also strongly evidenced in these curves. The particulate space charge tends to hide the presence of back corona at the lower voltages in the inlet electrical fields by removing the positive ions which are discharged from the collected ash layer. However, as more and more particles are removed from the gas stream, the evidence of back corona becomes more pronounced as is seen in the outlet voltage-current curve.

Factors Influencing Ash Resistivity[170]

Volume and Surface Conduction--

The electrical resistivity of a collected layer of fly ash varies with temperature in a manner illustrated in Figure 213. Above about 225°C, resistivity decreases with increasing temperature and is independent of flue gas composition. Below about 140°C, resistivity decreases with decreasing temperature and is dependent upon moisture and other constituents of the flue gases.

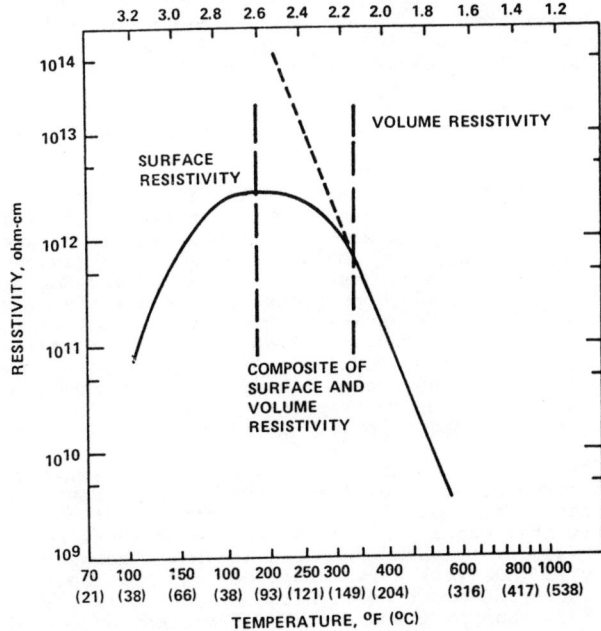

Figure 213. Typical temperature-resistivity relationship for flyash.

In analyzing the conduction process, it is convenient to consider the resistivity as involving two independent conduction paths, one through the bulk of the material (volume conduction) and the other along the surface of the individual particles, associated with an adsorbed surface layer of some gaseous or condensed material (surface conduction). Either of these paths may become the dominant conduction mode under conditions that exist in operating precipitators, or, as is the general case, both mechanisms may be important. The volume conduction is dependent upon the chemical composition of the particulate material, whereas surface conduction is controlled by the chemical compositions of both the particulate and the effluent gas stream.

The coal, ash, and flue gas compositions are very important in determining the mode or modes of conduction in a precipitated fly ash layer and the resistivity of the layer. Tables 11, 12, and 13 give data showing several representative coal, ash, and flue gas chemical analysis for coal-fired boilers followed by precipitators whose voltage-current characteristics were measured. In general, the measured voltage-current characteristics and resistivity can be correlated with the coal, ash, and flue gas compositions. Table 22 shows coal and flue gas compositions obtained from a large number of utilities in the U.S. From the tables containing the coal, ash, and flue gas compositions, it can be seen that a wide range of possible compositions exists. The wide range of possible coal, ash, and flue gas compositions is one of several factors which makes the prediction of fly ash resistivity difficult.

Factors Influencing Volume Resistivity--

Volume conduction in fly ash is an ionic process resulting from the migration of alkali metal ions. Whether the conduction takes place through the particles or along the particle surface has not been definitely established. The important distinction is that volume conduction, or volume resistivity, is governed only by the character and composition of the ash and is independent of gas composition.[170]

It has been shown[161,171] that lithium and sodium are principal ionic charge carriers in experimental environments excluding sulfuric acid vapor. Figure 214 shows the relations between the measured resistivity and the combined atomic concentrations of lithium and sodium for 33 fly ashes.[172] Eastern and Western ashes are indicated by closed and open symbols, respectively. These data were obtained from resistivity versus reciprocal absolute temperature plots for the individual ashes at $1000/T$ (°K) = 2.4 (144°C, 291°F). Prevailing test conditions included a simulated flue gas environment of nitrogen, 5% oxygen, 13% carbon dioxide, 9% water by volume and an electrical stress of 2 kV/cm. The flue gas environment contained no sulfur dioxide or sulfur trioxide.

In the upper, right corner of Figure 214, the expressions defining the curve produced by linear regression analysis are shown. One can either calculate the resistivity for the specific set of experimental conditions prevailing using these equations or read the resistivity value from the figure. The slope of approximately -2 indicates a two order of magnitude decrease in resistivity for a one order of magnitude increase in the atomic percentage of lithium plus sodium. A coefficient of correlation of 0.97 was determined. This coefficient defines the degree of fit between the data and the linear regression curve, and a value of 1.00 would define perfect correlation between the two factors.

TABLE 22. COAL AND FLUE ANALYSES OBTAINED FROM UTILITIES INDUSTRY SURVEY

E/W	Mois-ture	Proximate as Received			X10⁻³ BTU	Ultimate Dry								Flue Gas					Rank	Name
		Ash	Vola-tile	Fixed Carbon	Sul-fur		Car-bon	Hydro-gen	Nitro-gen	Chlor-ine	Sulfur	Ash	Oxy-gen	Water	Oxy-gen	Carbon Dioxide	Sulfur Dioxide	Sulfur Trioxide		
E	4.2	22.3		NA	2.4	11.3	64.8	4.5	1.0	NA	2.8	24.2	6.5	7.0	7.8	11.2	NA	NA	NA	1
E	3.5	13.9	28.6	53.1	1.0	12.7	NA	NA	NA	NA	NA	NA	NA	NA	3.8	15.2	NA	NA	B	2
E	6.4	10.9	31.1	51.6	1.1	12.4	75.1	4.9	1.5	NA	1.2	11.6	5.7	6.0	3.6	15.7	690	0.3	NA	3
E	7.5	8.3	NA	NA	1.3	12.5	NA	NA	NA	NA	NA	NA	NA	7.0	3.3	14.9	NA	NA	B	4
E	6.0	8.8	34.5	50.8	1.0	12.9	70.3	4.6	1.4	-	pyrite 0.38 organ 0.64 sulfate 0.00	9.0	7.9				645	13.2		5
E	4.9	9.5	32.7	52.9	1.0	13.1	73.2	4.8	1.5	-	pyrite 0.28 organ 0.71 sulfate 0.00		5.2				615	11.8		6
W	11.4	13.2	31.4	44.1	0.7	9.9	65.8	4.5	1.4	0.0	0.8	14.9	12.7	10.0	4.9	13.5	575	<0.5	SB	7
W	20.1	NA	NA	NA	0.2	8.1	59.7	3.8	0.8	NA	0.2	19.6	15.7	9.5	4.8	12.9	160	0	SB	8
W	24.2	NA	NA	NA	0.2	7.8	61.1	3.8	0.8	NA	0.3	18.3	15.7	11.3	4.4	11.3	190	0	SB	9
W	24.6	8.2	28.6	38.5	0.8	8.5	66.7	NA	0.9	NA	1.0	10.8	NA	10.7	5.0	12.5	650	2.0	SB	10
W	24.3	14.1	30.7	30.4	0.6	7.9	59.8	4.8	1.2	0.0	0.8	18.8	14.7	10.5	NA	NA	575	1.5	SB	11
W	23.3	10.0	32.4	34.3	0.7	8.7	56.2	4.5	0.9	0.0	0.9	13.1	14.4	NA	NA	NA	NA	NA	SB	12
W	14.1	11.2	32.2	40.9	2.5	10.6	69.4	4.9	1.1	NA	2.9	13.0	8.8	NA	NA	NA	NA	NA	B	13
W	23.6	4.3	33.0	39.1	0.4	9.6	72.5	5.1	0.9	NA	0.5	4.7	16.3	8.8	5.2	14.7	178	2.3	SB	14
W	16.3	8.9	37.0	37.8	3.7	10.6	67.7	5.5	1.4	NA	4.3	9.8	11.3	8.5	5.6	13.4	2343	24.0	B	15
W	21.2	6.0	30.3	44.5	0.5	8.5	65.9	4.6	1.1	NA	0.6	7.5	19.0	8.1	5.6	13.4	283	0.34		16
E	4.6	15.2	26.4	58.3	2.6	12.3	NA	NA	NA	NA	NA	NA	NA	NA	5.5	13.3	935	9.3		17
W	6.3	20.3	37.2	36.3	0.8	10.1	60.8	4.7	1.4	0.01	0.8	21.6	10.6	8.9	4.3	13.2	700	1.0	B	18
W	7.2	15.9	34.4	42.5	3.7	11.2	64.7	4.5	1.3	-	4.0	17.3	8.4	8.0	5.0	14.0	2775	27.0	B	19
W	7.9	16.3	30.8	45.0	0.8	11.0	67.7	4.4	1.3	-	0.9	17.7	8.0	8.0	5.0	14.0	600	6.0	B	20
W	6.9	20.6	30.3	42.1	2.3	10.6	62.2	4.2	1.3	-	2.5	22.1	7.7	8.0	5.0	14.0	1725	17.0	B	21
W	9.5	19.6	32.8	38.0	4.1	10.0	60.2	4.3	1.2	-	4.5	21.7	8.1	8.0	5.0	14.0	3075	31.0	B	22
W	7.7	15.4	31.7	45.2	1.8	11.3	67.4	4.5	1.3	-	2.0	16.7	8.1	8.0	5.0	14.0	1350	14.0	B	23
W	37.6	6.5	28.4	27.5	1.8	6.8	63.5	4.5	1.1	NA	1.1	10.4	19.4	NA	NA	NA	NA	NA	L	24
W	36.0	6.3	26.3	31.4	1.1	4.5	42.0	6.8	0.6	NA	0.7	6.3	43.8	NA	NA	NA	NA	NA	L	25
W	21.1	15.2	34.8	28.9	0.6	7.6	57.5	4.3	0.6	NA	0.6	18.2	17.9	9.5	5.6	13.5	465	<1.0	SB	26
W	18.9	9.7	29.4	42.0	0.6	9.5	68.5	4.5	1.3	0.02	0.7	12.0	13.0	9.1	7.8	11.6	375	3.3	SB	27
W	12.3	8.5	37.0	42.2	0.4	10.8	70.9	5.2	1.1	0.02	0.4	8.9	13.5	NA	NA	NA	NA	NA		28
E	8.5	8.5	38.0	45.0	2.8	11.7	72.9	5.1	1.3	NA	4.0	9.3	7.4	NA	NA	NA	NA	NA	B	29
E	9.4	10.4	NA	NA	1.3	11.6	NA	NA	NA	NA	1.4	11.5	9.0	NA	3.2	15.0	NA	NA	B	30
E	6.6	13.8	NA	NA	2.0	11.8	NA	NA	NA	NA	2.2	15.0	NA	6.5	4.1	13.6	556	NA	B	31
W	7.4	17.7	25.0	49.9	0.7	11.5	NA	NA	NA	NA	NA	NA	NA	NA	NA	NA	NA	NA	B	32
W	9.5	9.0	36.1	45.4	0.4	11.0	69.7	5.0	1.5	NA	0.5	9.9	13.4	NA	NA	NA	NA	NA	B	33
W	29.8	6.8	28.8	34.6	0.3	8.2	67.8	4.8	0.9	NA	0.5	9.7	16.4	8.0	3.5	16.3	475	<1.0	SB	34
E	9.6	13.8	31.2	45.4	0.6	11.2	70.4	4.5	1.4	0.4	0.7	15.3	7.3	10.3	3.7	13.8	380	NA		35

Analysis of Factors Influencing Performance 223

The high coefficient of correlation suggests that it is improbable that the relationship can be improved by examining these data as a function of the concentration of other chemical species appearing in the ash composition. Of course this statement may not be true if one subjectively selects a specific group of ashes from the larger universe of ashes shown in Figure 214.

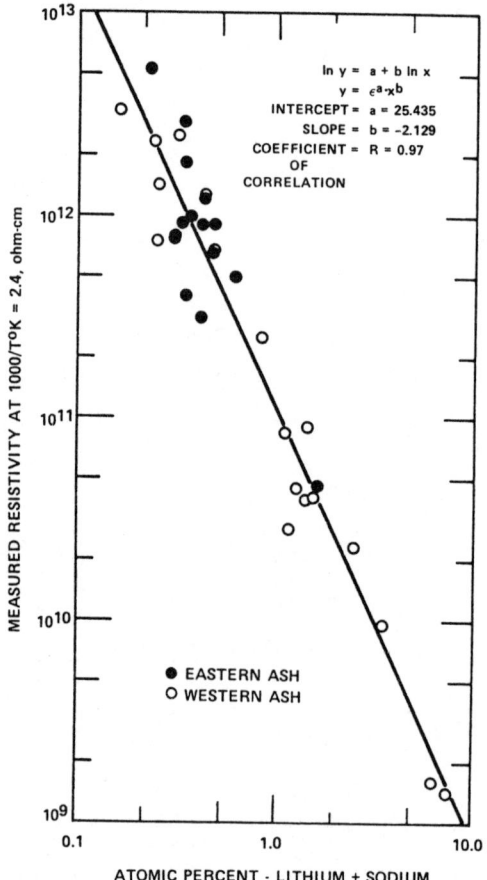

Figure 214. Resistivity as a function of combined lithium and sodium concentrations for a specific set of test conditions.[172]

Volume conduction in all dusts encountered in industrial gas cleaning is temperature dependent. In the case of ionic conduction, increased temperature imparts greater thermal energy to the structure of the material, allowing carrier ions to overcome adjacent energy barriers and to migrate under the influence of an electric field. Thus, for volume conduction, an increase in the temperature produces an increase in the number of carriers available to contribute to the conduction of the particulate layer.

Figure 215 shows the relationship between volume resistivity and temperature for two fly ash samples produced by combustion of coal.[173] The change of resistivity with temperature can be expressed in the form of an Arrhenius equation

$$\rho = \rho_0 \exp(Q/kT), \qquad (24)$$

where ρ is the resistivity, ρ_0 is a material constant. For the fly ash example shown in Figure 215, the material constant ρ_0 is different for fly ash with different sodium ion contents. Graphically, a shift in ρ_0 causes a parallel shift in the temperature-resistivity curve. The experimental activation energy Q is a rate phenomenon and represents the slope of the temperature-resistivity curve. The quantities ρ_0 and Q are useful in defining electrical conduction properties of solid or granular materials as a function of temperature.

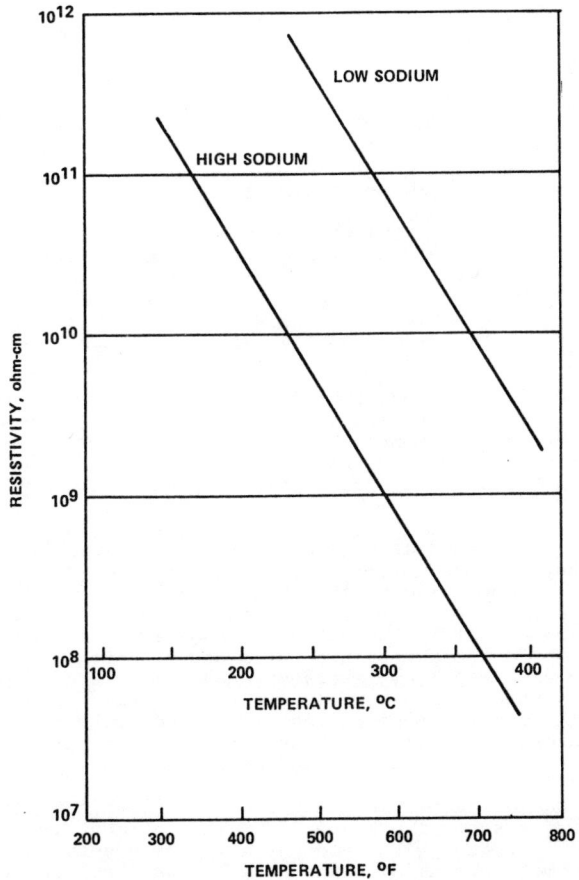

Figure 215. Resistivity vs. temperature for two flyash samples illustrating influence of sodium content.[173]

In some types of ashes, conduction may be electronic instead of ionic. Nevertheless, the Arrhenius equation applies, whether the conduction is electronic or ionic, the temperature-resistivity relationships are similar, differing only in the values of the constants in the Arrhenius equation.

Volume resistivity of a dust sample is also related to its porosity. Intuitively, one would expect a higher resistivity to be associated with a more porous dust layer due to the smaller quantity of material in a given volume.

For fly ash samples, a 25% change in specimen porosity causes a change of one decade in resistivity. A generalized relationship between specimen porosity and resistivity was found for fly ash to be

$$\log \rho_c = \log \rho_m + S(P_c - P_m) , \quad (25)$$

where

ρ_c = resistivity at porosity P_c (ohm-cm),

ρ_m = resistivity at porosity P_m (ohm-cm), and

$S = \Delta \log \rho / \Delta \%P = 0.04$.

Factors Influencing Surface Resistivity--

Surface conduction requires the establishment of an adsorbed layer of some material either to provide an independent conduction path or to interact with some component of the particulate material to provide a surface conduction pathway. If the effluent gas stream contains condensable material (e.g., water or sulfuric acid) and if the temperature is low enough that an adsorbed layer can form, then the surface conduction will become significant.

For temperatures below about 150°C (300°F), surface conduction occurs via the lower resistance path created by the adsorbed moisture or chemical components which occurs at these lower temperatures. Both moisture and chemically reactive substances such as sulfur oxides and ammonia are commonly present in many industrial gases.

Physical adsorption as well as condensation can be involved in surface conduction. At temperatures below the dew point, the rate of deposition on the surface of a dust would be high. However, for most circumstances the adsorbate is deposited on the dust surface and can provide a surface conduction pathway even at temperatures considerably above the dew point, as is shown in Figure 216.[172] The data were obtained from laboratory measurements on a particular fly ash sample under the same simulated conditions discussed earlier except that the water concentration was varied. The range of water concentration used was selected based on water concentration measurements made at several different power stations.

Another way of displaying the effect of water concentration on resistivity is shown in Figure 217. The attenuation of resistivity due to increased water concentration is observable at about 230°C and becomes very significant at the lower temperatures. At the higher temperatures the effect of water concentration on resistivity is not significant since the adsorption mechanisms needed for surface conduction are not present.

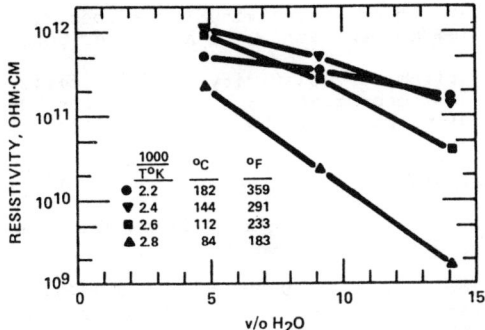

Figure 216. Flyash resistivity as a function of environmental water concentration for various test temperatures.[172]

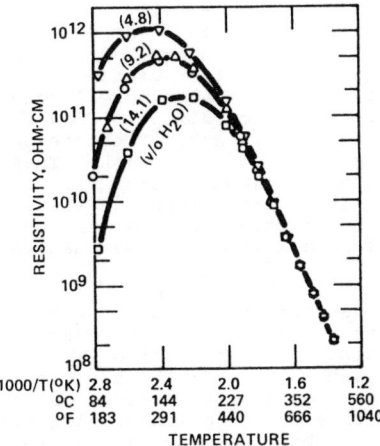

Figure 217. Typical resistivity-temperature data showing the influence of environmental water concentration.[172]

The data shown in Figure 216 are in a suitable form for use in the prediction of ash resistivity. In this interpretation, the resistivity data have been plotted as a function of water concentration for several isotherms. Expressions developed from data such as these can be used to correct the resistivity value predicted for a given set of baseline conditions to a value for some other set of conditions. For example, the average slope of the resistivity-water concentration curve at a temperature of $1000/T$ (°K) = 2.4 was -0.085. This is based on the data accumulated from 16 selected ashes used to evaluate the effect of water concentrations. A simple algebraic expression can be used to convert the resistivity value for 9% water shown in Figure 216 to the value for some other water concentration.[172]

$$\log \rho_{c,w} = \log \rho_c + (W_w - W_c)S_w \tag{26}$$

$\log \rho_{c,w}$: logarithm of resistivity for a specific lithium plus sodium concentration, c, and water concentration, W_w.

$\log \rho_c$: logarithm of resistivity for a specific lithium plus sodium concentration, c, and a water concentration of 9 volume percent. Value obtained from Figure 214.

W_w: volume percent water concentration to which the resistivity is to be corrected.

W_c: water concentration used in establishing Figure 214, 9 volume percent.

S_w: $\Delta \log \rho / \Delta\% \ H_2O$; -0.085 for $1000/T(°K) = 2.4$ and water concentrations between 5% and 15%.

In surface conduction, the mechanism of charge transport appears to be ionic; however, the migration species have not been identified. They could be ions extracted from or carried on the dust surface or those deposited from the gas stream.

An example of how surface resistivity of fly ash depends on the composition of the flue gas is the case of fly ash from coal-fired boilers burning sulfur containing coals. The burning of coal containing sulfur produces sulfur dioxide (SO_2) in quantities dependent on the sulfur content. Under normal conditions, about 0.5 to 1% of the SO_2 present is oxidized to SO_3, which serves to reduce the resistivity of the fly ash, if the temperature is low enough for the SO_3 to be adsorbed on the ash. Thus, high-sulfur coals tend to produce ash with lower resistivities than coals with lower sulfur contents. In general, lowering the flue gas temperature increases the SO_3 absorption, so that the resistivity of the fly ash can be controlled to some extent by changes in flue gas temperature.

The effect on ash resistivity of incorporating sulfur trioxide in an environment of water and air has been examined using a limited number of ashes and tests.[172] Figure 218 shows the results for six tests conducted on one ash to demonstrate the combined effect of sulfur trioxide concentration and temperature on resistivity. The circles represent data obtained in a linear flow electrode set while all other data were obtained in a radial flow electrode set.

Data obtained at 147-149°C using 2 kV/cm voltage gradient and a baseline environment of air containing 9 volume percent water are shown in Figure 219.[172] Eight ashes were used in conjunction with sulfur trioxide concentrations of nominally 2, 5, and 10 ppm. Since this data base is so small, it is not possible to quantify the effect of sulfur trioxide on ash resistivity. However, it is obvious that the effect can be dramatic in that the presence of 10 ppm of sulfuric acid can reduce the resistivity two or more orders of magnitude.

The influence of electric field on conduction in insulating materials has been well documented. In solid materials, increasing electric field permits a greater number of migrating ions to participate in the conduction process. In granular materials additional influences of electric field may become important. Possible effects are: an increase in temperature at the contact points between particles caused by joule heating, and an electric discharge in the dust layer due to the enhanced field near adjacent particles.

Analysis of Factors Influencing Performance   229

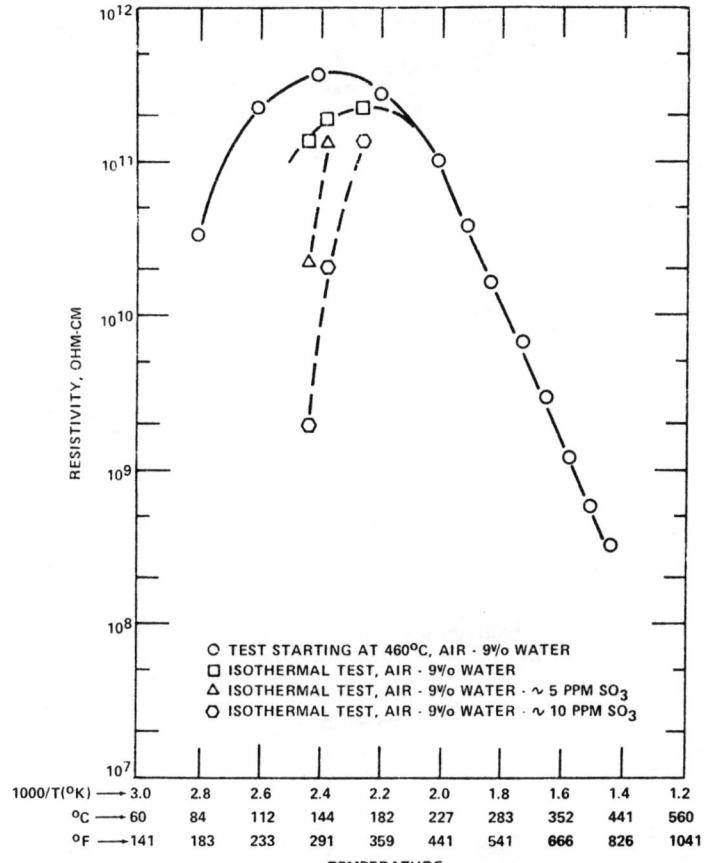

Figure 218. Effect of SO_3 on resistivity.[172]

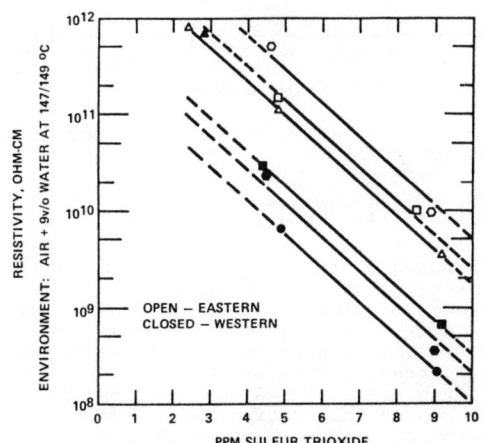

Figure 219. Resistivity as a function of environmental sulfur trioxide concentration for eight flyashes.[172]

Figure 220 shows relationships between in situ resistivity and electric field for fly ash (from coals with low and moderate sulfur contents).[174] The data were obtained using a point-to-plane probe and the parallel disc measurement method. The measurements were made at a temperature of 265°C (330°F) with a dust layer thickness of 1.0 mm.

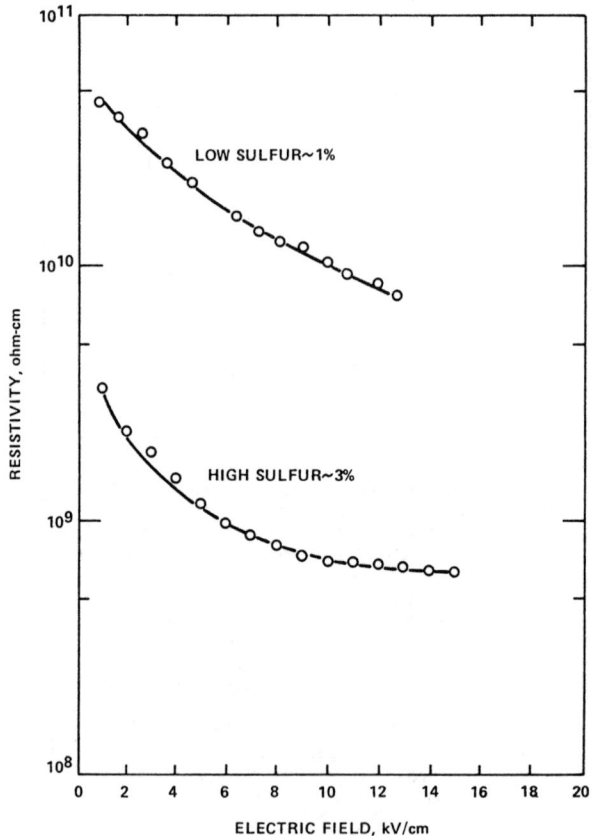

Figure 220. Variation in particulate in situ resistivity with electric field.

Laboratory investigations also show an effect of electric field on resistivity.[172,175] Figure 221 shows the effect of the electric field applied across the ash layer on the resistivity.[172] The upper curve illustrates the almost negligible effect experienced by a few fly ashes and the lower curve shows the average effect from the examination of 16 ashes. It should also be pointed out that the laboratory values of resistivity would increase at a significantly faster rate as a function of electric field as the electric field is decreased from 2 kV/cm to smaller values.[176] This is also indicated for the field data shown in Figure 220. The ASME PTC-28 code suggests that resistivity be determined just prior to dielectric breakdown. However, a research program involving many ashes and a multiplicity of test conditions can not afford to do this. Therefore, tests were conducted on a few ashes to establish a relationship between resistivity and electric field, and other data can be calculated from this relationship using an expression similar to equation (26).[172]

$$\log \rho_{c,w,e} = \log \rho_{c,w} + (E_e - E_c) S_e \quad (27)$$

$\log \rho_{c,w,e}$: logarithm of resistivity for a specific lithium plus sodium concentration, c, a water concentration W_w, and an applied voltage gradient E_e.

$\log \rho_{c,w}$: previously defined.

E_e: applied electric field to which $\log \rho_{c,w}$ is to be corrected.

E_c: applied electric field used in establishing Figure 221, 2 kV/cm.

S_e: $\Delta \log \rho / \Delta E$; -0.030 for $1000/T(°K) = 2.4$ and an applied voltage gradient range of 2 to 10 kV/cm.

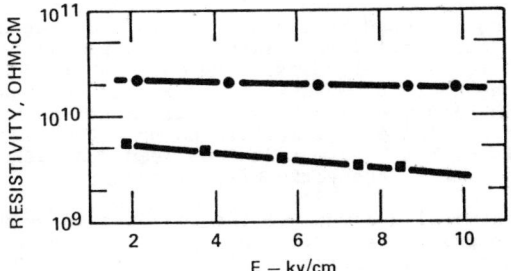

Figure 221. Typical resistivity values as a function of applied ash layer electric field.[172]

Combined Effects of Volume and Surface Conduction--

The initial evidence suggests that the presence of sulfuric acid in the environment provides an alternate conduction mechanism. Therefore, other than the effect of various ashes having different affinities for sulfuric acid vapor, there would seem to be no relationship between the acid and the ash composition with respect to conduction. It has been suggested that the effect of sulfuric acid can be combined with the other factors that influence resistivity by considering them as two independent conduction mechanisms and determining a resultant resistivity from the equation for parallel resistances.[172]

$$\rho_r = \frac{\rho_s \times \rho_{c,w,e}}{\rho_s + \rho_{c,w,e}} \quad (28)$$

ρ_r: resultant resistivity combining the effects of composition, water concentration, applied electric field, and sulfuric acid concentration.

ρ_s: resistivity resulting from the effect of environmental sulfuric acid concentration taken from Figure 219.

$\rho_{c,w,e}$: previously defined, equation (27).

Prediction Of Fly Ash Resistivity

Although little practical information exists concerning the prediction of fly ash resistivity, a recently proposed method of prediction appears very promising.[172] This method and a computer program used to perform the necessary calculations are described elsewhere.[138] Comparisons of the predictions of this method with in situ and laboratory measurements of resistivity on a limited number of fly ash samples have shown good agreement. A key feature of this method is that fly ash resistivity is predicted based on a coal sample. This method will be discussed briefly in the following paragraphs.

The information required to utilize the proposed technique for predicting resistivity is the as-received, ultimate coal analysis and the chemical composition of the coal ash. A stoichiometric calculation of the combustion products is made using 30% excess air to determine the concentration of sulfur dioxide and water. The quantity of excess air used in the calculation was established by comparing stoichiometrically calculated flue gas analyses with in situ analyses for several coals. The coal ash is prepared by first igniting the coal at 750°C in air, passing the ash through a 100-mesh screen, and then igniting the ash a second time at 1050°C ± 10°C in air for a period of 16 hours. Good agreement in chemical analyses has been obtained between coal ashes produced in this manner and their respective fly ashes.

The usual chemical analysis of the coal ash in weight percent expressed as oxides is performed. The analysis is converted from weight percent to molecular percent as oxides. The atomic percentage of the lithium and sodium is taken as 66.6% of the molecular percentage of the oxides. The sum of the atomic percentages of lithium and sodium is used to determine the resistivity value, ρ_c, from data similar to that shown in Figure 214 for various temperatures.

Using the concentration of water determined from the combustion products calculation and equation (26), the predicted resistivity in terms of ash composition and water concentration, $\rho_{c,w}$, is determined.

For the ash thickness used in the research to develop the predictive method, ~5 mm, it was found that dielectric breakdown generally occurred at applied electric fields of 8 to 12 kV/cm. Therefore, it was arbitrarily decided to use 10 kV/cm as the electric field at which the resistivity is predicted. Using equation (27) and E_e = 10 kV/cm, the predicted resistivity is put in terms of ash composition, water concentration and dielectric breakdown field, $\rho_{c,w,e}$. This value then is the predicted resistivity exclusive of the effect of sulfuric acid.

Using the information from a variety of field test programs for which flue gas data were available, it was observed that the average sulfur trioxide value was approximately 0.4% of the sulfur dioxide value at the inlet to cold-side precipitators. This factor is used to calculate the anticipated level of sulfur trioxide based on the amount of sulfur dioxide appearing in the stoichiometrically calculated flue gas. For example, a typical eastern coal can produce a flue gas containing 2000 ppm of sulfur dioxide which it is anticipated would yield 8 ppm of sulfur trioxide. Referring to Figure 219, a reasonable estimate of the resistivity, ρ_s, resulting from this sulfur trioxide concentration might be 2×10^{10} ohm-cm. One then determines the resultant resistivity, ρ_r, from equation (28) and the values for ρ_s and $\rho_{c,w,e}$.

Example of the Calculations Used to Predict Fly Ash Resistivity at 144°C (291°F)--

Step 1: Obtain an as-received, ultimate coal analysis and a coal ash analysis. Table 23 shows an as-received, ultimate coal analysis and coal ash analysis obtained for a particular coal. This information will be utilized in predicting the resistivity of the fly ash.

TABLE 23. AS-RECEIVED, ULTIMATE COAL ANALYSIS AND COAL ASH ANALYSIS USED IN PREDICTION OF FLY ASH RESISTIVITY

Coal Constituent	As-Received Ultimate Analysis lb/100 lb	Coal Ash Constituents	Coal Ash Composition (1050°C)
Carbon	57.21	Li_2O	0.04
Hydrogen	3.74	Na_2O	0.45
Oxygen	3.03	K_2O	3.7
Nitrogen	1.02	MgO	1.4
Sulfur	0.79	CaO	0.7
Water	8.41	Fe_2O_3	6.7
Ash	25.80	Al_2O_3	27.6
		SiO_2	58.2
		TiO_2	1.7
		P_2O_5	0.1
		SO_3	0.2

Step 2: Make a stoichiometric calculation of the combustion products using 30% excess air to determine the concentration of sulfur dioxide and water. Table 24 shows the steps in the calculation of stoichiometric flue gas from the coal analysis.[177]

Step 3: Determine atomic percentage of the lithium and the sodium from the chemical analysis of the coal ash. First, convert from weight percent to molecular percent as oxides as shown in Table 25. Second, calculate the atomic percentage of lithium and sodium as follows:

Atomic % of Li + Na = (.666)(0.096) + (.666)(0.518) = 0.409

Step 4: Determine the resistivity value due to ash composition at 144°C from Figure 214. The value of ρ_c is approximately 1×10^{12} ohm-cm.

Step 5: Include the effect of the calculated water concentration of 8.228% by volume by using equation (26).

$$\log \rho_{c,w} = \log (10^{12}) + (8.228 - 9.0)(-0.085)$$

$$= 12 + .0656 = 12.0656$$

$$\rho_{c,w} = 1.163 \times 10^{12} \text{ ohm-cm}$$

TABLE 24. CALCULATION OF STOICHIOMETRIC FLUE GAS FROM COAL ANALYSIS[177]

A. Calculation of combustion products, air, and O_2 for 100% combustion.

Coal con-stituent	As received, ultimate analysis lb/100* lb		Molecular weight		Moles per 100* lb fuel		Multipliers		Required for combustion moles/100* lb fuel at 100% total air	
									O_2	Dry air
C	57.21	÷	12.01*	=	4.764	x	1.00* and x	4.76*	4.764	22.677
H_2	3.74	÷	2.02*	=	1.852	x	0.50* and x	2.38*	0.926	4.408
O_2	3.03	÷	32.00*	=	0.095	x	−1.00* and x	−4.76*	−0.095	−0.452
N_2	1.02	÷	28.01*	=	0.036					
S	0.79	÷	32.06*	=	0.025	x	1.00* and x	4.76*	0.025	0.119
H_2O	8.41	÷	18.02*	=	0.467					
Ash	25.80		−		−					
Sum	100.00				7.239				5.620	26.752

B. Calculation of air and O_2 for 30%* excess air.

	Required for combustion moles/100* lb fuel at 30%* excess air	
	O_2	Dry air
O_2 and air x 130*/100 total	7.306	34.778
Excess air = 34.778 − 26.752	−	8.026
Excess O_2 = 7.306 − 5.620	1.686	··

C. Flue gas analysis calculation.

Flue gas constituent	Combustion / Fuel / Air			Products of combustion		
				Total moles/100* lb fuel	% by volume wet basis	% by volume dry basis
SO_2	4.764		=	4.764	12.864	14.018
H_2O	1.852 + 0.467 +	0.728[a]	=	3.047	8.228	−
SO_2	0.025		=	0.025	0.068	0.074
N_2	0.036 +	27.475[b]		27.511	74.288	80.948
O_2		1.686		1.686	4.553	4.961
Sum wet				37.033		
Sum dry = 37.033 − 3.047				33.986		

a. Moles H_2O in air = (34.778 x 29* x 0.013*) ÷ 18* = 0.728.
b. Moles N_2 in air = (34.778 x 0.79*) = 27.475.
* Constants in calculations. All other numbers depend on the particular calculation.

TABLE 25. CONVERSION OF WEIGHT PERCENT ANALYSES OF COAL ASH TO MOLECULAR PERCENT AS OXIDES

A. Calculate molecular weights of coal ash constituents.

Coal ash constituent	Calculation	Molecular weight
Li_2O	$(2)(6.939) + 15.9994$	= 29.8774
Na_2O	$(2)(22.9898) + 15.9994$	= 61.9790
K_2O	$(2)(39.1020) + 15.9994$	= 94.2034
MgO	$24.3120 + 15.9994$	= 40.3114
CaO	$40.0800 + 15.9994$	= 56.0794
Fe_2O_3	$(2)(55.8470) + (3)(15.9994)$	= 159.6922
Al_2O_3	$(2)(26.9815) + (3)(15.9994)$	= 101.9612
SiO_2	$28.0860 + (2)(15.9994)$	= 60.0848
TiO_2	$47.9000 + (2)(15.9994)$	= 79.8988
P_2O_5	$(2)(30.9738) + (5)(15.9994)$	= 141.9446
SO_3	$32.0640 + (3)(15.9994)$	= 80.0622

B. Calculate total number of moles per 100 grams of coal ash.

$$\text{Total \# of moles} = \frac{0.04}{29.8774} + \frac{0.45}{61.9790} + \frac{3.7}{94.2034} + \frac{1.4}{40.3114} + \frac{0.7}{56.0794} + \frac{6.7}{159.6922}$$

$$+ \frac{27.6}{101.9612} + \frac{58.2}{60.0848} + \frac{1.7}{79.8488} + \frac{0.1}{141.9446} + \frac{0.2}{80.0622}$$

$$= 0.000134 + 0.00726 + 0.03928 + 0.03473 + 0.01248 + 0.04196$$

$$+ 0.27069 + 0.96863 + 0.02128 + 0.00071 + 0.00250$$

$$= 1.40086$$

C. Calculate molecular percentages of coal ash constituents.

Coal ash constituent	Calculation	Molecular percentage
Li_2O	$\left(\frac{0.00134}{1.40086}\right) \times 100 =$	0.096
Na_2O	$\left(\frac{0.00726}{1.40086}\right) \times 100 =$	0.518
K_2O	$\left(\frac{0.03928}{1.40086}\right) \times 100 =$	2.804
MgO	$\left(\frac{0.03473}{1.40086}\right) \times 100 =$	2.479
CaO	$\left(\frac{0.01248}{1.40086}\right) \times 100 =$	0.891
Fe_2O_3	$\left(\frac{0.04196}{1.40086}\right) \times 100 =$	2.995
Al_2O_3	$\left(\frac{0.27069}{1.40086}\right) \times 100 =$	19.323
SiO_2	$\left(\frac{0.96863}{1.40086}\right) \times 100 =$	69.145
TiO_2	$\left(\frac{0.02128}{1.40086}\right) \times 100 =$	1.519
P_2O_5	$\left(\frac{0.00071}{1.40086}\right) \times 100 =$	0.051
SO_3	$\left(\frac{0.00250}{1.40086}\right) \times 100 =$	0.179

Step 6: Include the effect of the electric field across the ash layer by using equation (27). For purposes of illustration, assume dielectric breakdown occurs at 8 kV/cm.

$$\log \rho_{c,w,e} = \log(1.16 \times 10^{12}) + (8-2)(-0.03)$$

$$= 12.0656 - 0.18 = 11.8856$$

$$\rho_{c,w,e} = 7.684 \times 10^{11} \text{ ohm-cm}$$

Step 7: Include the effect of sulfur trioxide in order to obtain the final value of resistivity. The sulfur trioxide concentration is obtained by taking 0.4% of the sulfur dioxide concentration. In this example, the sulfur trioxide concentration is 0.004 x 680 ppm = 2.72 ppm. From Figure 219, a reasonable estimate of the resistivity resulting from this concentration of sulfur trioxide might be 1×10^{12} ohm-cm (low sulfur, eastern coal). The resulting resistivity due to all pertinent parameters is obtained using equation (28).

$$\rho_r = \frac{(1 \times 10^{12})(7.684 \times 10^{11})}{1 \times 10^{12} + 7.684 \times 10^{11}} = \frac{7.684 \times 10^{23}}{1.7684 \times 10^{12}} = 4.35 \times 10^{11} \text{ ohm-cm}$$

The most conservative estimate of ρ_r would have been obtained by taking $\rho_s \approx 1 \times 10^{13}$ ohm-cm as indicated by one of the data sets in Figure 219. In this case, $\rho_r = 7.14 \times 10^{11}$ ohm-cm.

Measurement Of Ash Resistivity

Factors Influencing Measurement of Resistivity--

Resistivity of a dust layer is determined experimentally by collecting a sample of the dust from a gas stream and measuring the current and voltage characteristics of a defined geometrical configuration of the dust. The method of forming the dust layer, and the conditions of measurement all influence the resistivity measurement.

Particle size distribution and porosity--For determination of the true particle size distribution, the sample should be taken from the gas stream in a manner (e.g., isokinetically) that insures that the sample is representative of the particle size distribution of the fly ash in the gas stream. However, due to problems of probe design, most of the resistivity probes either do not sample isokinetically or do not collect all the particles sampled.

Even if isokinetic sampling were used, the particle size distribution of the ash layer deposited in each field of a precipitator differs due to the variation in collection efficiency as a function of particle size. Consequently, in determining resistivity to correspond to that of each field of a precipitator, the particle size distribution associated with each field would have to be simulated. In general, such a procedure would be impractical, and some means of obtaining a reasonably representative sample is employed.

It has been shown that the resistivity of a fly ash layer depends on the particle size distribution in the layer.[171,176] Also, the effects of particle size distribution and porosity can not be considered independently since the particle size distribution will influence the porosity. Thus, depending on the conduction mode, effects on resistivity of different particle size distributions may be attributed to either particle size distribution or porosity. Laboratory measurements of resistivity as a function of temperature have been made on two fly ash samples under identical conditions except that the two samples differed in particle size distribution and porosity.[176] One sample had a MMD of 40 μm with a porosity of 54% while the other sample had a MMD of 2.7 μm with a porosity of 75%. These samples were obtained from a larger size fractionated sample by using the size fractions < 3 μm and > 25 μm. The resistivity versus temperature curves for these two samples crossed one another. The lower MMD sample had lower values of resistivity at the lower temperatures whereas the higher MMD sample had lower values of resistivity at the higher temperatures. These results are attributed to the greater specific surface area available in the lower MMD sample at temperatures where surface conduction is important and to the lower porosity of the higher MMD sample at temperatures where volume conduction is important.

Electric field--Since the resistivity of an ash varies with electric field, it is important that measurements be made at an electric field corresponding to that in the precipitator and/or that the value of the field at which the measurement is made be specified. In some resistivity measurement techniques the voltage is increased until the ash layer breaks down, and the resistivity reported is that corresponding to the condition just prior to breakdown. Other techniques impose a fixed voltage across a pair of electrodes to establish an electric field. Generally the magnitude of the field is very low, of the order of 1 kV/cm for this latter type of technique. The reported values of resistivity would be different depending upon whether the measurement was made at a low field or near breakdown.

Method of depositing ash layer--In an electrostatic precipitator, the ash layer is deposited electrostatically and the particles are aligned somewhat as the dust layer is built up. In some sampling probes the ash layer is deposited electrostatically, whereas in other probes the dust is collected by other means and allowed to fall into the measurement cell. In laboratory measurements the ash layer is mechanically deposited in the measurement cell.

The significance of the method of deposition has not been quantitatively determined. However, to the eye, dust layers deposited electrostatically appear denser than those established by free fall of the dust. In probes in which the dust is allowed to fall into the measurement cell, some attempt is made to vibrate the cell or otherwise establish a reproducible density of the deposited dust. In other probes, the measurement technique involves a disc placed on the dust surface. This disc provides some compaction of the dust layer.

The method of deposition of the ash layer may influence the porosity of the layer. Laboratory experiments have shown that the porosity of the layer will have a significant influence on the measured value of resistivity.[176] Laboratory measurements of resistivity as a function of temperature have been made on two fly ash samples under identical conditions except that the two samples differed in porosity. One sample had a porosity of 70% while the other sample had a porosity of 50%. The resistivity versus temperature curve for the higher porosity sample was above that of the lower porosity sample for all values of temperature.

This difference in porosity led to as much as a factor of 5 difference in the measured value of resistivity.

Thickness of ash layer--Limited laboratory experiments have been performed to examine the effect of the layer thickness on the resistivity measurement.[176] Laboratory measurements of dielectric breakdown strength have been made on three fly ash samples under identical conditions except that the three samples had three different thicknesses between approximately 3mm and 7mm. For all three samples, the applied voltage necessary to cause dielectric breakdown was essentially the same. Thus, the samples experienced dielectric breakdown at different values of average electric field. This suggests that the surface charge near one of the boundaries may be the important factor in determining dielectric breakdown of a fly ash layer and that the average electric field is not of significance.

Time of current flow--When voltage is applied across an ash layer, the magnitude of the current will initially be high and will then fall off, rapidly at first and slowly thereafter. The initial current surge is due to absorption current, which charges the capacitance associated with the ash layer. The subsequent decrease in current is due to depletion of the charge carriers or polarization at the ash-electrode interfaces. If the current is allowed to flow for a considerable time prior to making resistivity measurements, the value of current will be lower than that obtained immediately following application of a voltage.

Source variability--Another factor influencing resistivity measurement is source variability. In spite of attempts to obtain a uniform boiler fuel by blending the coal supply, the chemical composition of the coal will vary enough to be reflected in observable changes in the SO_2 level of the flue gases and in the chemical composition of the fly ash. Thus, to minimize errors due to source variability, resistivity measurements should be made on samples taken over a sufficiently long period of time, and the results should be averaged to obtain a representative value.

Methods For Measuring Ash Resistivity--

General considerations--The determination of the electrical resistivity of a fly ash layer is made indirectly. The resistivity is computed from the resistance of a sample of the fly ash with a known geometrical configuration. Typically, the geometry of the sample will be either a rectangular or cyclindrical solid, or the volume of space between concentric cyclindrical electrodes. In each instance, the relationship between the resistivity and resistance of what is considered to be a homogeneous material is given by

$$\rho = RA/\ell, \qquad (29)$$

where

ρ = resistivity (ohm-cm),

R = resistance (ohm),

A = cross sectional area (cm^2), and

ℓ = length (cm).

In each measurement device, the amount of material actually utilized for the measurement is on the order of one cubic centimeter or less. Layer thickness from one-half to five millimeters is common. Using this minute sample of material selected from the large quantities of fly ash generated during a measurement period raises serious questions as to just how representative of the total fly ash material this sample can be. This factor may, in part, explain the wide range of scatter actually observed in a resistivity measurement program.

Several techniques can be used for measuring the resistivity, and several types of equipment are available for this purpose, with no general agreement as to their relative merits. The choice of technique and equipment can be influenced by the intended use of the measured resistivity data.

One consideration is whether an absolute resistivity is to be made for scientific or engineering design purposes or whether a relative or rank ordering type of measurement is sufficient. If one is attempting to relate the behavior of an electrostatic precipitator to theoretically derived relationships, then it is important to attempt to evaluate the absolute resistivity of the dust. However, if one has accumulated a considerable quantity of resistivity data over a period of time with one type of device and in addition has similarly accumulated experience as to how a particular type of electrostatic precipitator behaves with the related particulate resistivity data, then the measured value of resistivity can be related to precipitator performance.

As discussed earlier, the measured value of resistivity is dependent upon a number of factors. If the measurements are contemplated for rank ordering or relative behavior, then wide latitude is allowed in the selection of a method. For the relative measurement type of investigation, it becomes important to merely assure that the measurement conditions are reasonably well duplicated for each condition, and the selection of method becomes of secondary importance. Either in situ or laboratory methods may be applicable to a study of this nature if the sample collection conditions, including temperature, are identical. However, if the purpose of the study is to evaluate how an electrostatic precipitator will behave with a new or significantly different type of dust under a given set of conditions, in situ measurements will probably be necessary.

For comparative evaluations, in situ and laboratory measurements must be made with the same instrumentation and technique. Extreme care must be exercised in attempting to compare resistivity data obtained with one device or technique with data obtained with another device or technique. This will become evident in the following discussions of the different measurement techniques.

Laboratory versus in situ measurements--The determination of whether the particulate resistivity should be measured in the laboratory or in situ is based on an evaluation of the significance of the surface conduction component. If the surface conduction is negligible because of high temperature (>200°C) or because of the absence of any reactive or condensable material (H_2O, SO_3, etc.) in the effluent gas stream, then laboratory measurements are appropriate.

However, if reactive constituents are present and if the temperature is in the vicinity of the dew point of the condensables, such that there is a reasonable probability that an adsorbed surface layer will exist, then it is important that both laboratory and in situ resistivity measurements be made for comparison.

It is also important to make measurements in the effluent gas stream in addition to the laboratory even though the chemical composition of the gas stream can be duplicated in the laboratory. The reason for this distinction is that as the particulate sample is collected, cooled and transported to the laboratory, there is a reasonable probability for chemical reactions to occur that would modify the particulate matter prior to measurement.

Laboratory measurements--Standard technique---The standard technique for conducting laboratory resistivity measurements is described in the American Society of Mechanical Engineers Power Test Code 28, Determining the Properties of Fine Particulate Matter.[178] This code was adopted by the Society in 1965 as a standard practice for the determination of all the properties of fine particulate matter which are involved in the design and evaluation of dust-separating apparatus. The tests include such properties as terminal settling velocity distribution, particle size, bulk electrical resistivity, water-soluble sulfate content, bulk density, and specific surface.

The document defines bulk electrical resistivity as the resistance to current flow, expressed in ohm-centimeters, through a dust sample contained in a cubic volume one centimeter on a side when exposed to an electrical voltage equivalent to 90% of the breakdown voltage of the sample, applied uniformly across two opposite faces of the cube. The code specifies that the property is to be determined at 150°C (300°F) and at a humidity of 5% by volume, unless otherwise specified.

Apparatus for standard technique---The basic conductivity cell is shown in Figure 222.[179] It consists of a cup which contains the ash sample and which also serves as an electrode, and an upper electrode with a guard ring. To conform with the code, the high-voltage conductivity cell must have the same dimensions as shown, and must use electrodes constructed from 25-micron porosity sintered stainless steel. The movable disk electrode is weighted so that the pressure on the dust layer due to gravitational force is 10 grams per square centimeter. The nominal thickness of the dust layer is 5 millimeters. The actual thickness is to be determined with the movable electrode resting on the surface of the dust. All electrode surfaces in the region of the dust layer are to be well rounded to eliminate high electric field stresses.

The controlled environmental conditions required for the standard measurement of resistivity in the laboratory can be achieved by an electric oven with thermostatic temperature control and with good thermal insulation to maintain uniform internal temperature, and a means to control humidity. Humidity may be controlled by any one of several conventional means, including circulation of preconditioned gas through the oven, injection of a controlled amount of steam, use of a temperature-controlled circulating water bath, or the use of chemical solutions which control water vapor pressure. It is desirable to circulate the humidified gas directly through the dust layer; hence the reason for the porous electrodes. Figure 223 illustrates a suitable set-up for standard resistivity measurements.[180]

Analysis of Factors Influencing Performance   241

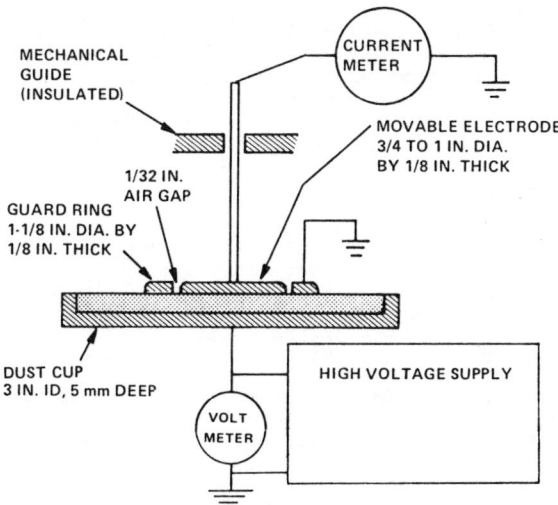

Figure 222. Bulk electrical resistivity apparatus, general arrangement.[179]

1. Pressure regulator
2. Constant temperature bath
3. Pump
4. Heater
5. Make-up water reservoir
6. Externally heated piping
7. PTC 28 apparatus
8. Environmental sampling port
9. Externally heated exit piping
10. Calibrated C/A thermocouple
11. Power source for oven
12. mV Potentiometer
13. Cold junction
14. Oven
15. Fritted disc
16. Environmental chamber
17. Fritted disc air bubbler
18. Bath water in overflow
19. Air flowmeter
20. Air tank

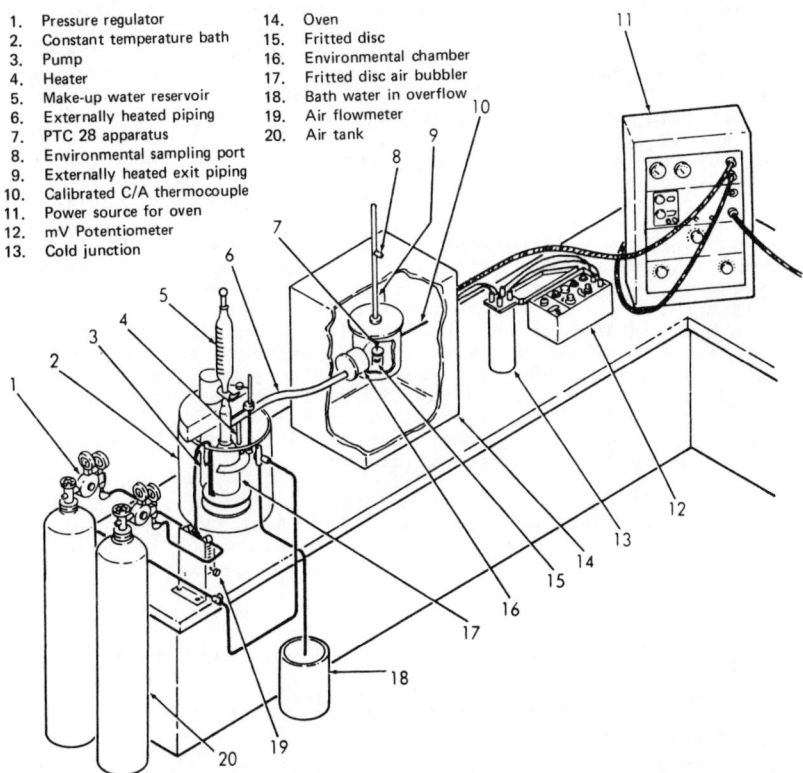

Figure 223. Schematic of apparatus setup for standard resistivity measurements.[180]

Experimental procedure for standard technique---The first problem encountered in making any resistivity measurement is obtaining an appropriate dust sample. The prescribed procedure for PTC-28 Code assumes that samples of gas-borne dust are taken from a duct in accordance with the Test Code for Determining Dust Concentration in a Gas Stream (PTC 27-1957). The PTC-27 Code involves isokinetic dust sampling at various points in the duct. It is recommended that samples should not be obtained from a large bulk of material in a hopper, silo, or similar location. If it is necessary that samples be obtained from such a location, procedures which will insure that the sample is representative of the whole must be used. For any resistivity test to be performed on a bulk sample, it is necessary that a random sample be obtained. This can be done by quartering the bulk sample to obtain the test sample. To break up agglomerates and to remove foreign matter, e.g., collection plate scale, the specimen can be passed through an 80-mesh screen.

The procedure for making the resistivity measurement according to Power Test Code 28 is as follows: (1) The sample is placed in the cup of the conductivity cell by means of a spatula. Then it is leveled by drawing a straight edge blade horizonally, across the top of the cup. (2) The disc electrode is gently lowered onto the surface. It should rest freely on the sample surface without binding on any supports. (3) The conductivity cell is mounted in the environmental chamber and equilibrium temperature and humidity are established. The Code specifies that a temperature of 150°C (300°F) and a humidity of 5% by volume are to be used for the test, unless otherwise specified. (4) A low voltage is applied to the cell and then gradually raised in a series of steps up to the point of electrical breakdown of the sample layer. Current transients will occur when the voltage is first applied or increased across the cell. A record of the current-voltage characteristic of the dust is obtained. Preferably using another sample, the above is repeated; when another sample is not available, the sample layer should be remixed and releveled after each run in order to break up any spark channels that may have been formed in the dust layer. A total of three runs should be made. The average breakdown voltage is then calculated. Before taking the samples to breakdown, it is necessary to determine whether the temperature and moisture content of the sample are in equilibrium with temperature and humidity of the controlled environment. A test for equilibrium is that the voltage-current measurements are reproducible to within 10% when determined by two successive measurements made 15 minutes apart. (5) The resistivity of the samples is then calculated in the range of 85 to 95% of the average breakdown voltage, using the corresponding currents from the previously recorded voltage-current characteristics.

Resistivity can be calculated in the following way. First, calculate the resistance of the dust layer R at the specified voltage.

$$R(ohms) = \frac{V(volts)}{I(amps)}$$

Then calculate the resistivity ρ at the specified voltage.

$$\rho(ohm\text{-}cm) = R(ohms) \frac{A(cm^2)}{l(cm)}$$

The moisture content of the air in the environmental chamber can

be determined by weighing a tube filled with calcium sulfate
(Drierite) before and after passage of a measured volume of air
through it. The volume of dry air passed through the tube is determined from the flow rate and the sampling time.

Variations for the standard technique used in laboratory
studies---Laboratory investigations using the PTC-28 or a similar
apparatus to study characteristics of ash resistivity usually
involve somewhat different procedures than that specified in the
standard technique. Usually it is not necessary or desirable to
determine the breakdown voltage of the ash layer. Hence, a fixed
potential prior to breakdown is applied across the cell, and then
the parameters under investigation are varied. Other laboratory
techniques may be desirable to determine certain electrical characteristics of the ash, for example, the method being used in
research on the resistivity of fly ash at elevated temperature.
The technique utilizes a self-supporting sintered disc of fly ash,
rather than loose powder. This technique is commonly used in the
electrical evaluation of ceramic insulators. It was selected for
the study of volume resistivity because it allows certain posttest analytical work to be done. The details of specimen preparation and measurement technique are given elsewhere.[162]

Another necessary refinement to the standard laboratory technique is based on the need to more nearly duplicate the gaseous
environment to which the ash is exposed. This refinement is
needed due to the strong influence on ash resistivity of the
various possible concentration levels of water and sulfur trioxide.

The different laboratories which make resistivity investigations of fly ash have developed their own measurement procedures
and techniques. Table 26 gives a comparison of the test procedures
utilized by several laboratories. As would be expected due to
their independent development, the procedures developed by the
different laboratories differ from one another to some extent.
The differences in the procedures are important because they may
influence the measured value of resistivity and because they make
it difficult to compare resistivity data from the different laboratories.

Other factors which influence the resistivity measurement
that are not addressed in Table 26 are the porosity of the ash
layer and the effect of sulfur trioxide on the measured value of
resistivity and the measurement technique. The value of porosity
at which the resistivity is determined is known to differ by as
much as a factor 1.5 between certain laboratories. In certain
systems which simulate the sulfur dioxide concentration, some of
the sulfur dioxide may oxidize to sulfur trioxide. The effects of
sulfur dioxide and sulfur trioxide on the measurement technique
will be discussed later.

Laboratory studies simulating flue gases containing SO_x[181]---
Experimental apparatus utilizing ASME, PTC-28, test cells----An
experimental arrangement was designed to determine resistivity for
four ash specimens simultaneously using ASME, PTC-28 test cells.
The test cells were contained in a 316 stainless steel chamber
that was housed in a high temperature oven. Simulated flue gas
environments were maintained in the test chamber under a small
positive pressure (2.54 to 5.08 cm of water). The electrical
circuit allowed the cells to be independently energized for resistivity measurements.

TABLE 26. RESISTIVITY TEST PROCEDURES COMPARISON OF CERTAIN FEATURES USED BY VARIOUS LABORATORIES

Laboratory	A	B	C
Resistivity Cell Design & Geometry	In House Guarded, Parallel Plate	In House Guarded, Parallel Plate	In House Unguarded, Parallel Plate
Environmental Containment	Environment contained within test cell	Test cell housed in an environmental chamber	Test cell housed in an environmental chamber
Standard Environment	Air – H_2O or N_2, O_2, CO_2, H_2O mixture	$N_2, O_2, CO_2, SO_2, H_2O$ mixture	Air – H_2O
Standard H_2O Concentrations	0,5,10,15 volume percent	0,3.2,7.8,15.2,22.1 volume percent	A constant value for each test
Ash Layer Thickness	0.25 – 0.30 cm	0.5 cm	0.3 cm
Usual Test Voltages	500,1000,1500,2000 volts	2,000 volts (can vary as desired)	1,000 volts (can vary as desired)
"Standard" E (kV/cm)	4	4	3.3
Time Voltage Applied Prior to Current Reading	20 – 40 seconds	3 – 5 minutes	30 seconds
Load of Electrode on Ash Layer	~17g/cm²	~12g/cm²	~6g/cm²
Test Temperature Range	120°C ascending to 400°C in 27°C increments	190°C descending to 90°C in 20°C increments	93°C ascending to 260°C in 27°C increments or 290°C ascending to 400°C in 27°C increments

(CONT'D)

TABLE 26. (CONT'D)

Laboratory	D	E	F
Resistivity Cell Design & Geometry	ASME, PTC-28 Guarded, Parallel Plate	In House Unguarded, Parallel Plate	In House Guarded, Parallel Plate
Environmental Containment	Test cell housed in an environmental chamber	Test cell housed in an environmental chamber	Environment contained within test cell
Standard Environment	Air – H_2O or N_2, O_2, CO_2, H_2O mixture	Air – H_2O	Air – H_2O or N_2, O_2, CO_2, H_2O mixture
Standard H_2O Concentrations	9 volume percent	A constant value for each test	0, 4.1, 8.2, 16.5, 32.9 volume percent
Ash Layer Thickness	0.6 – 0.7 cm	0.5 cm	0.5 – 0.6 cm
Usual Test Voltages	1,330 volts (can vary as desired)	500 volts	1000, 1500, 2000 volts
"Standard" E (KV/cm)	2	1	4
Time Voltage Applied Prior to Current Reading	60 seconds	?	5 minutes
Load of Electrode on Ash Layer	~10g/cm^2	?	~17g/cm^2
Test Temperature Range	460°C descending to 85°C continuously with readings taken periodically	5 points within temperature range of interest, ascending and then descending	110°C ascending to 260°C in 50°C increments

Figure 224 illustrates the physical arrangement of the apparatus.[182] Tank gases including commercially prepared and certified 1% SO_2 in N_2 were metered using precision rotameters to deliver the desired mixture at a total flow rate of 1.3 liters/minute at standard conditions. Depending on the temperature, this flow rate provided 5 to 10 volume changes per hour for the test chamber. The standard of baseline simulated environment contained by volume 5% O_2, 13% CO_2, 9% H_2O, 500 ppm SO_2 and the balance N_2.

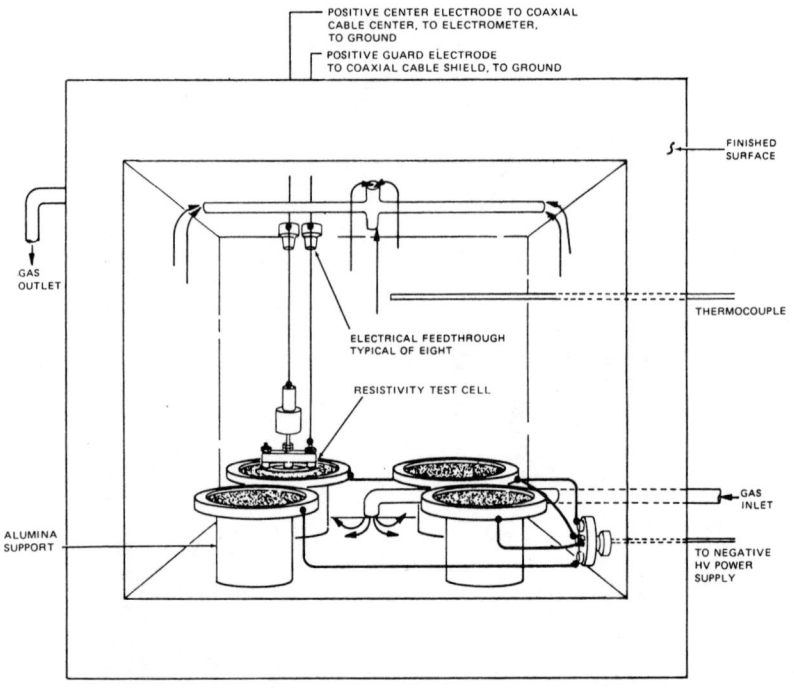

Figure 224. 316 stainless steel environmental resistivity chamber.[181]

The gases leaving the rotameters passed through a stainless steel manifold into a two liter stainless steel mixing vessel held at 200°C to preheat the gas. At the exit of this vessel an inlet was provided for the introduction of SO_3. The proper amount of SO_3 to be injected was governed by the temperature of the 20% sulfuric acid bath and the flow rate of the nitrogen used as a carrier.

A temperature of 160°C was maintained for the stainless steel tubing carrying the gas mixture to the oven. After entering the oven, the gas was passed through 7.62 m (25 feet) of tubing, maintained at the test temperature, before it entered the resistivity chamber. Gas exiting the chamber was passed through a bubbler external to the oven to provide visual evidence of the maintenance of a small positive pressure in the chamber.

Experimental procedure----Ashes were passed through an 80 mesh screen to remove any foreign material prior to being poured into the cup of the resistivity cell. While being filled, the cup was tapped to insure that ash bridging would be minimized. After the cell surface was leveled, the test cell was attached to the proper leads in the chamber, see Figure 224. The front piece of the chamber was sealed with C clamps after the four test cells were in position. Clamping together two finely machined surfaces was suitable for maintaining the small internal chamber pressure.

Nitrogen, passed through a drying column and the heated plumbing leading to the oven, was maintained in the test chamber overnight as the specimens were thermally equilibrated at 450-470°C. Prior to converting the environment to a simulated flue gas, the cell was tested by applying 1000 volts DC (5mm ash layer giving an E = 2 kV/cm) and determining the current one minute after the application of voltage.

After the environment was converted to the simulated flue gas, the current readings were repeated every 10 minutes until the current no longer increased with time. This usually took 20 to 40 minutes. At this point the oven was turned off, and current readings were taken periodically as the chamber temperature decreased. The cells cooled from 460°C to 145°C in about four hours and cooled further to 85°C in an additional two hours.

When it was of interest to determine resistivity as a function of ash layer field strength, the decreasing temperature was arrested at 162°C while the necessary measurements were made. Variation in water concentration was accomplished by changing the temperature of the water through which the nitrogen was bubbled prior to entering the 200°C preheating vessel. The nitrogen was valved so that it could be introduced dry or through the water bubbler. The water concentration was determined from an exit gas sample at least once during each resistivity test. Resistivity was calculated according to equation (29).

Problems encountered using SO_x.----It was stated above that the standard or baseline environment contained ~500 ppm of SO_2 and no injection of SO_3. Preliminary experiments had shown a small difference between resistivity data acquired using air-water environments versus the baseline simulated environment. At the time, it was believed that the small attenuation of resistivity was possibly due to the presence of SO_2.

The scope of research required the investigation of the effect of simulated environments containing 500, 1000, 2000, and 3000 ppm of SO_2. When the larger concentrations of SO_2 were incorporated, it was observed that resistivity values were significantly attenuated. Although one could not rule out the possibility that SO_2 affects ash resistivity, it seemed likely that large quantities of SO_3 were being generated and that the reduction in resistivity was due to the presence of sulfuric acid. Determination of SO_3 and SO_2 concentrations in the inlet and outlet gas samples when no SO_3 was being injected verified the presence of SO_3.

Several months were spent running ancillary experiments attempting to understand the problem and develop a way in which the existing equipment and test procedure could be utilized. When SO_2 was included in the environment, SO_3 was produced by catalytic oxidation of SO_2. A few ppm were produced even when oxygen was excluded. It was concluded that some oxygen was present as a trace impurity in other gases or that air diffused into the test chamber at the imperfect seal on the face. Furthermore, the amount of SO_3 catalytically produced was sensitive to the plumbing temperature and the temperature of the test chamber. When SO_2 was eliminated and SO_3 was injected, the difference in SO_3 concentration in the inlet and exhaust gas samples from the test chamber was sensitive to the chamber temperature. This indicated the chamber was capable of adsorbing a significant quantity of available SO_3 (H_2SO_4). Since temperature was one of the test variables and since it was desired to keep the SO_3 concentration constant during a specific test, the above observations indicated that the procedure and equipment utilized were not satisfactory for the evaluation of the effect of SO_3 on resistivity.

Experiments to develop apparatus and procedure to utilize environments containing SO_x.----A series of modifications took place in reaction to the observed test results. The first modification converted all plumbing and hardware from stainless steel to glass with the exceptions of electrical feedthroughs, test cells, lead wires, etc. This did not eliminate the formation of SO_3 from the SO_2 and O_2 present in the environment; however, the amount of SO_3 adsorbed by the system was decreased. It was then decided to convert to an environment of air, water vapor and injected SO_3 since no evidence was available to suggest a need for O_2, CO_2 and SO_2 to be present.

Under these conditions, the effect of 10 ppm of SO_3 on resistivity was not observed although a significant amount of SO_3 was removed from the environment as indicated by the measured SO_3 concentrations for chamber inlet and outlet gas samples. [This is in contrast to the observed reduction of resistivity reported for the stainless steel system. It has been rationalized that in the case of the earlier observations either a very great quantity of SO_3 had been generated and/or condensation of acid had taken place.] At this point the total environmental flow rate under standard conditions was increased from 1.3 liters/minute to 5.0 liters/minute, and the number of test cells were reduced from four to one. Under these conditions and with 25 grams of ash present in the single test cell, an injection rate of ∼10 ppm SO_3 could be maintained in both the inlet and outlet gas samples.

However, even overnight exposure to an environment consisting of air containing 9% water and 10 ppm of SO_3 did not produce a significant attenuation of resistivity. The resistivity cell was the type suggested in ASME PTC-28. The ash is held in a shallow dish having a porous, stainless steel bottom. The upper ash surface is exposed to the environment except where the measuring electrode and guard ring rest. Ash specimens were taken at various elevations between the exposed surface and the porous metal base at various positions exposed to the environment and beneath the measuring electrode. The amount of soluble sulfate was determined for each specimen as a measure of the penetration and adsorption of sulfuric acid from the environment. The results are shown in Figure 225 for an ash having a soluble sulfate value of 0.20 - 0.25% before testing.[183]

These data show that even after 24 hours of exposure at 145°C to an environment consisting of air, 9% water and 10 ppm of SO_3, a large concentration gradient of adsorbed acid (soluble sulfate) exists through the ash layer. The data show that in the area directly exposed to the environment the acid pickup was significant at the surface and a concentration gradient developed from position 1 to 3. Between the measuring electrodes there was little adsorption of acid. Therefore, no appreciable attenuation of resistivity was noted. Obviously even a thin ash layer (1-2 mm) between two parallel, porous electrodes would not be a successful test geometry under these conditions.

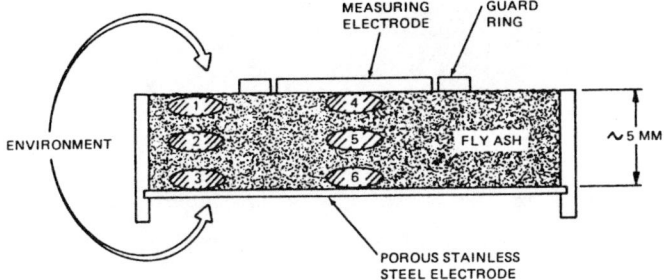

SAMPLE POSITION:	BLANK	1	2	3	4	5	6
% SOLUBLE SULFATE:	0.20/0.25	0.80	0.41	0.34	0.25	0.28	0.28

Figure 225. Weight percent soluble sulfate.[183]

Attempts to utilize vacuum to pull the environment through the electrodes and ash layer and other schemes to force it through under pressure failed. Besides the side effects of either compacting or fluidizing the ash layer, the concentration gradient of acid pickup expressed as soluble sulfate could not be eliminated.

The observations described above suggest that in addition to the ASME resistivity cell, other designs may be unsatisfactory for examining the effect of SO_3 on resistivity. Nevens, et al[184] recently evaluated three general types of laboratory resistivity test cells. Since these cells require the environment to permeate a porous stainless steel electrode and about 5 mm of ash, these designs are probably undesirable for environments involving SO_x.

Kanowski and Coughlin were successful in illustrating the effect of SO_3 on fly ash resistivity using a cell believed to be similar to that suggested by ASME PTC-28.[185] Although all apparatus and procedural details are not available, it would seem that the use of very high total environmental flow rates and the use of high concentrations (~30 ppm) of SO_3 contributed to this success. This approach was not attempted in the subject research, because the facilities limited the low rates available and interest was restricted to low SO_3 concentrations, <10 ppm.

Development of a radial flow test cell and procedure----Equipment-----The observation that the exposed ash surface adsorbed a significant amount of sulfuric acid (soluble sulfate), and the assumption that a thin layer of ash at the surface must become essentially "equilibrated" with the environment in a reasonable period of time led to the development of a test apparatus and technique that has provided useful laboratory resistivity data. Initial experiments showed that surface resistance readily reflected the effect of sulfuric acid in the environment. The test cell shown in Figure 226 was constructed to compare simultaneously a conventional test cell with a radial flow test cell using a 1 mm thick ash layer.[186] With this arrangement, one can alternately measure resistivity in the conventional parallel plate mode between electrodes 2 and 3 or in the radial flow mode between electrodes 1 and 2. The cell dimensions selected were based on the work of Amey and Hamburger regarding optimum geometries for surface and volume resistance measurements. Resistivity can be calculated for the radial flow cell from the expression:

$$\rho = \frac{2 \Pi c}{\ln(r_2/r_1)} \cdot \frac{V}{I} = \frac{1.56V}{I} \qquad (30)$$

where

ρ = resistivity, ohm·cm,

V = volts, applied between electrodes 1 and 2,

I = amperes, current flowing between electrodes 1 and 2,

c = 0.1 cm, thickness of electrodes 1 and 2,

r_2 = 1.90 cm, radius of I.D. of electrode 1,

r_1 = 1.27 cm, radius of electrode 2.

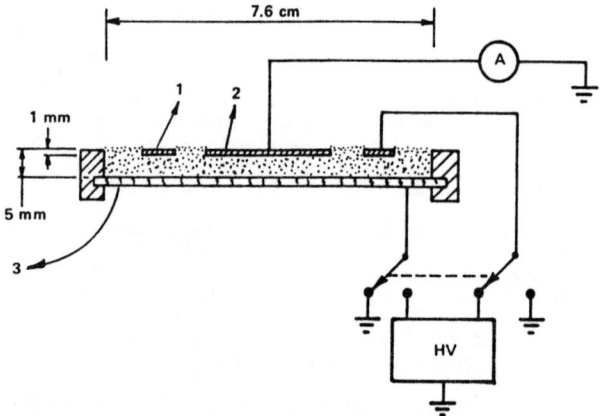

ELECTRODE 1 - 5.1 cm OD x 3.8 cm ID x 0.1 cm THICK, SOLID STAINLESS STEEL
ELECTRODE 2 - 2.54 cm OD x 0.1 cm THICK, SOLID STAINLESS STEEL
ELECTRODE 3 - 7.64 cm OD x 0.1 cm THICK, POROUS STAINLESS STEEL

Figure 226. Combination parallel plate-radial flow resistivity test cell and electrical circuit.[186]

Figure 227 shows a radial flow cell in the glass environmental chamber.[187] Figure 228 shows the comparative results for the two electrode geometries expressed as resistivity versus time of environmental exposure. For this experiment the apparatus shown in Figures 226 and 227 was used, and the ash was thermally equilibrated overnight in dry air at 145°C. Resistivity was determined, about 1.4×10^{13} ohm·cm with either electrode set, and the environment was changed to include 9% water at time = 0 hours. After 20 minutes, both electrode sets measured a resistivity of 2 to 3×10^{11} ohm·cm. This response time is typical. At this temperature, flow rate and chamber size, the time required to dilute a given environmental composition to 99% of a different composition was about six minutes. After the minimum resistivity due to water injection is reached, the resistivity gradually increases with time of exposure. Even though the injection of 10 ppm of SO_3 was started at time equal 30 minutes, the linear flow, parallel plate electrode set showed this increase in resistivity; i.e., the parallel plate electrode set did not respond to the presence of SO_3. However, the resistivity measured with the radial flow electrode set started to show the effect of SO_3 injection about 30 to 60 minutes after injection was started. After about two hours had elapsed, the attenuation of resistivity due to SO_3 injection was quite apparent and continued at a decreasing rate until a minimum value was attained about 24 hours after the start of the test. For this ash and set of conditions, it is assumed that a 24 hour exposure was required to "equilibrate" the 1 mm thick ash layer between electrodes 1 and 2, Figure 226, with the surrounding environment of air, water vapor and sulfuric acid vapor.

No effort has been made to formally evaluate the reproducibility of data using this cell; however, the cursory comparison of many pairs of tests would indicate the reproducibility is good. Also, no attempt has been made to evaluate the effect of variations in the test procedure on the data generated. It has been noted that the inlet and outlet SO_3 determinations indicate the environment is reproducible and that typically the inlet concentration is slightly greater than the outlet concentration for injections of <10 ppm SO_3.

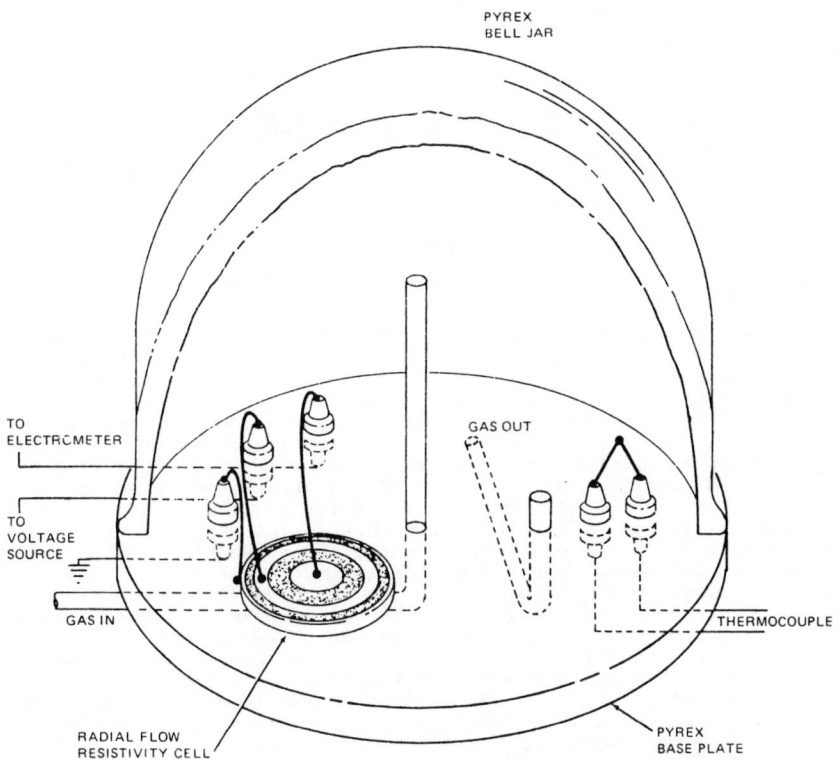

Figure 227. Glass environmental resistivity chamber.[187]

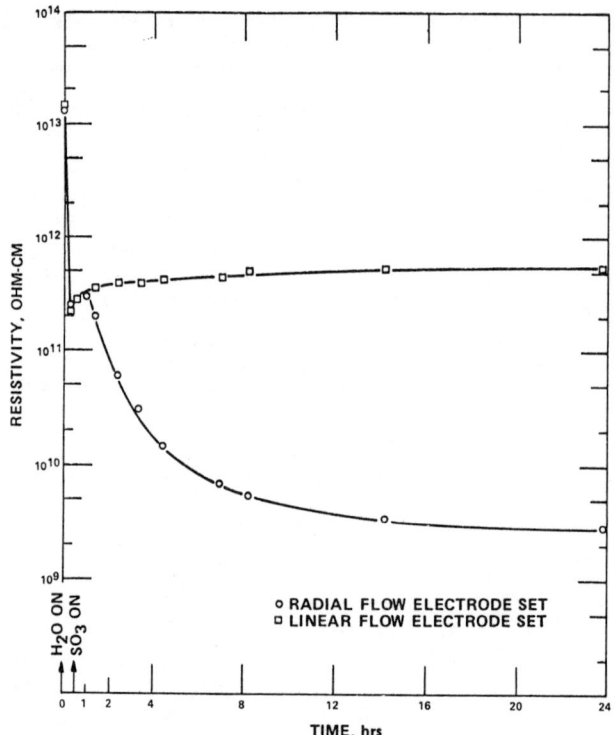

Figure 228. Resistivity vs. time of environmental exposure.

Test procedure-----The following test procedure was used to determine the resistivity for a number of ashes as a function of temperature and SO_3 concentration. This procedure is started at 11 am each day that a test is to be conducted: load cup of resistivity test cell with ash in the manner previously mentioned, place cup in chamber, attach lead wires and insert electrodes 1 and 2 by pressing them into the ash layer using a straight edge until ash slightly flows on to top of electrodes, cover chamber base plate with bell jar, start flow of dry air and turn on oven to desired set point, determine hot, dry resistivity at 2 pm and then divert dry air flow through controlled temperature water bubbler to introduce water vapor, determine resistivity at 2:15 and 2:30 pm and start nitrogen flow to inject desire concentration of SO_3, determine resistivity at 3:30 pm and take inlet and outlet gas samples for SO_3 determination, determine resistivity at 4:30 pm, determine resistivity at 8 am, take inlet and outlet gas samples for SO_3 determination and outlet gas sample for water concentration, determine resistivity at 9 am, convert environment to dry air and cool oven, open oven when cool and take ash sample for soluble sulfate determination, and begin new test at 11 am.

Resistivity is calculated using equation (30). Current is determined one minute after 1200 volts are applied to electrodes 1 and 2 in Figure 226.

The present apparatus has two severe disadvantages: only one test can be run per day and it is possible that in some cases the minimum resistivity could occur during the hours when no one is attending the apparatus or could require environmental exposure greater than 19 hours. While the radial flow cell and test procedure described have not been extensively evaluated and possess certain disadvantages, both the new cell and test procedure appear to provide a valid basis for determining ash resistivity in a simulated flue gas environment.

In situ measurements[169,188]--General considerations---Several decisions must be made in setting up and conducting in situ resistivity measurements. These decisions involve (1) device selection and operation, (2) site selection, (3) determination of the number of samples required to characterize the ash, (4) any auxiliary data required, and (5) necessary safety precautions. The selection of the device depends on a number of factors, including the availability of each device and the past experience of the intended user. However, selection should be based primarily on the operating characteristics of the various, available devices.

The first priority in selection of a sampling site is the location of a point in the operating system where the conditions of the gas and the gas-borne dust particles are representative of the environment for which resistivity is being determined. That is, the gas temperature, gas composition, and particle history must be the same as that found, for example, in the precipitator. Usually the inlet of the precipitator is selected as the point for making resistivity measurements. However, sampling at several points across the duct may be required to obtain a representative measurement where there are variations in temperature across the duct. Variations in gas flow velocity and dust loading in the duct must also be taken into account, since these conditions can result in nonrepresentative dust samples with some types of resistivity apparatus.

When selecting a site for the measurements, practical considerations must also be remembered. At the site location, sampling ports must exist or be installed. The normal practice is to use 4-inch pipe for the ports. Electrical power (117-120 VAC, 60 Hz), must be available at the site location for the operation of the measuring equipment. In many locations, adapters will be required for mating of plant electrical outlets with the standard three-prong plugs found on most laboratory equipment.

The determination of the number of individual measurements required to characterize the resistivity of the dust is related to the range of operating conditions anticipated and the variability in the particulate matter. It is desirable when designing a new precipitator installation that the worst operating conditions be covered in the test schedule.

The variability in plant operating conditions that is of the greatest concern is the variation in flue gas temperature throughout the year. The change in the ambient air temperature from winter to summer can cause the flue gas temperature to vary as much as 30°C (54°F) while the temperature variation across the duct downstream from a rotating (Ljundstrom) air heater may be 50°C (90°F). This combined temperature spread may cause a significant variation in the dust resistivity and care must be exercised to assure that the widest variation is covered.

The day-to-day variations in characteristics of the particulate matter may also cause significant variations in the particulate resistivity. This variability will show up as a considerable scatter in the measured value of resistivity over the measurement period. When this variation occurs, it becomes imperative to make a sufficient number of measurements at each temperature to obtain a statistically significant value for the resistivity.

The precipitator acts to smooth out short term variations in particulate resistivity. Dust layers ranging from perhaps one centimeter on the inlet plates to some lower value, perhaps only a millimeter, on the outlet plates build up during several hours of collection time. The average buildup rate on the precipitator plates is on the order of one millimeter per hour, exponentially distributed through the precipitator, such that the dust layer on the plates may represent an averaging of the instantaneous dust conditions of many hours of operation. Therefore, there is a rationale for averaging the measured values of resistivity for each temperature condition to arrive at the resistivity representative of the particular installation.

The determination of how many measurement points are required is therefore based on the variability of the source and the experience of the technician making the measurements. Typically, six to ten measurements each at intervals of 10°C (18°F) are sufficient if plant conditions are reasonably constant.

The auxiliary data required when conducting tests on an operating precipitator include: process samples for proximate and ultimate analysis, flue gas temperature and composition (including concentration of SO_3), precipitator voltage-current relationships, and particulate samples for laboratory analysis.

Extreme caution must be exercised when conducting measurements in ducts containing flue gas. Typically, the flue gas at temperatures exceeding 150°C (302°F) will contain a significant quantity of sulfur oxides and particles. If the access port has been covered for a period of time, significant amounts of particulate will accumulate in the port. Some ducts will be under a positive pressure of a few inches of water; in others, there exists the probability of "puffing". Therefore, extreme care must be exercised when opening ports and when inserting or extracting probes because of this presence of particulate and sulfur oxides in the gas.

Additional care must be exercised when utilizing resistivity probes with high voltages. Sufficient electrical grounds must be attached prior to handling any probe connected to an electrical supply.

A shock hazard also exists when inserting or extracting any ungrounded probe. An ungrounded probe inserted into a particulate-laden gas stream may become electrically charged by static electricity caused by particle impact. Therefore, probes should be grounded prior to insertion into a flue duct.

A hazard also exists because of the location of the sampling ports. Often, the ports were installed after the construction of the plant at locations remote from standard walkways. All scaffolds and walkways should be tested prior to use and all hazards that can be reasonably detected should be corrected.

A number of different instruments are available for making resistivity measurements. These instruments differ fundamentally in the method of sample collection, degree of compaction of the dust sample, and the values of the electric field and current density utilized for the measurement, as well as the method of maintaining thermal equilibrium and the method of deposition in the measurement cell. These differences in operation lead to differences in the characteristics of the sample and in the values obtained for the resistivity.

Instruments utilizing electrostatic collection and measurements on the undisturbed dust layer measure the resistance of a dust layer that was formed by collecting individual particles aligned by the electric field under conditions similar to those in a standard precipitator. This procedure leads to a compact dust layer with good interparticle contact. Those devices that utilize dust layers collected and redeposited will be operating on a disturbed and recompacted layer. This difference in operation may lead to differences in contact potential between the adjacent particles and to different porosity in the sample that may influence the value obtained for the resistivity.

In the remaining discussion of in situ measurements of resistivity, several devices and methods will be described and discussed. Particular emphasis will be placed on the point-to-plane probe since it is the most widely used probe in this country. The operating principles of these devices will be described. Also, the advantages and disadvantages of utilizing the different devices will be presented.

In situ resistivity probes---Point-to-plane probe----The point-to-plane probe for measuring resistivity has been in use since the early 1940's in this country. Two models of this device are shown in Figure 229.[189] The probe is inserted directly into the dust-laden gas stream and allowed to come to thermal equilibrium. The particulate sample is deposited electrically onto the measurement cell through the electrostatic action of the corona point and plane electrode. A high voltage is impressed across the point-to-plane electrode system such that a corona is formed in the vicinity of the point. The dust particles are charged by the ions and perhaps by free electrons from this corona in a manner analogous to that occurring in a precipitator.

The dust layer is formed through the interaction of the charged particulate with the electrostatic field adjacent to the collection plate. Thus, this device is intended to approximate the behavior of a full-scale electrostatic precipitator and to provide a value for the resistivity of the dust that would be comparable to that in a full-scale electrostatic precipitator.

In the point-to-plane technique, two methods of making measurements on the same sample may be used. The first is the "V-I" method. In this method, a voltage-current curve is obtained before the electrostatic deposition of the dust, while the collecting disc is clean. A second voltage-current curve is obtained after the dust layer has been collected. After the layer has been collected and the clean and dirty voltage-current curves obtained, the second method of making a measurement may be used.

In the second method, a disc the same size as the collecting disc is lowered on the collected sample. Increasing voltages are then applied to the dust layer and the current obtained is recorded until the dust layer breaks down electrically and sparkover occurs. The geometry of the dust sample, together with the applied voltage and current, provide sufficient information for determination of the dust resistivity.

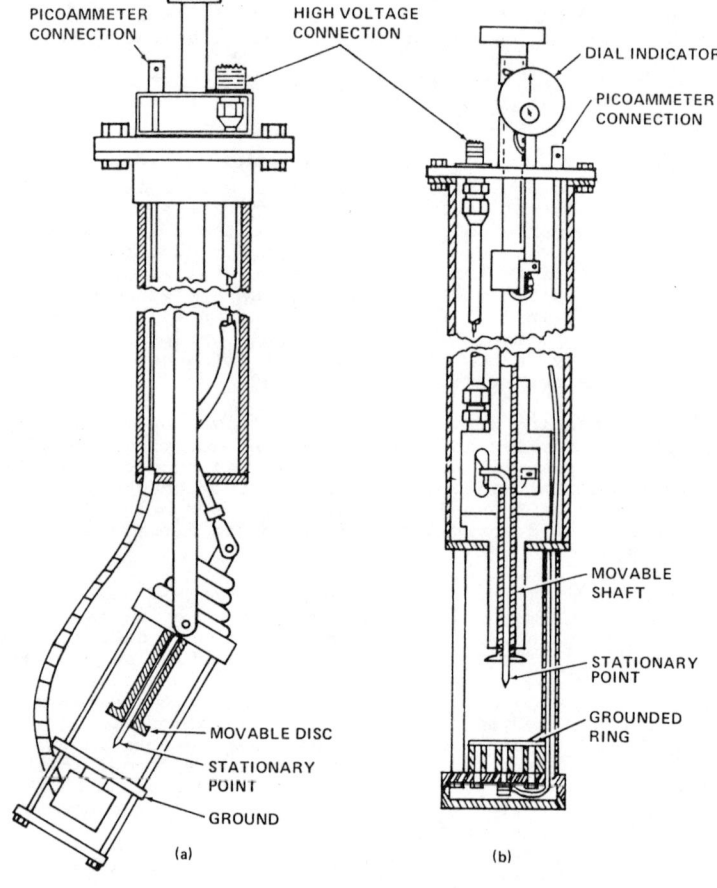

Figure 229. Point-to-plane resistivity probe.[189]

In the "V-I" method, the voltage drop across the dust layer is determined by the shift in the voltage vs current characteristics along the voltage axis as shown in Figure 230.[190] The situation shown is for resistivity values ranging from 10^9 to 10^{11} ohm-cm.

If the parallel disc (spark) method is used, dust resistance is determined from the voltage measured just prior to sparkover. In both methods the resistivity is calculated as the ratio of the electric field to the current density.

The practice of measuring the resistivity with increasing voltage has evolved because the dust layer often behaves as a non-linear resistor. As the applied voltage is increased, the current increase is greater than that attributable to the increase in voltage. Therefore, as described in the ASME Power Test Code No. 28 procedure, the resistivity reported is the value of resistivity calculated just prior to sparkover.

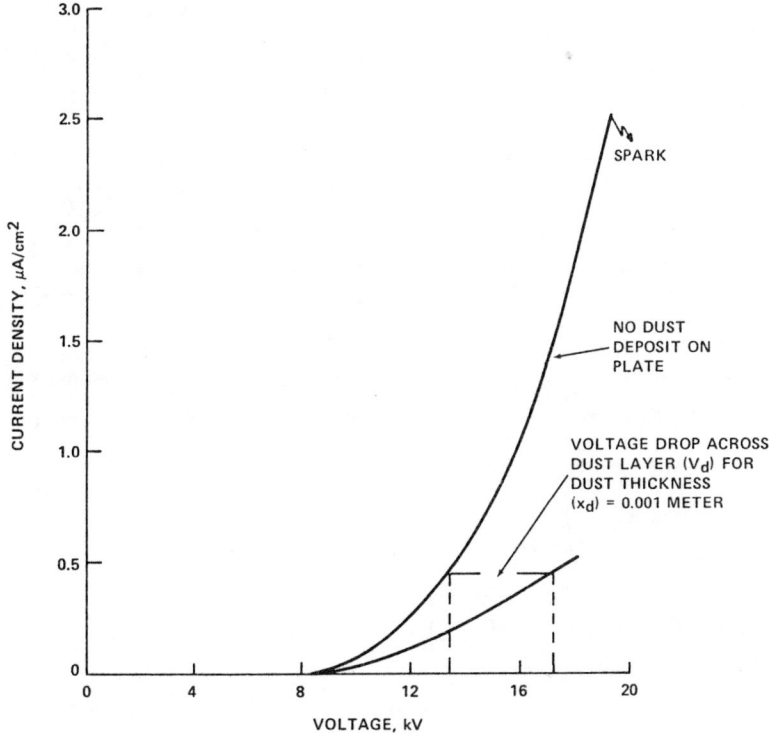

Figure 230. Typical voltage-current density relationships for point-to-plane resistivity probe.[190]

There is considerable justification for using the value of resistivity prior to electrical breakdown as the resistivity, since it is precisely at electrical breakdown that the resistivity causes problems within the precipitator. The electrical breakdown in the dust layer in the operating precipitator either initiates electrical sparkover or reverse ionization (back corona) when the resistivity is the factor limiting precipitator behavior. If neither of these events occur, the dust layer merely represents an additional voltage drop to the precipitator power supply.

Even though there are many similarities between the operation of the point-to-plane device and a full-scale precipitator, several problems also exist. The first problem encountered is the determination of the thickness of the dust layer. Some devices make use of a thickness measurement system built into the probe. In other devices, the instrument is withdrawn from the duct and the thickness of the layer is estimated visually by inspecting the dust layer. However, the dust layer is almost always disturbed by the air flow through the sampling port and extreme care is required to preserve the layer intact.

The advantages of utilizing the point-to-plane probe for in situ measurements are: (1) the particulate collection mechanism is the same as that in an electrostatic precipitator, (2) the

dust-gas and dust-electrode interfaces are the same as those in
an electrostatic precipitator, (3) flue gas conditions are preserved, (4) the values obtained for the resistivity are in general
consistent with the electrical behavior observed in the precipitator,
and (5) measurements can often be made by two different methods.

However, the following disadvantages exist: (1) the measurement of the dust layer thickness can be difficult, (2) high voltages
are required for collection, (3) considerable time is required for
each test, (4) a number of measurements are required for gaining
confidence in the measured value, (5) experienced personnel are
required for testing, (6) particle size of the collected dust is
not representative, (7) sample size is small, (8) carbon in the
ash can hamper resistivity measurements, and (9) length of probe.

Description of SoRI point-plane probe[188]----The SoRI resistivity probe system for making in situ resistivity measurements
includes a probe for insertion in the flue, a high voltage supply,
a voltmeter, an ammeter with overload protection and a temperature
indicator. A schematic diagram of the complete system is shown in
Figure 231.[191]

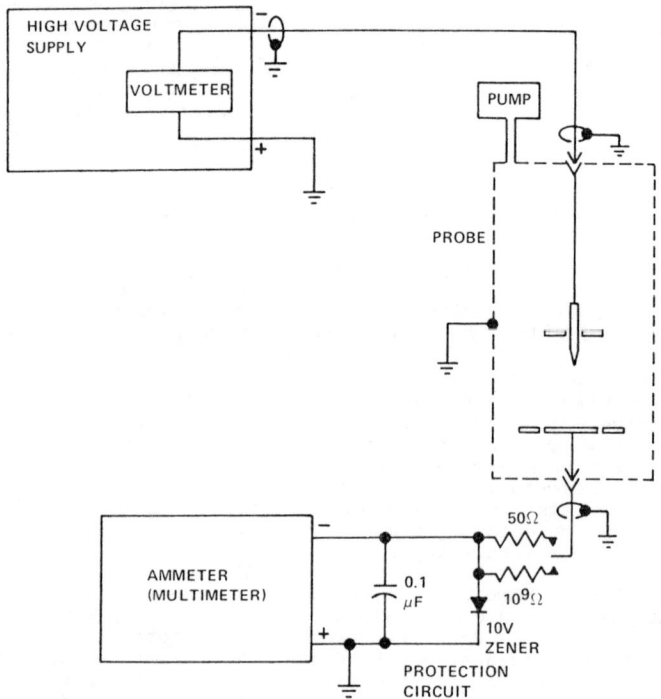

Figure 231. Schematic diagram of SoRI probe system.[191]

The power supply for the SoRI probe is a modified Spellman
Model UHR30N30 (30 kVDC Neg, 1 mA) with two voltage scales (0-30
kV and 0-3 kV). The ammeter is a Keithley digital multimeter model
150B (sensitivity to currents as low as 10^{-10} amps). The input to
the multimeter is protected from surge currents during sparkover

by a zener diode protective circuit. This circuit also contains a 10^9 Ω resistor for testing the probe.

The probe is equipped for collecting the dust, making electrical contact with the dust, and determining the dimensions of the collected dust layer, all without removing the probe from a sampling port. The particulate sample is collected by a point-plane corona discharge cell mounted in the end of the probe.

The corona point is located 5.72 cm from the 5.2 cm diameter collecting electrode. The collecting electrode consists of a guard electrode and a center disc electrode (diameter 2.52 cm, area 5.0 cm^2). The guard electrode is connected directly to ground. The center disc is isolated from ground by a machinable glass ceramic insulator and is connected to the external ammeter. A Chromel-Alumel thermocouple mounted in the back of the guard electrode is used for measurement of the duct temperature.

Electrical contact with the exposed surface of the collected dust layer is made by lowering a sliding disc electrode onto the collected dust. The thickness of the layer is determined by comparing the readings of a dial indicator connected to the sliding electrode. Readings are obtained when the electrode is lowered before and after the dust layer is collected. The sliding electrode is free to move up and down except for a lock clamp at the top of the probe and for an acme screw that engages before contact is made with the collecting electrode. This screw adjustment protects the dust layer from a sudden impact. The screw adjustment is also provided with a spring built into the sliding electrode push rod to limit the compression force applied to the dust layer.

The collection dust layer is removed by removing the probe from the flue and manually cleaning the electrodes.

General maintenance of SoRI point-plane probe----General maintenance of the probe requires that it be periodically disassembled and cleaned. Instructions for maintenance of the electrical equipment are given in the manuals supplied by the manufacturers.

To clean the probe, first remove all externally collected dust and the shield protecting the point-plane corona discharge assembly. Remove the high voltage and sliding electrode assembly from the probe by removing the bolts on the upper flange and the screws holding the high voltage junction block to the middle bulkhead (plate from which the corona point protrudes). Now slide this assembly out of the probe casing and clean.

The high voltage junction block consists of two concentric cylinders. It can be disassembled by removing the screws in the top of the junction block. Separating the cylinders exposes the high voltage connection and the sliding electrode contact. This area should be cleaned of any accumulated dust. The graphite contacts to the sliding electrode should be checked for electrical contact and for freedom of motion of the sliding electrode.

At the upper end of the high voltage and sliding electrode assembly is the dial indicator assembly, spring assembly, control mechanisms for lowering the sliding electrode, Swagelok quick connect connector, and the high voltage connector. The dial indicator mechanism has a tendency to corrode and should be lightly oiled. The vertical location of the dial indicator can be adjusted by loosing the locking screw and sliding the indicator up or down. When the probe is assembled, the dial indicator should be adjusted to read 5.00 when the sliding electrode is in contact with the collecting electrode. The spring assembly should be inspected to insure that the spring operates freely. If it does not, dust has probably accumulated in this assembly and it must be disassembled and cleaned. The electrode lowering control should be easy to turn and easy to move up and down when the acme screw is not engaged. The high voltage connector which was fabricated from alumina tubing, Swagelok connectors, and a banana plug, should be cleaned and the electrical continuity to the sliding electrode checked.

The collecting electrode electrical connections are accessible by removal of the flange from the bottom of the probe casing. The insulator isolating the center disc electrode from ground should be cleaned and the resistance to ground from the center electrode should be greater than 10^{12} ohms.

After reassembly, the probe electrode alignment must be inspected. The probe is designed to be self-aligning. In the lowered disc position, the sliding electrode should be parallel and in contact with the center disc electrode.

After assembly, the sliding electrode should move freely. When the sliding electrode is locked in the lower disc position, it should spring back into position if it is manually pushed into the probe casing and released.

Operation of the SoRI point-plane probe----Pre-field trip preparation-----Prior to use of the probe in the field, general maintenance should be performed to insure that the probe will operate properly. It is possible to bench test the probe using the 10^9 Ω resistor built into the spark protector box to simulate a collected dust layer.

Set up the probe system as described later in the operating instructions. (Lower the sliding electrode so that it makes contact with the collecting electrode and switch the control on the spark protector to the 10^9 Ω position.) Set the power supply for an output of 100 volts (V) and read the current (I) to the multimeter. Calculate the resistance (R) of the resistor in the protective box by Ohm's Law:

$$R = \frac{V}{I}$$

This value should be 1.00×10^9 Ω ± 5%. Electrical connectors and instrument calibration should be checked if the above value is not obtained.

A pre-field inspection check list is given in Table 27. (Some of the equipment listed here is not supplied with the probe and must be supplied by the user.)

Operating instructions-----At the site the equipment should be carefully unpacked and inspected. The electrical instrumentation package is not sealed to keep out moisture and must be located out of the weather but within 3 m (10 ft) of the sampling port. Connect the probe to ground. This is necessary to insure proper operation of the probe, and for operator safety. Before inserting the probe in the sampling port, lower the sliding electrode until it makes contact with the collecting electrode. If the metal shield for the corona discharge cell has been removed, replace it at this

Analysis of Factors Influencing Performance   261

time. (Between runs it is necessary to remove this shield to clean dust from the cell.) Adapters for 6" and 4" pipe nipples are supplied with the probe. For some sampling ports special adapting flanges must be made, or a temporary arrangement such as rags or other suitable sealing material, will have to be used. However, for strongly negative or positive pressure flues an airtight flange connector should be used.

The large cable (RG - 8/U) supplied with the probe is the high voltage cable. It connects the high voltage connector on the back of the power supply to the high voltage connector on the top of the probe. The 3 m cable (RG - 58/U) connects the BNC connector on the side of the probe to the input connector on the spark protector box. The coaxial cable with a double banana plug on one end connects the spark protector output to the multimeter.

TABLE 27.   RESISTIVITY PROBE
PRE-FIELD TRIP INSPECTION CHECK LIST

YES   NO

___   ___   1. Probe - including breakdown inspection and calibration - wiring - HV cable, etc.

___   ___   2. Power supply - inspect operation - general condition, wiring, calibration, etc.

___   ___   3. Multimeter - inspect operation - general condition, wiring, calibration, etc.

___   ___   4. Tool box - insure correct tools are in the box for in field breakdown repair and inventory spare parts.

___   ___   5. Power cords - insure operation of extension cord and box.

___   ___   6. Tent - covering for instruments.

___   ___   7. Field cleaning kit - insure rags, brushes, and dusters are included for on-site cleaning.

___   ___   8. Sample containers and data sheets - insure supply of bottles or plastic bags to collect ash samples and supply of data sheets is sufficient.

___   ___   9. Shipping boxes - insure boxes are serviceable and in condition to receive rough and abusive handling. Insure that instruments are sufficiently padded.

___   ___   10. Confirmation - insure unusual conditions at test site are accounted for: flue gas temperature, gas velocity, flue pressure, sampling port sizes, hot/cold weather conditions, etc.

Comments: _____
          _____
          _____

A suitable temperature indicator for a Chromel-Alumel thermocouple should be connected to the thermocouple output on the side of the probe using the supplied connector.

Plug the ac line from the instrumentation package into a 117-120 VAC line. The black clip lead on the power supply is an extra ground lead and should be attached to a good ground. Using their individual power switches, turn on the multimeter and the high voltage supply.

Assemble the probe completely and make all the necessary electrical connections. Place the spark protector box switch so that the 10^9 ohm resistor is in the circuit. Set the power supply for 100 V output. Lower the sliding electrode slowly while watching the current meter for a reading of approximately 100 nA. If this does not occur and the power supply current becomes excessive, the electrodes are misaligned and the probe must be disassembled and repaired. If all is normal, the initial test may begin.

Insert the probe in the flue with the 100 V applied as above. Insure that the holes in the screen are perpendicular to the gas flow. While the probe is warming up, it may expand sufficiently to cause a loss of electrical continuity indicated by reduced meter current, and 100 V indicated on the power supply. In this case lower the sliding electrode just enough to recontact the plate. If both the current and voltage drop to zero, the plate is misaligned and the test might have to be aborted. It is possible that the misalignment is due to uneven expansion and may return to normal when the probe reaches an even temperature.

Maintain the electrical contact of the sliding electrode and plate throughout the warm-up period. After the temperature as indicated on the thermocouple readout has equilibrated, a test may be started.

Check the current meter to insure that the sliding electrode is down in place. Adjust the dial indicator to have the pointer set at zero by loosening the lock nut on the dial face and rotating the scale to the proper location. Leaving the probe cover holes oriented perpendicular to the gas stream, unlock and unscrew the sliding electrode control and raise it to the up position. Lock in place. Now run a "clean-plate" V-I curve by placing the multimeter on the 100 nA scale and setting the power supply voltmeter switch to the high position. Check that the slide switch on the spark protector is in the normal position. Turn the high voltage on. The use of the high voltage supply is described in the manufacturer's manual. Adjust the OUTPUT control through its full range using the kV meter as a guide and make a current reading every 1000 volts until a spark level or the maximum output voltage is reached. Keep the multimeter within its range during these measurements to prevent excessive overranging. Record these readings on a data sheet and mark it "clean plate". Adjust the HV output control for a current of 1 µA and rotate the probe so the cover plate holes are in the gas stream.

A dust layer is then precipitated on the collecting electrode. The proper operating current density required for the type of ash being collected has to be experimentally determined. Thus the first test may not be useful for obtaining data. The current density normally used should fall somewhere in the range between 0.2 and 2.0 µA/cm^2 for this unit. If a high resistivity dust is encountered, reduced current densities may be necessary to obtain a good layer. Use of the V-I curves will be explained later to indicate how the proper current for precipitation may be found if the original selected value proves to be insufficient. A current of 1 µA, giving a current density of 0.2 µA/cm^2, is a good place to start the initial run. The voltage necessary to obtain this current is in the vicinity of 15,000 V. Depending on the resistivity of the dust being collected, the mass loading, and the current density selected, it will take from about thirty minutes to one hour to precipitate a sufficient sample of a thickness between 0.5 and 1.5 mm.

As a layer is being deposited, the current will begin to drop. This current drop may be used to estimate the collection time. When current drops significantly or if an hour has passed, whichever comes first, the test may be stopped. If an insufficient sample was collected on a short time run, run longer the next time no matter how the current happens to drop. After a sufficient sampling time has elapsed, turn the probe so that the holes in the cover plate are perpendicular to the gas stream. Now run a "dirty-plate" V-I curve using the same procedure as that for the "clean-plate" V-I.

After completing the "dirty-plate" V-I, turn the high voltage off by turning the control to zero and switching the power supply off. Place the switch on the spark protector box to the 10^9 Ω resistor in the circuit and protect the multimeter from an overload current when lowering the sliding electrode with the voltage on. Set the multimeter on the 100 nA range. Turn the voltage supply on and adjust for a 100 V output.

Unlock and very carefully and slowly lower the sliding electrode until the acme screw is engaged. Then turn the control lowering the electrode until the multimeter indicates that electrical contact with the dust layer has been made. Turn the control knob one-quarter turn and lock into position. If the dust resistivity is less than about 10^9 Ω-cm, the multimeter should read approximately 100 nA. For high resistivity dust smaller currents will be obtained, the exact current depending on the thickness of the dust layer and the resistivity. Now set the multimeter on the 1000 µA scale and switch the slide switch on the spark protector back to the normal position. If the power supply does not indicate an overload (1 mA), "direct contact" can be taken.* Increase the voltage across the dust layer in 100 V steps. Read and record the corresponding currents until a spark occurs across the dust layer. This will be indicated by the voltmeter jumping and an erratic reading on the multimeter.

Before starting another run the dust layer must be removed by mechanically removing the dust. Remove the metal cover from the discharge cell and clean the cell thoroughly. If saving the sample for chemical analysis or some other reason is desired, a sheet of paper placed under the disc will collect the sample when the operating rod is pulled back to its up and locked position.

At this time the dust layer thickness may be examined to insure the accuracy of the dial indicator. By utilizing an automotive

---

*Overloads frequently occur with high carbon content samples. The carbon particles or similar type conductors provide a conducting path between the disc allowing the full output current of the power supply to flow. If a short is encountered, it is impossible to obtain data for determining the resistivity of the layer between the parallel discs.

type metric feeler gauge the dust layer thickness may be estimated and compared to the dial indicator reading. The hole in the sliding electrode leaves an area of uncompressed dust that was protected from erosion when the probe was withdrawn from the flue and is an ideal point to gauge the thickness of the uncompacted layer.

After cleaning, replace the metal cover on the probe. Return the probe to the sampling port. While the probe is returning to the flue temperature make the calculations from the run just completed.

<u>Operating outline</u>-----The following outline summarizes the steps to be taken in operating the point-plane resistivity probe.

1. Prepare sampling port.
2. Clean and align cell.
3. Lower disc and lock.
4. Check current continuity, slide switch in 10^9 position.
5. Insert into flue, with inlet holes 90 degrees to flow, and bolt to flange.
6. Allow cell to reach flue temperature.
7. Zero dial indicator.
8. Raise operating rod.
9. Run "clean-plate" V-I, switch <u>normal</u> position.
10. Turn inlet holes into flow.
11. Apply necessary voltage to supply precipitating current.
12. After desired length of time turn probe so inlet holes are again 90 degrees to flow. (Leave high voltage applied so dust layer will not be shaken off in the turning process.)
13. Run "dirty-plate" V-I.
14. Lower disc, in 10^9 position, voltage 100 V.
15. Record thickness of dust layer.
16. Apply voltage in 100 V steps until sparkover occurs, switch in <u>normal</u> position.
17. Remove probe to remove collected dust.
18. Observe layer and save if needed.
19. Clean probe and check alignment.
20. Insert back into flue.
21. Make calculations.

Calculations-----A sample data sheet for a typical run is given in Figure 232. All the information necessary for making the resistivity calculations is given on this data sheet. The "clean" and "dirty" plate V-I information should be graphically plotted. The data on this data sheet is shown plotted in Figure 233.

The formula for calculating the resistivity is:

$$\rho = \frac{RA}{\ell}$$

or

$$\rho = \frac{\frac{V}{I} \times 5.00 \text{ cm}^2}{\ell \text{ (cm)}} \quad ,$$

where

ρ = resistivity (ohm-cm),

R = resistance $\frac{V}{I}$ (ohms),

ΔV = voltage across the dust layer (volts),

I = measured current (amps),

A = area of disc (5.00 cm^2),

ℓ = thickness of dust layer (cm).

The quantity A/ℓ is called the cell factor. This factor will remain constant for the V-I or spark calculation for each individual run. For different dust layers it is apparent that the cell factor will change.

SRI POINT PLANE PROBE DATA

Location - Power Plant Layer Thickness - 1.0 mm
Time - 0915 Data - 14 May 1973 Test No. - A-6 Temp. - 314°F (157°C)
Conditions - Normal, full load 56 MW
Unit 1, Port 3

V-I DATA			SPARK DATA		
KV	CLEAN	DIRTY	V	I	E
1			50	2.5 NA	500
			100	5.0 NA	1000
2			150	7.5 NA	1500
			200	10.0 NA	2000
3			250	13.5 NA	2500
			300	17.4 NA	3000
4			350	23.6 NA	3500
			400	29.0 NA	4000
5	1.0 NA	1.0 NA	450	39.5 NA	4500
			500	55.5 NA	5000
6	0.25 µA	0.1 µA	550	70.5 NA	5500
			600	96.7 NA	6000
7	0.65 µA	0.3 µA	650	0.14 µA	6500
			700	0.17 µA	7000
8	1.15 µA	0.5 µA	750	0.23 µA	7500
			800	0.36 µA	8000
9	1.8 µA	1.1 µA	850	0.46 µA	8500
			900	0.61 µA	9000
10	2.6 µA	1.65 µA	950	0.75 µA	9500
			1000	1.0 µA	10,000
11	3.2 µA	2.19 µA	1100	SPARK	
			1200		
12	4.3 µA	2.8 µA	1300		
			1400		
13	5.1 µA	3.7 µA	1500		
			1600		
14	6.2 µA	4.2 µA	1700		
			1800		
15	7.1 µA	4.8 µA	1900		
			2000		
16	8.2 µA	5.6 µA	2100		
			2200		
17	9.8 µA	6.25 µA	2300		
			2400		
18	11.1 µA	SPARK	2500		
			2600		
19	12.6 µA		2700		
			2800		
20	SPARK		2900		
			3000		

Figure 232. Sample data sheet for point-plane resistivity probe.

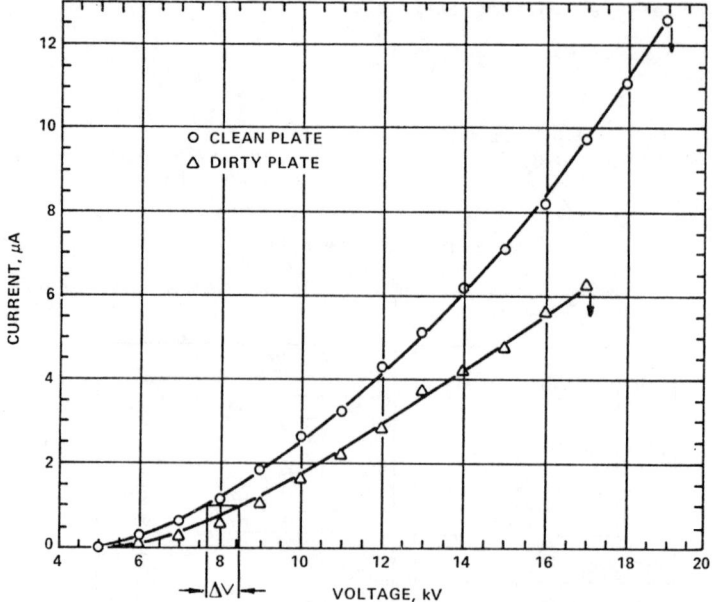

Figure 233. V-I data obtained from point-plane resistivity probe.

Example: Using the data from Figures 232 and 233, the following procedure shows how the resistivity is calculated. The following data was obtained from the V-I curve in Figure 233.

$$V = \Delta V = 850 \text{ V}$$

$$I = 1.0 \times 10^{-6} \text{ A}$$

$$\ell = 1.0 \text{ mm} = 1.0 \times 10^{-1} \text{ cm}$$

The value ΔV is the voltage drop across the dust layer as interpolated from the V-I curves at a current value of 1.0×10^{-6} A. Certain considerations must be taken into account when obtaining this voltage drop. The first is to look at the shape of the V-I curve. There are three basic shapes that may be encountered. Diagrams A and B in Figure 234 illustrate two of these shapes.

In Diagram A, the point x shows the voltage at which electrical breakdown occurs in the dust layer. This would show the onset of back corona, a characteristic of a high resistivity dust. It will be incorrect in this case to use any of the current and voltage relationships above the point x for calculating resistivity values.

This V-I curve may be used also to determine the operating point for the next run. If the point x is located at a lower current value than the one selected for collecting the sample, then there is a good chance that the sample was collected in a back corona situation. If this is the case then the current for the next run should be backed off to the value of current that corresponds to the point x. A more efficient collection should be found at this setting.

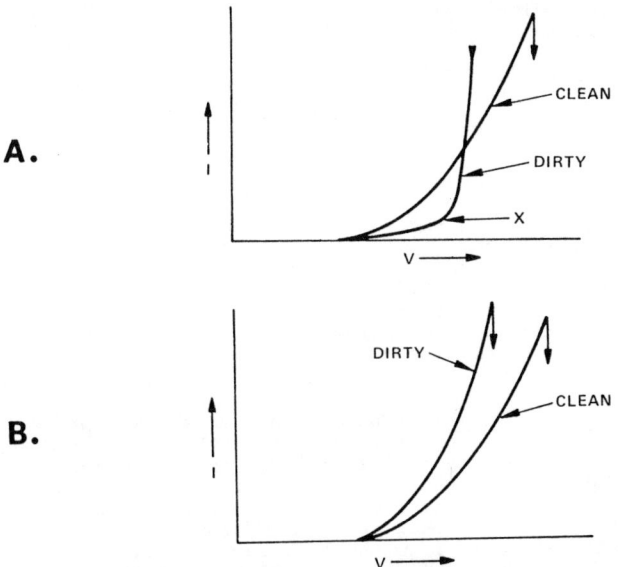

Figure 234. Two possible types of "dirty" V-I curves obtainable with a point-plane probe.

In Diagram B, the "dirty-plate" curve is on the left side of the "clean-plate" curve. This is a characteristic of either a very high or a low-resistivity dust. Since the ΔV taken from the curve will have a negative value, it will not be possible to use the V-I procedure for resistivity calculations in this case. Figure 233 is the third shape and it shows a standard curve.

The cell factor is the first calculation to be made. For the sample case, the cell factor is 50 cm and it comes from the term A/ℓ, where A is 5 cm^2 and ℓ is 0.1 cm. The next step is to find the resistance R of the dust layer. For this run, ΔV is equal to 850 V, this was taken from the V-I graph at a current of 1.0×10^{-6} amps. From this relation, $R = \Delta V/I$, the resistance is found to be 0.85×10^9 ohm. By multiplying the cell factor by the resistance, a resistivity of 4.2×10^{10} ohm-cm is obtained. This complete calculation is:

$$\rho = A/\ell \times \Delta V/I$$

or

$$\rho = \frac{5 \text{ cm}^2}{0.1 \text{ cm}} \times \frac{0.85 \times 10^3 \text{ V}}{1.0 \times 10^{-6} \text{ A}}$$

$$\rho = 4.2 \times 10^{10} \text{ ohm-cm}$$

After obtaining the resistivity from the V-I data, a check of this value may be obtained from the spark data information. The proper values to take from the spark data information are the last voltage and current reading before spark. In this case the layer broke down at 1100 volts and the last reading before breakdown was at 1000 volts with a current of 1.0×10^{-6} amps. Using the following formula the resistivity data may be obtained:

$$\rho = A/\ell \times V/I$$

$$\rho = 0.1 \text{ cm}^2 \times \frac{1.0 \times 10^3 \text{V}}{1.0 \times 10^{-6} \text{A}}$$

$$\rho = 5.0 \times 10^{10} \text{ ohm-cm}$$

The column labeled "E" on the spark data sheet is for the calculated electric field for the voltage applied and the thickness of the layer. In this example, breakdown of the layer occurred at an electric field of 10,000 volts/cm. When a series of measurements are made the resistivities should be calculated not only at sparkover for each run, but also at a fixed value of the electric field. This will eliminate the electric field dependence when comparing runs.

Cyclone resistivity probes----The cyclone resistivity probe measures the resistivity of a particulate sample that is extracted from the effluent gas stream by an inertial cyclone collector. The dust sample is deposited between two concentric cylindrical measurement electrodes. The dust-laden gas sample is extracted through a sampling nozzle by a pump into the cyclone separator where the collected dust falls into the measurement cell. The gas flow rate is adjusted to provide an isokinetic sample if desired. The collection characteristics of the cyclone are such that, even though the sampling system is operating isokinetically, the dust sample collected is not identical with that in the gas stream. Notwithstanding this, it is often desirable to use isokinetic conditions.

By applying a voltage across the cell and monitoring the current flowing through the cell, the filling of the cell can be observed by the increase in current through the cell. When the current levels off, the cell is full and the sampling is stopped. The current is then monitored until it stabilizes.

The resistivity of the sample is calculated from

$$\rho = KR , \qquad (31)$$

where R is the resistance of the dust layer (ohm) and K is a constant for any particular cell (cm). The constant K is defined by

$$K = \frac{2\pi L}{\ln (r_2/r_1)} , \qquad (32)$$

where

L = length of cell (cm),

r_1 = radius of inner electrode (cm), and

r_2 = radius of outer electrode (cm).

The Simon-Carves cyclone resistivity instrument, as described by Cohen and Dickinson, is one of the more widely used cyclonic devices.[192] The sample collection and measurement cell is located in a temperature controlled chamber as shown in Figure 235, external to the duct, with the sample extracted through a sample probe. The sampling line must be thermally controlled to preserve the flue gas condition. The dust sample is compacted into the measurement cell by the action of a vibrator.

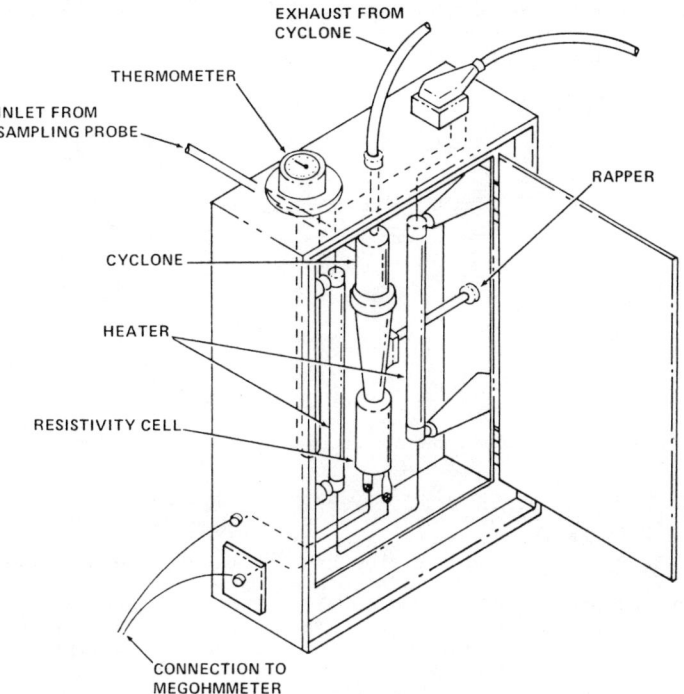

Figure 235. Resistivity apparatus using mechanical cyclone dust collector (from Cohen and Dickinson).[192]

A somewhat different design of this device is made to be inserted directly in the flue. The dust is collected and measured while the device is retained in the flue gas environment.

The probe is operated in the following manner: it is inserted into the flue and permitted to come to thermal equilibrium with the flue gas. A sample is then drawn through the apparatus by a pump, and the gas flow measured. Isokinetic sampling can be achieved by adjusting the flow so that the inlet velocity of the gas to the probe and the flue gas velocity are the same. A vibrator attached to the probe is used to keep dust from collecting on the walls of the probe and to give uniform compaction. Figure 236 shows a schematic of this instrument.[193]

The advantages of utilizing the cyclone probe for extractive or in situ measurements are: (1) low voltage instrumentation may be used, (2) dust layer thickness is fixed by cell geometry, and (3) the electric field is easily duplicated from test to test.

However, the following disadvantages exist: (1) the cylindrical cell yields a nonuniform electric field, (2) the electrical

noise is unusually high, (3) it is difficult to determine when
the sample cell is full, (4) compaction of the dust layer is not
reproducible, (5) the thermal control of the external model is
difficult, and (6) the values of resistivity obtained are un-
realistically high for electrostatic precipitator applications,
(7) particle size of dust is not representative, and (8) the dust
layer in the cell is not electrostatically deposited.

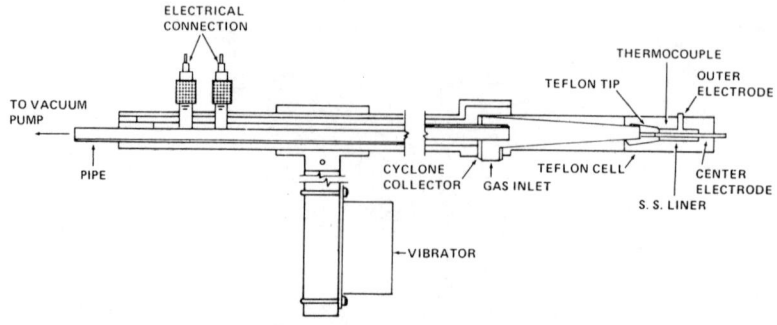

Figure 236. Cyclone probe inserted in duct.[193]

Kevatron electrostatic precipitator analyzer----The Kevatron
resistivity device is designed to simulate in situ measurements
in an external thermally controlled cell.[194] The sampling probe
is inserted directly into the flue gas for extracting an isokinetic
sample. The sampling line leads to a miniature wire-pipe type of
electrostatic precipitator, where the particulate material is collected on the surface of the pipe. The collected dust layer is
removed from the pipe and deposited in a concentric cylindrical
measurement cell by removing the electrical energization and
applying an acceleration to the pipe. A schematic drawing of
the system is shown in Figure 237.

The particulate matter is in the flue gas environment throughout
the entire measurement period. The flue gas flows through the sampling
lines and wire-pipe precipitator and exhausts to the atmosphere.
Provisions must be made to preserve the thermal conditions in the
flue duct through the sampling line to avoid upsetting the chemical
equilibrium conditions in the flue. Without this precaution, a
temperature drop in the sampling line may lead to an increased
absorption for any naturally occurring conditioning agents such
as sulfur trioxide and moisture in the effluent gas stream.

The instrument is designed to internally compute the resistivity of the dust in the measurement cell, when used with the
graph paper supplied. The system projects a spot of light on the
graph grid, thus eliminating the computation of resistivity that
is required for other instruments. The measurement is conducted
with applied voltage of 3, 30, or 300 volts across an electrode
spacing of 0.2 cm for electric fields of 15, 150, or 1500 volts
per centimeter, respectively.

The advantages of utilizing the Kevatron probe for resistivity
measurements are: (1) the resistivity is internally computed,
eliminating field calculation, (2) clean electrode and dust covered
electrode voltage-current curves can be obtained, and (3) some
variation in electric field is allowable in the measurement.

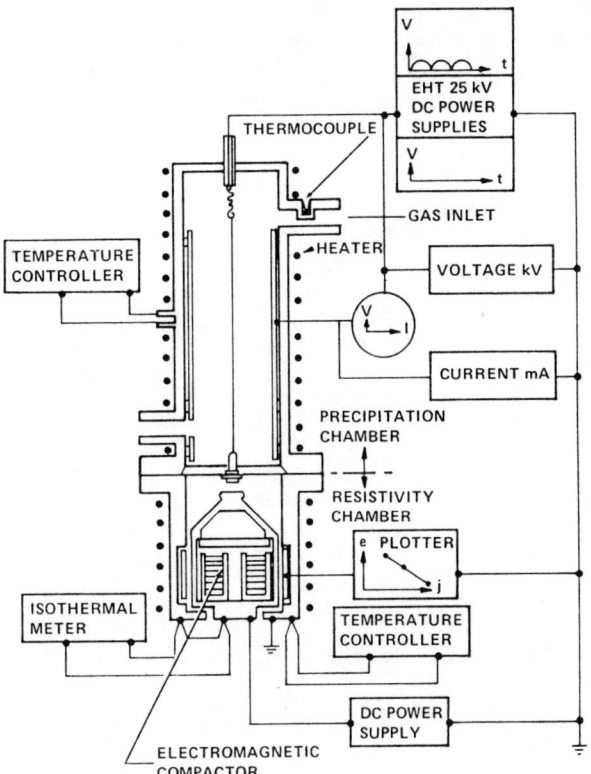

Figure 237. Kevatron resistivity probe (from Tassicker, et al).[194]

However, the following disadvantages exist: (1) the equipment is very heavy and bulky, difficult for field work, (2) sampling lines require temperature control, (3) mirror alignment in resistivity computation is critical, (4) particle size of the dust is not representative, (5) density of dust in the cell is not reproducible, (6) dust is not deposited in the cell electrostatically, and (7) resistivity values can be unreasonably high.

Lurgi electrostatic collection resistivity device----The Lurgi Apparatebau-Gesellschaft mbh in Frankfurt, West Germany, developed an in situ resistivity probe described by Eishold, consisting of two corona wire electrodes equally spaced from an interlocking comb arrangement as shown in Figure 238.[195] This device is inserted either directly into the flue duct for in situ measurements or into a thermally and environmentally controlled chamber for simulated in situ laboratory measurements.

The dust is collected on the interlocking comb structure by electrostatic forces. The dust layer forms on the surface of the comb structure and fills the region between the two comb segments. After the sample is collected, a potential is applied across the dust layer. The configuration of the cell (the cross-sectional area and spacing between the electrodes) is such that the resistivity of the sample is ten times the measured resistance. This factor of ten is based on neglect of any electrical fringing through the adjacent fly ash. The measurements are made using an ohm-meter without specifying the electric field at which the measurements are made.

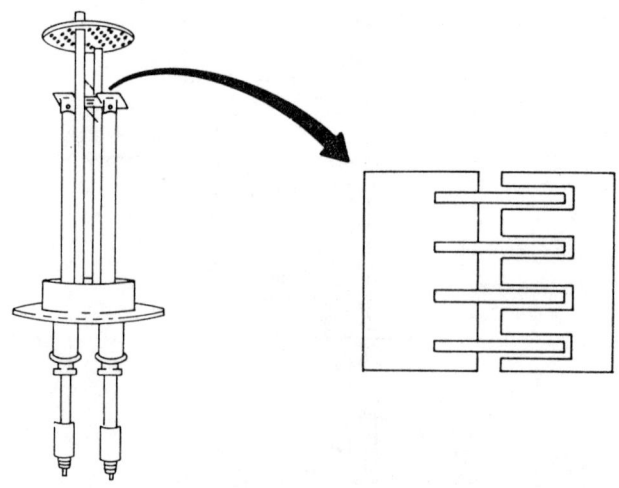

Figure 238. Lurgi in situ resistivity probe.[195]

Comparison of in situ resistivity probes----The resistivity probes previously described differ primarily in the manner of collection of the dust particles from the gas stream, the manner of dust deposition in the measuring cell, the cell geometry, and the electrical conditions during measurement.

Because of the nature of the collection devices, the size distributions of the particles in the samples are not representative of the size distribution of the dust particles in the duct. Neither the cyclone nor the electrostatic devices are efficient collectors of fine particles, so the particle size distribution in the resistivity sample is biased toward the larger particles. This condition can cause some variation in the results obtained with different devices.

A second difference in the resistivity probes is the manner of depositing the dust in the measuring cell. The point-plane probes and the Lurgi probe deposit the dust electrostatically onto the surface of the measuring cell. Consequently, some alignment of the dust particles occurs and in general the deposited dust layer is more dense than that in the other types of measurement apparatus. The effect of alignment on dust resistivity has not been quantitatively determined. However, variations in density can influence resistivity values by as much as 10-fold, as reported by Cohen and Dickinson.[192]

A third difference in the resistivity probes is the value of the electric field at which resistivity is measured. Standard procedures for the Kevatron and Simon-Carves probes are to measure resistivity at relatively low electric fields. By contrast, the procedure for the point-plane probe is to measure the resistivity at a field near breakdown. As a consequence, the values of resistivity as measured by the different methods vary by as much as an order of magnitude due to electric field differences. The combined effect of these variables is that the resistivity values reported by investigators using different techniques vary widely. Upper values of resistivity measured by a point-plane probe in the vicinity of 10^{12} to 10^{13} ohm-cm have been reported, whereas upper values of 10^{14} to 10^{15} ohm-cm have been reported by other techniques.

There have been no definitive studies to compare results of resistivity measurements by the various devices. However, limited studies have been conducted at electric power generating plants using the instack cyclone, Kevatron, and point-plane probes.[196] Resistivity values measured by these probes are compared in Figures 239 and 240. Figure 239 shows the settled-out cyclone data plotted against the point-plane data, using the point-plane data at 2.5 kV/cm, which corresponds to the field in the cyclone apparatus.

Figure 240 shows the peak values of resistivity from the Kevatron and cyclone probes plotted against point-plane data from the same (2.5 kV/cm) field. In this case, much better agreement is obtained between the cyclone and point-plane data. The Kevatron data are still higher than the average of the cyclone or point-plane data, although there are statistically insufficient data to draw firm conclusions regarding the Kevatron values. The logic of comparing the peak values of resistivity from the cyclone with the point-plane data can be rationalized to some extent by the fact that fresh dust is being deposited on the surface during the precipitation process. In view of the scatter of the data obtained with any one probe, the discrepancies shown in Figures 239 and 240 are not unexpected.

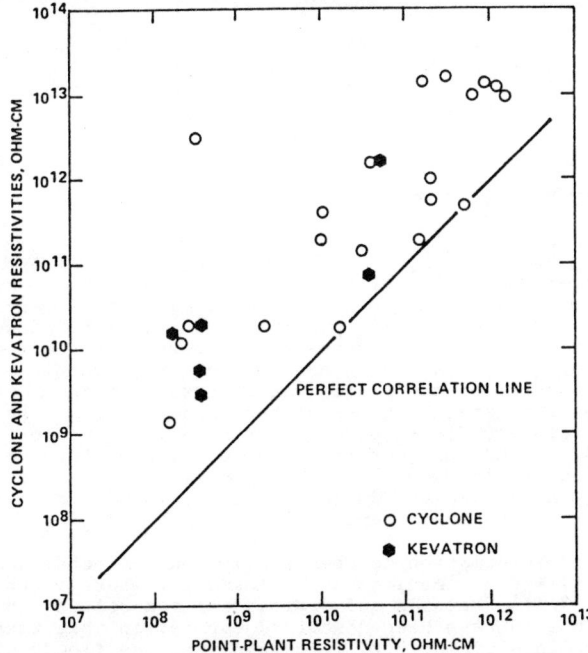

Figure 239. Comparison of Kevatron and cyclone resistivities with point-plane resistivities at an electric field of 2.5 kV/cm. Settled values for cyclone peak values for Kevatron.[196]

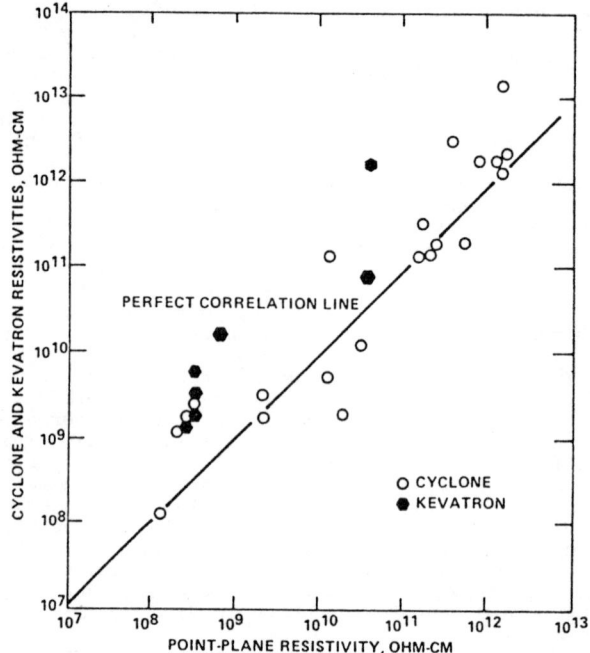

Figure 240. Comparison of Kevatron and cyclone resistivities with point-plane resistivities at an electric field of 2.5 kV/cm. Peak current values used for Cyclone and Kevatron.[196]

LIMITATIONS DUE TO NON-IDEAL EFFECTS

Gas Velocity Distribution

General Discussion--

Nonuniform gas velocity distributions result in reduced precipitator performance due to (1) uneven treatment of particles in different velocity zones, (2) possible reentrainment of collected particles from the plate surfaces and hoppers in regions of high gas velocity, and (3) a possible nonuniform particulate mass loading distribution entering the precipitator, resulting in excessive dust accumulation in certain regions of the precipitator. The uniformity of the gas velocity distribution entering a precipitator is influenced by (1) the configuration and location of turning vanes, (2) the location and types of diffuser elements, such as grids and perforated plates, (3) the ductwork design, and (4) coupling of the precipitator to the draft fan.

Detailed information on the description, effects, and control of the gas flow distribution can be found elsewhere in the literature.[197,198,199] Methods and devices for controlling the gas flow distribution have been discussed earlier in this text. Now, some major points of interest concerning the gas flow distribution will be discussed. These include criteria for determining a good flow distribution, measurements of gas flow distributions associated

Analysis of Factors Influencing Performance   275

with full-scale precipitators, and the effect of gas flow distribution on precipitator performance.

Criteria for a Good Gas Flow Distribution--

Good uniformity of the gas velocity distribution must be achieved in order to attain the present requirements of high collection efficiencies (99.5-99.9%) with a minimum in precipitator size. To be meaningful, the criteria for an acceptable gas velocity distribution must be stated in terms of measurable quantities. In 1965 a definition of an acceptable deviation from an ideal gas distribution was introduced by the Industrial Gas Cleaning Institute (I.G.C.I.), which states:

> "Uniform gas distribution shall mean that a velocity pattern five feet or less ahead of the precipitator inlet flange shall have a minimum of 85% of the readings within $\pm$ 25% of the average velocity in the area with no reading varying more than $\pm$ 40% from the average."[200]

The above criteria are the most widely used at the present time. However, some power companies have specified even more stringent criteria for an acceptable gas distribution at the inlet of a precipitator; for example:

> "A minimum of 8% of the readings within $\pm$ 10% of the average velocity and no reading varying more than $\pm$ 20% from the average."[199]

At the present time, I.G.C.I.'s Committee on Gas Flow Model Studies is in the process of preparing a new more detailed set of criteria for an acceptable gas velocity distribution. These criteria include a restated velocity distribution pattern, an R.M.S. deviation criteria, and limitations on gas velocity deviations between individual chambers of large precipitator installations.

Field Experience with Gas Flow Distribution--

A particular case history which has been reported demonstrates many of the important aspects associated with gas flow distribution and precipitator performance.[198] In this case, an electrostatic precipitator installed on a 500 MW tangentially-fired steam generator burning coal was to collect 99.5% of the fly ash entrained in the flue gas emanating from the combustion process. The installations reported had the following specifications: collection efficiency of 99.6%, treated gas volume flow of 723.5 m^3/sec (1,530,000 acfm) at 126°C (260°F), collecting plate area of 25,154 m^2 (270,400 ft^2), specific collection area of 35 $m^2/(m^3/sec)$ [178 ft^2/1000 acfm], and coal with an ash content of 12% dry basis and with a sulfur content of 3.65% as-fired. The efficiency achieved during the first three years of operation was measured several times and ranged from 98.8 to 99.1%. Mechanical remedies, electrical remedies, and gross gas flow corrections were attempted without improving the performance. Finally, an in-depth study of the gas flow distribution revealed serious problems which were limiting performance.

Figure 241 is a side elevation of the entire precipitator complex for Unit A. Gas leaves the Ljungstrom^R air preheater and is divided between the two precipitators of the double deck installation. During initial operation, gas-flow traverses were conducted to determine the gross division of gas between the precipitators. Detailed velocity traverses were also conducted in the vertical outlet flue leaving the upper precipitator, and in the inlets to the i.d. fans. The gas flow passing through the lower precipitator was determined by subtracting the measured gas flow leaving the upper precipitator from the measured gas flow entering the induced draft (i.d.) fan inlets. These initial tests showed that approximately 54.5% of the gas was going through the lower precipitator with the remainder going to the upper precipitator. Based on the recommendation of a model study, a perforated plate was installed in the vertical portion of the flue just before the turn into the lower precipitator. The turning vanes (Figure 241) shown in the inlet to the upper and lower precipitators and in the outlet of the upper precipitator also were installed based on recommendations from this same model study.

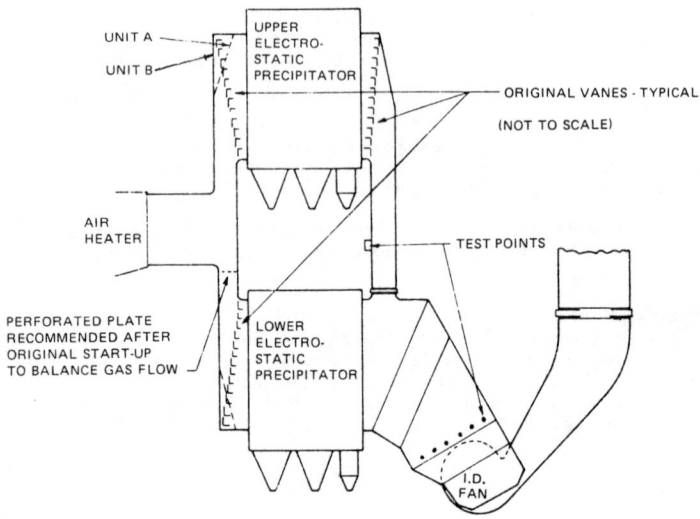

Figure 241. Side elevation of electrostatic precipitator.[198]

The velocity traverses conducted at the inlets to the i.d. fan also revealed a lateral imbalance of gas flow across the precipitators. Figure 242 shows the results of these tests. The north i.d. fan was receiving 9% more flow than the south but, more importantly, the inboard legs of each fan received more flow than the outboard legs. Finally, when dust samples were taken in the inlet to each i.d. fan to check performance, it was found that 88% of the total dust collected in each inlet was collected in sample port #1 as noted in Figure 243.

Based on this history of gas-flow related problems, a decision was made to conduct detailed field evaluations. As shown in Figure 244, four 20 cm (eight-inch) diameter observation ports were installed in the roof and side wall of the lower precipitator on the north side of the unit. The system was operated at full load and high intensity lights were used to illuminate the gas flow zones of interest through these ports. These observations pointed out dramatically the effects of poor gas flow distribution on precipitator performance. Although initial short term observations showed no apparent problems, extended observations revealed that huge clouds of dust would suddenly appear in the lower precipitator outlet. Careful observation of this phenomenon revealed that these eruptions were occurring only in limited areas of the precipitator, and usually occurred when one or more collecting electrodes in these areas were being rapped. At first, it was thought that plate rappers were occasionally rapping entire precipitator lanes at once, but this proved not to be the case. The dust eruptions would occur only when the plates in the immediate vicinity of either of the i.d. fan inlet boxes were rapped.

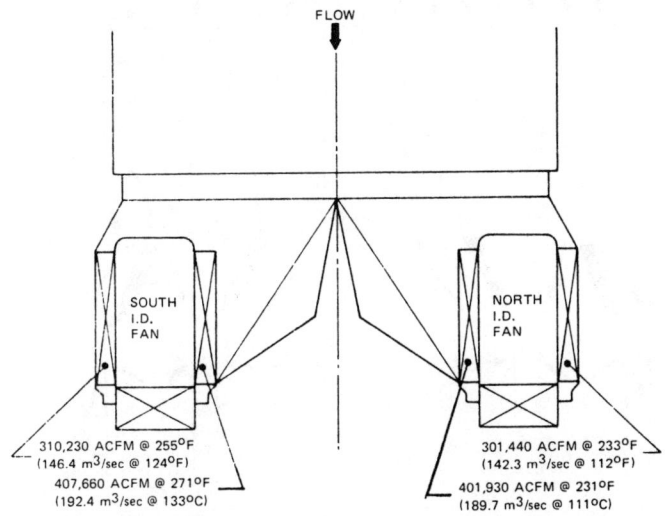

Figure 242. Gas-flow imbalance, outlet flues and i.d. fans (Unit A).[198]

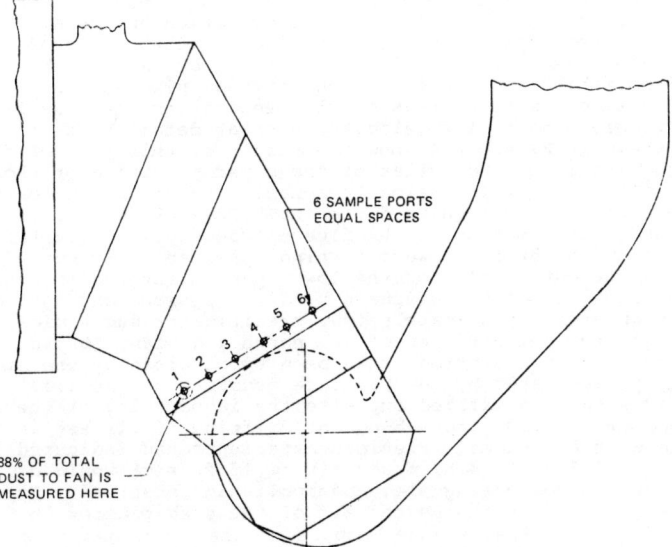

Figure 243. Side elevation of i.d. fans (Unit A).[198]

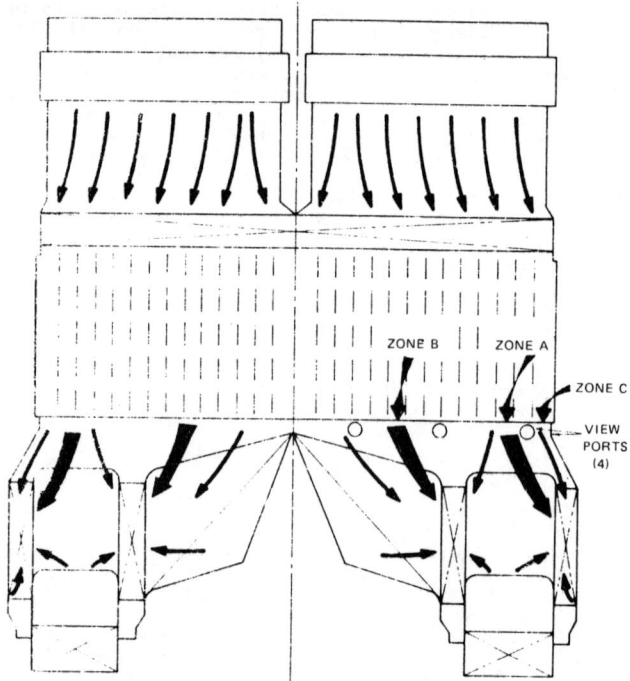

Figure 244. Gas-flow patterns, plane view of outlet flues (Unit A).[198]

To further define the problems observed through the observation ports, the unit was taken out of operation and detailed internal inspections of both the inlet and outlet flues of each precipitator was made. A skilled observer, by careful observation of polishing and deposition on internal pipe struts, vanes, and dampers, can define areas of flow separation, reverse flow, and extremely high or low velocity in great detail. The flow arrows shown in Figure 244 show the result of this type of flow mapping. The inlet and outlet of the upper precipitator showed no unusually high or low velocity zones. The situation for the lower precipitator was quite different. Several feet of fly ash were found in the bottom of the flue entrance of the lower precipitator, with the two lowest turning vanes actually buried in fly ash. The outlet flue of the lower precipitator also exhibited areas of high velocity (evidenced by dust erosion) and dust dropout. In an area approximately four precipitator ducts wide, adjacent to the outboard leg of the north i.d. fan, the surfaces of the collecting eleectrodes had been swept clean by the high-velocity jets created by the pressure gradient of the i.d. fan. A similar situation existed opposite the inboard leg of the same fan. Also, hopper sweepage and subsequent drifting of fly ash into the outlet flue were evidenced. Previous experience had indicated that velocities of 3.05 to 4.57 m/sec (10 to 15 ft/sec) would be required to produce the collecting electrode polishing observed. These phenomena were repeated in the south half of the precipitator, but the problems appeared less severe because of the lower gas flow through that half of the installation.

Based on the results of the on-line observations and off-line inspection, it was obvious that the gas flow problems in this unit were a major contributing factor to its deteriorated performance. Since the original model study of this installation did not reveal any of the problems just described, it was decided that a complete velocity traverse of the inlets to both the upper and lower precipitators would be conducted. This information could then be used to check the as-built model results to insure an accurate representation of the problem. Because of limited unit availability, the field velocity traverses could not be conducted on Unit A. They were, however, conducted on Unit B, a duplicate of installation A, which also had experienced performance problems of the same nature as Unit A. A quick walk-through of Unit B was conducted to ensure that the problems observed in Unit A were evident in Unit B. Unit B was then thoroughly cleaned before attempting to perform the field velocity traverses so that the traverses would be indicative of a new system.

A heated thermocouple anemometer was used to obtain velocity data. The anemometer was traversed down the first two discharge electrodes of every fourth precipitator duct. Selected traverses were also obtained in the outlet of each precipitator. Figure 245 shows a sample of the data obtained from one precipitator duct.

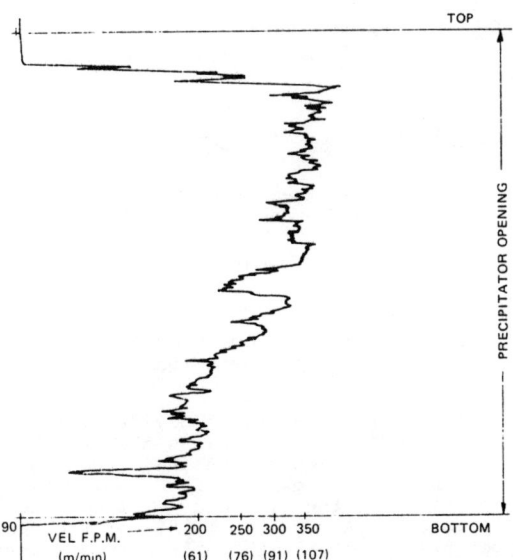

Figure 245. Lower precipitator inlet velocity profile duct 68 as measured with continuous traverse (Unit B).[198]

The unit was operated on cold air at approximately 60% of design velocity. This provided a Reynolds Number approximately equal to that which would be seen under actual full load operation. Figure 246 is an example of a typical field velocity profile after the velocity had been corrected back to a linear scale. Once all the data curves had been linearized, they were reduced to numerical form. An overlay grid was prepared of twenty equally spaced lines representing precipitator elevations. The overlay was placed over each linearized velocity profile and the value of the velocity profile at each evaluation was recorded as a point velocity. These velocity data points were then numerically averaged to establish an average vertical and horizontal velocity profile for each precipitator.

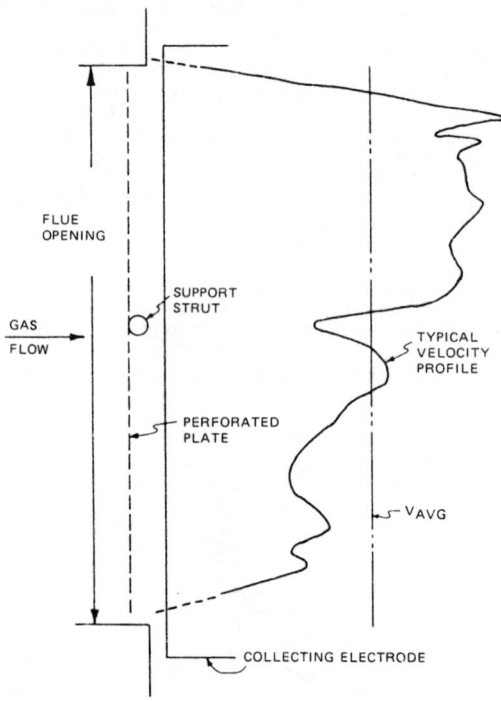

Figure 246. Typical measured velocity profile, as installed -- lower precipitator inlet (Unit B).[198]

Figure 247 illustrates a simplified side elevation view of the upper and lower precipitators showing the average vertical inlet velocity profile for each as obtained from the field test. It is important to note the skewness of the velocity profile in the lower precipitator and the imbalance of flow between the upper and lower precipitators. Approximately 58% of the gas was passing through the upper precipitator with the remainder passing through the lower. It should also be noted that this imbalance was not completely detrimental since previous field tests had indicated that 80 to 90% of the dust went to the lower precipitator. If the design velocity had actually been met, the high velocity zones in the lower precipitator would have further reduced the efficiency of the overall system.

Figure 248 demonstrates the dramatic effect that the outlet flue has on the velocity profile leaving the lower precipitator. This points out the condition that had to be eliminated if re-entrainment and hopper sweepage in the lower precipitator were to be eliminated.

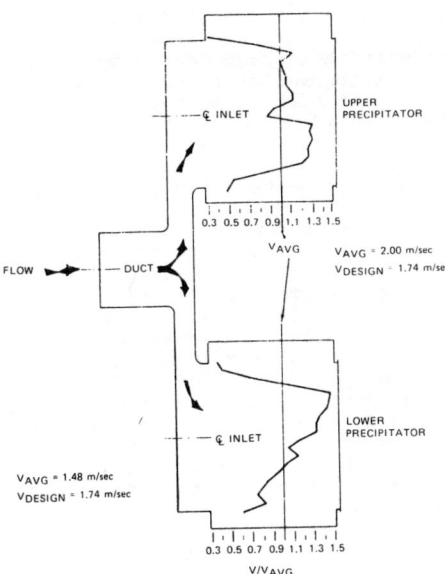

Figure 247. Average inlet velocity side elevation profiles -- as installed (Unit B).[198]

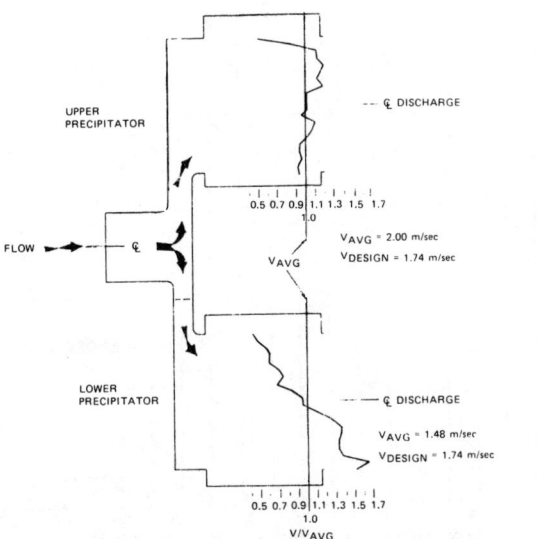

Figure 248. Average outlet velocity side elevation profiles -- as installed (Unit B).[198]

Figures 249 and 250 detail the statistical distribution of the data points taken in the upper and lower precipitators and compare these results with those recommended by the IGCI. The vertical bars of these histograms represent the percentage of the data points occurring at each velocity. The actual velocity values have been normalized (divided by the average velocity), following standard practice. As can easily be seen, neither precipitator meets the IGCI requirements, with the upper precipitator approximately two times better than the lower precipitator.

A model study in conjunction with the field data resulted in the correction of several mechanical defects in the existing flow devices and in the addition of new flow control devices. The precipitator and flue inlet perforated plates had been installed in panels 91.5 cm (3 feet) to 122 cm (4 feet) wide by 305 cm (10 feet) high. These panels had been clipped together for alignment to maintain the effect of a large single-piece perforated plate. Some clips had not been installed while others had broken loose permitting the adjacent plates to buckle over their 305 cm (10 ft) height. Where a 2.54 cm (1 inch) gap had been desired, gaps of 10.16 to 20.32 cm (4 to 8 inches) were found. Gaps of this type were found in both the inlet flue and precipitator perforated plates. This is not uncommon; this particular type of erection defect has been found in many installations. The gaps were oriented such that they accounted for the high flows measured in several of the ducts.

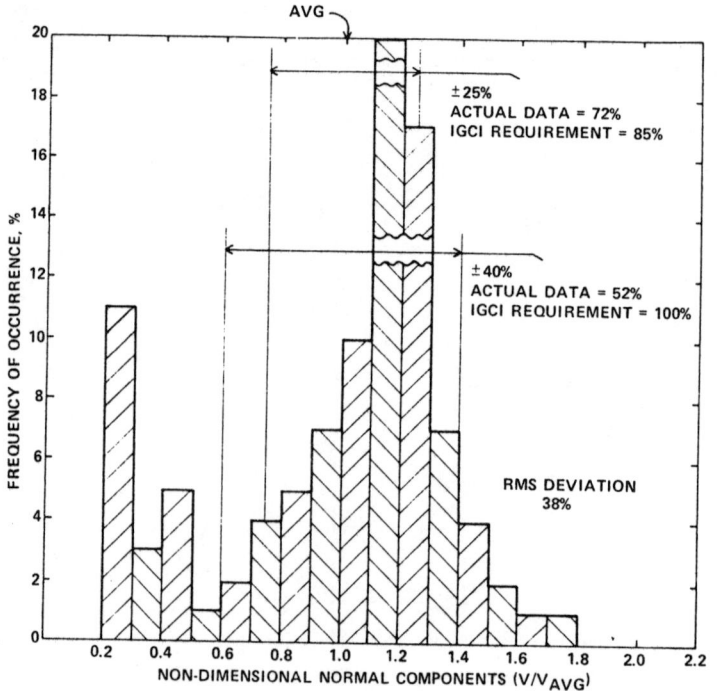

Figure 249. Histogram analysis of upper precipitator inlet velocity measurements (Unit B).[198]

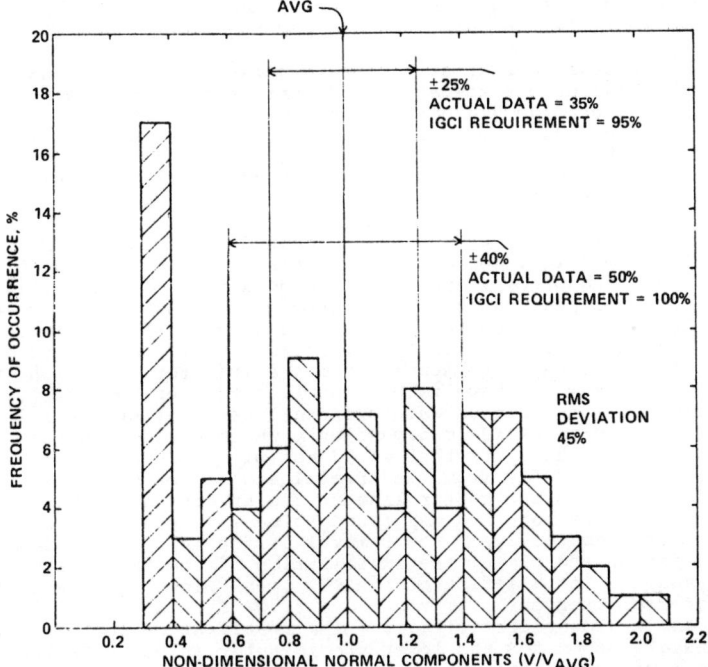

Figure 250. Histogram analysis of lower precipitator inlet velocity measurements (Unit B).[198]

A gap in the flue perforated plate created a jet that moved northward between the two sets of perforated plates. This jet then responded to a combination of a gap in the precipitator inlet perforated plate and the effects of the north i.d. fan to create a high velocity in one of the ducts. Similarly, high velocities measured in three other ducts were created by a 10.16 cm (4 inch) gap between the end of the flue and precipitator perforated plate and the north wall of the flue. This gap was the result of cumulative inaccuracies in hanging the plate panels.

All the perforated plate panels, both flue and precipitator inlet, were rehung, aligned, and clipped so as to present a flat plate structure to the gas flow. These corrections improved the gas flow distribution and brought field and model data into agreement. Based on the agreement between field and model data, a model study could then be performed to determine the design of new flow corrective devices and produce an optimized flow field in the precipitators.

Because of the very close coupling between the inlet-flue expansion turn and the precipitator, it was decided that "ladder vanes" would be used to replace the inlet radius vanes. Ladder vanes are a series of flat surfaces that are oriented perpendicular to the direction of the turn inlet gas flow. The optimum positioning of these vanes can only be done under actual flow conditions or in a model. The model study also indicated that the floor of the lower precipitator inlet flue would be subject to potential fly ash dropout. It was, therefore, recommended that a dust blower be installed in this area to keep the flue clean.

A major problem that still remained was the correction of the lower precipitator outlet-gas-flow distribution. The upper precipitator outlet flue did not have to be changed once the inlet flue was corrected. The lower precipitator outlet was still experiencing both vertical and lateral gas-flow problems. It was again confirmed that this was the result of the close coupling of the lower precipitator to the i.d. fans.

It was felt that, if a pressure-drop device could be placed at the lower precipitator outlet, a satisfactory decoupling of the i.d. fan and the precipitator could be obtained. The installation of a perforated plate at the lower precipitator outlet was rejected. It was known that perforated plates installed at the precipitator outlet tended to plug due to the electrically grounded plate collecting the residual dust leaving the precipitator. Completely new rapping systems would have to be installed to keep this perforated plate clean. The solution was to install vertical structural shaped channels of standard dimensions, which would then form continuous vertical slots that would not plug. This satisfactorily decoupled the i.d. fan from the precipitator. The vertical slots were lined up with the center line of the precipitator ducts. The net free area required was found to be 15% open. The net result of the above, i.e., the removal of the inlet flue flow biasing perforated plate, the installation of the inlet ladder vanes, and the installation of the 15% open "picket" fence at the lower precipitator outlet, produced a flow distribution slightly biased to the lower precipitator. The resultant corrected flow patterns are shown for the lower precipitator inlet in Figure 251 and the lower precipitator outlet in Figure 252. The gross improvement is noted when compared to Figures 247 and 248.

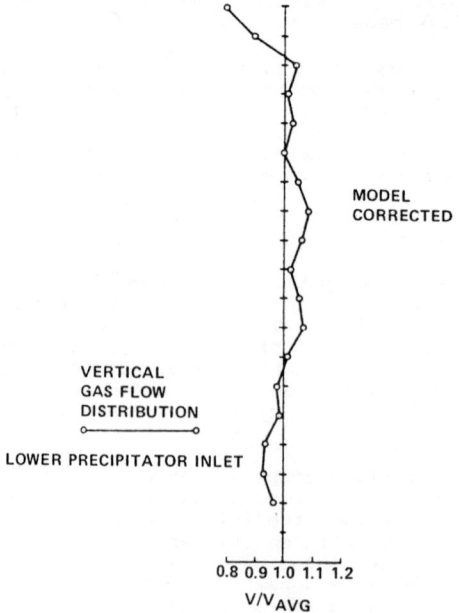

Figure 251. Vertical gas flow distribution lower precipitator inlet -- model corrected.[198]

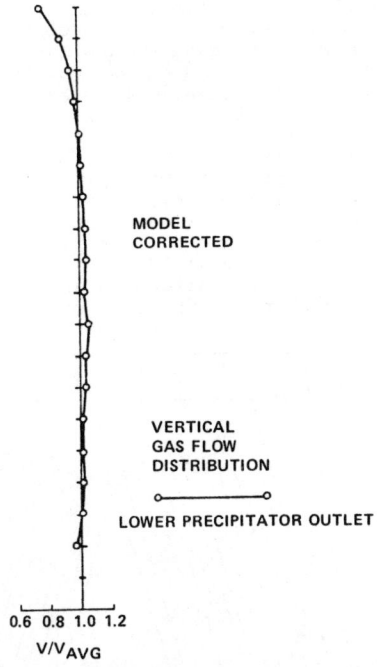

Figure 252. Vertical gas flow distribution lower precipitator outlet -- model corrected.[198]

Because of the favorable results obtained from the model study and from mathematical predictions of an improvement in precipitator efficiency to 99.76%, the full-sized flues were modified in accordance with the model recommendations. The corrections were first made on Unit B, including the complete rehanging of the inlet perforated plates. Once this was completed, a fans-running walk-through inspection was performed. No evidence of high-velocity jets or hopper sweepage could be found.

Mass efficiency tests were performed on Unit A. The unit had been permitted to operate for at least one month after the flow device modifications were made before testing. Three performance tests were run. All three tests produced results equal to or better than dust collection efficiencies which were required.

The case history just described points out the severe effect that poor gas flow distribution can have on precipitator performance, several possible causes of poor gas flow distribution, devices utilized to control the gas flow distribution, and remedies for certain specific problems. The studies performed demonstrate that, if poor gas flow distribution is a contributing factor to poor precipitator performance, the extent of the problem can readily be evaluated and appropriate remedial actions taken. The studies also point out the need for (1) careful design in integrating flues, flow control devices, precipitators, and i.d. fans, (2) careful mechanical construction, and (3) model studies of gas flow to assist in design and troubleshooting.

Figure 253 illustrates the direction of gas flow and precipitator arrangement (chevron) at a second installation where gas velocity distribution measurements have been made and analyzed.[138] Each precipitator consists of two collectors in series, each of which has 144 gas passages, with 0.229 m plate to plate spacing (9 in), 9.14 m high plates (30 ft), and 5.45 m in length (18 ft). Thus, each precipitator consists of 144 gas passages 9.14 m high (30 ft), 10.97 m long (36 ft), for a total collecting area of 28877 m^2 (311,000 ft^2) per precipitator. The precipitators each have twelve electrical sections arranged in series with the gas flow, such that the individual sections power 1/12 of the plate area and 1/12 of the length. Gas flow at full load (~700 MW) for each precipitator is about 520 m^3/sec (1.1 x 10^6 cfm) at 149°C (300°F). The specific collecting area at these conditions would be 55 m^2/(m^3/sec) or 283 ft^2/1000 cfm.

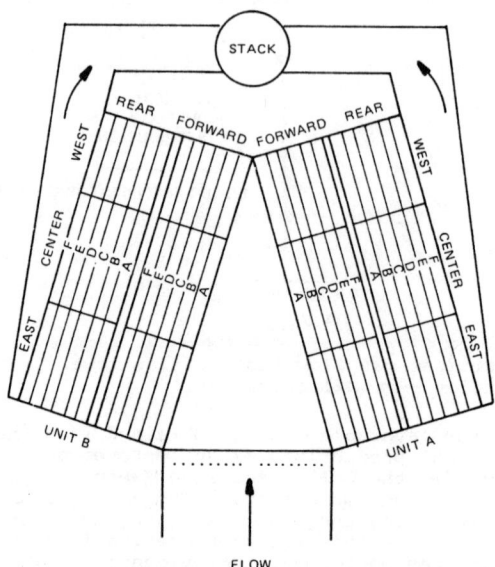

Figure 253. Precipitator layout for installation with chevron arrangement.[138]

Velocity measurements were obtained from Unit A between the third (C) and fourth (D) sections from an I beam located approximately 2.44 m (8 ft) from the top of the precipitator housing. Air flow was set at 2.6 x 10^6 kg/hr (5.7 x 10^6 lb/hr), which corresponds to full load conditions. Although the unit had been washed, considerable amounts of suspended and attached dust were present when the measurements were made. The velocity measurements were made with a thermal anemometer. The anemometer was calibrated to the "T" position frequently, but the dust concentration may have been sufficient to influence somewhat the data obtained with this instrument. The anemometer probe was maintained perpendicular to the gas stream by an aluminum guide which was held in position by the collecting electrodes. Considerable short term variation of velocity with time was noted at each point, and an attempt was made to obtain the time-averaged value at each point. The time of observation at each point was generally less than 30 seconds.

Table 28 shows the data obtained from measurements in several lanes (ducts). Standard statistical calculations on the data from the rectangular traverse from 0.92 m (3 ft) to 9.2 m (30 ft) from the baffle on lanes 1, 12, 24, 36, and 47 are shown in Table 29. Note however that the maximum and minimum velocities occur 15.24 cm (6 in) and 0 cm (0 in) from the baffle and therefore are not included in the standard traverse. Thus, the variance is possibly worse than calculated. Also note that the high velocities are at the top and bottom, thus inducing greater gas sneakage than might otherwise occur.

As a cross check, the profiles were plotted with an intention of using planimeter integrations to determine a better mean velocity than the arithmetic mean of a rectangular traverse. In the raw data, there are some anomalous points that make it difficult to draw a consistent curve for each lane. However, by averaging at each elevation across all lanes and omitting the most severely anomalous points, it was possible to obtain an average profile for the left, center, and right sections as well as for the entire unit.

Similarly, the squares of the profiles were constructed and their mean values determined by the use of the planimeter. Then the standard deviations were calculated as follows:

$$\sigma = \sqrt{\frac{\Sigma x^2}{N} - \left(\frac{\Sigma x}{N}\right)^2} = \sqrt{\frac{\int_a^b x^2}{b-a} - \left(\frac{\int_a^b x}{b-a}\right)^2}, \quad (33)$$

where

σ = standard deviation,
x = value of variable at a point,
N = total number of measurements,
a = lower limit on region containing the variable, and
b = upper limit on region containing the variable.

The analysis obtained with the planimeter is also shown in Table 29.

By comparison, the arithmetic means lie within 1% of the planimeter means and the coefficients of variance lie within 50% of those determined by planimeter. Note that the determination of the mean square by planimeter is not very accurate because the averaging over each elevation is poor (the square of the average is not the average of the squares) and because of the subjective nature of the fit of the curve to the points. The standard deviation and coefficient of variance can be expected to be small for the average because averaging reduces the effect of peaks and valleys. However, the standard deviation of two of the three groups of data are higher in the calculation by planimetry because the high values at 15.24 cm (6 in.) from the baffle are included.

Although some improvement could be obtained through corrective measures, this gas velocity distribution is fairly good. Since the precipitator could achieve a collection efficiency in excess of 99.5% with an average current density of 20 nA/cm^2 (21.5 µA/ft^2), the gas velocity distribution was probably not a serious factor

TABLE 28. VELOCITY DISTRIBUTION FROM UNIT A OF CHEVRON ARRANGEMENT

(Obtained on I Beam between 3rd and 4th field in direction of flow)

VELOCITY FT./MIN (m/min)

LOCATION FROM TOP OF BAFFLE, FT. (m)	LANE #1L	LANE #12L	LANE #24L	LANE #36L	LANE #47L	LANE #1R	LANE #12R	LANE #24R **
0 (0)	18(5.5)	15(4.6)	18(5.5)	18(5.5)	18(5.5)	35(10.7)	20(6.1)	15(4.6)
1/2 (.15)	60(18.3)	210(64.)	260(79.3)	150(45.8)	350(106.8)	450(137.2)	410(125.)	350(106.8)
3 (.92)	250(76.2)	185(56.4)	220(67.1)	240(73.2)	300(91.5)	350(106.8)	330(100.6)	300(91.5)
6 (1.83)	300(91.5)	185(56.4)	265(80.8)	250(76.2)	340(103.7)	370(112.8)	290(88.4)	320(97.6)
9 (2.75)	255(77.8)	160(48.8)	230(70.2)	185(56.4)	250(76.2)	80(24.4)	240(73.2)	280(85.4)
12 (3.66)	183(55.8)	150(45.8)	170(51.8)	140(42.7)	270(82.4)	340(103.7)	200(61.)	210(64.)
15 (4.58)	146(44.5)	135(41.2)	135(41.2)	125(38.1)	240(73.2)	320(97.6)	170(51.8)	175(53.4)
18 (5.49)	135(41.2)	130(39.6)	150(45.8)	155(47.3)	215(65.6)	300(91.5)	180(54.9)	180(54.9)
21 (6.41)	155(47.3)	175(53.4)	175(53.4)	190(58.)	200(61.)	280(85.4)	240(73.2)	190(58.)
24 (7.32)	135(41.2)	148(45.1)	50(15.2)*	240(73.2)	200(61.)	280(85.4)	240(73.2)	190(58.)
27 (8.24)	140(42.7)	170(51.8)	170(51.8)	200(61.)	205(62.5)	200*(61.)	250(76.2)	250(76.2)
30 (9.15)	155(47.3)	190(58.)	170(51.8)	240(73.2)	230(70.2)	320*(97.6)	300(91.5)	320(97.6)

* Change to Lane 2R

** Sneakage above the plate hangers 25 ft./min (7.63 m/min) 45° from normal flow, one foot above wire hanger

(CONTINUED)

TABLE 23. (CONTINUED)

VELOCITY FT./MIN (m/min)

LOCATION FROM TOP OF BAFFLE, FT. (m)	LANE #36R	LANE #47R	LANE #LC1*	LANE #LC12	LANE #LC24	LANE #LC36	LANE #LC48
0 (0)	20(6.1)	20(6.1)	35(10.7)	15-20(4.6-6.1)	15(4.6)	20(6.1)	100(30.5)
1/2 (.15)	400(122.)	70(21.4)	300(91.5)	300(91.5)	340(103.7)	450(137.2)	400(122.)
3 (.92)	300(91.5)	310(94.6)	230(70.2)	250(76.2)	270(82.4)	310(94.6)	280(85.4)
6 (1.83)	280(85.4)	280(85.4)	240(73.2)	230(70.2)	280(85.4)	290(88.4)	270(82.4)
9 (2.75)	180(54.9)	270(82.4)	220(67.1)	190(58.)	230(70.2)	200(61.)	270(82.4)
12 (3.66)	220(67.1)	220(67.1)	195(59.5)	190(58.)	200(61.)	190(58.)	250(76.2)
15(4.58)	200(61.)	190(58.)	200(61.)	170(51.8)	140(42.7)	190(58.)	260(79.3)
18(5.49)	180(54.9)	180(54.9)	190(58.)	170(51.8)	170(51.8)	180(54.9)	280(85.4)
20(6.10)						250(76.2)	
21(6.41)	180(54.9)	230(70.2)	190(58.)	160(48.8)	40(12.2)		280(85.4)
24(7.32)	190(58.)	230(70.2)	170(51.8)	150(45.8)	230(70.2)	230(70.2)	300(91.5)
27(8.24)	190(58.)	250(76.2)	180(54.9)	210(64.)	240(73.2)	280(85.4)	290(88.4)
29(8.85)						300(91.5)	
30(9.15)	330(100.6)	230(70.2)	220(67.1)	250(76.2)	300(91.5)		370(112.8)

*Sneakage above electrode hanger 35 ft./min., 8 rows down 150 ft./min.
LC30 150 ft./min., LC36 45 ft./min.

HOPPERS-RIGHT CHAMBER
Row 12 3 ft. (0.92 m) below plate 40 ft./min. (12.2 m/min)
Row 24 3 ft. (0.92 m) down   40 ft./min. (12.2 m/min)
Row 36 3 ft. (0.92 m) down   50 ft./min. (15.2 m/min)
Row 48 3 ft. (0.92 m) down   40 ft./min. (12.2 m/min)

in limiting precipitator performance. The only bad feature of the distribution is the location of the peak velocities at the top and bottom baffles. This might tend to increase gas sneakage past the electrified regions.

TABLE 29. STATISTICAL EVALUATION OF VELOCITY DISTRIBUTION FROM UNIT A OF CHEVRON ARRANGEMENT

Standard Statistical Calculations

	Mean	Standard Deviation (S.D.)	Coefficient of Variance S.D./Mean
Left	192.6	54.7	.284
Center	227.5	56.5	.249
Right	246.7	61.5	.249
Average	222.3	61.5	.277

Planimeter Analysis

	Mean	Standard Deviation (S.D.)	Coefficient of Variance S.D./Mean
Left	191	36.6	.191
Center	229	88.5	.386
Right	245	79.4	.324
Average	221	44.1	.200

Figure 254 illustrates the direction of gas flow and precipitator arrangement at a third installation where gas velocity distribution measurements have been made and analyzed.[201] The electrostatic precipitator consists of three fields in the direction of gas flow. The precipitator is physically divided into two collectors (A & B). The test program was performed on the "A" side of the precipitator. The total collecting area for the "A" side is 7,374.4 m^2 (79,380 ft^2), 2458.13 m^2 (26,460 ft^2) per field. This gives a specific collection area of 34.475 m^2/(m^3/sec) (175 ft^2/1000 cfm) for the design volume flow of 213.82 m^3/sec (453,000 acfm) per collector. Each collector has three double half-wave transformer rectifiers, one per field. The precipitator has 27.94 cm (11 in.) plate spacing and operates at approximately 149°C (300°F). The emitting electrodes are square twisted wires with an approximate diameter of .419 cm (.165 in.) and are 10.0 m (32' 9 3/4") long. There are 12 wires per lane per field for a total of 1512 wires. The discharge electrodes are held in a rigid frame, and each frame holds 4 wires.

Figure 255 shows the gas velocity distribution obtained under air load conditions at the face of the first field of the precipitator. These measurements were obtained using a thermal anemometer after the precipitator was washed during an outage. The average velocity and the square of the average velocity for all the passages on which measurements were obtained are plotted as a function of vertical position. The average velocity and the average of the velocity squared were obtained by planimetry. The average velocity obtained was 1.74 m/sec (5.71 ft/sec), and the standard deviation was 0.955 m/sec (3.13 ft/sec), or 55% of the average velocity. This distribution is undesirable because of the large standard deviation and the location of the highest velocities in the region near the bottom of the precipitator. However, at the outlet sampling plane, the flow distribution was changed such that the highest velocities occurred in the upper portion of the duct. Flow distribution plates located at the precipitator outlet offered more flow resistance at the bottom than at the top and thus are probably responsible for the change in relative flow pattern.

Figure 256 illustrates the direction of gas flow and precipitator arrangement at a fourth installation where gas velocity distribution measurements have been made and analyzed.[202] A mechanical collector, which was reported to have been reworked when the precipitator was installed, precedes the electrostatic collector at this installation. The precipitator consists of four fields in the direction of gas flow and is physically divided into two collectors (A and B). The test program was conducted on the "B" side of the precipitator. The total collecting area for the "B" side is 5900.64 m^2 (63,516 ft^2) with 1475.16 m^2 (15,878 ft^2) per field. This gives a specific collection area of 43.48 m^2/(m^3/sec)(220.9 ft^2/1000 cfm) for the design volume of 135.70 m^3/sec (287,500 acfm) per collector. The precipitator has six full-wave transformer rectifiers; each transformer rectifier has an "A" and "B" bushing. The precipitator has 30.5 cm (12 in.) plate spacings and operates at approximately 160°C (320°F). The emitting electrodes are rigid "barbed" electrodes which are 0.502 m (1' 7 3/4") apart in the direction of gas flow.

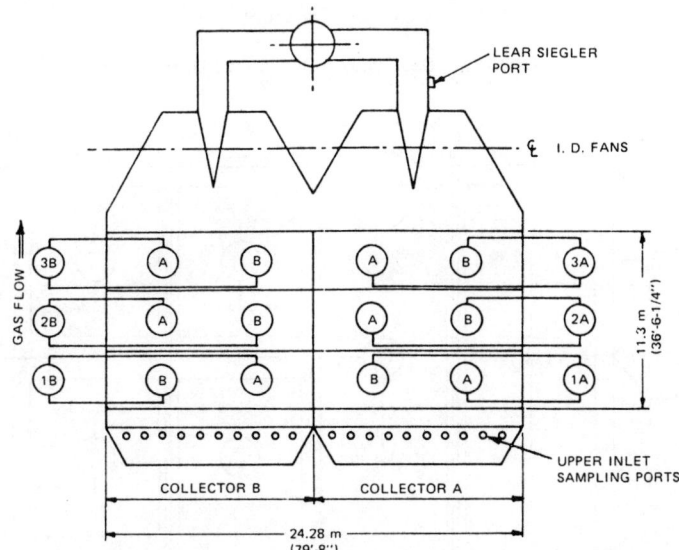

Figure 254. Precipitator layout for third gas velocity distribution analysis.[201]

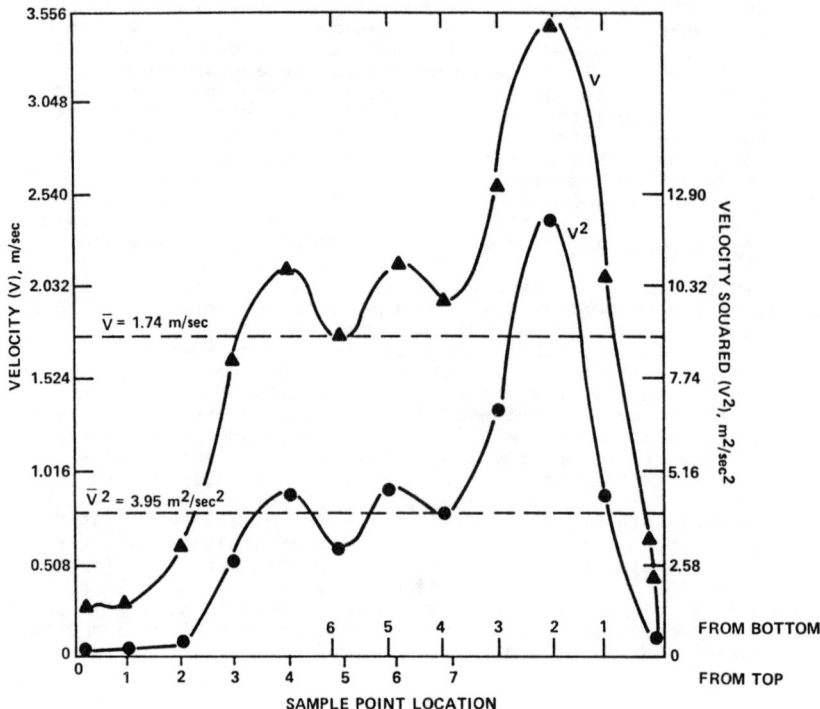

Figure 255. Gas velocity distribution.[201]

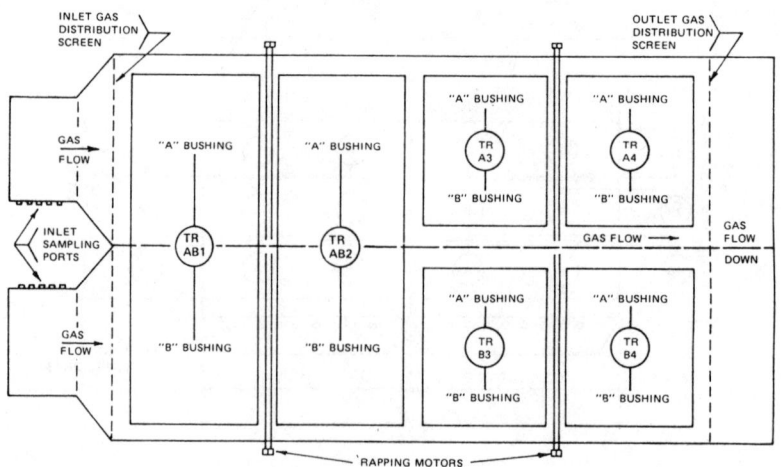

Figure 256. Precipitator layout for the fourth gas velocity distribution analysis.[202]

The gas velocity distribution inside the precipitator was measured at the leading edge of the second field. The gas velocity was measured at 132 points as indicated by the black dots on the isopleth of the velocity distribution shown in Figure 257. This isopleth was constructed from air load data obtained with the F.D. and I.D. fans operating with current settings corresponding to full load operation. The mean velocity was 1.51 m/sec (4.95 ft/sec) with a standard deviation of 0.47 m/sec (1.54 ft/sec) or 34% of the mean velocity. The isopleth shows that the velocity is higher in the top of the unit than in the lower portion. These data also show that the upper diagonal support braces in the unit produced regions of higher than normal gas velocity.

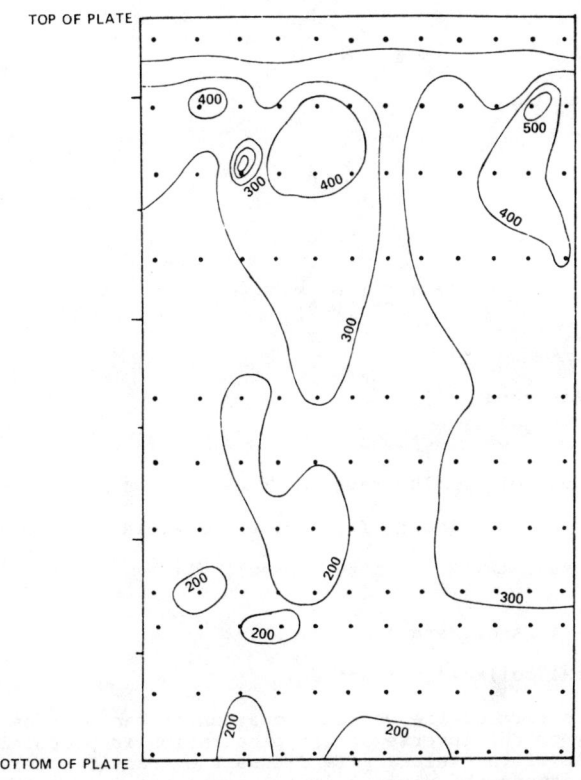

Figure 257. Gas velocity distribution (ft/min).[202]

Correlation of Collection Efficiency with Gas Velocity Distribution--

At the present time, no exact methods exist for correlating precipitator collection efficiency with gas velocity distribution. However, several approaches have been proposed that demonstrate the general trends to be expected due to a nonuniform gas velocity distribution.[198,199,203] All these approaches utilize equation (2) or one that is similar in form. Thus, a reduced gas flow in a finite section of the precipitator results in an increased collection efficiency whereas an increase in gas flow will result in a decrease in collection efficiency.

In order to demonstrate the general considerations to be made in accounting for the effects of a nonuniform gas velocity distribution on collection efficiency, one[203] of the previously referenced approaches will be developed here. It will be assumed that Equation (2) as written applies to each particle size with a known migration velocity, w, and that the specific collection area and size of precipitator are fixed.

Given:

$$\eta = 1 - e^{-\frac{A_p w}{Q}}.$$

It can be seen that

$$1 - \eta = e^{-\frac{A_p}{A_1}\frac{w}{u_a}}, \qquad (34)$$

and

$$\ln\left(\frac{1}{1-\eta}\right) = \frac{A_p}{A_1}\frac{w}{u_a} = \frac{k}{u_a}, \qquad (35)$$

where

A_p = plate area (m²),

A_1 = inlet cross sectional area (m²),

Q = inlet volume flow rate (m³/sec),

w = migration velocity for a given particle size (m/sec),

u_a = average inlet velocity (m/sec),

$k = \frac{A_p w}{A_1}$ (m/sec), and

η = ideal collection fraction.

From this form of the Deutsch equation it can be seen that the logarithm of the inverse of the penetration is proportional to the inverse of the velocity (and thus the transit time). The precipitator can now be divided into a number of imaginery channels corresponding to pitot traverse points. Using the altered form of equation (2), the losses for all the channels can be summed and averaged to obtain the mean loss in the precipitator using an actual velocity distribution instead of an assumed uniform distribution. This can be accomplished as follows:

(1) Calculate constant k from the efficiency predicted under ideal conditions:

$$k = u_a \ln \frac{1}{1-\eta}$$

(2) Calculate the mean penetration:

$$p = \frac{1}{Nu_a} \sum_{i=1}^{N} u_i(1-\eta_i) , \qquad (36)$$

or

$$p = \frac{1}{Nu_a} \sum_{i=1}^{N} u_i e^{-\frac{k}{u_i}} , \qquad (37)$$

where

N = number of points for velocity traverse,

u_i = point values of velocity (m/sec), and

η_i = point values of collection fraction for the particle size under consideration.

Note that the average penetration is a weighted average to include the effect of higher velocities carrying more particles per unit time than lower velocities.

For any practical velocity distribution and efficiency, the mean penetration obtained by summation over the velocity traverse will be higher than the calculated penetration based on an average velocity. If an apparent migration velocity for a given particle size is computed based upon the mean penetration and equation (2), the result will be a value lower than the value used for calculation of the single point values of penetration. The ratio of the original migration velocity to the reduced "apparent" migration velocity is a numerical measure of the performance degradation caused by a non-uniform velocity distribution. An expression for this ratio may be obtained by setting the penetration based on the average velocity equal to the corrected penetration obtained from a summation of the point values of penetration, and solving for the required correction factor, which will be a divisor for the migration velocity.

The correction factor "F" may be obtained from:

$$\exp\left(-\frac{k}{F} \cdot \frac{1}{u_a}\right) = \frac{1}{Nu_a} \sum_{i=1}^{N} u_i \exp(-k/u_i) = p . \qquad (38)$$

Therefore,

$$F = -\frac{k}{u_a (\ln p)} . \qquad (39)$$

Whether the quantity F correlates reasonably well with statistical measures of velocity non-uniformity is yet to be established. A limited number of traverse calculations seem to indicate a correlation between the factor F and the normalized standard deviation of the velocity traverse. Figure 258 shows F as a function of the ideal efficiency for several values of gas velocity standard deviation. These curves were obtained by computer evaluation of equation (39), and the data on which the calculations are based were obtained from Preszler and Lajos.[18] The standard deviations have been normalized to represent a fraction of the mean. The overlapping of the curves for standard deviations of 1.01 and 0.98 indicates that the standard deviation alone does not completely determine the relationship between F and collection efficiency.

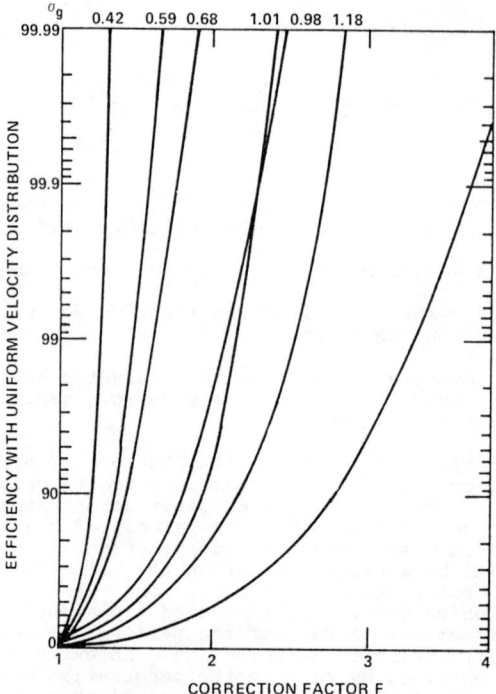

Figure 258. "F" as a function of ideal efficiency and gas flow standard deviation.

The data in Figure 258 were used to obtain the following empirical relationship between F, the normalized standard deviation of the gas velocity distribution (σ_g), and the ideal collection predicted for the particle size under consideration:

$$F = 1 + 0.766\ \eta\sigma_g^{1.786} + 0.0755\ \sigma_g \ln\left(\frac{1}{1-\eta}\right), \qquad (40)$$

where

$$\sigma_g = \frac{\sqrt{\frac{1}{N}\sum_{i=1}^{N}(u_a - u_i)^2}}{u_a} \qquad (41)$$

This relationship is based on a pilot plant study, and should be regarded as an estimating technique only. If it is desirable to simulate the performance of a particular precipitator, the preferred procedure would be to obtain the relationship between F, η and σ_g

for the conditions to be simulated from a velocity traverse at the entrance to the unit.

Gas Sneakage--

Gas sneakage around the precipitation zones may occur at the bottoms of the plates, at the tops of the plates, and on the outsides of the plates adjacent to the precipitator shell. Gas sneakage oocurs because of the pressure drop across the precipitator, flow separation, and in some cases by aspiration effects. Adequate measures exist to prevent significant gas sneakage. Gas sneakage can be reduced by frequent baffles which force the gas to return to the main gas passages between the collection plates, subdivision of collecting zones into several series sections, and maintenance of good gas flow conditions to and out of the precipitator. The use of baffles has been discussed earlier in this text.

If there were no baffles, the percent sneakage would establish the minimum possible penetration because it would be the percent volume having zero collection efficiency. For example, if 5% of the gas volume bypasses the precipitation zones, the collection efficiency can be no better than 95%, even though all other factors are perfect. Gas sneakage can be an especially serious problem for precipitators designed for very high collection efficiencies because only a small percentage of gas bypassage may be sufficient to prevent the attainment of the desired performance. With baffles, the gas sneakage remixes with part of the main gas flow and then bypassage occurs again in the next unbaffled area. The limiting penetration due to gas sneakage will therefore depend on the amount of sneakage gas per baffled section, the degree of remixing, and the number of baffled sections.

Gas sneakage results in undesirable gas flow and eddy formation inside and above the hoppers. This can result in considerable particle reentrainment back into the main gas stream due to hopper sweepage and hopper boil-up after a section of collection plates is rapped. Baffles and hoppers must be designed to minimize these reentrainment effects due to gas sneakage. Hopper designs have been discussed earlier in this text. Even with good baffling, some gas flow will travel through the regions inside and above the hoppers. Thus, hopper design must be such as to minimize the effects of gas sneakage. Effective hopper design must consider several aerodynamic effects, including Bernoulli's principle, flow separation, and vortex formation. Actual designs are best determined by model studies and observations on full-scale precipitators.

If we make the simplifying assumption that perfect mixing occurs following each baffled section, an expression for estimating and demonstrating the effect of gas sneakage may be derived as follows:[204]

Let S = fractional amount of gas sneakage per section,

η = collection fraction of a given size particle obtained with no sneakage for total collection area,

η_j = collection fraction per section of a given particle size = $1 - (1 - \eta)^{1/N_s}$,

N_s = number of baffled sections, and

p_j = penetration from section j.

Then the penetration from section one is given by:

$$p_1 = S + (1 - n_j)(1 - S) \quad , \tag{42}$$

and from section 2

$$\begin{aligned} p_2 &= Sp_1 + (1 - n_j)(1 - S)p_1 \\ &= p_1 [S + (1 - n_j)(1 - S)] \\ &= [S + (1 - n_j)(1 - S)]^2 \quad , \end{aligned} \tag{43}$$

and from section N_s (the last section),

$$\begin{aligned} p_{N_s} &= [S + (1 - n_j)(1 - S)]^{N_s} \\ &= [S + (1 - S)(1 - n)^{1/N_s}]^{N_s} \quad . \end{aligned} \tag{44}$$

Figure 259 shows a plot of the degradation of efficiency from 99.9% design efficiency versus percent sneakage with number of baffled sections as a parameter. For high efficiencies, the number of baffled sections should be at least four and the amount of sneakage should be held to a low percentage. With a high percentage of sneakage, even a large number of baffled sections fails to help significantly. This graph can also be applied to reentrainment due to hopper sweepage and hopper boil-up as will be described later when discussing particle reentrainment.

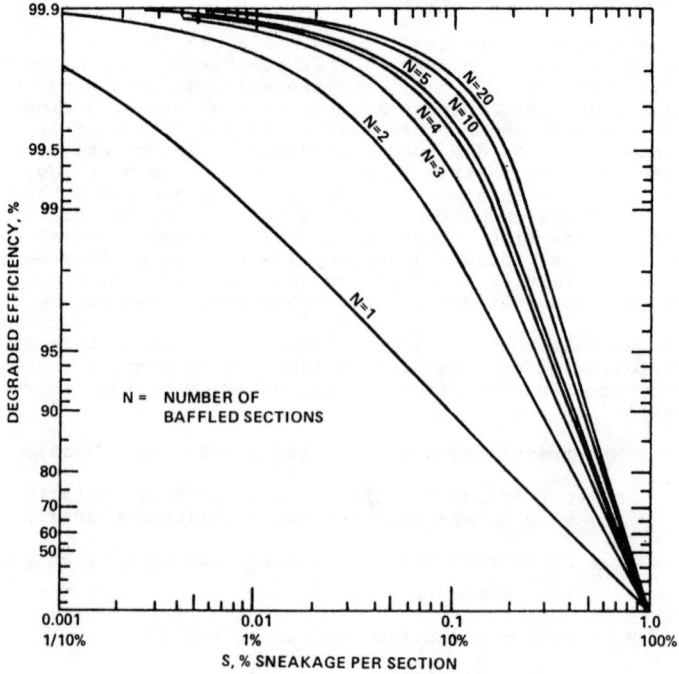

Figure 259. Degradation from 99.9% efficiency with sneakage.

We can define a bypass or sneakage factor, B, analogous to the gas flow quality factor F, in the form of a divisor for the migration velocity in the exponential argument of equation (2):

$$B = \frac{\ln(1-\eta)}{N_s \ln[S + (1-S)(1-\eta)^{1/N_s}]} \quad . \quad (45)$$

Figure 260 shows a plot of the factor versus sneakage for a family of ideal efficiency curves for five baffled sections. Similar curves can easily be constructed for different numbers of sections.

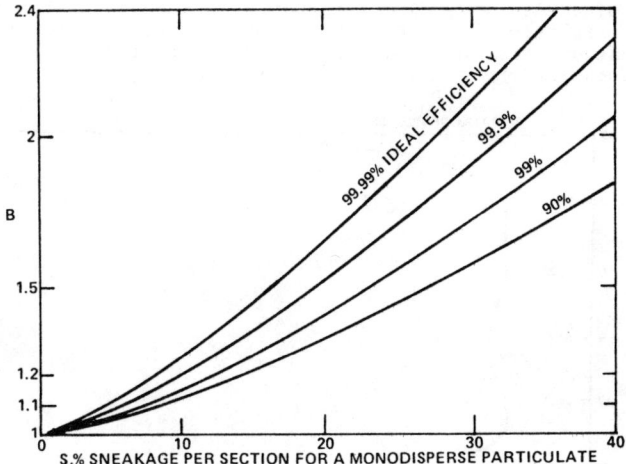

Figure 260. Correction factor for gas sneakage when $N_s = 5$.

The foregoing estimation of the effects of sneakage is a simplification in that the sneakage gas passing the baffles will not necessarily mix perfectly with the main gas flow, and the flow pattern of the gas in the bypassage zone will not be uniform and constant. The formula is derived to help in designing and analyzing precipitators by establishing the order of magnitude of the problem. Considerable experimental data will be required to confirm the theory and establish numerical values of actual sneakage rates.

A rough estimate of the gas sneakage occurring at the second installation discussed earlier (shown in Figure 253) has been made based on velocity measurements made above and below the collection plates (see Table 28). Calculations were performed based on

$$\% \text{ Sneakage} = 100 \times \frac{\sum(\text{Peak-Average})}{N \times \text{Average}} \quad , \quad (46)$$

where the peak values recorded in the standard traverse are utilized and N is the total number of measurements. The calculations yield the following results:

	Right	Center	Left	Average
% Top Sneakage	4.2	2.3	5.5	3.9
% Bottom Sneakage	3.4	6.2	2.4	4.0
Total	7.6	8.5	7.9	7.9

Therefore, a rough estimate of the gas sneakage is 8%.

A profile of the hopper sneakage occurring at the fourth installation discussed earlier was also measured. This profile is shown in Figure 261. The sharp increase in air flow observed at the bottom of the hopper occurred where the center baffle of the hopper terminated. This flow probably does not occur when the hopper is partially full. Comparison of Figures 257 and 261 indicates that gas sneakage through the hopper regions is significant at this installation.

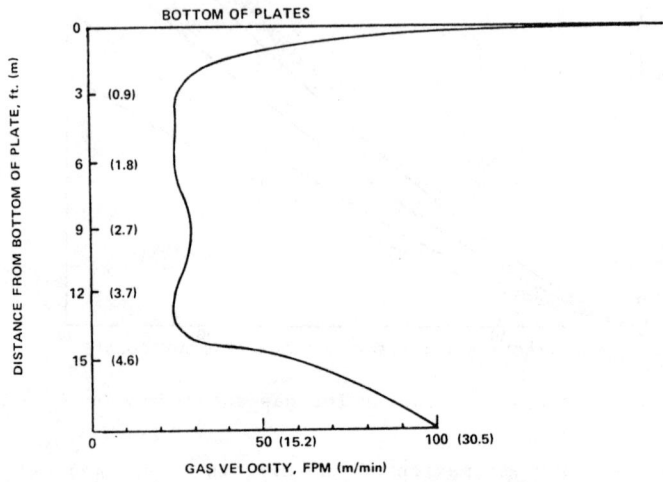

Figure 261. Velocity profile in hopper.[202]

Air Flow Model Studies--

Basis for model studies--Although the first precipitator flow model studies were performed in the early fifties, a widespread use of flow modeling techniques was not made until the late sixties. With the ever increasing size of thermal power stations, uniformity of gas flow, dust distribution, and the gas temperature profile at the inlet of the precipitator become of prime importance. With prior attention focused primarily on structural and space problems, negligence of proper ductwork, flow control device, and hopper design resulted in poor performance of the equipment, excessive pressure losses, large dust accumulations, and corrosion due to uneven gas temperature distribution. In larger boiler units for modern thermal power plants, one inch w.g. of pressure drop can be evaluated as an annual operating cost of $40,000 or more.

The main purposes of a model study are to determine the location and configuration of gas flow control devices, such as vanes, baffles, perforated plates, to satisfy the contractural requirements on gas velocity distribution in the inlet and outlet of the electrostatic precipitator, and to minimize the pressure drop through the complete system. Also, accurate flow model studies offer the potential of detecting and correcting flow problems in the design stage. Even if only qualitative results can be obtained in a model study, these can be extremely useful in providing recognition of potential problem areas. Also, it is the opinion of some that proper attention to details (in both the model and prototype) will produce a one-to-some correlation between the model and the field.[198] This was demonstrated quite convincingly in one field and model study where detailed velocity traverses for both configurations were in good agreement.[198] Since the investment in a flow model study is relatively small when compared to the total precipitator investment and possible financial losses due to poor flow design, this type of study is probably justified when designing most new precipitator installations.

Similarity of fluid flows[199]--The gas velocity distribution in the electrode system of the electrostatic precipitator is analyzed using accepted procedures based on similarities of fluid flows in the three-dimensional scale model and the full-size system (prototype).

The similarity of the fluid flow conditions between the model and the prototype are dependent on matching some or all of a series of dimensionless parameters, which describe the characteristics of the prototype and model, as well as those of the flows, as ratios of the fluid forces.

In general, three similarities must be satisfied to obtain valid results with the fluid flow models. These similarities are: (1) Geometric Similarity, (2) Kinematic Similarity, and (3) Dynamic Similarity. Any of two flow systems satisfy geometric similarity, if all dimensions have the same scale factor,

$$\ell_1/\ell_2 = \text{constant}, \tag{47}$$

where

ℓ = typical length (indices 1 and 2 distinguish between the two flow systems).

Kinematic similarity requires, in addition, that any two flow systems have the same relative velocities and accelerations throughout such that

$$v_1/v_2 = \text{constant} \tag{48}$$

$$b_1/b_2 = \text{constant} \tag{49}$$

where

v = typical velocity and

b = typical acceleration.

It is normally not too difficult to match all of these requirements in a precipitator model.

The third requirement of a dynamic similarity requires, in

addition, the similarity of pressures at corresponding points of model and prototype such that

$$p_1/p_2 = \text{constant}, \qquad (50)$$

where

p = typical dynamic pressure.

To completely satisfy the similarity of dynamic pressures, a number of ratios of forces in both fluid systems, such as Reynolds. Froude, Weber, Euler, and Mach Number have to be identical.[205] Most of the time, this condition can not be met in model study work.

A different approach would be to maintain equality of Reynolds and Froude numbers only. Scales for the model can be developed based on gas viscosities and densities of both systems.

Scales for model and prototype can be developed by using Reynolds and Froude Numbers of both systems, resulting in a length scale of:

$$\lambda = \ell_1/\ell_2 = (\nu_1/\nu_2)^{2/3} \qquad (51)$$

a time scale of:

$$\tau = t_1/t_2 = (\nu_1/\nu_2)^{1/3} \qquad (52)$$

and a general force scale of:

$$\Xi = p_1/p_2 = \rho_1/\rho_2 \, (\nu_1/\nu_2)^2 \qquad (53)$$

The general force scale can be extended to cover scales for inertial, frictional, and gravitational forces with ν representing gas kinematic viscosity and ρ gas density.

Scale factors for other units of measurements can be calculated from these basic scales; for example, for gas velocity:

$$v_1/v_2 = \lambda/\tau = (\nu_1/\nu_2)^{1/3} \qquad (54)$$

and for gas volume:

$$Q_1/Q_2 = \lambda^3/\tau = (\nu_1/\nu_2)^{5/3} \qquad (55)$$

The general use of these ratios in flow model studies of electrostatic precipitators would require rather large models. For example, for flue gas with a temperature of 180°C in the prototype and an air temperature of 20°C in the model, the length scale would be 1 to 1.6; the velocity scale 1 to 1.3; and the volume scale 1 to 3.2.

A different approach for model studies, where a significant decrease in size of the model is intended, would be to arbitrarily select a scale factor for a typical length; for example, 1 to 16, and to match the Reynolds number of the prototype by increasing the fluid velocity in the model or changing the fluid properties.

Increasing the system velocity creates significantly larger pressure losses and requires higher head fans. The gas velocity of the prototype system mentioned earlier may be 1.2 m/sec. To match the Reynolds Number using air as model fluid in a 1 to 16 scale model would require an air velocity of 9.6 m/sec., an increase by a factor of eight. The system pressure loss would, thus, increase by a factor of 64; for example, from a design pressure loss of 250 mm H_2O to 16,000 mm H_2O.

To match the Reynolds Number of the prototype, the model fluid properties could be adjusted by changing the fluid temperature or using a fluid other than air, but neither of these approaches is very practical.

The Reynolds Number is the ratio of inertial to viscous forces. When the inertial forces predominate, flow separation from the critical surfaces occurs and is principally a function of the geometry of the system. If the value of the Reynolds Number is well within the turbulent range (Re > 3×10^3, for example), the behavior of the fluid can be successfully modelled at a Reynolds Number other than that of the prototype system.

The model flow pattern observed at reduced Reynolds Number levels will be identical to the full size system, and the model pressure drop will be only slightly higher due to the influence of the Reynolds Number on frictional pressure losses.

In industrial flue systems, which are usually designed to connect major pieces of equipment, many conditions will establish flow separation and induce turbulence. Therefore, the calculated value of Re is no indication of the quality of flow or the state of turbulence. Once the condition of flow separation is established (inertial forces predominating) the flow pattern tends to remain the same over a wide range of calculated Re values. That is, kinematic similarity is established in industrial flues substantially independent of variations in average velocity or model factors.

Therefore, the model study does not have to match the full size Re value. It suffices that:

$$\frac{v_1 \ell_1}{\nu_1} \geqslant 3 \times 10^3 \tag{56}$$

and

$$\frac{v_2 \ell_2}{\nu_2} \geqslant 3 \times 10^3 \ . \tag{57}$$

However, Reynolds Number must be considered when the conditions of pressure drop and dust drop-out are studies in the model. It has been shown that the boundary-layer thickness of gas on any surface is an inverse function of Re. In the usual industrial scale model study, Re will be proportionately low due to the scale factor, and the boundary-layer will be too thick. This condition tends to give a conservative estimate of pressure drop and dust drop-out.

Model studies involving two-phase fluid systems or airborne particulates influenced by gravitational forces, require the adherence to a constant Froude Number, i.e.:

$$\frac{v_1^2}{\ell_1 g} = \frac{v_2^2}{\ell_2 g} \ . \tag{58}$$

Another approach, which is frequently used, consists of using a 1:16 scale model with the collecting surface plates installed in a 1:8 scale, and, thus, test at a flow condition characterized by a Reynolds Number in the turbulent range, closer to the Reynolds Number of the prototype.

The fluid velocity level in a model should be selected to be in a range which can be easily and accurately measured (velocity head above 10 mm H_2O) but low enough to be incompressible (Mach Number below 0.2). As a result, the fluid velocity in a duct will normally range from 10 to 20 m/sec, and in the precipitator model itself, from 0.5 to 3.0 m/sec; the latter being measured with a hot wire anemometer.

If smoke is used to visualize the flow pattern, the fluid velocity should not exceed 10 m/sec to maintain visibility of the smoke pattern.

It is recommended to use a fan with a variable speed drive or to have air dampers between flow model and fan to be able to reduce the air flow through the model to one-half or one-third of the design flow volume during the test program.

Flow model construction[199]--Three-dimensional scale models have become the most widely used means for a fluid velocity distribution analysis. For purpose of convenience, some of the models made in early gas flow studies were constructed on a scale of 1.9 cm (3/4 in.) in the model being equivalent to .3048 m (one foot) in the prototype; that is, the model was 1/16 of actual size. This scale became common and is widely used in the industry, although for more demanding work, especially for predictions of pressure drop and dust fallout, a scale of 1 to 12 or even 1 to 10 could become increasingly more necessary.

The model is a precise replica of the entire gas cleaning system and includes items, such as air preheaters, steam generator economizers, flues, flow control devices, precipitators, fans, stacks, etc. All of the model or parts of it are made of transparent plastic to make it possible to observe flow indicators, such as cotton tufts or smoke and dust fallout. Internal parts of the ductwork, such as flow control devices, may be constructed of light gauge sheet metal.

The precipitator model has sidewalls, hoppers, box girders, and roof made out of transparent plastic. Collecting surface plates are made out of flat sheets of plastic or metal and are hung between the box girders or plate supports. Normally, only the first and last electrical fields need to be equipped with collecting surface plates. Walkways, horizontal and vertical baffles are included in the model, as well as hopper partitions. The discharge system is normally not included in the model.

Inlet and outlet nozzles are also made out of transparent plastic. Perforated plates or similar devices used for gas distribution are selected with equal opening ratios as those to be used in the prototype.

The air preheater is modeled as exact as possible, complete with transitions between the round arc of the wheel and the rectangular outlet flanges, as well as the wash-out hopper underneath the air preheater outlet duct.

The model is set up on the suction or pressure side of a fan with a suitable gas volume, normally following the configuration used in the prototype.

Larger models need a separate support structure and access platforms next to the test ports.

Instrumentation[199]--Air velocity distributions in the ductwork of the model can be measured with a calibrated standard pitot tube; for example, Dwyer 0.32 cm (1/8 inch) diameter, with an inclined water manometer; for example, Meriam Model M-173-FB with a range of 0-15.2 cm (0-6 inches) and minor graduations of 0.03 cm (0.01 inch).

Static pressures in ductwork can be measured with a calibrated standard pitot tube connected to an inclined water manometer; for example, Meriam Model HE 35 WM with a range of 0-35.6 cm (0-14 inches) and minor graduations of 0.03 cm (0.01 inch).

The air velocity distribution in the model of the precipitator chamber is measured with a hot-wire linear flow flowmeter; for example, Datametrics (Gould) model 800-LV with a model U-25 probe.

Particle Reentrainment

Rapping Reentrainment--

Background--Rapping reentrainment is defined as the amount of material that is recaptured by the gas stream after being knocked from the collection plates by rapping or vibration. With perfect rapping, the sheet of collected material would not reentrain, but would migrate down the collection plate in a stick-slip mode, sticking by the electrical holding forces and slipping when released by the rapping forces. However, the rapping forces are necessarily large to overcome adhesion forces, and a significant percentage of the material is released into the gas stream as sheets, agglomerates, and individual particles. Most of the material is recharged and recollected at a later stage in the precipitator.

The purpose of an electrode rapping system is to provide an acceleration to the electrode which is sufficient to generate inertial forces in the collected dust layer that will overcome those forces holding the dust to the electrode. Electrode rapping systems have been described earlier in this text. A successfully designed rapping system must provide a proper balance between electrode cleaning and minimizing emissions resulting from rapping reentrainment. Presently, two approaches are prevalent with regard to the removal and transfer of the particulate from the collecting plates. One approach is to rap often and to provide maximum rapping acceleration to the plates during each rap in an attempt to minimize the thickness of the residual ash layer. The other approach is to vary the intensity and frequency of the rapping in an attempt to minimize the quantity of material reentrained. A determination of the best rapping technique for a specific application depends on an understanding of the mechanisms by which ash is actually removed and transferred from the collection plates during a rapping sequence and of the effects of residual ash layers.

The mechanics of the ash removal process vary with the properties of the ash, precipitator operating conditions, and rapping parameters. Ash properties and precipitator operating conditions affect the adhesion and cohesion of the ash layer. The adhesion and cohesion of ash layers depend upon particle-to-particle forces.

According to Tassicker,[206] the component forces are: London-van der Waals, triboelectric, capillary, surface dipole, and electric-field corona forces. These component forces are influenced by the following: particle diameter, porosity and compaction of the layer, complex dielectric constant, humidity in the gas, adsorbed surface dipolar molecules, work-function interfaces on the material, and the electric field and current density in the ash layer. The above considerations point out the difficulty and complexity which would be involved in predicting ash removal properties.

A relationship for the electrostatic force which acts upon the ash layer as a whole has been presented earlier in equation (23). In most practical applications, the net electrical force should be in the direction that forces the dust layer on to the collection surface. However, in certain cases, the net force can be such as to pull the ash layer off the collection surface. This can occur for low resistivity ashes or for low operating current densities, as indicated in Figure 211.

An elementary theory of dust removal which considers only the tensile strength of the dust layer and the acceleration normal to the plate has been developed by Tassicker.[207] The theory predicts that the dust layer is removed only when

$$a_n > \frac{P}{\delta \ell} = \frac{P}{(M/A)} \quad , \qquad (59)$$

where

a_n = acceleration normal to the plate (m/sec^2),

P = tensile strength of the dust layer (nt/m^2),

δ = bulk density of the dust (kg/m^3),

ℓ = dust layer thickness (m), and

M/A = mass per unit area (kg/m^2).

According to this relationship, for removal of a given dust thickness, the rapping intensity must be of sufficient magnitude to produce an acceleration greater than the ratio of the tensile strength of the ash layer to the mass per unit area. For a given normal acceleration, the dust layer is removed only when

$$M/A > P/a_n \quad ; \qquad (60)$$

that is, when the mass per unit area (dust surface density) is greater than the ratio of dust layer tensile strength to the normal plate acceleration. Since the mass per unit area depends on the dust layer thickness, which in turn is related to collection time between raps, the time interval between the raps is directly related to the efficiency of dust removal from the plates. As collection time between raps is increased, the mass per unit area is increased, and the acceleration required for removal is decreased. Experimental data obtained by Sproull[208] and by Penney and Klingler[169] show that the requirements for removal of a precipitated dust layer are in basic agreement with Tassicker's elementary theory for dust removal.

Sproull[208] has conducted a series of experiments which illustrate the effect of dust composition, corona forces, accelerations, and temperature on the removal of dust layers from collection electrodes. Figure 262 presents some of Sproull's data to illustrate the relative effects of these parameters as a function of the maximum shear acceleration of the collecting electrodes in multiples of "g". A comparison of these curves indicates that, under the conditions of the experiments, the cement dust was more difficult to remove than fly ash, even though the particle size distributions of the two dusts were similar, presumably as a result of differences in composition. It is also clear that the electrical holding force was acting to retain the dusts on the collection electrode surface. Similar data were obtained for acceleration perpendicular to the electrode plate produced by a normal rap. Lower values of acceleration were required for removal of difficult-to-remove dust with normal rapping than was the case for shear rapping.

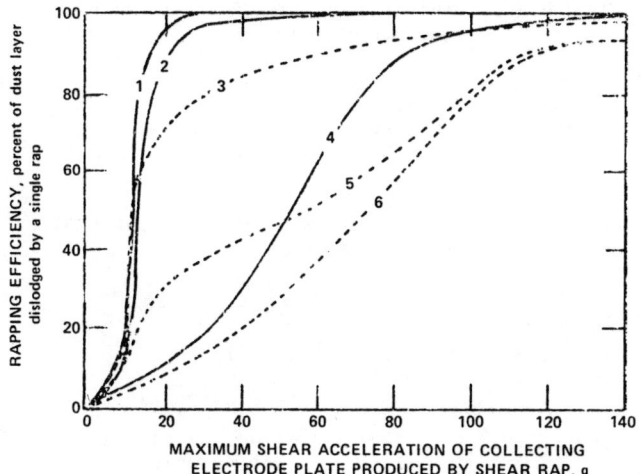

Figure 262. Shear (parallel) rapping efficiency for various precipitated dust layers having about 0.2 grams of dust per square inch as a function of maximum acceleration in multiples of "g". Curve (1) fly ash, 70° to 300°F, power off. Curve (2) fly ash, 300°F, power on. Curve (3) cement kiln feed, 70°F, power off. Curve (4) cement kiln feed, 200 or 300°F, power on. Curve (5) fly ash, 70°F, power on. Curve (6) cement kiln feed, 70°F, power on.[208]

Figure 263 (also from Sproull) illustrates the effect of temperature on the removal efficiency of a precipitated layer of copper ore reverberatory furnace dust. These data indicate that the net holding force on the dust layer decreases with increasing temperature until softening or partial melting occurs, excluding the cases in which the dust temperature falls below the dew point of the surrounding gases.

Particle reentrainment is influenced by factors concerning the design and operation of the precipitator as well as the physical and chemical properties of the dust. White[209] has summarized the particle properties and precipitator design factors which affect reentrainment and these are presented in Table 30.

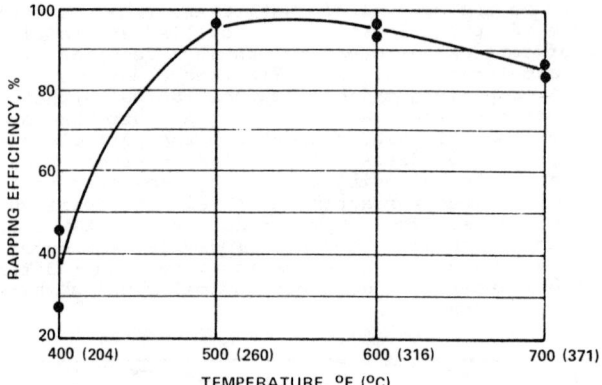

Figure 263. Rapping efficiency for a precipitated layer of copper ore reverberatory furnace dust, rapped with a ballistic pendulum having an energy of 0.11 foot-pound, at various temperatures.[208]

TABLE 30. PARTICLE PROPERTIES AND PRECIPITATOR DESIGN FACTORS WHICH AFFECT REENTRAINMENT[209]

PARTICLE SIZE	PRECIPITATOR FACTORS
1. Size Distribution	1. Gas Velocity
2. Shape	2. Gas Flow Quality
3. Bulk Density	3. Collecting Electrode Configuration and Size
4. Adsorbed Moisture and Other Vapors	4. Electrical Energization
5. Environment-Gas Temperature	5. Rappers: Type, Number, and Amplitude
6. Resistivity	6. Hopper Design
	7. Air In-leakage into Hoppers or Precipitator Proper
	8. Dust Removal System Design and Operation
	9. Single Stage or Two Stage

Although hopper design and ash removal system operation do not influence the manner in which particles are directly reentrained as a result of rapping, improper operation of the ash removal system can increase emissions through hopper boil-up resulting from rapping or as a result of gas circulation through the hoppers.

Sproull[210] has reported that optimum rapping conditions are achieved when the collected dust layer is permitted to accumulate to a reasonable thickness and then rapped with sufficient intensity to progress down the plate in a slip-stick mode. This procedure has the advantage of resulting in the deposition of only a portion of the dust on the lower portion of the collecting plate into the hoppers at any one time. These circumstances would minimize the disturbance of previously deposited dust since the velocity of the falling layer would be relatively low.

The foregoing considerations illustrate that it is desirable to vary both rapping intensity and rapping interval in order to optimize the performance of a dust removal system. Since the mass rate of dust collection varies with length through a precipitator, it follows that rapping frequency variations between the inlet and exit fields would be expected to yield the best rapping conditions. If a precipitator consists of four fields in the direction of gas flow and exhibits a no-reentrainment efficiency of 99%, the rate of build-up in the first field would be about 30 times that in the outlet field, again neglecting reentrainment effects. However, the optimum rapping intervals for these fields would not be expected to correspond to the dust collection rate ratios.

Recently, work has been performed to develop models for describing dynamically the vibrational modes and accelerations produced in a full-scale collection plate with an attached dust layer by a rapping force of a given intensity.[211,212] Although these studies are still in the elementary stage and are not of practical value as of yet, they have the potential for answering several questions pertaining to what is the best method of removing the dust layer from the collection electrode. For example, are high frequency, small amplitude vibrations or low frequency, large amplitude vibrations more effective in removing a dust layer? Also, what is the relative importance of normal and shear forces in removing a dust layer?

Emissions due to rapping--Emissions due to rapping and their dependence on rapping parameters have been reported by Sproull,[210] Plato,[213] Sanayev and Reshidov,[214] Schwartz and Lieberstein,[215] and Nichols, Spencer, and McCain.[216] Some of the results of this work has been discussed above. In summary, these workers have observed that (1) reducing the intensities of the raps lead to a reduction in rapping emissions,[210] (2) vertical stratification of the emissions occurred during rapping, with higher concentrations in the lower portion of the precipitator,[210] and (3) improvements in the performance of full-scale precipitators occurred when the time intervals between raps were increased.[213,214,215,216] Although these studies have added to the understanding of rapping reentrainment and some of the variables affecting the emissions due to rapping reentrainment, they do not provide quantitative data on the amounts of emissions and particle size distributions due to rapping reentrainment.

One study on a pilot plant[20] and another study on six full-scale precipitators[19] have yielded quantitative information on the emissions due to rapping reentrainment. A complete characterization of rapping reentrainment requires the measurement of a large variety of variables. A block diagram of an experimental layout for the pilot study is shown in Figure 264. In addition to the data that are obtained with this arrangement, a complete characterization utilizes the precipitator design data.

The field experiments included a similar set of measurements to those made during the pilot studies. However, sampling view ports for photographing rapping emissions and for determining the vertical stratification of the rapping emissions were not available in the full-scale units nor were load cells for measuring the quantity of fly ash collected on the collection plates. Hence these measurements were not included in the field tests.

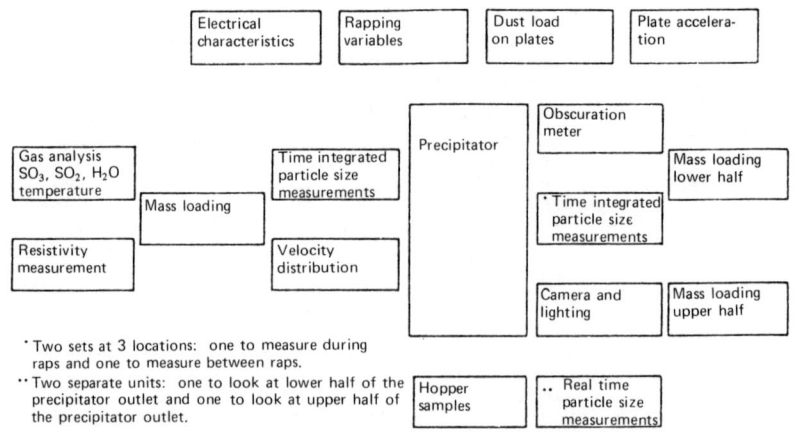

* Two sets at 3 locations: one to measure during raps and one to measure between raps.
** Two separate units: one to look at lower half of the precipitator outlet and one to look at upper half of the precipitator outlet.

Figure 264. Block diagram of experimental layout for a rapping reentrainment study.[20]

The quantification of rapping reentrainment requires methods of measuring the mass and particle size distribution of particulate exiting the precipitator with and without rapping. During both the pilot and full-scale precipitator test programs, an optical real-time system and integrating mass systems were used. For the full-scale tests, particle size measurements were obtained using a method based on electrical mobility analysis for particle diameters between 0.01 µm and 0.3 µm.

Mass measurements were obtained with in-stack filters. The sampling probes used at the inlet and outlet were heated and contained pitot tubes to monitor the velocity at each sampling location for the full-scale tests. Glass fiber thimbles were used at the inlet to collect the particulate and Gelman 47 mm filters were used at the outlet. Different procedures were employed at the pilot unit compared to the full-scale units.

At the pilot plant facility, two outlet sampling trains were used: (1) the upper sampling train for the upper 68% of the precipitator outlet and (2) the lower sampling train for the lower 32% of the precipitator. The outlet sampling locations were about 1 meter from the plane of the outlet baffles, and only one lane of the precipitator was sampled. Both outlet mass trains were modified to consist of two systems: one of which was used to measure emissions between raps and the other was used to measure emissions during raps. Each outlet sampling probe consisted of a 2.5 cm pipe, to the end of which two 47 mm Gelman filters with 1.25 cm nozzles pointed 110° apart were attached. Separate copper tubes were run to each filter from a three-way valve. The valve was used to connect the appropriate filter to the metering box. Sampling rates at each traverse point were based on velocity traverses made prior to the sampling.

One of the two filters on each of the two outlet probes was designated the between rap sampler and the other the rapping puff sampler. After stable conditions were obtained, the between rap sampling systems were started. Before rapping the plates, sampling was discontinued and the probes were rotated so that both nozzles on each probe pointed downstream. The dust feed was turned off, and after a clear flue was obtained, the second filter was rotated into the gas stream. Sampling was resumed and the plates were rapped. When dust had settled, sampling with this second set of filters was discontinued and the nozzles to the filters were again pointed downstream. The dust feed was then turned on and the sampling was resumed again with the between rap system.

Data obtained with the between rap system were handled in the usual manner and were used to calculate steady-state mass emission rates. Data from the second set, or "rap" set of filters were used to calculate emission rates from the rapping puffs independently of the between-rap emissions. These emission rates were calculated from

$$E_p = \frac{M_p A_s N_H}{A_N N_s}, \qquad (61)$$

where

E_p = emission rate from rapping puffs (kg/hr),

M_p = mass collected by the filter while sampling the flue gas during rapping (kg),

A_s = cross-sectional area of precipitator sampled by the probe (m²),

A_N = cross-sectional area of nozzle (m²),

N_H = number of raps per hour (#/hr), and

N_s = number of raps sampled.

The emission rates between raps and from raps were combined to obtain the overall hourly emission rate.

For a full-scale precipitator installation one would expect to be able to measure rapping reentrainment simply by obtaining data with either a mass train or an impactor sampling system, with a rapping system energized and subsequently de-energized and then comparing these measurements. When utilizing these integrating or time-averaging, inertial systems for the measurement of rapping reentrainment, a sampling strategy must be developed which will differentiate between steady-state particulate emissions and those which result from electrode rapping. At the first full-scale installation (Plant 1) tested, the strategy employed consisted of sampling on subsequent days with the rapping system energized and subsequently de-energized while an attempt was made to maintain boiler operating parameters as constant as was practical. The precipitator was characterized by high collection efficiency (99.9%), which required extended sampling times to obtain meaningful mass measurements. However, it was found that the sensitivity of the electrostatic precipitator to changes in resistivity and other process variables could mask the differences in total emissions caused by energizing and de-energizing the rappers. The variation in precipitator performance caused by the resistivity and other process variable changes made it impossible to determine rapping reentrainment losses from a direct comparison of data obtained one day with rappers in the normal mode and rappers de-energized on subsequent days.

In order to minimize the above difficulty, a revised sampling strategy was adopted for the remaining installations. The revised strategy consisted of sampling with mass trains and impactors dedicated to designated "rap" and "no-rap" periods. Data with a rapping system energized and de-energized were obtained by traversing selected ports with dedicated sampling systems in subsequent periods on the same day. This procedure, while necessarily distorting the frequency of the rapping program being examined, minimized the effects of resistivity and other process variable changes.

The use of the alternating sampling strategy leads to at least three possible procedures for calculating the fraction of losses attributable to rapping reentrainment. The first procedure consists of the calculation of the ratio of emissions obtained with rappers off to rappers on and subtracting from unity. The emissions data utilized in this procedure were obtained during the time in which alternating sampling periods for rap and no-rap sampling trains were employed. The second procedure consists of subtracting the mass emissions obtained with the rappers de-energized from those of the previous day with normal rapping, and dividing by the emissions obtained with the rappers operating normally. The data obtained from the "rap" period will be approximately equal to that obtained during other test periods in which the rappers are operating in a normal fashion if: (1) the distortion of the rapping frequency does not significantly influence emissions during the "rap" period and (2) there are no other variations in parameters affecting the precipitator performance.

A third possible procedure consists of the use of a weighted time average emission during the rap-no-rap periods as an approximation to the normal emission rates, subtracting the no-rap emission from the weighted time average, and dividing the difference by the weighted time average to obtain the fraction of emissions attributable to rapping. This procedure provides an estimate of rapping reentrainment with the effective intervals which result from the alternating sampling periods. All of the above calculation procedures were used when applicable to analyze emissions data from the six installations tested.

The three size selective sampling systems which were used in the measurement programs consisted of a large particle sampling system (LPSS) containing an optical single particle counter, an ultrafine particle sampling system containing an electrical aerosol analyzer (EAA), and cascade impactor sampling systems. The operating principles of these sampling systems have been discussed earlier in this text. The optical and electrical sampling systems provided real-time data obtained from an extracted gas sample while the inertial sampling systems provided time-integrated in situ data. The large particle sizing system (diameter range $\overline{0.6\text{-}2.0}$ μm) was employed only for outlet measurements to provide qualitative information on the relative fractions of the emissions that could be attributed to rapping losses in the precipitator. In addition, this system also provided data on particulate concentration changes with time. The ultrafine particle sampling system (0.01 μm to 0.3 μm) was employed at both the inlet and outlet of the full-scale precipitators for purposes of providing fractional efficiency data and to give quantitative information on the contribution of rapping, if any, to emissions in this particle size range.

The pilot-scale rapping tests were conducted on a nearly full-scale pilot precipitator owned and operated by FluiDyne Engineering. Figures 265 and 266 illustrate the features of the test facility. This pilot unit effectively represents one electrical section in a full-scale precipitator. The plate height is 6 meters, and the plate length is 2.7 m. The total collecting area is 167 m^2, and wire-to-plate spacing is 11 cm. In the original design, the plates were constructed from expanded metal. For this rapping reentrainment study, three of these plates were replaced to provide two lanes with solid plates on each side of the lane. Outlet sampling was confined to the lanes with solid plates. The plate rappers are of the single shot pneumatic type. The rapper weight is supported in a cylinder by low pressure compressed air. When a rap is desired, a signal to a solenoid valve pressurizes the other side of the cylinder and forces a weight down on top of a rod that transmits the force to a plate support beam.

Dust feed is supplied from a dust dispersion system which has an adjustable feed capability. Three oil burners are available to heat the gas stream to the desired temperature level. A water injection system consisting of three atomization nozzles, each with a capacity of eleven liters of water per minute, is available to supply the desired humidity. The water is atomized by compressed air and is vaporized by the burners that heat the system gas flow to the design temperature.

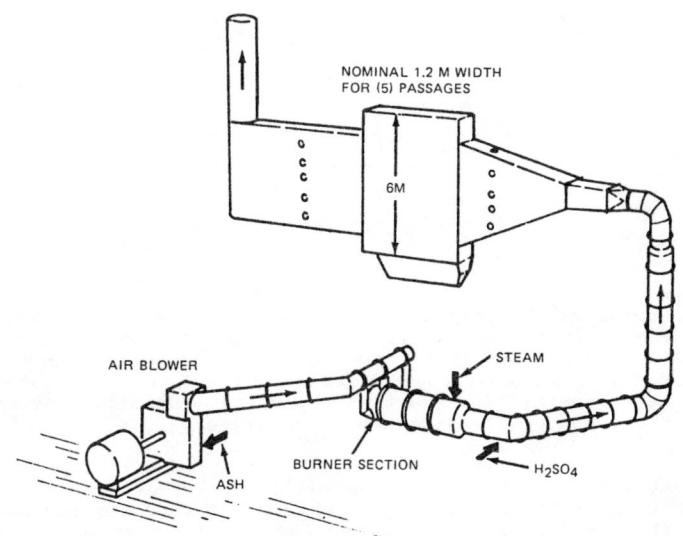

Figure 265. Near full-scale pilot precipitator at FluiDyne Engineering.[20]

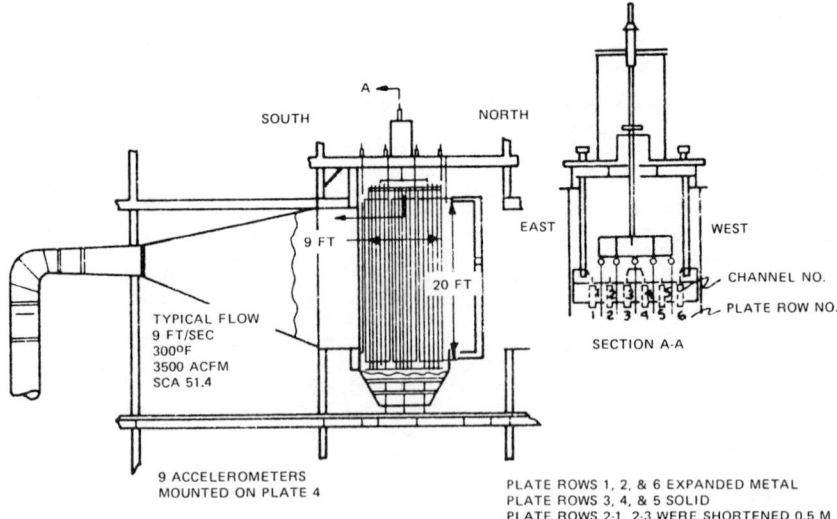

Figure 266. FluiDyne pilot precipitator.[20]

Table 31 presents a summary of results obtained from the experiments on the FluiDyne Pilot Unit. These results indicate that rapping emissions decreased with increasing time between raps.

TABLE 31. RESULTS FROM PILOT-SCALE RAPPING EXPERIMENTS

Type of test	Plate acceleration G's x,y,z axis	Rap intervals, min	Gas velocity, m/sec	Avg. plate current density, nA/cm²	Total penetration, %	Penetration due to rapping reentrainment, %
Rap	11 16 15	12	0.87	23.3	11.4	53
Rap		32			7.6	32
Rap		52			6.1	18
Rap		150			6.9	25
No Rap		--			5.2	--

Figure 267 shows the effect of rapping interval on efficiency. The percentage of the collected dust removed from the collecting electrode also increased with increased time between raps, as Figure 267 illustrates. These results are consistent with the theory of dust removal which indicates that the product of the normal plate acceleration and the dust surface density must be greater than the tensile strength of the layer.

Figure 268 also illustrates the build-up of a residual dust layer that was not removed with normal plate accelerations on the order of 11 Gs. There are several possible causes for the development of the residual layer. For one, the dust layer directly in contact with the collection plates has a much higher tensile strength than the remainder of the layer. Estimates for removal of the layer called for accelerations greater than 10^3 Gs (9.8×10^5 cm/sec²). Consolidation of the dust that remains on the plate after a rap also aids in producing residual layers. The vibrations

during a rap can have the effect of compacting the dust layer if
it is not removed making it more difficult to remove. A third
possible cause of the residual layer is the removal of patches of
dust only from selected locations on the collection plates where
the removal criteria are met. Dust can be removed from one location during one rap and from another location on the next rap due
to changes in distribution of the dust surface density. This
results in a nonuniform dust layer and the presence of a residual
layer. This is often the result of nonuniform plate accelerations.
At one location where plate accelerations were on the order of
only 4 to 5 Gs (3.4-4.9×10^3 cm/sec^2), residual dust layers as
thick as 2 cm were observed in the vicinity of plate baffles where
the plate accelerations are dampened. Between the baffles, the
residual layers were only 1 to 2 mm thick.

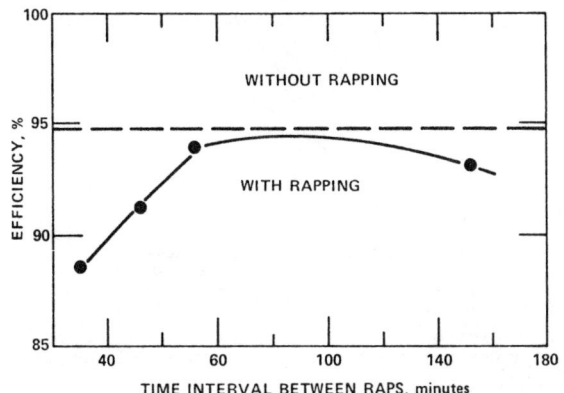

Figure 267. Average efficiencies for FluiDyne pilot precipitator for various rapping intervals.[20]

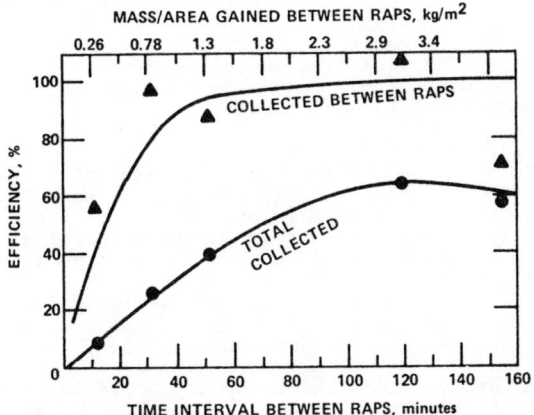

Figure 268. Dust removal efficiency versus the time interval between raps.[20]

Figure 269 presents particle size distribution of rapping puffs for the indicated rapping interval. These data suggest that thicker dust layers produce larger reentrained particles upon rapping. An inspection of the impactor substrates at the outlet sampling locations 2 and 3 revealed that the majority of the large particles in the rapping puffs were agglomerates. Producing relatively large agglomerates instead of individual particles is desirable because the larger agglomerates are recollected faster than discrete particles or smaller agglomerates.

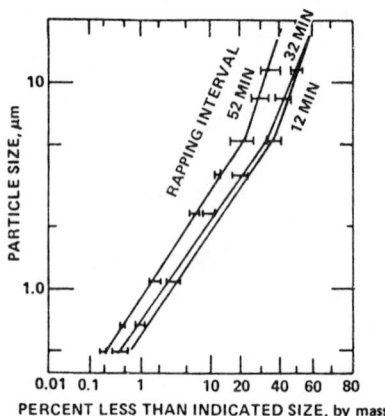

Figure 269. Cumulative percent distribution for rapping puffs, rapping intervals of 12, 32, and 52 minutes, pilot test.[20]

In the FluiDyne pilot plant study, it was evident that "boil-up" from the hoppers comprised a significant portion of the reentrainment. The measurement of the vertical distribution of the rapping loss at the FluiDyne Pilot unit indicated that 82% of the rapping emission occurred in the lower 32% of the precipitator. This effect was apparently due to both hopper boil-up and gravitational settling of the reentrained material. Figure 270 illustrates the vertical stratification as a function of particle size. All of the particle size bands show a decrease in concentration with increasing distance from the bottom baffle.

Rapping puffs observed in the lower portion of the precipitator occurred in two bursts for both upstream and downstream raps as shown in Figure 271. The first burst lasted 2-4 seconds. This burst was interpreted as being the result of particulate reentrained directly in the gas stream and being carried out of the precipitator at the velocity of the gas through the unit. The longer lasting second burst, which for the larger particles was a series of puffs, can be interpreted as resulting from hopper "boil-up". These data indicate that hopper "boil-up" contributes significantly to rapping reentrainment emissions.

Motion pictures of the dust removal process in the Southern Research Institute (SoRI) small-scale precipitator and the FluiDyne pilot precipitator have produced several observations relating to the dislodgement of dust after a rap and to the reentrainment of dust due to the rap. Motion pictures (32 frames/sec) of the removal of a dust layer (2-3 mm thick) by rapping in the SoRI small-scale unit show the dust layer fracturing along lines of discontinuity in the dust surface. The resulting fractured sheet of dust starts to fall as separate sheets which break up as they encounter other falling sheets and patches of unremoved dust. The dust appears to fall without being recollected and to become turbulently mixed as it falls. The motion pictures show the majority of the dust dropping into the hoppers from which a portion boils up and becomes reentrained into the gas stream. Motion pictures taken in the large pilot precipitator at FluiDyne Engineering showed similar behavior.

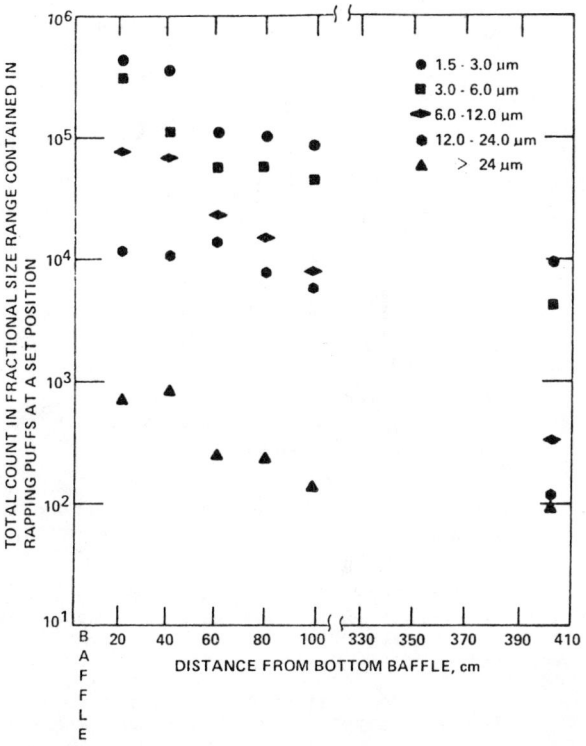

Figure 270. Spatial distribution of particles in rapping puff.[20]

In terms of location in the power plant system and type of fuel burned in the boiler, the six full-scale installations studied may be classified as follows:

    Plants 1 and 5 - Cold-side ESPs collecting ash from low sulfur western coals,

    Plant 6   - Hot-side ESP collecting ash from low sulfur western coal,

    Plant 4   - Hot-side ESP collecting ash from low sulfur eastern coal,

    Plants 2 and 3 - Cold-side ESPs collecting ash from high sulfur eastern coals.

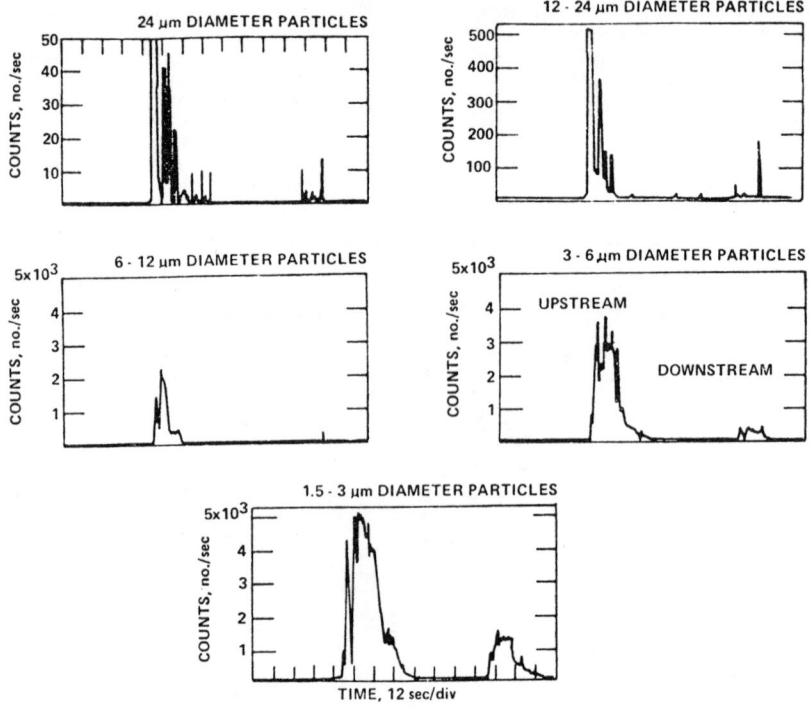

Figure 271. Rapping puffs at the exit plane of the pilot precipitator, upstream and downstream raps.[20]

Table 32 summarizes the important design parameters and the results obtained for the six installations.[217] A mechanical collector preceded the precipitator at Plant 1 and Plant 3. The installations were characterized by relatively high overall mass efficiency. Rapping losses as a percentage of total mass emission ranged from over 80% for one of the hot-side units to 30% for the cold-side units. The high rapping losses at Plant 4 are probably due both to reduced dust adhesivity at high temperatures and the relatively short rapping intervals.

Table 33 lists the rapping intervals for each field at the various installations.[217] Also shown are the effective rapping intervals resulting from the alternating sampling schedules which were used to obtain the rap-no rap data. To the extent allowed by process variations, the range of emissions attributable to rapping should be established by the calculations using (rap-no rap) and (normal-no rap) data sets. However, the time weighted average (TWA) calculation is of interest in that it indicates the change in rapping emissions caused by the effective increase in time intervals between raps. With the exception of the normal current density data set at Plant 2, the time weighted average calculation gives the lowest percentage emissions due to rapping of the three calculation methods. Table 34 provides typical flue gas and fly ash compositions obtained at the test sites.

TABLE 32. SUMMARY OF RESULTS FROM EPRI TESTS[217]

Plant	1	2	3	4	5	6
Number of Electrical Fields in Direction of Gas Flow	6	3	4	4	5	6
Plate-to-Plate Spacing, cm	30.48	27.94	25.4	22.86	24.76	22.86
Emitting Electrode Design	Mast with Square Twisted Wires	Mast with Square Twisted Wires	Rigid Barbed Wires	Hanging Round Wires	Electrode Frame With Spiral Wires	Hanging Round Wires
Rapper Design	Drop Hammer	Drop Hammer	Tumbling Hammers	Magnetic Drop Hammer	Tumbling Hammers	Magnetic Impulse Hammers
Portion of ESP Tested	Total	1/2	1/2	1/2	1/6	1/16
Boiler Load During Test, MW	128	160	122	271	508	800
Gas Flow During Test, am^3/sec	330.2	155.2	117.2	203.9	149.4	126.8
Temperature During Test, °C	152.2	155	157.2	321.1	106.1	358.9
SCA During Test, m^2/(m^3/sec)	113.5	47.6	50.4	76.8	117.9	55.4
Measured Efficiency, %	99.92	99.55	99.80	99.64	99.85	98.98
Dust Resistivity at Operating Temp, Ω-cm	1.4x10^{11}	1.7x10^{10}	2x10^{10}	3.2x10^{10}	4.6x10^{11}	1.5x10^{9b}
% of Mass Emissions Attributed to Rapping[a]	31	65-33	30	85	36-29	63-44

[a] Indicating range of values from two methods of calculation.
[b] Laboratory measurement.

TABLE 33. SUMMARY OF REENTRAINMENT RESULTS [217]

Plant	1		2			3			4			5			6	
	Raps/Hr		Raps/Hr Rap - No-Rap			Raps/Hr			Raps/Hr			Raps/Hr			Raps /73 Min	Raps/Hr Rap-
Field	Normal	Normal	Normal Current Density	One-Half Normal Current Density		Normal	Rap- No-Rap		Normal	Rap- No-Rap		Normal	Rap- No-Rap		Normal	No-Rap
1	6	10	4.29	3.75		10	1.67		30-60	12.5-25		10	4.17		8	2.74
2	6	6	2.57	2.25		10	1.67		30-60	12.5-25		5	2.08		8	2.74
3	3	1	0.43	0.38		5	0.83		30	12.5		5	2.08		3	1.03
4	3	-				5	0.83		30	12.5		2	0.83		3	1.03
5	1	-				-			-	-		1	0.42		1	0.34
6	1	-				-			-	-		-	-		1	0.34
Rapping Losses, % of Emissions																
Rap-No Rap/Rap	-		65	55		30			85			29			44	
Normal-No Rap/ Normal	31		33	82		-			85			36			63	
T.W.A.-No Rap/ T.W.A.	-		45	38		18			71			15			24	

TABLE 34. TYPICAL FLUE GAS AND ASH COMPOSITIONS[217]

Plant	1	2	3	4	5	6
Date	8/7/75	1/16/76	2/25/76	4/28/76	10/6/76	1/31/77
Flue Gas						
Temp., °C	164	154	155	333	106	346
SO_2, ppm by vol.	282	3200	2430	750	470	355
SO_3, ppm by vol.	6.5	12	8.3	2.7	<0.5	<0.5
H_2O, vol. %	8.2	7.2	8.2	7.4	8.7	9.6
Fly Ash						
Ash Source	Hopper 1	High Vol. Sample	High Vol. Sample	High Vol. Sample	High Vol. Sample	High Vol. Sample
Date	8/7/75	1/15/76	3/2/76	4/27/76	10/5 & 10/6/76	1/31/77
Wt. % of[1]						
Li_2O	0.02	0.02	0.03	0.04	0.02	0.013
Na_2O	0.9	0.54	0.67	0.43	1.38	1.52
K_2O	1.72	2.49	2.12	3.5	0.54	1.4
MgO	3.61	0.95	1.00	1.3	1.1	1.8
CaO	8.71	4.73	4.95	1.1	5.8	6.0
Fe_2O_3	5.49	22.72	13.13	7.2	6.1	5.0
Al_2O_3	24.64	18.52	21.76	28.4	13.2	24.3
SiO_2	50.55	45.69	50.23	53.8	70.8	57.6
TiO_2	1.22	1.45	1.96	1.8	0.87	2.1
P_2O_5	0.50	0.30	0.78	0.23	0.05	0.32
SO_3	0.55	2.77	2.29	0.50	0.50	0.54
LOI[2]	0.61	5.72	10.92	3.5	1.0	0.11

[1] Chemical analyses obtained from ignited samples

[2] Loss on ignition

Figure 272 shows the time variations over the test period at Plant 1 in boiler load, precipitator power, dust resistivity and relative particle concentrations in two size bands (0.6 to 1.8 μm and 1.5 to 3 μm). August 5 and 6 were "normal" rapper operation test periods, whereas August 7 and 8 were "no-rap" test periods. It is readily apparent that, on August 7, changes in variables other than rapper energization caused exit particulate concentration changes which masked the effect of rapping system de-energization. The LPSS system, however, was able to detect rapping puffs, as described below.

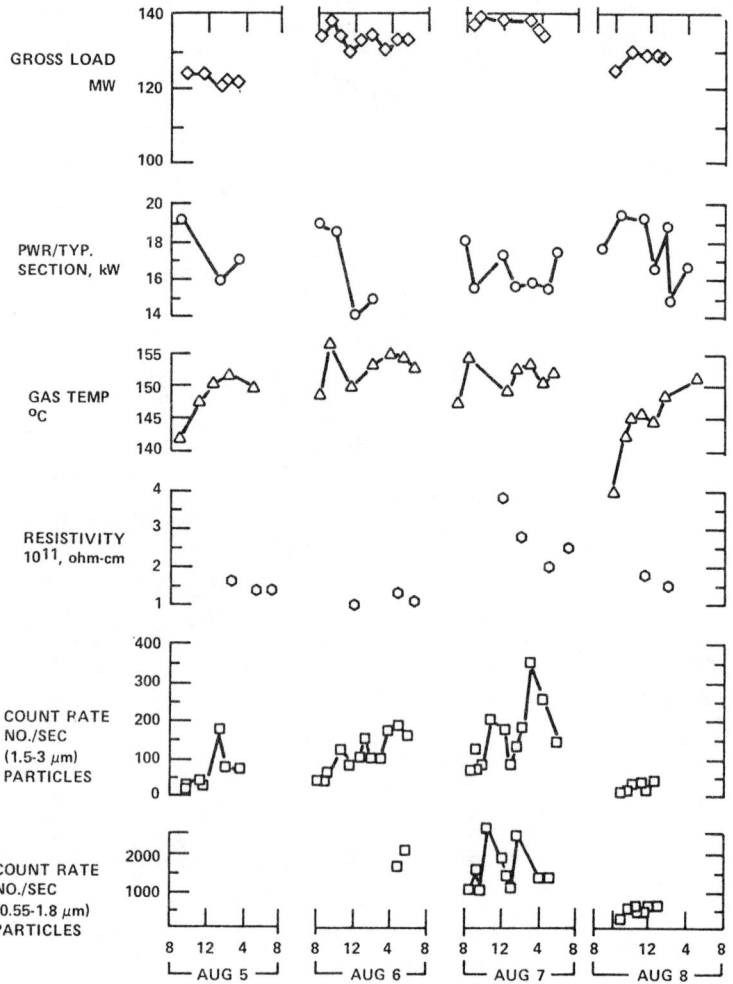

Figure 272. Events for test period Plant 1.[217]

Figures 273 and 274 show the number of 6-12 and 12-24 μm diameter particles counted in 10 minute intervals through one day of testing with rapping and one day of testing without rapping, respectively. Cyclic concentration variations with a period of one hour were expected when the rappers were on and are fairly apparent in the data shown in Figure 273. No such cyclic pattern is apparent in the data shown in Figure 274 which were obtained with the rappers de-energized. Note the obvious effect of losing power to one of the TR sets. The average counting rate was much reduced in the 6-12 and 12-24 µm channels with the rappers turned off as can be seen by comparison of Figures 273 and 274.

As indicated previously, the attempt to determine rapping losses at Plant 1 by comparison of mass train and impactor data sets from normal and no-rap periods was not successful due to other factors influencing outlet emissions. However, an estimate of the contribution of rapping losses to total mass emissions was made from data from the LPSS and outlet impactor systems. The estimate is that 30% of total outlet mass emission during normal rapper operation can be attributed to rapping reentrainment. Figure 275 shows the rap-no-rap data for the EAA system and the rap and no-rap impactor derived efficiencies. The estimated no-rap efficiencies are based on the data from the LPSS system and these are subject to large uncertainties because of the poor counting statistics for the larger particles coupled with the limited time span over which the data were taken. Fifty percent confidence intervals are shown for the impactor and EAA data. Even with the existence of the indicated uncertainties, it is apparent that very high collection efficiencies are achieved in the particle diameter range 0.05 to 20.0 µm. The minimum collection efficiency is approximately 99.2% at 0.20 µm diameter.

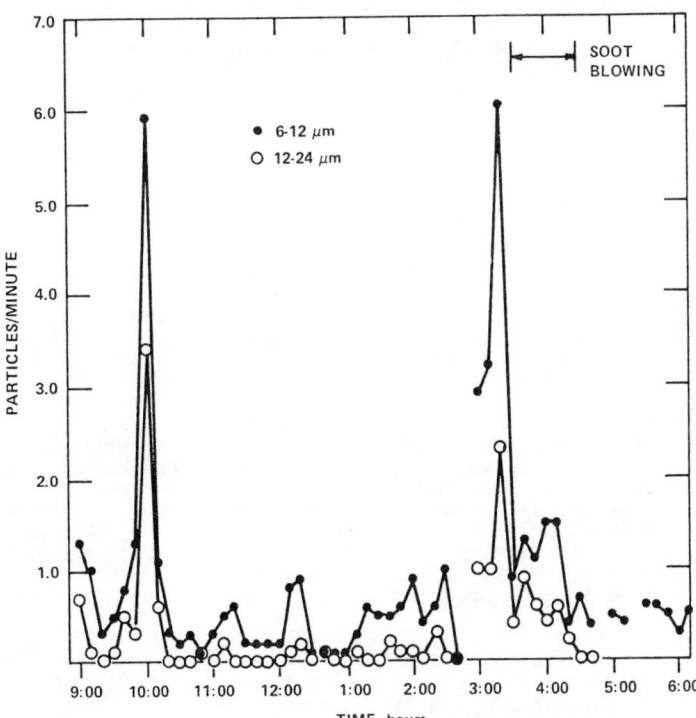

Figure 273. Particles per minute vs. time for large particle system on August 6, 1975 -- rappers on (Plant 1).[217]

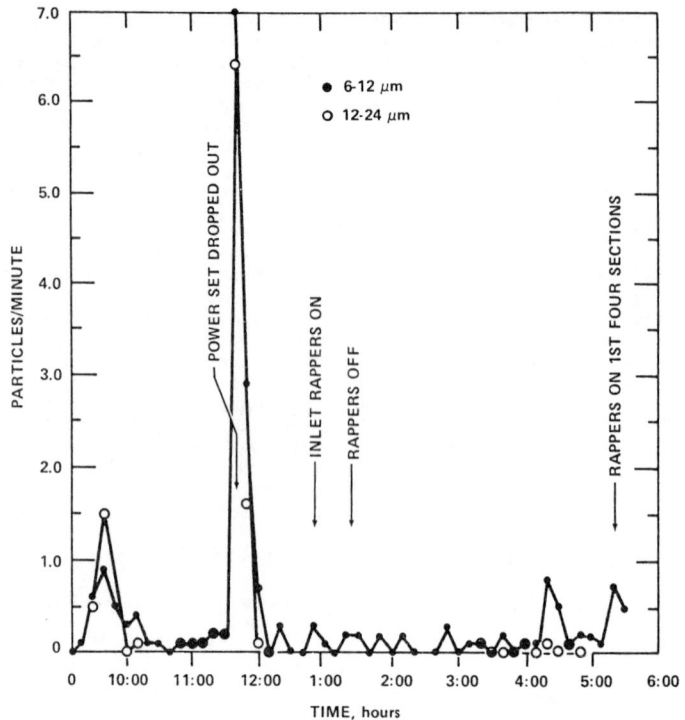

Figure 274. Particles per minute vs. time for large particle system on August 7, 1975 -- rappers off (Plant 1).[217]

The alternating sampling strategy with impactors and mass trains was successfully employed at Plant 2 and subsequent test sites to differentiate between reentrainment resulting from rapping and steady-state emissions. Figure 276 presents rap and no-rap data from Plant 2 from the EAA and the impactor sampling system. The large error bars (50% confidence intervals) on data obtained from the ultrafine particle system are a reflection of difficulties encountered with condensation of sulfuric acid, which created an interferring aerosol in the ultrafine size range. The data were screened and those results which were felt to be non-representative were discarded. It is apparent that rapping losses become significant only for particle diameters larger than 1 to 2 μm. The presence of significant large particle emissions in the absence of rapping is also indicated by Figure 276, and was confirmed by data obtained from the LPSS. These emissions apparently resulted from sparking or voluntary reentrainment. Plant 2 was operating with a high sulfur eastern coal which produced a fly ash with low electrical resistivity.

Figure 277 illustrates the large particle losses (on a relative basis) measured at Plant 4, which is a hot-side installation, using the impactor and ultrafine sampling systems with the rap-no-rap sampling sequence. The data obtained with normal rapper operation (not shown) show reasonable agreement for sizes greater than 1.0 μm diameter, indicating the alternating sampling strategy did not significantly distort the results obtained. As with the previously discussed data, the results indicate that rapping reentrainment does not cause a significant change in fine particle emissions.

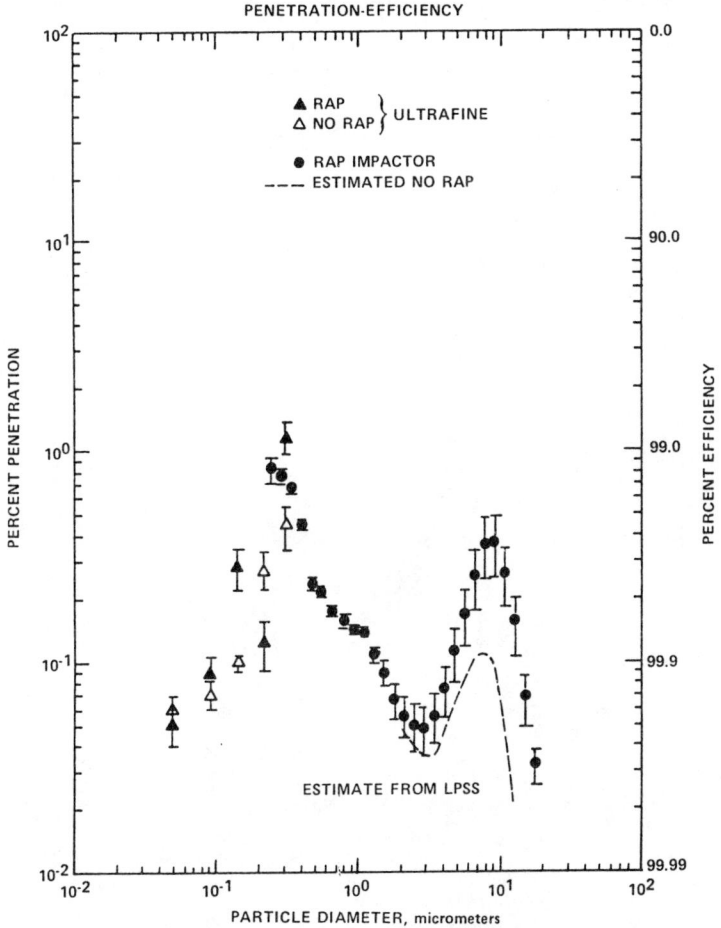

Figure 275. Plant 1 rap-no rap fractional efficiency including ultrafine and impactor measurements.[217]

The rapping emissions obtained from the measurements on the six precipitators are graphed in Figure 278 as a function of the amount of dust calculated to have been removed by the last field. The dust removal in the last field was approximated by applying the relation

$$\eta'/\text{section} = 1-\exp(-X'/N_E), \qquad (62)$$

where

$\eta'/\text{section}$ = overall mass collection fraction per section,

$$X' = -\ln(1-\eta_0), \qquad (63)$$

η_0 = overall mass collection fraction determined from mass train measurements under normal operating conditions, and

N_E = number of electrical sections in series.

These data suggest a correlation between rapping losses and particulate collection rate in the last field. Data for the two hot-side installations (4 and 6) which were tested show higher rapping losses than for the cold-side units. This would be expected due to reduced dust adhesivity at higher temperatures. Data 2a and 2b are for a cold-side unit operating at normal and approximately one-half normal current density, respectively. The decrease in current density at installation 2 resulted in a significant increase in rapping emissions due to the increased mass collected in the last field and smaller electrical holding force for the same rapping intensity.

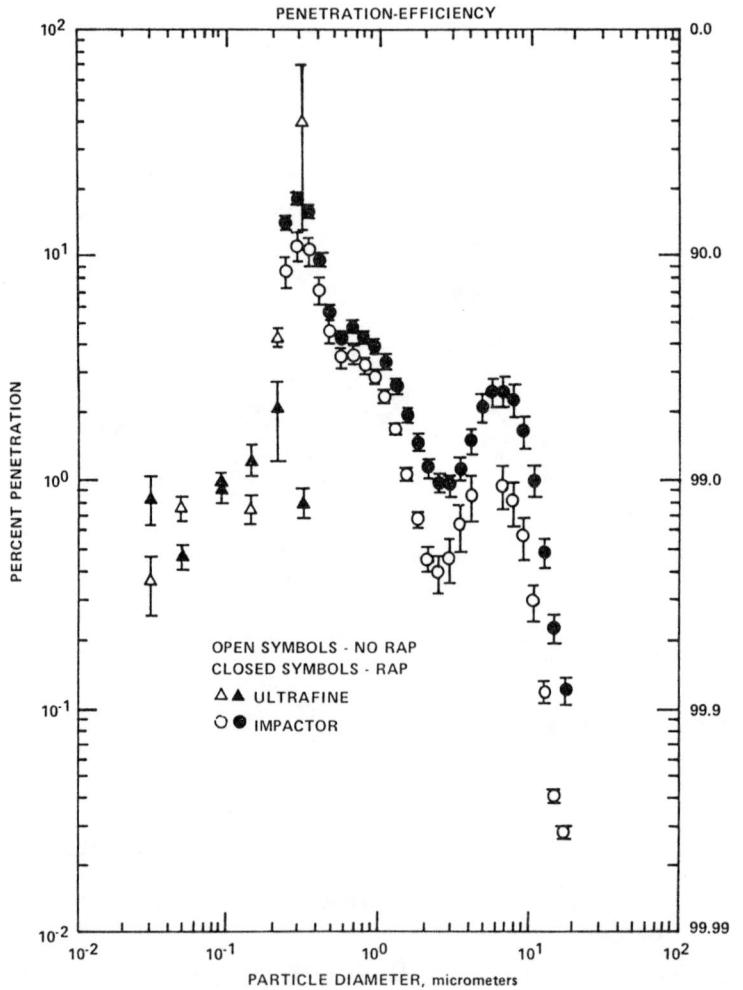

Figure 276. Rap-no rap ultrafine and impactor fractional efficiency. Normal current density, Plant 2.[215]

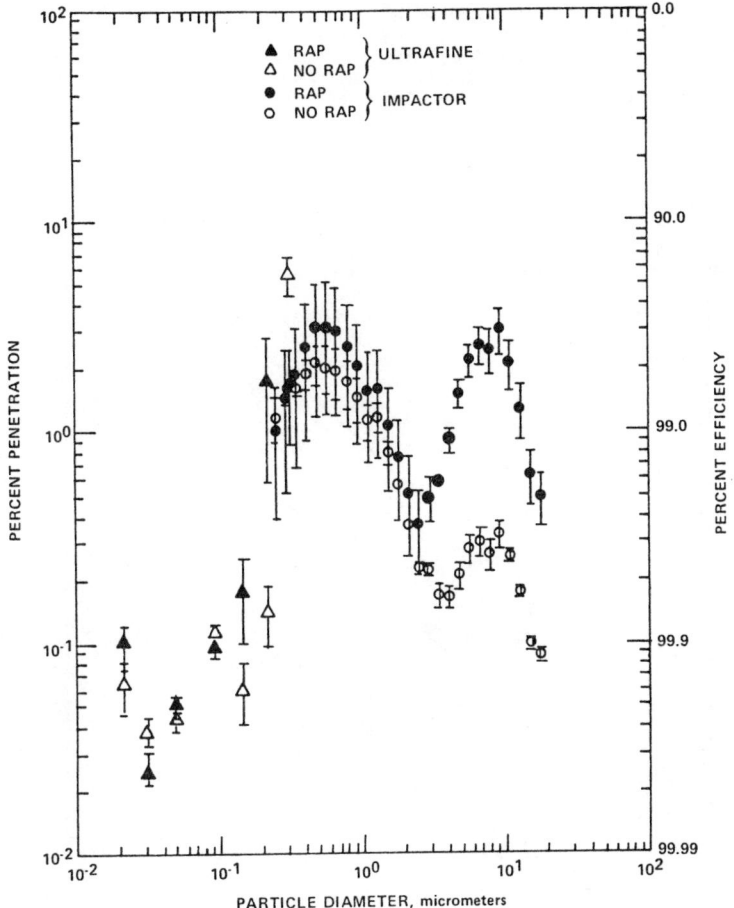

Figure 277. Ultrafine and impactor rap-no rap fractional efficiencies, Duct Bl., Plant No. 4, with 50% confidence intervals.[217]

The simple exponential relationships

$$y_1 = (0.155) x^{0.905} \qquad (64)$$

and

$$y_2 = (0.618) x^{0.894} \qquad (65)$$

can be used for interpolation purposes in determining the rapping emissions (mg/DSCM) for a given calculated mass removed by the last field (mg/DSCM) for cold- and hot-side precipitators, respectively. Figure 278 was constructed using the calculated mass removed in the last field determined by the measured overall mass collection efficiency during normal operation of the precipitator. This was done because complete traverses were made by the mass trains during the normal tests whereas this was not the case for the measurements made during the no-rap tests. In principle, the no-rap efficiencies should be used to calculate the mass removed in the last field. Obviously, the limited amount of data obtained thus far is not sufficient to validate in general the approach presented here. However, this approach gives reasonable agreement with the existing data and offers a quantitative method for estimating rapping losses.

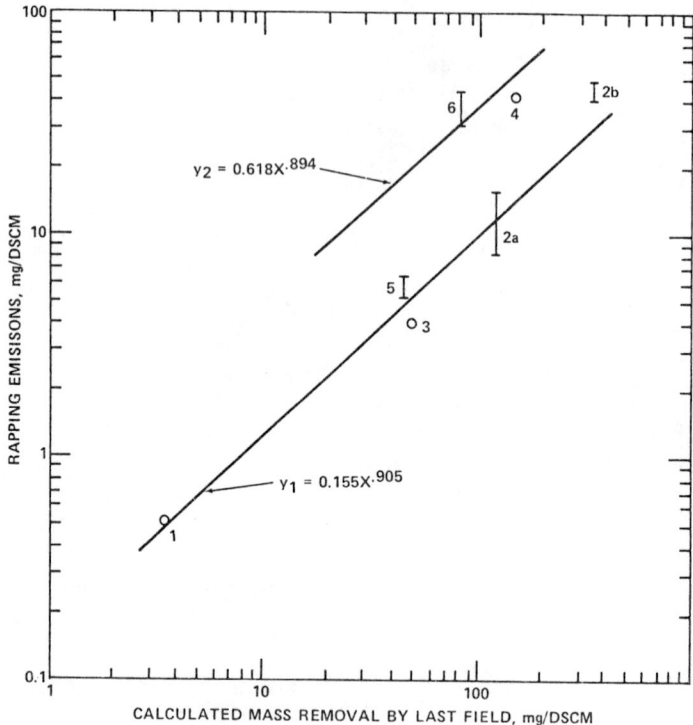

Figure 278. Measured rapping emissions versus calculated particulate removal by last field.[19]

The apparent size distribution of emissions attributable to rapping at each installation was obtained by subtracting the cumulative distributions during non-rapping periods from those with rappers in operation, and dividing by the total emissions (based on impactor measurements) resulting from rapping in order to obtain a cumulative percent distribution. Figure 279 contains the results of these calculations. Although the data indicate considerable scatter, an average size distribution has been constructed in Figure 280 for use in modeling rapping puffs.

Analysis of Factors Influencing Performance 329

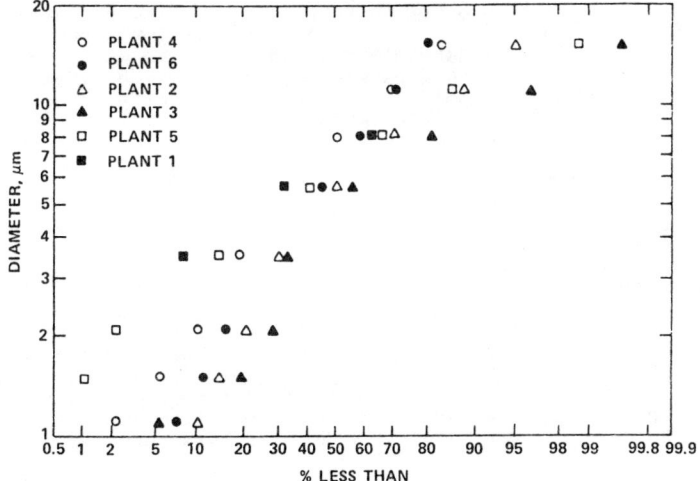

Figure 279. Apparent rapping puff size distribution for six full-scale precipitators.[19]

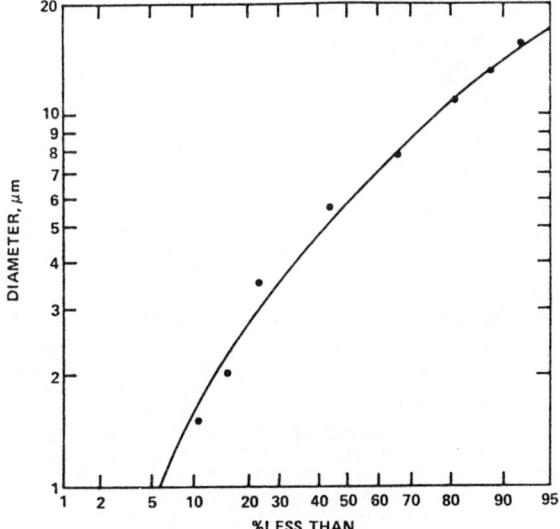

Figure 280. Average rapping puff size distribution for six full-scale precipitators.[19]

Summary of the results of rapping studies--Pilot plant studies indicate that rapping emissions decrease with increasing time between raps. Also, the percentage of the collected dust removed from the collecting electrode increases with increased time between raps. The buildup of a residual dust layer on the collecting electrodes that could not be removed with the maximum, available normal plate acceleration has been evidenced. By varying the rapping frequency, the penetration due to rapping reentrainment could be varied from 18 to 53% of the total penetration. These results point out the need for a flexible rapping system in which the rapping frequencies for the different sections can be varied and in which the rapping intensities can be varied. With this type of system, the rapping function can be optimized for specific precipitator operating conditions and ash properties in order to minimize the penetration of particulate out of the precipitator due to rapping reentrainment.

The pilot plant data suggest that thicker dust layers produce larger reentrained particles upon rapping. The majority of the particles in the rapping puffs were agglomerates. These data would indicate that as one proceeds from the inlet section to the outlet section increased times are needed between raps in order to obtain dust layers with sufficient thickness to produce large reentrained agglomerates during rapping. The large agglomerates can be easily recollected by the precipitator. Depending on the rapping frequency, typical mass median diameters for reentrained particulate from an inlet section would range from approximately 10 µm to 20 µm with very little of the mass less than 2.0 µm.

The pilot plant studies also showed significant vertical stratification of particulate matter reentrained as a consequence of rapping. All particle size bands showed a decrease in concentration with increasing distance from the bottom baffle. This was attributed to both gravitational settling and hopper boil-up. Emissions due to hopper boil-up were observed at some time after observation of emissions due to particulate matter reentrained directly into the gas stream from the collection electrodes. In these studies, hopper boil-up contributed significantly to the reentrainment emissions. This points out the need for adequate design of hoppers and hopper regions to prevent excessive hopper boil-up.

The data obtained from the six full-scale precipitators showed that rapping losses as a percentage of total mass emissions ranged from over 80% for one of the hot-side units to 30% for the cold-side units. The high rapping losses for the hot-side unit were probably due both to reduced dust adhesivity at high temperature and relatively short rapping intervals. It was also found that reduction of the operating current density at Plant 2 resulted in increased emissions due to rapping. This was due to increased mass collected in the last field and reduced electrical holding force.

Measurements of fractional efficiency with and without electrode rapping showed that losses in collection efficiency due to rapping occur primarily for particle diameters greater than 2.0 µm. The available mass emission data suggest a correlation between the dust removal rate in the last rapped section of the precipitator and the emissions due to rapping. Apparent rapping puff particle size distributions measured at the outlets of the full-scale precipitators had mass median diameters ranging from approximately 6.0 µm to 8.0 µm. Real-time monitoring of outlet emissions also revealed sporadic emission of particulate matter due to factors other than rapping.

Reentrainment from Factors other than Rapping--

Although it is difficult to quantify the complex mechanisms associated with particle reentrainment due to (1) the action of the flowing gas stream on the collected particulate layer, (2) sweepage of particles from hoppers caused by poor gas flow conditions or air inleakage into the hoppers, (3) bouncing of particles following impaction on the collection surface, (4) impaction of large particles with small particles previously deposited on the collection electrode, and (5) excessive sparking, the effect of these nonideal conditions on precipitator performance can be estimated if some simplifying assumptions are made. If it is assumed that a fixed fraction of the collected material of a given particle size is reentrained and that the fraction does not vary with length through the precipitator, an expression can be derived which is identical in form to that obtained for gas sneakage:[218]

$$P_{N_R} = [R + (1-R)(1-\eta_i)^{1/N_R}]^{N_R}, \qquad (66)$$

where P_{N_R} is the penetration of a given particle size corrected for reentrainment, R is the fraction of material reentrained, and N_R is the number of stages over which reentrainment is assumed to occur.

Since equations (44) and (66) are of the same form, the effect of particle reentrainment without rapping can be expected to be similar to the effect of gas sneakage, provided that a constant fraction of the collected material is reentrained in each stage. It is doubtful that such a condition exists, since the gas flow pattern changes throughout the precipitator and different holding forces and spark rates exist in different electrical sections. However, until detailed studies are made to quantify the losses in collection efficiency as a function of particle size for these types of reentrainment, equation (66) provides a means of estimating the effect of particle reentrainment without rapping on precipitator performance.

Several things should be done in order to minimize the particle reentrainment due to factors other than rapping. The gas velocity distribution should meet IGCI criteria as a minimum and should have an average value of 1.5 m/sec (5 ft/sec) or less. Hoppers should be designed with proper baffling to prevent excessive flow in the ash holding regions and should have no air inleakage. Excessive sparking should be avoided.

Nonuniform Temperature And Dust Concentration

Nonuniform temperature and dust concentrations may exist in a precipitator and may result in adverse effects. A nonuniform temperature may result in variations in the resistivity of the collected dust layer, variations in the electrical properties of the gas, and corrosion in low temperature regions. The first two effects may lead to excessive sparking in certain regions of the precipitator. A nonuniform dust concentration may result in excessive buildups of dust on corona wires, collection plates, beams, etc. and excessive sparking due to particulate space charge effects. Excessive dust buildups and possible "doughnut" formations on corona wires tend to suppress the corona and to cause uneven corona emission. Excessive dust buildups on the collection electrodes between raps may result in significant particle reentrainment, undesirable electrical conditions, and reduced cross-section for gas flow.

The effects of nonuniform temperature and dust concentration on precipitator performance have not been analyzed or studied extensively. Therefore, at the present time, these effects can not be quantified. Generally, it is assumed that if a good gas flow distribution exists, then the temperature and dust distributions will also be good. This may be a poor assumption for many precipitator arrangements that are commonly employed.

Section 7

Emissions from Electrostatic Precipitators

PARTICULATE EMISSIONS

The data required for determining the mass efficiency of control devices collecting fly ash are obtained by sampling the flue gas upstream and downstream of the pollution control device. Mass concentrations of particulate matter in flue gas are measured by drawing a sample of gas through a probe and filter and weighing the collected material.

Methods For Determination Of Overall Mass Efficiency

Various organizations have proposed specific procedures and sampling train designs for mass concentration measurements. The Environmental Protection Agency's Method 5 specifies the use of an extractive sampler.[219] Sampling trains constructed to meet Method 5 specifications were initially designed to operate at flow rates up to one cubic foot per minute (28.3 liters/min). Recently, a four cubic feet per minute (113 liters/min) extractive sampler has been developed which is claimed to comply with the requirements of Method 5. The proposed EPA Test Method 17 specifies the use of in situ sampling.[220] The American Society of Mechanical Engineers (ASME) Performance Test Code 27 specifies the use of either an in situ or extractive sampler.[221] The ASME will soon be releasing a new Performance Test Code 38 which will supercede the Performance Test Code 27. The Industrial Gas Cleaning Institute (IGCI Publication No. 101) and Western Precipitation Co. (Bulletin WP 50) have also suggested sampling methods. The American Society of Mechanical Engineers (ASME) Performance Test Code 27 specifies the use of either an in situ or extractive sampler.

EPA Test Method 5--

Official performance testing of stationary sources for particulate emissions from coal-fired power plants must be conducted with the EPA Test Method 5 "Determination of Particulate Emission from Stationary Sources".[219] Method 5 relies on the removal or extraction of a dust-laden gaseous sample from the duct or stack followed by removal of the particles by a filter while monitoring sample volume. With this method one obtains a measure of the average particulate mass concentration for the cross-sectional area of the duct during the time of sampling. There is some difference of opinion as to how the results should be interpreted, especially in regard to condensation of vapors in the probe and filter box which contains the condensers. Originally the Environmental Protection Agency proposed that any material collected in the condenser portion of the sampling train (shown in Figure 281)[222] must be added to that of the dry collector (filter) portion. After

Emissions from Electrostatic Precipitators   333

numerous objections from people in the field, the proposed method was altered so that compliance now is based only upon material collected in the filter and in the probe preceding the filter. Hemeon and Black contend, however, that even this modification is not valid since condensation and chemical reaction occurs in the probe prior to the filtering of the sample.[223,224] Therefore, the SO_2 in the gas forms sulfates which later are collected on the filter. However, one might argue that such reactions, if they occur, would also occur in the atmosphere and should be included as particulate matter. Some investigators conducting performance tests of control devices on emission sources prefer to use a sampling train that differs from Method 5 in that the filter for collecting particulate matter is located in the stack instead of outside the stack at the end of the sampling probe (ASME Performance Test Code 27).

With EPA Method 5, one obtains a sample from the duct by using a prescribed traversing procedure which involves isokinetic extraction from different points within the duct. This procedure yields, in effect, an approximate integration of collected mass and sample volume over the cross-sectional area of the duct. Before sampling, the number of traverse points must be determined using EPA Test Method 1, "Sample and Velocity Traverse for Stationary Sources". The EPA sampling train consists of a thermally controlled probe, with a variety of sampling nozzles and a pitot tube assembly, which is connected to a sampling case containing a heated filter assembly housing, filter, and a number of impingers located in an ice bath (Figure 281). The control console contains the flow meters, pressure gauges, thermal control systems, timer, and vacuum pump required for sampling.

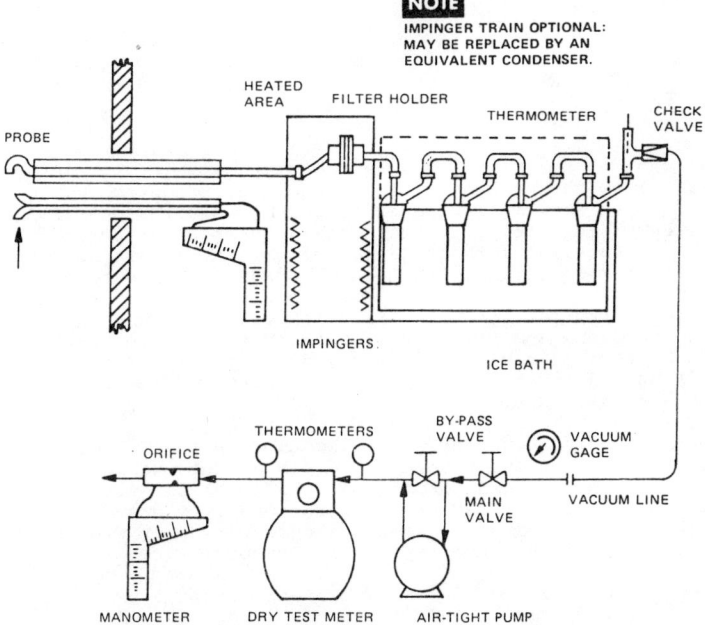

Figure 281. The EPA Method 5 particulate sampling train.[222]

DESCRIPTION OF COMPONENTS

The nozzle removes the sample from the gas stream and should disturb the gas flow as little as possible. This means a thin wall and sharp edge. The major requirement of the probe, which removes the sampled stream from the stack, is that it does not significantly alter the sample from stack conditions. The sample temperature should be maintained at 120°C ± 14°C (248°F ± 25°F) or at such other temperature as specified by an applicable subpart of the standards or approved by the Administrator of the EPA for a particular application. Glass probe liners are desirable over metal probe liners, but steel probes are allowed for probe lengths over 2.5 meters. New regulations require a thermocouple to be attached to the probe end for monitoring the stack gas temperature.

Pressure drop, generated by the gas velocity in the duct, is monitored by an S-type pitot tube to insure isokinetic sampling velocities. The glass fiber filter should be at least 99.95% efficient in collecting 0.3 micron dioctylphthalate smoke particles.

An optional cyclone type of collector precedes the filter and results, when used, in the removal of larger particles. The four impingers in the train remove water, gases, vapor, and condensable particulate matter. The EPA and some states do not require the measurement of the condensable particulate fraction and hence the impingers are not specifically required. The impinger train may be substituted by any type condenser such as a piece of coiled tubing immersed in an ice bath. The condenser should be followed by a silica gel drying tube to collect the remaining moisture and protect the vacuum pump and dry gas meter. The sampling box holds the probe, the filter holder, and the impinger train and its ice bath.

The filter holder is contained in a heated area of the sampling box and the temperature of this area should be maintained at 120°C ± 14°C. Where the condensable particulate fraction is not required by state regulation or is of no interest, the sampling box can be simplified.

The control box contains a vacuum pump capable of maintaining isokinetic flow during heavy filter loadings, a control valve to vary the sample stream flow rate, a vacuum gauge for measuring the sample stream pressure, a dry gas meter equipped for determining the sample volume, a calibrated orifice meter which is used to monitor the sample stream flow rate, a pressure gauge to measure the pitot tube pressure drop, a pressure gauge to measure the orifice meter pressure drop, a variable voltage power supply to maintain the probe and filter box at their respective temperature by means of their individual heaters, and a pyrometer or potentiometer calibrated for thermocouple measurements of the duct and filter box temperature.

Calibration requirements are discussed in the EPA maintenance procedures.[225] Critical laboratory calibrations include the orifice meter, dry gas meter, and pitot tube. Calibration of the orifice meter and dry gas meter requires the use of a wet gas meter. Various other common laboratory instruments are required for the maintenance and calibration of the other system components.

Many commercial models for conducting Method 5 tests are available and a list of some manufacturers is given in Table 35.[226]

TABLE 35. SAMPLING SYSTEMS FOR TESTING
BY EPA METHOD 5[226]

Company	Address and Telephone Number
Aerotherm-Acurex	485 Clyde Avenue Mountain View, California 94042 (415) 964-3200
Glass Innovations, Inc.	Post Office Box B Addison, New York 14801
Joy Manufacturing Co.	Commerce Road Montgomeryville, Pennsylvania 18936 (215) 368-6100
Lear Siegler, Inc. Environmental Technology Division	74 Inverness Drive East Englewood, Colorado 80110 (303) 770-3300
Misco International Chemicals, Inc.	1021 South Noel Avenue Wheeling, Illinois 60090 (312) 537-9400
Research Appliance Co.	Pioneer and Hardies Road Gibsonia, Pennsylvania 15044 (412) 443-5935
Scientific Glass & Instruments, Inc.	7246 Wynnewood Houston, Texas 77001
Lace Engineering Co.	8829 North Lamar Post Office Box 9757 Austin, Texas 78766 (512) 836-5606
Bendix Corporation Environmental & Process Instruments Division	1400 Taylor Avenue Baltimore, Maryland 21204 (301) 825-5200

ASTM - Test Method (Figure 282)[227]

Both the ASTM and the ASME provide specifications for in situ samplers. The ASTM Method is similar to the EPA Test Method 5. The main difference is the use of an instack filter with no restrictions on the sampling flow rate used. However, the filter should be preheated by being allowed to reach temperature equilibrium in the process stream for at least thirty minutes prior to sampling. When inserting the filter for preheating, the nozzle must be pointed in the downstream direction of the gas flow to prevent accumulation of fly ash in the nozzle. Also, when inserting the filter into a duct which is not under ambient pressure, the sampling lines must be closed to prevent undesirable gas flow through the filter.

ASME Performance Test Code 27

The ASME Performance Test Code provides for the use of a variety of instruments and methods.[221] Since testing experience has not been uniform enough to permit standardized sampler design, this code merely gives limiting requirements which past experience has shown gives the least sources of error. The Code is designed as a source document which provides technically sound options to be selected and agreed upon by the contractor and the contractee performaning the sampling. According to ASME Performance Test Code 27, the sampling device shall consist of a tube or nozzle for insertion into the gas stream and through which the sample is drawn, and a filter (thimble, flat dish, or bag type) for removing the particles. For the purpose of the Power Test Code, 99.0% collecting efficiency by weight is satisfactory, and the filter may be made of cotton, wool, filter paper, glass wool, nylon, or orlon. The filter arrangement may be extractive or in situ.

The main advantage of in situ sampling over extractive sampling is that substantially all of the particulate matter is deposited directly on the filter, which means that only a small area other than the filter contains particulate matter and requires washing. Also, since the filter is maintained at the stack gas temperature, auxiliary heating of the filter is not needed. The main disadvantage of the in situ sampler over the extractive sampler is the fact that the in situ sampler is limited to process streams where temperatures do not exceed the limit of the filter medium and holder. In fact, thermal expansion of the filter holder may create gas leakage problems. Of course, the instack filter system cannot yield data on condensable particulate matter in the plume.

Another difference between the filtration methods is the sampling flow rate used in each method. Sampling trains constructed to meet EPA Method 5 specifications were initially designed to operate at flow rates up to 28.3 ℓ/min (1 ft^3/min); recently a 113 ℓ/min (4 SCFM) sampler has been developed which complies with EPA Method 5 specifications. ASTM and ASME Methods do not define a flow rate range. Some high volume trains can operate at flow rates up to 1.98 m^3/min (70 ft^3/min).

The main advantage in the use of a high flow rate sampler lies in the fact that the amount of time required to sample a given volume of stack gas is small compared to the time required to sample with a low flow rate sampler. In a process stream where the mass concentration is constant, the time required for sampling is markedly reduced.

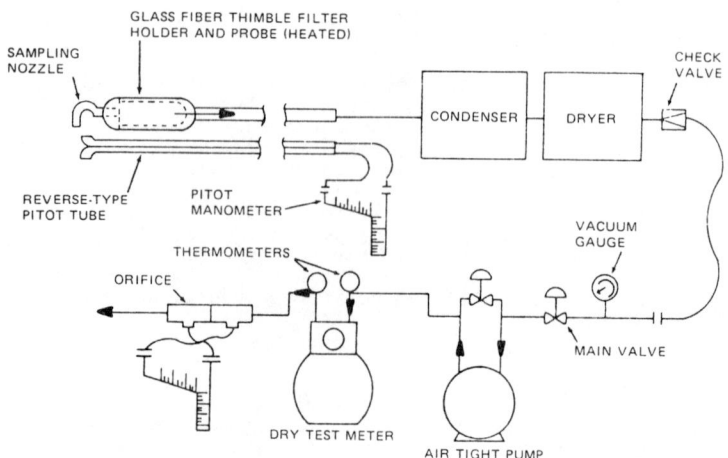

Figure 282. ASTM-type particulate sampling train.[227]

In a process stream where the mass concentration is highly variable, a large number of high volume runs would be required to obtain a value representation of the same average mass concentration obtainable from one run of the low volume run. Statistically, it is more desirable to obtain several samples of a value than just one sample. For stable streams this will give additional information revealing the precision with which the method has been applied. When using high flow rate extractive samplers the high ratio of sample gas flow rate to probe wall area minimizes errors due to loss of particulate matter on the tubing walls between the nozzle and the filter, minimizes heat losses, and thus helps to prevent the condensation of vapors in the train. The high ratio also can be a disadvantage when cooling of the sample gas stream is required to protect the equipment since auxiliary cooling equipment may be needed.

STATUS OF RULES AND REGULATIONS GOVERNING PARTICULATE MATTER, SULFUR OXIDE, NITROGEN OXIDE, AND OPACITY FOR COAL-FIRED POWER BOILERS IN THE UNITED STATES

Background[228]

The Clean Air Act of 1970 gave the Environmental Protection Agency (EPA) the responsibility and authority to control air pollution in the United States and its territories. In 1971 EPA issued National Ambient Air Quality Standards for six pollutants -- sulfur dioxide, nitrogen dioxide, particulate matter, carbon monoxide, hydrocarbons, and photochemical oxidants. For each pollutant both primary and secondary standards were issued. Primary standards were set at levels necessary to protect the public health and were to be met no later than three years from the date of promulgation (subject to limited extensions of up to three years). Secondary standards were designed to protect the public from adverse effects to their welfare. Each state was required to adopt and submit to the Environmental Protection Agency a plan for attaining, maintaining, and enforcing the standards in all regions of the state. The State Implementation Plans specified all details necessary to insure attainment and maintenance of the standards. Most of the state implementation plans were approved by the Environmental Protection Agency in 1972.

In addition to the state implementation plans, new source performance standards were issued by the Federal Government. New sources include newly constructed facilities, new equipment which is added to existing facilities, and existing equipment which is modified in such a way that results in an increase of pollutant emissions. New source standards limit specific pollutant emissions from categories of sources (such as fossil fuel-fired steam generators) which are determined to contribute significantly to the endangerment of public health and welfare.

Current Status Of Emission Regulations

According to the Environmental Protection Agency, particulate and opacity standards for new coal-fired power boilers of 25 MW or more are 0.05 g/10^6 cal (0.03 lb/million Btu) and 20% (on a six minute average), respectively.[230,231] Also, final sulfur standards just released by EPA indicate a "sliding" standard that requires scrubbing of 70 to 90 percent of the sulfur from the flue gas, depending upon the sulfur content of the coal.[232] For coal with a sulfur content that would cause an emission, uncontrolled, of less than 3.6 g/10^6 cal (2 lb/

million Btu), only 70 percent of the sulfur dioxide need be removed from the flue gas. For uncontrolled emission levels from 2 lb up to 6 lb the desulfurization must be sufficient to bring the controlled emission level down to 0.27 kg (0.6 lb). For coal-sulfur levels from 2.72 kg (6 lb) to 5.45 kg (12 lb) the control efficiency must be 90 percent. Above 5.45 kg (12 lb), the degree of desulfurization must be enough to bring the emission down to no more than $2.16 \text{ g}/10^6$ cal (1.2 lb/million Btu), which was the old limit. The nitrogen oxides standard is $0.90 \text{ g}/10^6$ cal (0.50 lb/million Btu) from subbituminous coal, shale oil, or any solids, liquids, or gaseous fuel derived from coal.

Table 48 in the Appendix C gives a compilation of emission limits for particulate matter, sulfur oxide, and nitrogen oxide limits for coal-fired power boilers for every state in the United States. Table 48a gives emission limits for California. California's counties each have separate rules and regulations. Therefore emission limits were obtained from most of the counties in an SoRI survey. Table 49 in Appendix C gives a compilation of opacity limits as they apply to those power plants which come under the "existing source" category of each state's opacity regulations. New source limits for opacity were not compiled since they generally follow the present Federal limit of 20%.

Performance Evaluation

To evaluate the performance of new stationary sources, the Environmental Protection Agency has specified reference methods for the manner in which tests must be conducted at each plant. The Code of Federal Regulations 40, Part 60-Standards of Performance for New Stationary Sources, Appendix A - Reference Methods, contain the reference methods to be used to check performance standards. Method 9 is the reference method for visual determination of the opacity of emissions from stationary sources. This method is basically a visual determination by a qualified observer. There are also performance specifications and test procedures for transmissometer systems which are used to continuously monitor opacity of stack emissions. These specifications are found in Appendix B of the Code of Federal Regulations 40, Part 60. Where disagreements occur between a qualified visual observer's determination (Method 9) and a transmissometer, Method 9 takes precedence in the opinion of the Environmental Protection Agency.[230] Method 5 is the reference method for performance testing of stationary sources for particulate emissions. Method 5 relies on the removal or extraction of a dust-laden gaseous sample from the duct or stack followed by removal of the particles on a filter while measuring sample volume. Methods 6 and 7 in Appendix A describe the reference methods for determination of sulfur dioxide and nitrogen oxide emissions from stationary sources, respectively. In Method 6 a gas sample is extracted from the sampling point in the stack. The acid mist, including sulfur trioxide is separated from the gaseous sulfur dioxide. The sulfur dioxide fraction is then measured by the barium-thorin titration method. In Method 7 a grab sample is collected in an evaporated flask containing a dilute sulfuric acid-hydrogen peroxide absorbing solution, and the nitrogen oxides, except nitrous oxide, are measured colorimetrically using the phenoldisulfonic acid procedure. Performance specifications and specification test procedures for monitors of SO_2 and NO_x are given in Appendix B, Performance Specification 2.

A helpful procedure for planning and implementing tests for control device evaluation can be found in a recent SoRI publication.[54]

Discussion And Definition Of Opacity

   Suspended particles in an aerosol will scatter and absorb radiation from a beam passing through it; the remaining portion is transmitted. The transmittance, T, of a fluid medium containing suspended particles is defined as the ratio of the transmitted radiation intensity to the incident radiation intensity. T is given by the Bouguer, or the Beer-Lambert, law:

$$T = \exp(-EL) \quad (67)$$

where L is the path length of the beam through the aerosol medium and E, the extinction coefficient of the medium, is a complicated function of the size, shape, total projected area, refractive index of the particles, and the wavelength of the radiation. Sometimes the measured transmittance is expressed in terms of optical density defined as

$$O.D. = \text{Log}(1/T) \quad (68)$$

instead of the transmittance. Consequently, instruments and methods for aerosol measurement based upon light transmission principles have been referred to as transmissometers, smoke density meters, photo-extinction measurements, or turbidimetric measurements.

   While transmittance is defined as the ratio of light transmitted through the aerosol to the incident light, opacity is defined as the ratio of the light attenuated from the beam by the aerosol to the incident light (i.e., opacity = 1-T). Aerosols which transmit all incident light are invisible, have a transmittance of 100%, and an opacity of zero. Emissions which attenuate all incident light are totally opaque, having an opacity of 100% and a transmittance of zero.

   Many versions of transmissometers, or smoke meters, are available as stack emission monitors. If the transmissometer is used to measure instack opacity for purposes of compliance to federal regulations, it must meet the EPA requirements for opacity measurement systems as specified in the Federal Register of September 11, 1974. The use of visible light as a light source is required because the response of the instrument is supposed to match that of the human eye (photopic response). The angle of view and the angle of projection is specified, for compliance, as no greater than 5° (see Figure 283).[233]

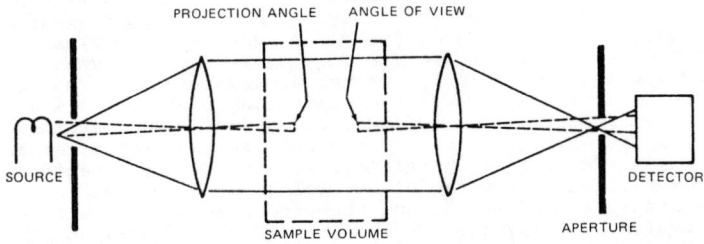

SCHEMATIC OF A TYPICAL TRANSMISSOMETER SYSTEM

Figure 283. Schematic of a transmissometer showing projection and view angles which must be no greater than 5° for EPA compliance.[233]

To obtain true transmittance data the collimation angles (angles of view and projection) for the transmitter and receiver must be limited to reduce the sensitivity to stray light scatter (see Figure 283). A zero degree angle is the ideal collimating angle, whereas a non-zero angle will introduce a systematically low reading of opacity. However, a compromise is necessary, since as a zero degree collimation is approached, instrument construction costs, operating stability, and optical alignment problems increase. A transmissometer having a 5° collimating angle applied to the emissions of a pulverized coal-fired steam generator gave an opacity measurement that was about 5% low relative to the 0° value.[234]

The error in the transmissometer measurement due to the use of different light detection angles has been analyzed theoretically by Ensor and Pilat and shown to be a function of detection angle and particle size.[235] They showed that, in general, the error associated with a given detector viewing angle increases with an increase in the particle mean diameter.

All transmissometers require purge air systems to protect the optical windows or reflectors. Still, regular cleaning is required with the accumulation rate varying widely from one location to another. Most commercial instruments have automatic zero and span checking capabilities to verify proper functioning and calibration between cleanings.

Transmissometers can be used to measure the instack opacity in order to obtain an estimate of the plume opacity for compliance testing; or they can be used to measure the in situ opacity for process control or as an estimate of mass concentration.

When the required measurement is the opacity of the emissions at the exit of the stack, a measurement at any other location in the stack has to have its optical path length adjusted to the exit diameter. The calculation for this adjustment can be found in the Federal Register.[236] Figure 284 gives the relationship of effluent transmittance at the stack exit as a function of instack transmittance for various ratios of stack exit diameter to transmissometer optical path length.[237]

As opacity, 1-T, approaches zero the relative error in its measurement with a transmissometer becomes unavoidably large. For example, a two per cent error in the transmittance measurement gives a 100 per cent error in an opacity of two per cent. In such cases, important during diagnostic studies of control devices, a nephelometer as used by Ensor,[238] may be a more accurate measure of opacity although it requires a probe and sampling traverses. This instrument when used as an opacity monitor attempts to determine E, the extinction coefficient, through a measurement of the scattering coefficient alone where E = scattering coefficient + absorption coefficient. This is performed using a predetermined relationship between E and the instrument response for a calibration aerosol. The errors in this type of opacity measurement depend upon the variation of the ratio, aerosol absorption coefficient to the scattering coefficient and the errors associated with extraactive sampling. This ratio varies from zero for non-absorbing particles to about one for highly absorbing aerosols giving possible errors in opacity of ≈ 100 per cent depending upon the calibration aerosol. However, if the calibration aerosol is chosen judiciously (i.e., with optical properties close to those of the sample aerosol) and the opacity is low, the nephelometer errors are much smaller than those obtained with the transmissometer at low opacities.

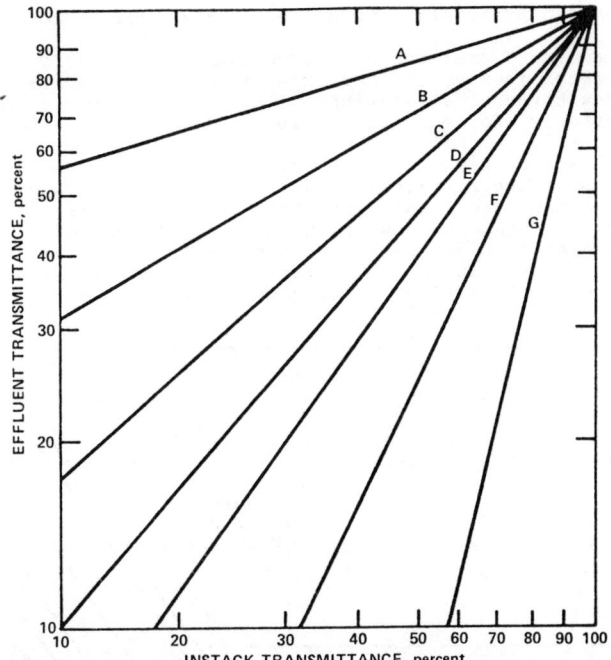

Figure 284. Effluent transmittance vs. instack transmittance for varying ratios of stack exit diameter to in stack path length: A = 1/4, B = 1/2, C = 3/4, D = 1, E = 4/3, F = 2, G = 4.[237]

Relationship Between Opacity And Mass Concentration And Particle Size

Theoretical Relationship--

Because of the interrelation between particle size distribution in a stack and the opacity, it is possible to meet mass emission standards and still have an opacity problem. In fact, some changes in flue gas streams causing a reduction in mass emissions have produced an increase in opacity. The relevance of this particular aspect of opacity is described below.

The dependence of opacity upon the total mass concentration, size distribtuion, and particle composition is given by

$$O = 1 - I/I_0 = 1 - \exp(-W \cdot L/\rho \cdot K) \tag{69}$$

where

O = opacity,

I = intensity of transmitted light,

I_0 = intensity of incident light,

W = total particulate mass concentration,

L = illumination path length or diameter of plume,

ρ = particle density, and

K = specific particulate volume/extinction coefficient ratio.

The parameter K, related to the volume/surface ratio of the aerosol, is determined by the particle size distribution and refractive index through calculations using the Lorentz-Mie theory of light scattering for each size class. Illustrative calculations of K assuming a log-normal size distribution and various refractive indices have been carried out by Ensor and Pilat.[239] The results for two values of the refractive index are given in Figures 285 and 286. It can be seen that K and thus opacity is very sensitive to MMD, geometric standard deviation, and refractive index. Since opacity increases as K decreases the minimum occurring around 0.5 to 0.1 µm in diameter is of particular interest. This light scattering theory is based on a homogeneous sphere model for the particles.

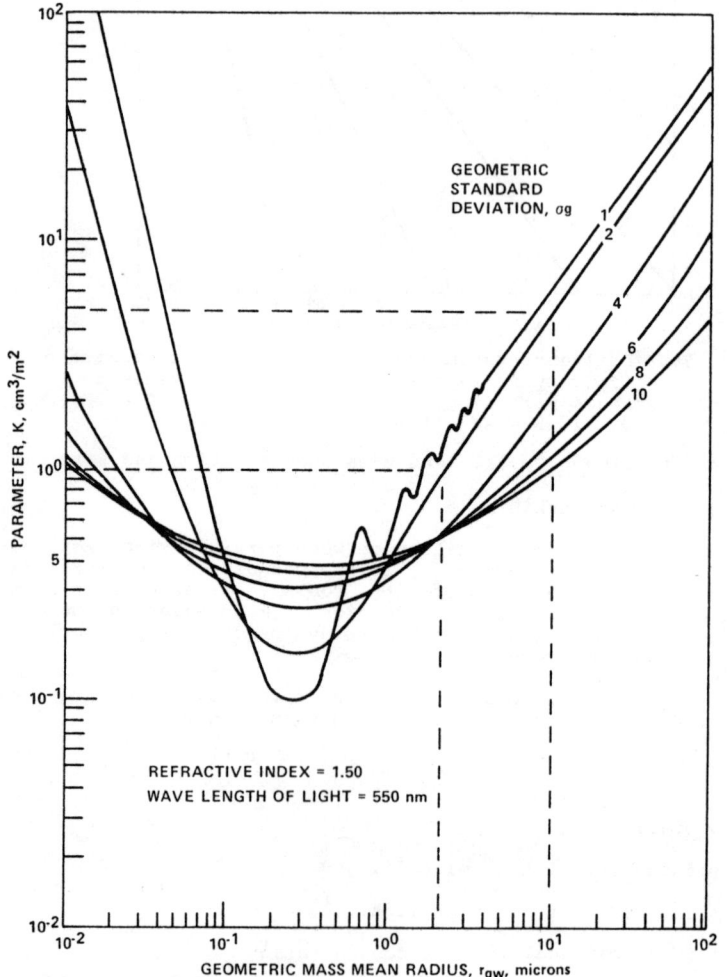

Figure 285. Parameter K as a function of the log-normal size distribution parameters for a white aerosol after Ensor and Pilat.[239]

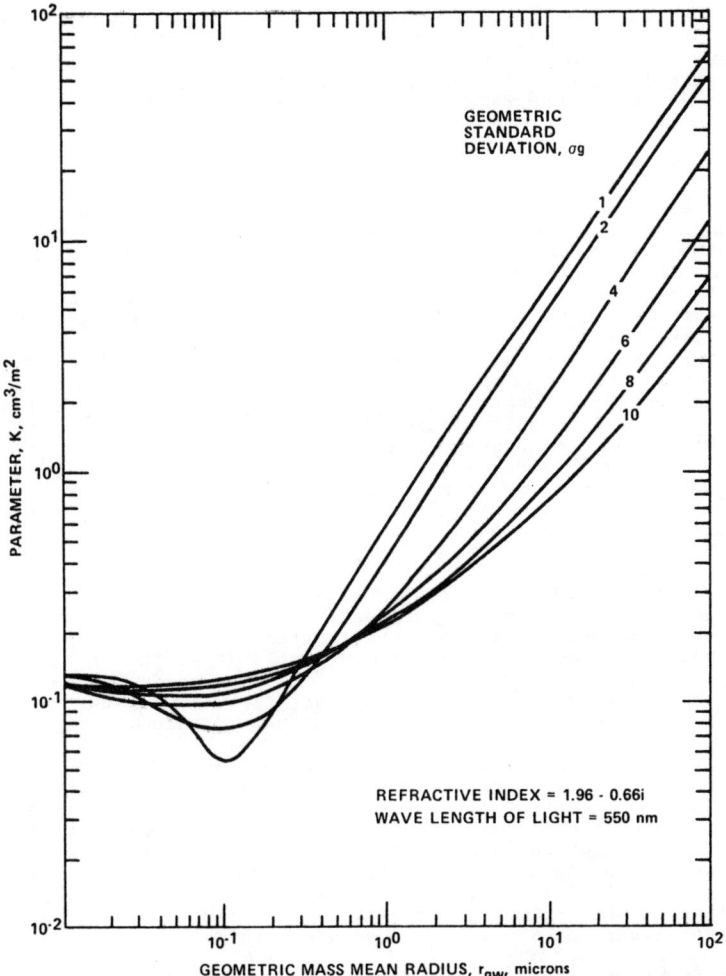

Figure 286. Parameter K as a function of the log-normal size distribution parameters for a black aerosol after Ensor and Pilat.[239]

Since control devices generally reduce the MMD while removing particles a reduction in the total emitted mass will not effectively reduce the opacity if the inlet and outlet MMD's are to the right of the minima in Figure 285.

For example, if an aerosol originally had an MMD of 10 μm and a geometric standard deviation of 2 (shown in Figure 286), and a control device removed 80% of the mass from the aerosol while reducing the MMD to 2 μm, then there would still be no change in opacity. On the other hand, if the inlet MMD is close to the minimum then a further reduction in total mass and/or MMD will be much more effective at reducing opacity. Figures 285 and 286 with equation

(69) show that the change in opacity for a given change in total mass requires knowledge of the aerosol size distribution and refractive index. While the size distribution is of greatest importance in determining opacity, the differences in Figures 285 and 286 show that refractive index (determined by the composition of the particles) is also important.

Observed Relationship--

Several plants with which SoRI has had experience demonstrate the importance of particle size distribution to opacity. A power plant in Wyoming has a cold-side electrostatic precipitator with an SCA of about 98.5 $m^2/(m^3/sec)$ (500 $ft^2/1000$ cfm). This plant is near the particulate emission standard but does not meet the opacity standard. Three other western plants which have hot precipitators with SCA's in the 59.1-69 $m^2/(m^3/sec)$ (300-350 $ft^2/1000$ cfm) range have the same problem. This can be attributed in large part to the generally fine particle size distribution of ash obtained from burning western coal. (See discussion of Figure 285.) Another interesting case in point is a northern utility which was burning an eastern coal at one of its plants equipped with a normal cold-side electrostatic precipitator. This plant was meeting the opacity standard but not the emission standard. After switching to a western coal, the plant was able to meet the mass emission standard but could no longer meet the opacity requirement.

Even more dramatic is the situation at Southwest Public Service, Harrington Station. This plant burns low sulfur coal and uses an electrostatic precipitator/scrubber system to meet the particulate standard. Measured emissions are 19.4 ng/J (0.45 $lb/10^6$ Btu) and the opacity is around 38%. Sparks[239a] has analyzed this case and concluded that the high opacity was primarily due to the fine aerosol produced by the precipitator/scrubber system.

For a transmissometer to be useful as a monitor of the mass concentration, the properties of the particles (other than mass) being monitored must remain fairly constant over the monitoring period. Experimental data are available showing that good opacity-mass concentration calibration can be obtained on some sources. The sources that have been evaluated include coal-fired power plants;[240,241,242] lignite-fired power plants;[243] a cement plant;[244] a Kraft pulp mill recovery furnace;[245] petroleum refinery; asphaltic concrete plant; and a sewage sludge incinerator.[246]

Nader reported tests that were performed over one 3-month interval and two 2-month intervals representing different seasons of power plant operation.[247] Emissions were increased at various times by cutting off one or more electrostatic precipitator stages. Correlation curves were essentially the same for the three different time periods with coefficients of 0.93, 0.98, and 0.99. The coefficient for the composite correlation curve for the data for all three time intervals is 0.97 (see Figure 287). Mass concentration ranged from 55 to 360 mg/m^3. No problem with window contamination occurred with continuous operation of the transmissometer spanning the one year period.

For an emission source with high efficiency particulate control equipment, the size distribution of the emitted particulate matter may be relatively constant. Therefore, emission sources with variable emission and low efficiency particulate control equipment (i.e. cyclone and low energy scrubbers) can be expected to provide poorer correlation of instack plume opacity to particle mass concentration. Transmissometers may be useful indicators of mass emissions, once calibrated, on sources where the aerosol properties remain constant.

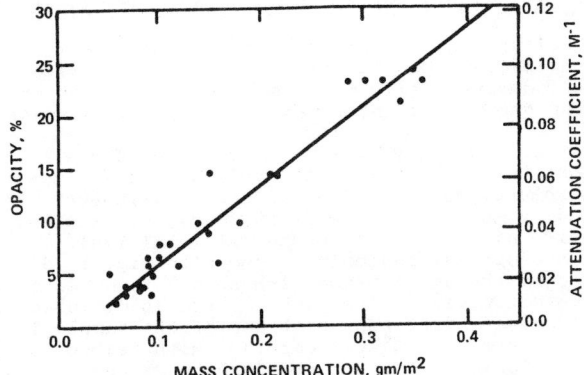

Figure 287. Correlation data between opacity and mass measurements of particulate matter in emissions for a coal-burning power plant. After Nader.[247]

Example Of Modeling Of Opacity Versus Mass At The Exit Of An Electrostatic Precipitator

The SoRI-EPA mathematical model of electrostatic precipitation has been used with certain modifications to simulate the operation of a power plant precipitator collecting fly ash from the burning of coal under test conditions. Based on the simulation of test conditions, the model has been employed to estimate the performance of the precipitator as a function of current density, specific collection area, inlet particle size distribution, and inlet mass loading. Performance of the precipitator has been determined in terms of both overall mass collection efficiency and opacity.

The set of parameters used in the simulation of the test conditions yielded an overall mass efficiency of 88.75%, opacities in the range from 39 to 49%, and an outlet size distribution with a mass median diameter (MMD) of 2.35 μm and a geometric standard deviation (σ_p) of 2.91. The above values compare favorably with the measured values. The simulation of the test conditions was based on an inlet size distribution with an MMD of 4.0 μm and σ_p of 2.45, a normalized standard deviation of the gas velocity distribution of 0.25, 5% gas sneakage per stage, a rapping loss size distribution with an MMD of 4.5 μm and a σ_p of 2.8, and 35% of the mass collected in the last field being reentrained in the outlet emissions. The rapping emissions constituted approximately 40% of the total outlet emissions for the simulation. Although the parameters characteristic of the rapping losses will vary with current density, specific collection area, and inlet mass loading and particle size distribution, they were held fixed in making projections since these dependences can not be quantified at the present time.

The results of this particular application of the precipitator model for design purposes in control of opacity are encouraging.

It appears that inlet and outlet size distribution and opacity measurements along with precipitator operating parameters will provide enough information to predict the necessary modification to the precipitator to achieve a given level of opacity.

Measurement Of Relative Stack Emission Levels And Opacity

A number of optical techniques are used to determine relative stack emission levels. Usually these techniques involve a determination of the degree of light transmittance or light scattering. Some of the representative instrumentation used is discussed below:*

Nephelometers, devices that attempt to measure all of the scattered light, have recently been applied to stack monitoring. One such instrument, call the Plant Process Visiometer (PPV), has been developed by Meteorology Research, Inc., 464 West Woodbury Road, Altadena, California 91001, telephone (213) 791-1901.[248,249,250] A diagram of the optical assembly is shown in Figure 288. The sample, extracted through a probe with no dilution, is passed through the detector view. The light source is diffused so that light rays illuminate different portions of the sample in a wide range of angles from near 0° to near 180° with respect to the detector view. During operation the detector signal is calibrated with an opal glass calibrator which has been adjusted to give a certain scattering coefficient which corresponds to an opacity of 5.4 percent assuming no light absorption. This device gives an acceptable measure of mass concentration if calibration is performed against a direct mass technique and if the size distribution and composition of the aerosol remain nearly constant.

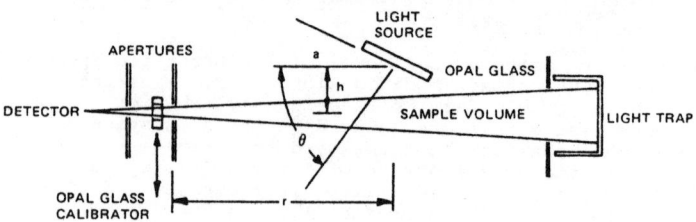

Figure 288. Optical assembly diagram of a nephelometer used in stack monitoring. The scattering angle θ, for any light ray from the source, is the angle between the ray and the horizontal line a. From Ensor and Bevan.[248]

An *in situ* monitor has been developed that is based on the measurement of backscattered light.[251] This instrument, called PILLS V, was developed by Environmental Systems Corporation, Post Office Box 2525, Knoxville, Tennessee, 37901, telephone (615) 637-4741, and uses a laser as the light source. As shown in Figure 289, both the light source and detector are located within the same enclosure.[252] One of the features of the PILLS V is its ability to determine mass concentration. The instrument optically defines a sample of 12 cm³ (0.73 in³) at 10 cm from the end of the probe within the process stream. Detection of the scattered light

---
*Southern Research Institute and the Environmental Protection Agency bear no responsibility for the promotional claims of these companies.

at angles greater than 160° relative to the beam produces an electrical signal that is proportional to the mass contained within the sample volume. Since the sample volume is a constant, the mass concentration is read directly from an appropriately labeled scale on the instrument meter. The instrument does not possess the capability to traverse large stacks in order to obtain multi-point measurements. Since the particulate mass concentration is frequently not uniform across the entire cross-sectional area of the stack, the use of such a small sampling volume and the inability to traverse creates a problem when trying to obtain data that is representative of the actual total mass concentration present within the stack.

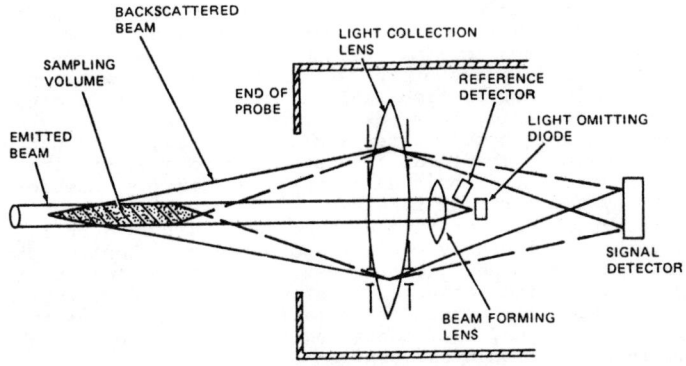

Figure 289. Optical diagram of the PILLS V instrument. From Schmitt, et al.[252]

An improved version of PILLS V, the model P-5A, has been developed. This instrument has the following specifications: a measurement range of 0.001 to 10 grams/ACM, response that is proportional to particle mass concentration and is relatively independent of the particle size in the range of approximately 0.1 to 8 μm, a process gas pressure limit of +5 inches of water from ambient (higher limits are optional); a process gas temperature limit of 260°C (500°F) (negative pressure streams permit use at higher temperatures), an instrument response that is independent of gas velocity, an optional automatic zero and span calibration at preset intervals without removal from the stack, and a light source consisting of a highly collimated beam of monochromatic laser light whose wavelength is 0.9 μm.

A backscattering instrument, called an LTV monitor, has been used in making mass measurements, but a commercial model is not available.[253] This device, illustrated in Figure 290, utilizes a high intensity argon or xenon laser and a television camera with telephoto lens. The camera optics image the backscattered light of 175° from the focused view volume, intersecting the laser beam. Particles that produce illumination above the sensitivity threshold can be resolved as distinct flashes and the intensity of each can be measured.

A portable opacity measurement system called RM41P has been developed by Lear Siegler, Inc., Environmental Technology Division, 74 Inverness Drive, E., Englewood, Colorado 80110, telephone (303) 770-3300. This system includes a transmissometer to measure light transmittance through an optical medium such as fly ash. The transceiver unit contains the light source, the detector, and electronic circuitry. The retroreflector is housed in the end of a slotted probe which is attached to the transceiver and is inserted into a stack or duct through a conventional stack sampling port. The probe causes negligible flow disturbance, and air flushing keeps the optical window and retroreflector free of dust and dirt deposits. The transceiver output is transmitted to a portable control unit that simultaneously provides an indication of optical density and opacity corrected to stack-exit conditions. There is a switch activated, self-contained, calibration checking of transceiver zero, instrument (with probe) zero, and instrument span. Automatic, electronic compensation of instrument zero output is provided whenever zero calibration is activated. The standard stainless steel probes will withstand stack temperatures up to 1200°F, though to minimize thermal conduction into the transceiver, care must be exercised to limit exposures at extreme temperatures. Some of the other features of the system are as follows: optical density output for correlation with particulate grain loading, opacity output corrected to stack-exit conditions to comply with emission standards, choice of ten measurement ranges and outputs, chopped light source for total insensitivity to ambient light, dual-beam measurement technique for maximum accuracy, double-pass measurement system for high sensitivity and easy calibration, probe inserts into stack or duct through a conventional 3½ inch I.D. sampling port, continuously variable adjustment on control panel to correct opacity outputs to stack-exit conditions for any stack or duct, choice of interchangeable one meter or five foot probe lengths, provision for permanent installation when so desired, and manually activated, self-contained transceiver zero, probe zero, and instrument span calibrations.

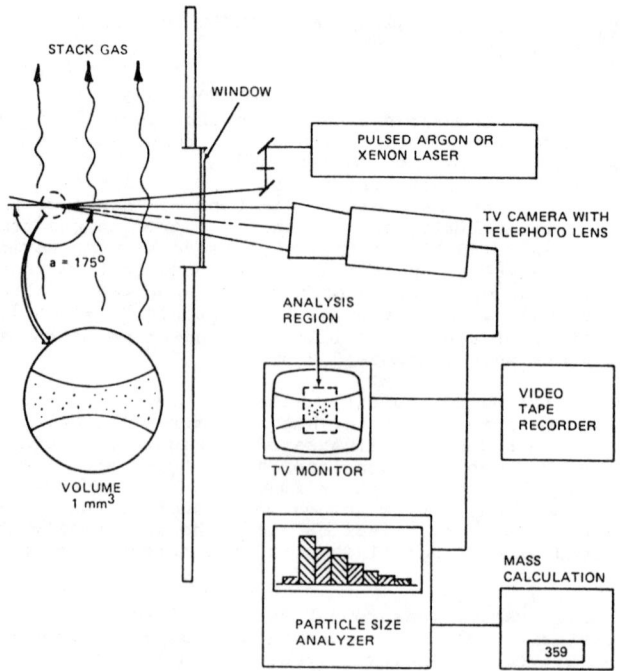

Figure 290. Schematic of Laser-TV monitor. After Tipton.[253]

Another Lear Siegler, Inc. product is the RM41 Visible Emission Monitoring System which is being used successvully to measure opacity and amount of particulate matter in effluent from large industrial stacks. The instrument performs automatic calibration and zero correction, and offers a wide choice of built-in measurement ranges and status indicators on the remote control unit to maximize system performance and operator effectiveness. Unattended operation can be expected for three to six months. The system contains a transmissometer consisting of an optical transceiver mounted on one side of a stack and a reflector mounted on the other, a forced-air purge system, and a control room unit. Containing only the essential optics and electronics required to implement the dual-beam measurement technique, the transceiver incorporates automatic continuous correction for variations in ambient temperature, line voltage, lamp aging, detector drift, and associated changes in component characteristic. Output from the transceiver is interconnected to a remote control unit, which provides simultaneous readings of opacity, corrected to stack exit conditions, and optical density, indicating actual two-pass conditions. There is an optical density output for correlation with particulate grain loading and determination of mass emission flow rates. In typical applications the standard system can be used with stack temperatures up to 316°C (600°F).

The RM7A Opacity Monitor by Lear Siegler, Inc. is a transmissometer consisting of a transceiver mounted on one side of a stack and a reflector mounted on the other side. The transceiver unit contains a light source, dual photocell detectors, and electronic measuring circuitry. A special corner-cube retroreflector is housed in the reflector unit. Both units contain provisions for optical alignment verification and correction. Zero and alarm-level adjustments are built into the transceiver. A manual zero-calibration reflector assembly and storage container are attached to the transceiver. This system is used on small or medium sized industrial facilities.

The Model 1100 Double Pass Opacity Monitoring System is manufactured by Dynatron, Inc., Barnes Industrial Park, Wallingford, Connecticut 06492, telephone (203) 265-7121. The system works by measuring variations in "double pass" light transmittance. The light source and two photo detectors are mounted on one side of the stack and a retroreflector is mounted on the other side. The light source projects a collimated beam of light which is split by a beam splitter into a reference beam and a transmitted beam. The reference beam is directed to the reference detector. The transmitted beam is projected to a "double pass" across the stack to a retroreflector which reflects it back across the stack to the measurement detector. The measurement detector working on a ratio basis with the reference detector generates an output signal directly related to smoke opacity. Some of the features of the system are: 100% solid state design, a restriction of ambient light interference, flexible air line which supplies clean filtered air, and alignment viewing port to allow a visual check by the operator.

The Model 301 Opacity Monitor by Dynatron is a rugged economical monitoring system utilizing a single pass transmissometer which enables the operator to meet opacity monitoring regulations and optimize combustion efficiency. Each system includes the following design features as standard: an analog panel meter which indicates single pass opacity at the transmissometer in 2% increments from 0 to 100% opacity, an optional digital panel meter is available with an easy to read numeric display, and a fuel saving early warning system which alerts the operator prior to a violation.

The following list gives a number of other suppliers of smoke measuring instruments and supplies:

Bailey Meter Company
Beltram Associates, Inc.
W. N. Best Combustion Equipment Company
Catalytic Products International, Inc.
Cleveland Controls, Inc.
De-Tec-Tronic Corporation
E. I. duPont deNemours & Company, Inc.
Dwyer Instruments, Inc.
Electronics Corporation of America
Environmental Data Corporation
GCA Technology Division
Horiba Instruments, Inc.
Institute for Research, Inc.
International Biophysics Corporation
ITT Barton
Jacoby-Tarbox Corporation
Leeds & Northrup Company
Milton Roy Company
NAPP, Inc.
Photobell Company, Inc.
Photomation, Inc.
Preferred Instruments
Process & Instruments Corporation
Reliance Instrument Manufacturing Corporation
Research Appliance Company
Royco Instruments, Inc.
Von Brand Filtering Recorders
Robert H. Wager Co., Inc.
Westinghouse Electric Corporation, Computer & Instrumentation Div.

Section 8

Choosing an Electrostatic Precipitator: Cold-Side vs Hot-Side Conditioning Agents

ADVANTAGES AND DISADVANTAGES OF THE DIFFERENT PRECIPITATOR OPTIONS

General Discussion

There are presently three accepted methods of utilizing electrostatic precipitators for the collection of fly ash. These methods include cold-side operation (120-180°C), hot-side operation (315-480°C), and chemical flue gas conditioning (CFGC). Whether or not one of these methods is preferable to the others depends primarily on the type of ash to be collected, the space available for control equipment, and economic considerations. Depending on the circumstances, each of these methods may have certain advantages and disadvantages. In this section, the advantages and disadvantages of the three precipitator options are discussed. Also, the precipitator requirements and economics which would be necessary to achieve a given high level of collection efficiency for high resistivity ashes are estimated for the three options.

Cold-Side Electrostatic Precipitator

Cold-side electrostatic precipitators provide the most economical and reliable option for providing high collection efficiency of fly ash with low-to-moderate resistivity ($0.1 - 5 \times 10^{10}$ ohm-cm). The low pressure drop across the precipitator, relatively low gas volume to treat on the cold-side of the air preheater, and good electrical operating conditions provide significant advantages. Figure 101 shows measured fractional efficiency data obtained from a cold-side precipitator collecting fly ash with a measured resistivity of approximately 2.2×10^{10} ohm-cm.[254] This unit operated with an average applied voltage of 51.0 kV and average current density of 38.0 nA/cm^2. A relatively high overall mass collection efficiency of 99.6+% was measured with a relatively low specific collection area of 43.5 m^2/(m^3/sec)(221 ft^2/1000 ACFM). This precipitator was preceded by a mechanical collector and was treating particulate with an inlet mass median diameter of approximately 10 µm.

The use of a cold-side precipitator becomes questionable when the resistivity of the fly ash is high (greater than 10^{11} ohm-cm). Due to the poor electrical conditions that will be experienced with a high resistivity fly ash, a cold-side precipitator has to be very large in size in order to achieve high collection efficiencies. Although there may be economic and practical drawbacks, large cold-side precipitators have been utilized successfully to collect high resistivity fly ash. Figure 89 shows measured fractional efficiency data obtained from a cold-side precipitator collecting fly ash with a measured resistivity of 1.8×10^{11} ohm-cm.[255] This unit operated with an average applied voltage of 40.9 kV and average current density of 12.1 nA/cm^2. A very high overall mass collection efficiency of 99.9+% was measured with a relatively high specific collection area of 99.2 m^2/(m^3/sec)(504 ft^2/1000 ACFM).

For sufficiently high values of fly ash resistivity, the size of a cold-side precipitator that can attain high collection efficiencies becomes excessively large. The large precipitator size needed for high efficiency collection of high resistivity ash results in large precipitator costs, increased space requirements, and possible impracticality of enlarging an existing precipitator which was originally designed to collect a low resistivity fly ash. Also, for very high values of resistivity (greater than 10^{13} ohm-cm), accurate cold-side precipitator design is probably not possible due to uncertainties regarding the attainable electrical operating conditions and useful operating voltage and current.

In addition to excessive precipitator size, there are other possible disadvantages of cold-side collection of high resistivity ash that must be considered. Due to the tendency of high resistivity ash to adhere tenaciously to the collection electrodes, high intensity impact rappers are required (120-200 g) to remove the ash from the collection electrodes. To withstand these higher rapping forces, more costly rigid electrode frames are desirable. The high rapping forces increase the possibility of ash reentrainment, structural collection electrode failures, and more difficult equipment maintenance.

Hot-Side Electrostatic Precipitator

The motivation for locating the precipitator on the hot gas side of the air preheater where temperatures are in the neighborhood of 371°C (700°F) rests entirely on data which show that ash resistivities should be very favorable. As discussed earlier, the controlling conduction mechanism in the precipitated ash layer at this temperature is intrinsic or volume conduction, instead of the surface conduction mechanism which predominates on the cold gas side of the air preheater. Thus, the fly ash resistivity at high temperature is not sensitive to the SO$_3$ or moisture content of the flue gas. Most published resistivity data indicate that resistivities below 2×10^{10} ohm-cm will occur above 600°F. Therefore, high temperature operation should offer an alternative approach for achieving high collection efficiency of fly ash which would have a high resistivity under cold temperature operation.

Another advantage of high temperature operation is that fouling of the air preheater by fly ash is reduced. However, in installations burning high sulfur coal with a basic fly ash, it is probable that removal of this ash ahead of the air preheater would result in increased corrosion rates of air preheater cold end elements. For installations in which coal and oil firing are employed, high temperature operation minimizes oil ash handling problems.

The decrease in precipitator size that can be achieved by hot-side collection of a fly ash which would have a high resistivity at cold-side temperatures is moderated by two factors. First, a higher gas volume must be treated due to the higher temperature. The increase in gas volume dictates that the precipitator be increased in size by approximately 50% in comparison to a cold-side precipitator operating at the same applied voltage and current in order to achieve the same collection efficiency. Second, the decreased gas density results in lower operating voltages and electric fields prior to sparkover than in the case of a cold-side precipitator. Thus, additional precipitator size is needed to compensate for the reduced operating voltages.

Certain economic disadvantages are associated with a hot-side precipitator. Special expansion provisions, increased insulation, increased draft fan requirements, and additional ductwork in an unconventional configuration add increased costs as compared to a cold-side precipitator. In addition, the hot-side operation reduces boiler efficiency due to heat loss through the precipitator.

Recently, it has been found that hot-side precipitators may be sensitive to the composition of the ash.[256] This sensitivity is manifested in voltage-current characteristics which are abnormal and unfavorable for electrostatic precipitation. Figures 203 and 204 show abnormal voltage-current characteristics obtained from a hot-side precipitator which responded unfavorably to fly ash deposits on the collection electrodes. These curves should be compared to those in Figures 200, 203, and 204 for normal hot-side precipitator operation. The steep voltage-current curves and low maximum applied voltages shown in Figures 203 and 204 are not expected at the elevated temperatures and result in decreased precipitator performance. In addition, the abnormal electrical conditions could not be attributed to ash resistivity since both in situ and laboratory measurements indicate a value of less than 10^{10} ohm-cm. However, these measurements were made over a relatively short period of time, and there is reason to believe that the resistivity of the collected dust layer may increase with time. Due to the above discussion, the most serious disadvantage of a hot-side precipitator is the unpredictability of the electrical conditions. Although adequate electrical conditions may be obtained with certain fly ashes, inadequate electrical conditions may result due to other fly ashes. This makes the design of a hot-side precipitator extremely difficult and makes hot-side operation less attractive as an option.

Figure 104 shows measured fractional efficiency data obtained from a hot-side precipitator collecting fly ash from a low sulfur eastern coal.[257] This unit had normal hot-side voltage-current characteristics and operated with an average applied voltage of 31.7 kV and average current density of 35.6 nA/cm^2. A relatively high overall mass collection efficiency of 99.6+% was measured with a moderate specific collection area of 76.8 m^2/(m^3/sec) (390 ft^2/1000 ACFM).

Figure 110 shows measured fractional efficiency data obtained from a hot-side precipitator collecting fly ash from a low sulfur western coal.[258] This unit had anomalous hot-side voltage-current characteristics and operated with an average applied voltage of 25.1 kV and average current density of 32.2 nA/cm^2. An overall mass collection efficiency of 98.5% was measured for the entire unit with a specific collection area of 57.1 m^2/(m^3/sec)(290 ft^2/1000 ACFM). The poor performance of this unit could be attributed primarily to the low operating voltages, especially in the outlet electrical fields.

Cold-Side Electrostatic Precipitator With Chemical Flue Gas Conditioning

Possible Advantages of Chemical Flue Gas Conditioning--

There are several attractive features and possible benefits

of adding chemical conditioning agents to the gas stream on the cold or hot gas side of the air preheater and upstream from a cold-side precipitator. First, certain chemical conditioning agents can be used to lower the resistivity of unconditioned ash from high values to values which are favorable for electrostatic precipitation. One manufacturer of conditioning systems will guarantee that the resistivity of SO_3 conditioned fly ash will not exceed 4×10^{10} ohm-cm.[259] Second, certain chemical conditioning agents can be used to increase the cohesiveness of the precipitated fly ash.[166,260] This capability can be utilized to reduce emissions due to particle reentrainment caused by rapping, high gas velocities, or hopper boil-up. Conditioning can cause particulate reentrained due to rapping to consist of large agglomerates which can be easily recollected. Third, certain chemical conditioning agents can be used to introduce a beneficial space charge effect in the precipitator.[164] With a beneficial space charge effect, higher applied voltages can be obtained at a given current density than in the unconditioned gas. The increase in applied voltage can be large enough to make a significant improvement in precipitator performance. The three effects just described have been substantiated and discussed earlier in this text. Fourth, certain chemical conditioning agents can be used to increase the resistivity of unconditioned ash from extremely low values (less than 10^8 ohm-cm) to values (approximately 10^{10} ohm-cm) which are more favorable for electrostatic precipitation.[156,260] The increase in resistivity reduces particle reentrainment due to scouring and rapping by increasing the electrical forces holding the ash layer to the collection electrode. In addition, if the low value of resistivity is due to an excess of SO_3 caused by burning high sulfur coal, the conditioning agent added in a hot section of the boiler may remove excess SO_3 by neutralizing reactions on the surfaces of the particles.[260] This is significant because high exit gas temperatures are maintained in order to prevent condensation of excess SO_3 from the flue gas which could result in corrosion and air preheater pluggage. This method of operation not only reduces boiler efficiency, but also increases the gas volume and velocity through the precipitator, thus reducing the precipitator performance. Fifth, there have been claims that certain chemical conditioning agents can favorably modify the fly ash particle size distribution by causing agglomeration of particles.[261] However, this effect has not been substantiated. If significant agglomeration of fine particles can be produced, a larger particle size distribution which can more easily be collected would be produced. Due to the wide applicability of chemical conditioning agents, one manufacturer of conditioning systems is now offering a performance guarantee that its system will reduce emissions in excess of compliance levels by a minimum of 60%, regardless of type of coal, boiler, or precipitator.[262] In order to take advantage of the multiplicity of mechanisms of fly ash conditioning, the technique of dual injection can be utilized.[260] This technique involves the application of one additive into a hot section of the boiler, followed by injection of the same or a different additive into a relatively low temperature zone, usually after the air heater.

In addition to offering improved precipitator performance, chemical flue gas conditioning has several favorable economic aspects. First, the capital costs of a new precipitator installation can be greatly reduced by using a conditioning system in conjunction with a relatively small cold-side precipitator. Second, less space is required when conditioning is used. Third, the retrofitting of existing precipitators can be accomplished relatively quickly and with little or no loss in power generating capacity.

Properties and Utilization of Well-Known Conditioning Agents--

Compounds which have been examined for use as conditioning agents in cold-side precipitators include sulfur trioxide, ammonia, sulfonic acid, sufamic acid, ammonium sulfate, ammonium bisulfate, sodium carbonate, triethylamine, and several proprietary agents.[156,164,263,264] Table 36 gives the names, chemical formulas, and physical properties of some of the conditioning agents which have been studied.[263] Some are vapors or liquids that can be volatilized without much difficulty. Others are solids that may or may not be liquified or volatilized without decomposition. All of the compounds listed are highly soluble in water. For those that are not readily volatilized, aqueous solutions provide a convenient method for injection into a flue-gas stream.

The best known conditioning agent is sulfur trioxide or the chemically equivalent compound sulfuric acid. One of the significant properties of sulfuric acid in flue gas is its tendency to undergo condensation from the vapor to the liquid state, the latter consisting of a mixture of sulfuric acid and water. The dewpoint curve given by Verhoff and Banchero[265] for sulfuric acid in flue gas containing 10% of water vapor is shown in Figure 291. If the gas stream is at a given temperature, it can contain no more vapor than is indicated by the appropriate point on this curve. At 138°C (280°F), for example, the maximum vapor concentration that can exist is 10 ppm.

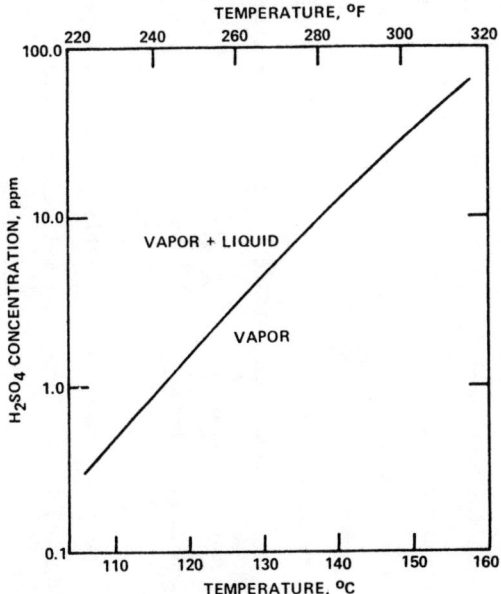

Figure 291. Dewpoint curve for sulfuric acid in the presence of 10% water vapor.

TABLE 36. PHYSICAL PROPERTIES OF CONDITIONING AGENTS

Agent[a]	Formula	State at 21°C(70°F)	Mp, °C	Mp, °F	Bp, °C	Bp, °F
Sulfur Trioxide	SO_3	Liquid	17	(62)	45	(113)
Sulfuric Acid	H_2SO_4	Liquid	10.6	(51)	326	(619)
Ammonia	NH_3	Gas	-78	(-108)	-33	(-28)
Ammonium Sulfate	$(NH_4)_2SO_4$	Solid	Dec.[b]		—	—
Triethylamine	$(C_2H_5)_3N$	Liquid	-114	(-174)	89	(193)
Triethylammonium Sulfate	$[(C_2H_5)_3NH]_2 SO_4$	Solid	—	—	—	—
Sulfamic Acid	HO_3S-NH_2	Solid	205	(401)	Dec.[b]	
Sodium Carbonate	Na_2CO_3	Solid	851	(1564)	Dec.[b]	

a. All compounds are highly soluble in water and some are used in aqueous solution.
b. Dec. signifies thermal decomposition.

Once condensed, sulfuric acid conducts electricity readily. Thus, if it condenses on fly-ash nuclei, it provides a conductive surface film. If absorbed on fly ash particles under conditions that do not allow condensation, it may again provide a conductive surface film. Actually, little is known about the chemistry and physics of adsorbed sulfuric acid, but there is evidence that part of the adsorbed material may react chemically with ash constituents to form non-conductive sulfate salts (such as calcium sulfate) but that part retains its integrity as a conductive acid.[164]

All available data indicate that SO_3 conditioning will significantly lower the resistivity of an unconditioned, high resistivity ash. In this case, SO_3 conditioning will result in improved electrical operating conditions and increased collection efficiency. The effects which can be expected from adding the other compounds mentioned are not so well defined. The realized effects, if any, appear to depend strongly on the gaseous environment and the chemical composition of the ash. In a certain application, one of these compounds may improve precipitator performance by one or more of the mechanisms discussed earlier whereas, in another application, it may have a different or no effect. A data base which is much larger than that existing is needed in order to establish the effects on precipitator operation resulting from adding the various possible conditioning agents to flue gases of differing gaseous composition and containing particles of differing chemical composition.

Unlike sulfur trioxide and sulfuric acid, ammonia is not recognized as an important naturally occurring constituent of flue gas. The distinguishing feature of ammonia vapor in flue gas is its behavior as a base. At temperatures that are not too high--say around 149°C (300°F)--it is capable of combining with sulfuric acid vapor to form ammonium sulfate, as shown by the following reaction:

$$2NH_3(g) + H_2SO_4(g) \rightarrow (NH_4)_2SO_4(s)$$

There are other acidic gases in flue gas--sulfur dioxide and carbon dioxide--but, even though they are present at much higher concentrations than sulfuric acid, they are unable to react with ammonia.

The addition of triethylamine to flue gas can be expected to lead to similar reactions, for this compound is also a base. It is stronger as a base than ammonia, however, and thus it may combine with sulfuric acid at higher temperatures or it may even react with some of the other acidic gases in flue gas.

Comparatively little is known about the chemical behavior of addition compounds of sulfur trioxide and ammonia that are used as conditioning agents. Such compounds as sulfamic acid and ammonium sulfate are frequently added at temperatures around 1100 or 1200°F. It is claimed by vendors who sell proprietary blends of these agents that injection at high temperatures is needed to decompose the agents to other products that are engaged in the actual conditioning process. Knowledge of what decomposition processes occur at high temperatures or what reactions of the decomposition products occur as the gas temperature is lowered is not complete. However, the following equations give a fairly realistic estimate of reactions that may be expected at high injection temperatures:[264]

$$HO_3S-NH_2(s) \rightarrow SO_3(g) + NH_3(g)$$

$$(NH_4)_2SO_4(s) \rightarrow SO_3(g) + H_2O(g) + 2NH_3(g)$$

Reversal of these reactions may then occur as the temperature is lowered.

Sodium compounds may be injected into the boiler along with coal.[266] In such an event, decomposition will occur:

$$Na_2CO_3(s) \rightarrow Na_2O(l) + CO_2(g)$$

The sodium oxide is incorporated in the fly ash and increases the sodium content of the ash. Sodium compounds may also be injected into the gas stream near the temperature of the electrostatic precipitator.[267] In this event, no chemical change is to be expected, and solid particles of the added compound are subject to co-precipitation with the ash.

Utility Utilization and Capital and Operating Costs of Conditioning Systems--

Capital and operating costs for cold-side conditioning systems will depend primarily on the type of conditioning agent and the system used to inject the agent. One company which makes SO_3 conditioning systems estimates the capital costs to be between $2.00 to $2.50 per KW with operating costs of $0.02 to $0.03 per ton of coal burned.[262] As of December, 1978, this company had 85 CFGC systems on stream, under construction, or on order, at 13 utilities, serving more than 16,000 MW of generating capacity. Another company which makes conditioning systems for injecting proprietary compounds has a system installed with capital costs of approximately $0.45 per KW and operating costs of less than $0.50 per ton of coal burned.[262] As of December, 1978, this company had CFGC systems at 18 utilities with the vast majority of the units in the range of 200 to 700 MW.

Recently, it has been reported that chemical conditioning agents can be utilized to improve the performance of poorly operating hot-side precipitators.[163,259,260] Laboratory studies have been conducted to evaluate the effectiveness of several different conditioning agents in improving poor, hot-temperature voltage-current characteristics which result when certain types of ashes are deposited on the collection electrodes of a precipitator.[163] With respect to effectiveness in improving the voltage-current characteristics, $NaHSO_4$, Na_2SO_4, $NaOH$, Na_2HOP_4, KOH, $KHSO_4$, and Na_2CO_3 were evaluated as good, $NaCl$ and $NaHCO_3$ were evaluated as moderate to good, NH_3 was evaluated as moderate, triethylamine and ferrous sulfate were evaluated as moderate to poor, and SO_3, $NH_3 + SO_3$, $(NH_4)_2SO_4$, and TiO_2 were evaluated as poor. All these conditioning agents were in the solid form except NH_3, SO_3, and $NH_3 + SO_3$. It has been reported that conditioning with sodium carbonate and certain proprietary compounds has been successful in improving the performance of full-scale, hot-side precipitators.[259,260] This offers another possible option for upgrading existing hot-side precipitators which are not performing adequately. A particular sodium based conditioning system has been installed with capital costs ranging between $1.75 to $2.00 per KW and operating costs between $1.00 to $1.20 per ton of coal.[259]

Possible Disadvantages of Chemical Flue Gas Conditioning--

Although chemical flue gas conditioning offers several attractive, potential benefits, there are several possible disadvantages which must be considered. First, a chemical injection system must be operated and maintained. Second, certain chemical compounds which are effective in improving precipitator performance are hazardous. Third, the effects that conditioning with certain chemical compounds will have on precipitator performance cannot always be predicted in advance. Fourth, in certain cases, the injection of chemical conditioning agents has resulted in an ash which was very sticky. If this situation results, the rapping forces might not be sufficient to remove the material collected on the discharge and collection electrodes. In addition, if the conditioning agent is injected on the hot gas side of the air preheater, pluggage and fouling of the air preheater would result. Fifth, operating costs associated with certain chemical conditioning agents can be significant. Sixth, possible future regulations concerning the emissions of chemical conditioning agents may make flue gas conditioning more difficult to implement and less effective. Future regulations appear plausible since certain agents may be potentially hazardous. It is inconsistent to regulate the emissions of certain gases such as SO_2 while allowing similar injected gases such as SO_3 to escape in significant quantities. Sufficiently high concentrations of SO_3 at sufficiently low temperatures will produce a highly visible blue plume due to the condensation of H_2SO_4. It has already been emphasized by an EPA official that any emissions of sulfuric acid, SO_3, or ammonia resulting from chemical treatment should not exceed a combined total of 10 ppm.[262] Also, it should be pointed out that only a few parts per million of certain conditioning agents contain a significant amount of mass. For example, 5 ppm of SO_3 is equivalent to 20 $\mu g/m^3$ (about 0.01 gr/ft^3). Thus, the possibility exists of treating the emissions due to conditioning on a mass basis and adding this to the mass due to fly ash emissions in order to obtain the total particulate emissions. This type of treatment of emissions of chemical conditioning agents would require that a high percentage of the injected agent be adsorbed on the surfaces of the fly ash particles.

Precipitator Requirements and Economic Comparisons--

Precipitator requirements and economic comparisons for the different precipitator options can be estimated by using the projections obtained from a mathematical model of electrostatic precipitation.[137,152] Figure 292 shows projected curves for overall mass collection efficiency as a function of specific collection area for several cases where the different precipitator options can be compared. The curve for an ash resistivity of 4×10^{10} ohm-cm at 148°C (300°F) corresponds to an ash with a favorable resistivity without conditioning or to an ash with an unfavorable resistivity that can be conditioned to a guaranteed resistivity of 4×10^{10} ohm-cm. The curves for ash resistivities of 1×10^{11}, 5×10^{11}, and 1×10^{12} ohm-cm at 148°C (300°F) correspond to cold-side precipitator operation without conditioning. The curve for hot-side precipitator operation with normal voltage-current characteristics was obtained based on electrical operating conditions demonstrated in Figure 200. The curve for hot-side precipitator operation with anomalous voltage-current characteristics was obtained based on electrical operating conditions demonstrated in Figures 203 and 204 and an adjustment to these conditions, as described elsewhere,[256] in order to obtain agreement with measured data.

All the curves were generated for an electrode geometry consisting of plate-to-plate and wire-to-wire spacings of 22.86 cm (9 in) and a corona wire diameter of 0.277 cm (0.109 in). The cold-side precipitator calculations, the maximum allowable current density for a given value of ash resistivity, was estimated by using the experimental data shown in Figure 208. Although these values of current density are probably somewhat conservative for the higher values of ash resistivity since higher useful currents might be obtained with the presence of limited back corona, it is best to be conservative in design due to the lack of predictive capabilities concerning back corona. Operating current densities for the resistivities of 4×10^{10}, 1×10^{11}, 5×10^{11}, and 1×10^{12} ohm-cm were chosen to be 22.0, 8.9, 1.7, and 0.9 nA/cm^2, respectively. The applied voltage in each electrical section for a specified current density was estimated by using the experimental voltage-current curves shown in Figure 196. These data are representative of a full-scale, cold-side precipitator treating an ash with a resistivity of approximately 2×10^{10} ohm-cm. The calculations were based on a precipitator with four electrical sections in the direction of gas flow. An applied voltage for use in the second and third electrical sections of the specified precipitator was obtained by averaging the values from the experimental inlet and outlet curves.

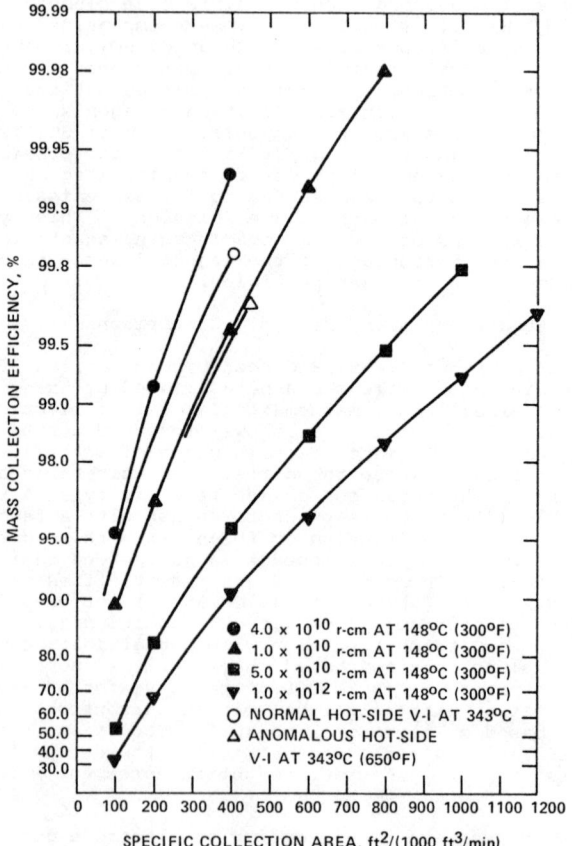

Figure 292. Effect of specific collection area on overall mass collection efficiency (curves based on a fractional gas sneakage of 0.05 and a normalized standard of deviation of gas velocity distribution of 0.25).

For all the curves, specific collection area was varied in the calculations by changing the gas volume flow and holding the plate area fixed. Although the voltage-current characteristics will change to some extent with changes in gas volume flow, it was assumed that they remain constant in making the calculations. The number of baffled sections for gas flow redirection was increased appropriately with increasing specific collection area in order to account for increased precipitator size.

The measured inlet mass loading and particle size distribution used in the calculations are typical of fly ash generated by coal-fired boilers. The inlet mass loading was 5.7 gm/m^3 (2.5 gr/acf). The log-normal fitted inlet particle size distribution had a mass median diameter of 25.5 µm with a geometric standard deviation of 5.1. To account for the effect of particle size distribution, especially in the fine particle range (0.25-3.0 µm), the measured particle size distribution was divided into size intervals with midpoints of 0.2, 0.4, 0.7, 1.1, 1.6, 2.5, 3.5, 4.5, 6.0, 8.5, 12.5, 20.0, and 27.5 µm.

The curves were generated based on a fractional gas sneakage and particle reentrainment without rapping per baffled section of 0.05 and a normalized standard deviation of the gas velocity distribution of 0.25. These values are typical of a precipitator which is in good mechanical condition. All overall mass collection efficiencies have been corrected for rapping reentrainment using an empirical procedure based on field test data from full-scale precipitators as discussed earlier.[19,137]

The curves in Figure 292 do not address the problems of (1) opacity, (2) variations in the significant parameters influencing precipitator performance, and (3) outage of electrical sections. Therefore, these curves are intended only for use in making relative comparisons of the different precipitator options for treating ashes with different resistivities and should not be used per se for design purposes. Problems (2) and (3) can be conservatively accounted for by designing the precipitator with more collection area than that needed to achieve the desired collection efficiency. However, problem (1) requires a somewhat extensive analysis to determine if the opacity standard will be met and to determine what safety margins should be included in the precipitator design to account for normal variations in precipitator parameters that would cause an increase in opacity. In many cases, the mass emissions standard will be attained at collection efficiencies well below that needed to meet the opacity standard.

The curves in Figure 292 can be used to make a relative economic comparison of the different precipitator options in terms of total fixed (capital) investment for an 800 MW unit. As an example, the total fixed investment for each of the precipitator options can be determined based on a required overall mass collection efficiency of 99.5%. The design parameters for the different precipitator options, cold-side ash resistivity values, and possible hot-side electrical conditions are given in Table 37.

Estimated costs for cold-side, hot-side, and conditioned precipitators for use on an 800 MW unit have been published recently.[259] These estimates will be used here for comparing the relative capital costs of the various precipitator options. In this particular analysis, cold-side, hot-side, and conditioned precipitators would cost $14.82, $15.65, and $16.62 per square foot of collection plate area, respectively.

TABLE 37. DESIGN PARAMETERS FOR DIFFERENT PRECIPITATOR OPTIONS AND OPERATING CONDITIONS ON AN 800 MW UNIT

	Cold ESP $\rho=4\times10^{10}$ Ω-cm (conditioned or unconditioned)	Cold ESP $\rho=1\times10^{11}$ Ω-cm	Cold ESP $\rho=5\times10^{11}$ Ω-cm	Cold ESP $\rho=1\times10^{12}$ Ω-cm	Hot ESP Normal V-I	Hot ESP Anomalous V-I
Gas Volume Flow m³/min (1,000 ACFM)	78.4 (2,800)	78.4 (2,800)	78.4 (2,800)	78.4 (2,800)	114.5 (4,089)	114.5 (4,089)
Gas Temperature °C (°F)	149 (300)	149 (300)	149 (300)	149 (300)	343 (650)	343 (650)
Collection Efficiency (%)	99.5	99.5	99.5	99.5	99.5	99.5
Collecting Surface Area 1,000 m² (1,000 ft²)	61.1 (658)	98.8 (1,064)	212 (2,282)	289 (3,108)	114 (1,227)	152 (1,636)
Specific Collection Area m²(m³/sec) (ft²/1,000 ACFM)	46.3 (235)	74.9 (380)	160.6 (815)	218.7 (1,110)	59.1 (300)	78.8 (400)

The quoted costs include the following items:

1. A gas volume flow for the cold-side precipitator systems which includes 9% leakage at the air heater.
2. Base equipment.
3. Flues which are sized to provide a gas velocity of 18.3 m/sec (60 ft/sec).
4. Plenums.
5. Necessary expansion joints for thermal motion and dampers for isolation and gas distribution.
6. Accessories which include safety interlocks, internal walkways, hopper heaters, hopper level indicators, remote controls, transformer-rectifier removal systems, weather enclosures, gas distribution devices, facilities, and typical instrumentation.
7. Support structures.
8. Erection.
9. Insulation.
10. SO_3 gas conditioning system in the case of the cold-side precipitator with conditioning.
11. Ash handling system at $5,000 per hopper.
12. Capacity charge at $800/KW.
13. Required land at $10,000/acre.

Based on the above considerations, Table 38 gives a comparison of the different precipitator options under different operating conditions in terms of total fixed investment. The comparisons in Table 38 and Figure 292 show several points of interest. First, an unconditioned, cold-side precipitator is the most economically effective option for ash resistivities of 4×10^{10} ohm-cm or less and, in addition, should be considered seriously until the ash resistivity is greater than 1×10^{11} ohm-cm. Second, for ash resistivities greater than 1×10^{11} ohm-cm, flue gas conditioning and hot-side operation with normal voltage-current characteristics become attractive options from an economic standpoint when compared to unconditioned, cold-side operation. However, at best, hot-side operation will be a factor of 1.76 times as costly as cold-side operation with conditioning. Third, if a hot-side precipitator is sized to account for the possibility of anomalous voltage-current characteristics, then it will cost a factor 1.33 times that of a hot-side precipitator with normal voltage-current characteristics. This would make the hot-side option extremely unfavorable when compared to flue gas conditioning and would make it competitive with unconditioned, cold-side operation treating ashes with resistivities near 5×10^{11} ohm-cm or less.

Since annual operating and overhead costs will be dominated by amortization of the debt (including interest, taxes, and insurance at approximately 20% of the total fixed investment), the relative comparison of these costs between the different precipitator options should parallel that of the total fixed investment analysis. The operating costs include (1) heat loss for the hot-side options, (2) an energy charge for all the options that depend on power input to the transformer/rectifier sets and pressure drop across the precipitator, (3) cost of the conditioning agent for the flue gas conditioning option, and (4) maintenance. These operating costs are small compared to the amortization and at most will probably not exceed 25% of the amortization. The heat loss penalty for the hot-side option will probably make the estimate of its operating costs somewhat higher than the other two options when all three options are evaluated for the collection of high resistivity ash. Of course, the cost of the conditioning agent can vary widely, depending on the type of agent and the supplier. Finally, the estimation of maintenance costs is difficult and would vary significantly from one type of precipitator to another.

Due to the uncertainties involved in estimating the operating costs for the different options, this type of analysis will not be presented here. However, estimated operating costs can be found elsewhere.[259]

TABLE 38. TOTAL FIXED INVESTMENT OF PRECIPITATOR OPTIONS UNDER DIFFERENT OPERATING CONDITIONS FOR AN 800 MW UNIT ($1000)

	Cold ESP $\rho=4\times10^{10}\,\Omega\text{-cm}$ (unconditioned)	Cold ESP $\rho=4\times10^{10}\,\Omega\text{-cm}$ (conditioned)	Cold ESP $\rho=1\times10^{11}\,\Omega\text{-cm}$	Cold ESP $\rho=5\times10^{11}\,\Omega\text{-cm}$	Cold ESP $\rho=1\times10^{12}\,\Omega\text{-cm}$	Hot ESP Normal V-I	Hot ESP Anomalous V-I
Total Investment	9,752	10,936	15,769	33,819	46,061	19,203	25,603
Relative Investment Ratio	1.00	1.12	1.62	3.47	4.72	1.97	2.63

Section 9

Safety Aspects of Working with Electrostatic Precipitators

RULES AND REGULATIONS

The only regulations specified by OSHA as being applicable to safety practices around an electrostatic precipitator are (1) the National Electrical Code - found in 29 Code of Federal Regulations 1910 Subpart S, and (2) Occupational Health and Environmental Control - found in 29 Code of Federal Regulations 1910.1000 Air Contaminants. Table 7-3 of the CFR gives exposure limits to silica and coal dust, and Table 7-1 of the CFR sets an exposure limit for ozone, which is produced during electrical discharge, and for sulfur dioxide, which results from coal combustion.

HAZARDS[268,269,270,271]

Since the operation of an electrostatic precipitator involves high voltage, extreme caution should be taken when inspecting and troubleshooting to avoid electrical shock. Also, serious fires and explosions have occurred, resulting in large losses and long shut-downs. Other hazards one encounters while inspecting precipitator internals involve toxic gases, especially ozone and sulfur oxides, sudden accidental activation of rapping equipment, possible burns and heat exhaustion from working inside the shell, eye and lung contamination from foreign particles, especially fly ash, and the possibility of falling from areas being inspected. These hazards and preventive measures will be discussed in detail below.

Fire And Explosion Hazards[269,272]

Combustion may be defined as the rapid chemical combination of oxygen with the combustible elements of a fuel. There are just three combustible chemical elements of significance - carbon, hydrogen, and sulfur. Sulfur is of minor significance as a source of heat. Carbon and hydrogen when burned to completion with oxygen unite as shown below:

$$C + O_2 = CO_2 + 14,100 \text{ Btu/lb of C}$$

$$2H_2 + O_2 = 2H_2O + 61,100 \text{ Btu/lb of } H_2$$

Excess air, blown into the primary furnace of the steam generator, is the usual source of oxygen for boiler furnaces. The objective of good combustion is to release heat while minimizing losses from combustion imperfections and superfluous air. Adequate combustion then requires temperatures high enough for ignition, turbulence or mixing, and sufficient time. These factors are known as the "three T's" of combustion. If one of these requirements is deficient incomplete combustion occurs with its resultant unburned carbon constituents. Fires can quickly become a problem with the presence of combustibles, oxygen, and source of ignition (high voltage sparking).

Some of the areas in which fires have occurred due to poor combustion are in the electrostatic precipitators themselves, air heaters, flues, ducts, coal pipes, and precipitator hoppers. In one case where improper combustion occurred, there was excessive air in-leakage between the primary furnace and the precipitator. This air leakage, together with unburned carbon and arcing in the precipitator, caused a fire. In another case the formation of clinkers (large carbonaceous ash masses which adhere to tube surfaces) produced plugged secondary superheaters which allowed more fuel to carry over to the precipitator.

In summary, poor maintenance and poor operating practices at the plant facility are the major causes of fires and explosions in electrostatic precipitators. Poor combustion due mainly to improper amounts of excess air appears to be the major operating practice leading to fires.

Electrical Shock Hazards[269]

Electrical shock to operators of precipitators is due to the failure, misuse, or faulty condition of electrostatic precipitator equipment and may cause the following conditions: painful shock from sudden contact, resultant action from shock contributing to a secondary hazard (falling, dropping tools, etc.), flesh burns at points of contact, and death if the victim cannot release himself from the energized conductor within a reasonable period of time (this factor depends greatly upon one's physiological condition, amount of current, resistance and path of current flow through the body, and type of electrical energy in question).

Pure direct current produces a steady sensation of intense heating and burning along the current path with only slight muscular contraction. A direct current flow of ten milliamperes through the body causes little or no sensation, but secondary hazards, such as falls, are possible. At about 60 to 80 milliamperes the sensation becomes unbearably painful, with no tissue damage. However, the muscular reactions due to breaking contact may be sufficient to throw a person bodily. With higher currents the above effects are increased and serious burns may be encountered. Fibrillation appears in the range of 500 to 2000 milliamperes on contacts that exceed a quarter of a second.

Large, high voltage electrostatic precipitators usually have double interlocking safety controls to prevent electrical shock accidents. These safety controls prevent entrance into the electrostatic precipitator unless the unit has been deenergized. If the primary safety control fails and the access door is opened while the precipitator is in operation, the secondary control immediately grounds out the transformer and the unit is deenergized. Sometimes, however, maintenance and operation personnel do not want to take the time necessary for proper shutdown and bypass the interlocking safety controls. When safety controls are misused in this way, accidents often result. Another potential problem with the safety controls occurs when they are not inspected and maintained periodically. An actual case of electrocution occurred when safety controls, which operated in a corrosive atmosphere,

corroded to the point of not functioning. When a worker entered
the unit, thinking it would be deenergized, electrocution resulted.

Poking the precipitator collection hoppers with long poles
to facilitate the flow of bridged fly ash is a common practice.
Obviously a non-conductive pole, never a metal pole, should be
used. If a metal pole makes electrical contact between the
energized parts of the unit and the hopper, electrocution could
result.

Toxic Gas Hazard

Purging the inside of the electrostatic precipitator with air
is necessary before allowing personnel to enter because of the
presence of toxic gases. Sulfur oxides and ozone are two gases
which can be present in concentration great enough to cause a
health risk. Sulfur dioxide and sulfur trioxide are common gaseous emissions when burning sulfur-containing coal. Sulfur
trioxide (SO_3) is not likely to be present in large quantities
(a few parts per million) but it readily combines with water
vapor to form sulfuric acid mist and can be dangerous. Sulfur
dioxide (SO_2) could be present in several hundred or even several
thousand parts per million inside the precipitator depending upon
the sulfur content of the coal. The taste threshold for SO_2 is
about 0.3 ppm, and SO_2 is a very unpleasant experience at 1 ppm.
A level of 5 ppm of SO_2 causes respiratory irritations and even
spasmodic reactions in some sensitive individuals.[273] Ozone is
produced by the discharge of high tension electrical current inside the precipitator. The body is very sensitive to ozone,
detecting its odor as low as 0.02 ppm. Nasal and throat irrita-
tion occur at 0.3 ppm. At 1 ppm, severe restriction of respiratory
passages occurs and many persons cannot tolerate higher concentrations. Ozone appears to damage lung tissue by accelerating the
aging process, making it more susceptible to infection.

Other More Minor Hazards[270]

The rapping area contains rotary equipment which is deenergized when the weather enclosure door is open. However, if the
door is closed the equipment may operate if a padlock is not used
to lock open the disconnect on the panel feeding the rappers.

Heat exhaustion and/or severe burns can result from entering
the precipitator too soon after shut-down since the steel takes a
very long time to cool down.

Eye protection should be worn to protect eyes from fly ash
and other foreign particles.

There are areas within the precipitator from which one could
fall. Ladders should be properly secured and safety belts may be
appropriate.

Section 10

Maintenance Procedures

Proper maintenance precautions and procedures can make the difference between an electrostatic precipitator which operates satisfactorily and one which is continually beset with operational difficulties. Most of an installation's problems are mechanical in nature and, though many of the breakdowns can be traced to poor structural design or poor installation, poor maintenance is the cause often enough to merit a detailed discussion. Two general categories of precipitator maintenance problems exist: those problems due to lack of proper preventive maintenance and those problems associated with failure or breakdown of precipitator components. A careful, step-by-step start-up procedure is an invaluable preventive maintenance aid, and a typical start-up procedure and inspection is given in Table 39.[274] After start-up preventive maintenance schedules should be established to conform to the requirements for the particular installation. A typical maintenance schedule for an electrostatic precipitator is given in Table 40.[274,275,276]

TABLE 39. INITIAL ELECTROSTATIC PRECIPITATOR START-UP PROCEDURE AND INSPECTION[274]

Ducting

1. Check all ducting for foreign material.
2. Check all welded joints for leakage.

Internals

1. Check collecting plates for straightness and flatness and give tolerances.
2. Check spacing of collecting plates and give tolerances.
3. Check pendulum movement of collecting plates.
4. Check rappers for freedom of movement and alignment.
5. Check the spacing between the plates and discharge wires and give tolerances.
6. Spot check discharge wires for proper tension.
7. Check for foreign material clinging to discharge wires, collecting plates, precipitator chamber bottom, and hopper area.
8. Check all welds on high voltage frames.
9. Check all motors, bearings, reducers, etc. for proper lubrication.
10. Check all motors for direction of rotation.
11. Check underside of insulators for cleanliness, foreign material, and position of high voltage hanger rods.

(CONTINUED)

TABLE 39 (CONTINUED)

Insulator Compartment

1. Check insulator compartment for debris.
2. Check insulator for cracks.
3. Check installation of high voltage hanger.
4. Check welds on high tension hangers.
5. Check for dryness.

Access Doors

1. Insure that door is free swinging.
2. Check latches for tightness when door is shut.
3. Check gasket for gas tight seal.

Rapper Drives

1. Check alignment of all rapper shafts.
2. Check for proper installation of insulators through casing wall for high voltage rapper shafts.

Hopper Conveyors

1. Check rotation of screw conveyors if used.
2. Check for binding.

Safety Interlock System

1. Check to insure that all keys are in master keyboard.
2. Check all key-locks to insure that safety lock is operating.

Electrical

1. Inspect the control panel and insure that all motor and heater control circuits, inter-locking arrangements, and remote controls function properly.
2. Arrange that all time relays, end position switches, rotation guards, etc. be set properly and that the function of all alarm signals be checked.
3. Check that all electric heaters function and set the thermostats correctly.
4. Inspect the rectifier units with regard to oil level, etc. (follow rectifier manufacturer's instructions).
5. Inspect all transformer rectifiers.
6. Check all electrical wiring to precipitator.
7. Check wiring between control cabinet and high tension transformers to be certain control cabinet is actually connected to the proper high voltage transformer and that interlocks are in the proper sequence. Check ground wiring.
8. Connect one rectifier unit at a time corresponding to the emitting system of the precipitator.
9. Start the rectifier.
10. Check current and voltage at different settings and test the signals circuits.

TABLE 40. TYPICAL MAINTENANCE SCHEDULE[274,275,276]

1. Check for drift of meter readings away from baseline values established when ESP was installed. Record readings for each control unit.

(CONTINUED)

TABLE 40 (CONTINUED)

2. Keep an accurate log of all aspects of precipitator operation. In addition to the electrical data, record changes in rapper and boiler operation and variations in fuel quality.

3. Check insulator heaters for operation mode and record ammeter readings of each insulator heater.

4. Check all "Push to Test" lights on panel and replace as necessary.

5. Check all rapper timers for operation.

6. Test annunciator panel for operation and replace any bad lights.

7. To warn of hopper ash buildup and ash conveyor stoppage, check skin temperature of hopper.

8. Check operation of rapper and vibrator controls.

9. Check oil level of all transformer-rectifier units and record oil temperature.

10. Note and report any leaks on tank of transformer-rectifier.

Weekly

1. Make visual inspection of rapper action and check vibrator operation by feel.

2. Check control sets internally for deposits of dirt that may have penetrated the filter. Accumulation of dirt can cause false control signals and can be destructive, particularly to large components such as printed circuits.

3. Clean all insulators.

4. Check access doors for tightness.

Monthly

1. Shut down unit, tag switches, apply ground protectors, and proceed with inspection and maintenance.

2. Using low pressure air, blow out rectifier compartments and control cabinets.

3. Clean with carbon tetrachloride and check for chips and arc tracks the following:

    a. transformer bushings
    b. stand-off insulators
    c. potheads
    d. rectifier rotor and cross arms
    e. rectifier tubes

4. Clean or change ventilating fan air filters.

5. Check rotor and stator shoes for wear and proper adjustment.

(CONTINUED)

TABLE 40 (CONTINUED)

6. Inspect on the drag motor the foundation bolts, alignment, and rotor end play.

7. Inspect on the screw conveyor motor the foundation bolts.

Quarterly

1. Clean inside all panels.

2. Check all electrical components for signs of overheating.

3. Clean and dress electrical distribution contacts, surfaces, and lubricate pivots.

4. Check vent fan for operation and check clearances between blades and shroud.

5. Install new filters in control panel.

6. Routine inspection, cleaning, and lubrication of hinges and test connections.

7. Exterior inspection for corrosion, loose insulation, exterior damage, and loose joints.

Annually

1. Remove dust buildup on wires and plates, and adjust intensity of rappers and vibrators if necessary.

2. Inspect perforated diffuser screen and breeching for dust buildup.

3. Perform maintenance and lubrication of pressurized fans and check for leaks in pressurized system.

4. Check for loose bolts in frames, verify that suspension springs are in good order, and examine wearing parts.

5. Inspect discharge wires for tightness and signs of burning and measure to see if they hang midway between plates.

6. Check plates for alignment and spacing.

7. Check insulators for cracks.

8. Drain oil, wash out, and refill gear boxes.

9. Check transformer fluid and dielectric strength.

Several surveys have been conducted in an effort to identify the major sources of operating malfunctions most commonly encountered with electrostatic precipitators.[277,278,279,280] A survey conducted by the Industrial Gas Cleaning Institute in 1969 identified problems in the order listed in Table 41.[279] The number identified with each problem is a percentage of the respondents identifying the particular component as a maintenance problem. Results of a 1974 Air Pollution Control Association (APCA) survey of electrostatic precipitator maintenance are similar (Table 42).[279]

TABLE 41

MOST COMMON MAINTENANCE PROBLEMS[279]

Component	Percent
Discharge Electrode Failure	68
Rapper Malfunctions	40
Insulator Failures	28
General Dust Buildup Causing Shorts	28
Hopper Plugging	24
Transformer Rectifier Failures	20

TABLE 42. POWER PLANT ELECTROSTATIC PRECIPITATOR MAINTENANCE PROBLEMS[279]

| Component | Major Maintenance Problem, % | Component Failure Frequency, % | | |
		Frequent	Infrequent	Very Seldom
Discharge Electrodes	35.2	29.5	38.6	28.4
Dust Removal Systems	31.8	36.4	42.0	20.5
Rappers or Vibrators	5.7	9.1	38.6	47.7
Collecting Plates	13.6	4.5	7.9	68.2
Insulators	1.1	8.0	34.1	48.9

Discharge electrode failures are typically caused by electrical arcing, corrosion, and fatigue. When a wire breaks an electrical short circuit often occurs between the high-tension discharge wire system and the grounded collection plate. The short trips a circuit breaker, disabling a section of the precipitator until the discharge wire is removed or replaced. Some of the more common specific causes of discharge wire breakage are:[277]

(1) Inadequate rapping of the discharge wire which eventually allows arcing to occur.

(2) Improperly centered wires leading to sparking at those points too near the bracing.

(3) Clinker or a wire that bridges the collection plates and shorts out the wire.

(4) Ash buildup under the wire, causing it to sag and short out.

(5) Corrosion caused by condensation.

(6) Excessive localized sparking leading to wire erosion.

(7) Fatigue leading to wire breakage, especially at those points where wires are twisted together.

(8) Fly ash buildup in certain spots which leads to a clinker and burns off the wire.

Continuous sparking at any one location along a discharge wire will ultimately lead to wire failure since small quantities of metal are vaporized with each spark. Localized sparking can be caused by misalignment of the discharge electrodes during construction or by electric field variations caused by "edge" effects where the discharge and collection electrodes are adjacent to each other at the top and the bottom of the plates. Mechanical fatigue often occurs when the discharge wire is twisted around the support collar at the top of the discharge electrode.

Since the existence of temperatures below 121°C (250°F) may lead to excessive corrosion and fouling of the cold-end elements of the air heater and corrosion of cold-side precipitator elements, the topics of corrosion and fouling are of considerable importance and deserve proper attention. However, since proper design should result in temperatures above 121°C (250°F) and since an adequate coverage of the topics of corrosion and fouling requires extensive text, a discussion of low temperature corrosion and fouling is given in Appendix D instead of in the main text. Appendix D includes discussions of (1) sulfuric acid occurrence in flue gas based on SO_x, H_2O, and H_2SO_4 equilibria, determination of the sulfuric acid dew point, and condensation characteristics, (2) factors influencing corrosion rates such as acid strength, acid deposition rate, fly ash alkalinity, and hydrochloric acid, (3) fouling of low temperature surfaces, (4) laboratory corrosion studies, and (5) power plant data.

Problems with the dust removal systems are caused primarily by hopper plugging, followed by screw conveyor and dust valve deficiencies. Improper adjustment of hopper vibrators or complete failure of the ash conveyor are common causes of hopper overflow. Heaters and/or thermal insulation for the hoppers to prevent ash agglomeration may be helpful in some cases.

Rapping is required for both discharge and collection electrodes. A number of different rapping systems are used but those rapping systems using vibrators, either pneumatic or electric, appear to require more maintenance than impulse-type systems. Failures of support insulators are caused primarily by arc-overs from accumulations of dust or moisture on the surface of the insulator. These failures are often caused by inadequate pressurization of the top housing of the insulators.

Other problems which cause difficulty, but to a lesser extent, are dust buildup in the upper outside corners of hoppers, corrosion in the less accessible parts of the precipitator such as around the access doors and frames, box girders, and housing, plugging of gas distribution plates, problems with rapping system drives, wear of rappers and bushings, and problems of wear and movement occurring at points of impact.

Another point of inquiry in the APCA survey involved overall experience with electrostatic precipitators from operational and maintenance standpoints. The utilities' responses were:

Utilities - Operation of Precipitators

Excellent	Good	Fair	Poor
14.8%	45.5%	29.5%	10.2%

Utilities - Precipitator Maintenance

Excellent	Good	Fair	Poor
13.6%	52.3%	13.6%	20.5%

Some of the data reported represent precipitator installations that have been in service for many years and often these installations have not received proper attention.

Proceedings from a recent specialty conference on the operation and maintenance of precipitators would be extremely useful to users who experience many of the problems discussed in this section.[281]

Section 11

Troubleshooting

DIAGNOSIS OF ESP PROBLEMS

Causes for an electrostatic precipitator to fail to achieve its design efficiency can be due to poor maintenance as discussed in the previous section, or they can be due to inadequate design, electrical difficulties, improper gas flow, inadequate rapping, installation problems, electrode misalignment, or improper operation.

Structural engineering and design considerations are frequently overlooked by the engineer who specifies and buys electrostatic precipitators, for he often assumes that the manufacturer's experience and engineering capability is sufficient. In the competitive atmosphere which exists among precipitator manufacturers, a manufacturer normally proposes only the equipment and features absolutely necessary to meet contract requirements.[282] Any deviations from a manufacturer's standards would increase costs and possibly cost him his competitive advantage. An example of one of the structural problems which has occurred is the lack of provision for expansion, possibly stemming from a temperature assumption that allows no margin, thus causing excessive deflection of the substructure or the interior precipitator beams and columns.[282] Other structural problems arise from insufficient attention to fabrication and erection tolerances, which result in misalignment and operating difficulties.

Indications of electrical difficulties can usually be observed from the levels of corona power input. Efficiency is generally related to power input, and if inadequate power densities are indicated, difficulties can usually be traced to:[275]

(1) high dust resistivity,
(2) excessive dust accumulations on the electrodes,
(3) unusually fine particle size,
(4) inadequate power supply range,
(5) inadequate sectionalization,
(6) improper rectifier and control operation,
(7) misalignment of the electrodes.

Because of the importance of resistivity in the precipitation process, in situ resistivity measurements should be one of the initial trouble shooting steps. If resistivity exceeds 10^{10} ohm-cm, the resistivity may be the blame for most of the difficulty.

Other electrical problems encountered with electrostatic precipitators are shorting of the high tension frame by dust accumulation in the hoppers, broken wires, insulator bushing leakage, and leaking or broken cables.

Quality of gas flow can be determined by measurement of a gas flow distribution profile at the precipitator inlet. The IGCI recommends a gas quality such that 85% of the local velocities is within 25% of the mean with no single reading more than 40% from the mean. Poor gas flow often results from dust accumulation on turning vanes and duct work and plugging of distribution plates.

Gas "sneakage", a term describing gas flow which by-passes the effective precipitator section, can also be a problem. "Sneakage" can be identified by measurement of gas flow in the suspected areas (the dead passages above the collection plates, around the high tension frame, or through the hoppers) during a precipitator outage with the blowers on. Also, problems of reentrainment of dust from the hoppers because of air inleakage or gas "sneakage" can often be identified by an increase in dust concentration at the bottom of the exit to the precipitator. Corrective measures usually involve baffling to redirect gas flow into the electrified region of the precipitator.

Improper rapping is usually the cause when excessive dust deposits occur on the discharge and collection electrodes. Adequacy of rapping can be measured by accelerometers mounted on the electrodes. One should carefully adjust the rapping intensity and cycle to maintain a practical thickness of dust deposit without excessive reentrainment.

Most problems associated with hopper and ash removal systems are usually due to improper adjustment of the hopper vibrators or failure of the conveyor system. In some instances heat and/or thermal insulation for the hoppers to avoid moisture condensation may be necessary.

Severe difficulties with electrostatic precipitators are usually caused by inadequate electrical energization or excessive reentrainment. The following is a rather general guide which may be useful in pinpointing the causes of severe precipitator problems:[275]

(1) Measure the high tension voltage, current, and spark rate.

(2) Measure gas flow distribution.

(3) Observe collecting plates for evidence of back corona.

(4) Use an oscilloscope to record the high tension voltage to determine the duration of the corona current.

(5) Observe the collection plates for evidence of excessive reentrainment (this requires construction of a glass plate and wiper for an access port and a means for illumination of the interelectrode space).

(6) Examine alignment and condition of the hoppers, insulators, and other components.

(7) Measure rapping accelerations.

Table 43 is a trouble shooting chart for use in determining the cause of common electrostatic precipitator malfunctions, with suggestions for remedying these problems.[277]

TABLE 43

TROUBLESHOOTING CHART[277]

Symptom	Probable Cause	Remedy
(1) No primary voltage No primary current No precipitator (ESP) current Vent fan on	DC overload condition Misadjustment of current limit control Overdrive of rectifiers	Check overload relay setting Check wiring and components Check adjustment of current limit control setting Check signal from firing circuit module
(2) No primary current No precipitator current Vent fan off Alarm energized	Fuse blown or circuit breaker tripped Loss of supply power	Replace fuse or reset circuit breaker Check supply to control unit
(3) Control unit trips out an overcurrent when sparking occurs at high currents	Circuit breaker defective or incorrectly sized Overload circuit incorrectly set	Check circuit breaker Reset overload circuit
(4) High primary current No precipitator current	Short circuit condition in primary system Too high precipitator voltage for prevailing operating conditions High voltage circuit shorted by dust buildup between emitting and collecting electrodes Slack or broken emitting electrode wire shooting the high "V" circuit Circuit component failure Trouble in ESP: (1) Dust buildup in hopper; check meters: —Ammeter very high —KV meter very low (½ normal) —Milliamperes very high (2) Metallic debris left in unit during shutdown for maintenance (3) Unhooked collecting plate touching emitting frame (4) Broken support insulator (5) Excessive dust buildup on hopper beams or cross member	Check primary power wiring Lower the precipitator voltage Remove dust buildup Deenergize precipitator and remove or replace broken or slack wire Check transformer-rectifier and precipitator: Ground T-R high "V" connector to precipitator Clean off dust buildup Deenergize ESP and remove Repair Repair Clean
(5) Low primary voltage High secondary current	Short circuit in secondary circuit or precipitator	Check wiring and components in high voltage circuit; check ESP for: interior dust buildup, full hoppers, broken wires, ground switch left on, ground jumper left on, broken insulators, foreign material on high voltage frames or wires
(6) Abnormally low ESP current and primary voltage with no sparking	Misadjustment of current and/or voltage limit controls Misadjustment of firing circuit control Heavy coating on emitting electrode wires Stream of cold air entering ESP from defective door gasket duct opening, inlet gas system rupture-condensation Wet dust clinging to wires causes extremely low millampere readings Severe arcing in the ESP without tripping out the unit	Check settings of current and voltage limit controls Turn to maximum and check setting of current and voltage limit controls Check emitting frame vibration and emitting vibration shaft insulator Repair Eliminate source of condensation Eliminate cause of arcing
(7) Spark meter reads high-off scale	Continuous conduction of spark counting circuit	Deenergize, allowing integrating capacitor to discharge and reenergize

(continued)

TABLE 43 (CONTINUED)

Symptom	Probable Cause	Remedy
Low primary voltage and current; no spark rate indication	Spark counter counting 60 cycles peak	Readjust controls
(8) Spark meter reads high primary voltage and current very unstable	Misadjustment Loss of limiting control	Readjust Replace control
(9) No spark rate indication; voltmeter and ammeter unstable indicating sparking	Failure of spark meter Failure of integrating capacitor Spark counter sensitivity too low	Replace spark meter Replace capacitor Readjust sensitivity
(10) No response to voltage limit adjustment Does respond to current adjustment	Controlling on current limit or spark rate	None needed if unit is operating at maximum current or spark rate Reset current and spark rate adjustment if neither is maximum
(11) No response to spark rate adjustment Does respond to other adjustment	Controlling on voltage or current	None needed if unit is operating at maximum voltage or current Reset voltage and current adjustment if neither is at maximum

AVAILABLE INSTRUMENTATION FOR ELECTROSTATIC PRECIPITATORS

Spark Rate Meters

The term "spark rate" refers to the number of times per minute that electrical breakdown occurs between the corona wire and the collection electrode. A spark-rate controller establishes the applied voltage at a point where a fixed number of sparks per minute occur (typically 50 - 150 per corona section). The sparking rate is a function of the applied voltage for a given set of precipitator conditions. As the spark rate increases, a greater percentage of input power is wasted in the spark current, and consequently less useful power is applied to dust collection. Continued sparking to one spot will cause errosion of the electrode and sometimes mechanical failure. Therefore, to meet rapid or periodic changes in the gas and ash composition, the rectifier should be fitted with a spark rate controller which can automatically adapt the current to the changing operating conditions. The precipitator is thus supplied with a maximum of current at all times.

The spark rate meter may be supplied as a self-contained unit or built into the automatic voltage control system. Some of the companies which supply the spark rate meter and/or total voltage control system are given below:

- Environecs
  1654 Babcock Street
  Costa Mesa, California 92627
  (714) 631-3993

The Environecs spark rate meter circuit is a standard part of their total automatic voltage control system (Figure 293[283]). Other standard features of this system (see Figure 294[283]) other than the spark rate meter are: (1) Electronic Current Limit, which prevents drift in the current setting; (2) Soft Start, which prevents high in-rush current to the high voltage power supply at start-up; (3) Recovery Control, which adjusts the rate at which voltage recovers from the zero level after a spark back up to the setback point; (4) Setback Control, which determines the reduction of output voltage after a spark is detected; (5) Hold Control, which holds the voltage at the adjusted setback

level for a short period of time, allowing the precipitator to
stablize; (6) Rise Rate, which determines the rate at which the
output power increases to the current limit setting or until a
spark is detected; (7) Spark Detection, which senses the spark
on the first half cycle, allowing the control logic circuits to
adjust the precipitator power immediately following the spark;
(8) Automatic/Manual Control with Bumpless Transfer, allows the
operator to select the optimum operating point of the precipitator
in the manual control mode of operation and then switch to the
automatic position and have the thyristor control automatically
start operating at the same output level selected in the manual
mode, (9) Arc Quench circuit, is an added safety feature to insure
against power arcs; (10) Under-Voltage Relay, monitors the AC
voltage across the primary of the high voltage power supply and
can be a useful device for indicating potential problems when
properly adjusted for a plant's particular operation.

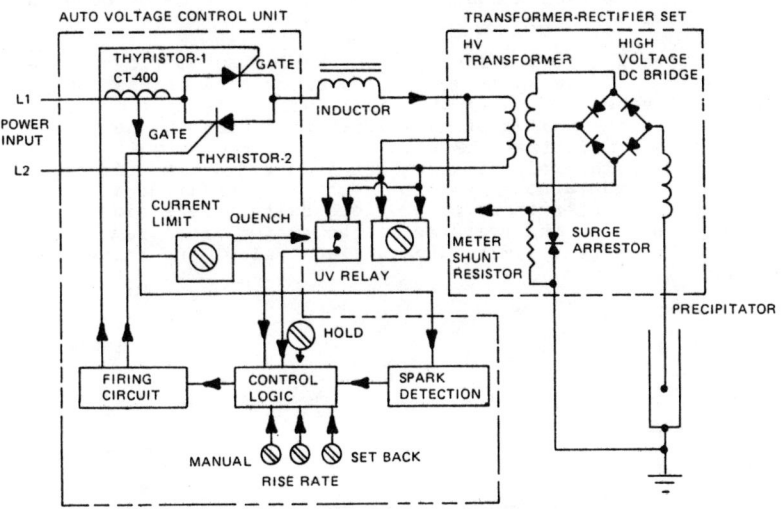

Figure 293. Schematic of Environecs Automatic Voltage Control Unit.[283]

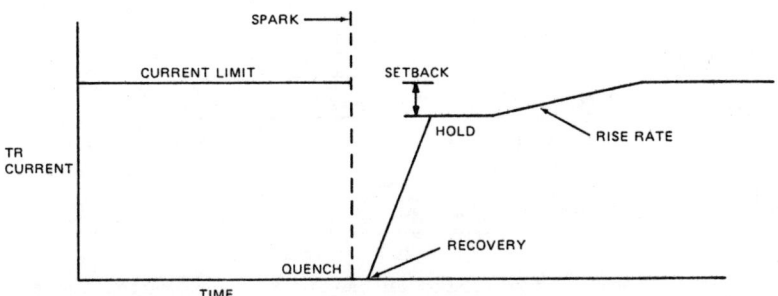

Figure 294. Typical response to spark.[283]

- Wahlco, Inc.
  3600 West Segerstrom Avenue
  Santa Ana, California 92704
  (714) 979-7300

The Wahlco Spark Rate Meter is designed for installation in conjunction with new or existing precipitator controls. The unit is self-contained requiring 120 VAC input for powering and the signal input is derived from the ground leg resistor of the transformer rectifier set. All detecting and conversion components are solid state. The only mechanical component is the meter movement. The solid state system takes the steep wavefront of the spark signal, integrates this over a time base, and delivers an analog signal into the meter movement. The spark sensing input signal is fed through a full wave bridge rectifier to eliminate polarity sensitivity. The unit has multipole filters enabling it to respond quickly and yet follow a spark signal without the meter bouncing objectionably.

In Figure 295 is a diagram of the Wahlco automatic voltage control unit.[283] The spark detector's circuit memorizes the peak amplitude attained by the input signal during one half cycle, compares it to the peak amplitude attained during the next half cycle, and then memorizes the value of the latter signal. From the controller standpoint a spark has occurred if the signal is at least 25 instantaneous peak volts and its amplitude is at least 5 volts greater than the previous half cycle's signal peak amplitude. Some of the features of the system are: ramp rate and set-back, current limit, undervoltage relay, and recovery time control.

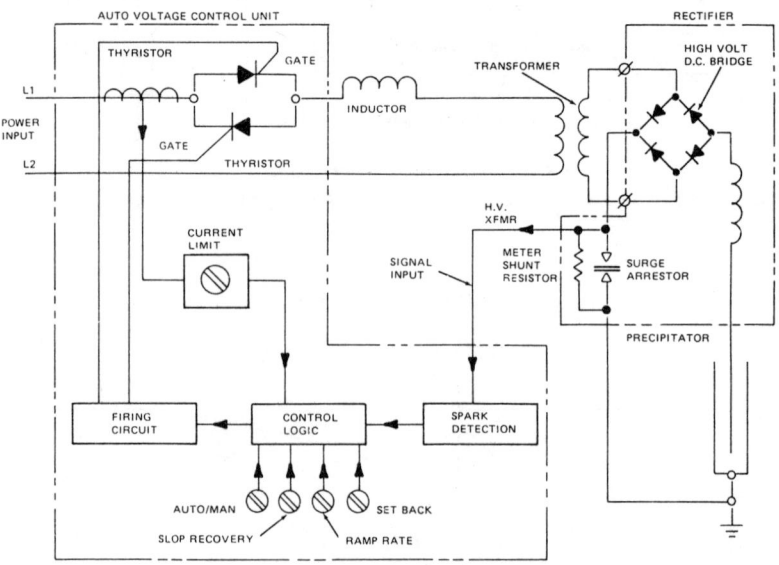

Figure 295. Diagram of a Wahlco automatic voltage control unit.[283]

- A.V.C. Specialists, Inc.
  2612 Croddy Way, Suite 1
  Santa Ana, California 92704
  (714) 540-2321

Figure 296 is a connection diagram for the external connections to the A.V.C. self-contained spark rate meter.[283] This unit can be added to any TR set controller providing that the input power and spark signal are made available. The meter mounts in the hole pattern for General Electric "Big Look" meters, 3½ inch type 162 (AO/DO91). Depth behind the panel is 4½ inches maximum, and an additional ½ inch minimum should be allowed for clearance at the terminals.

A.V.C. Specialists concentrates on providing voltage controls for precipitators, both new and existing. Much of their business is upgrading existing units to achieve better electrical performance, better collection efficiency, more reliable operation, the elimination of maintenance problems caused by non-responsive "automatic" controls. Some of the important standard features of the automatic voltage controllers are: ramp rate control, set back control, quench control, current limit control, fast acting overload protection, and manual control mode.

There are two types of voltage controls that A.V.C. Specialists, Inc. has developed for electrostatic precipitators:

(1) Saturable Core Reactor Type Controller, which is designed to drive the D.C. control winding of a saturable core reactor. (See Figure 297[283]);

(2) Thyristor (SCR) Type Controller, which controls the phase angle of firing of two SCRs in order to control the output of the TR set (See Figure 298[283]).

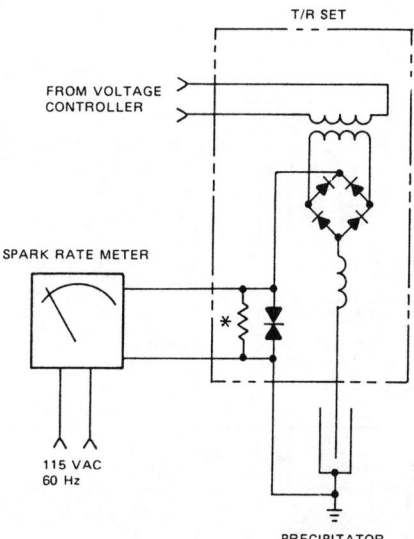

Figure 296. Connection diagram for the external connections to A.V.C. self-contained spark rate meter.[283]

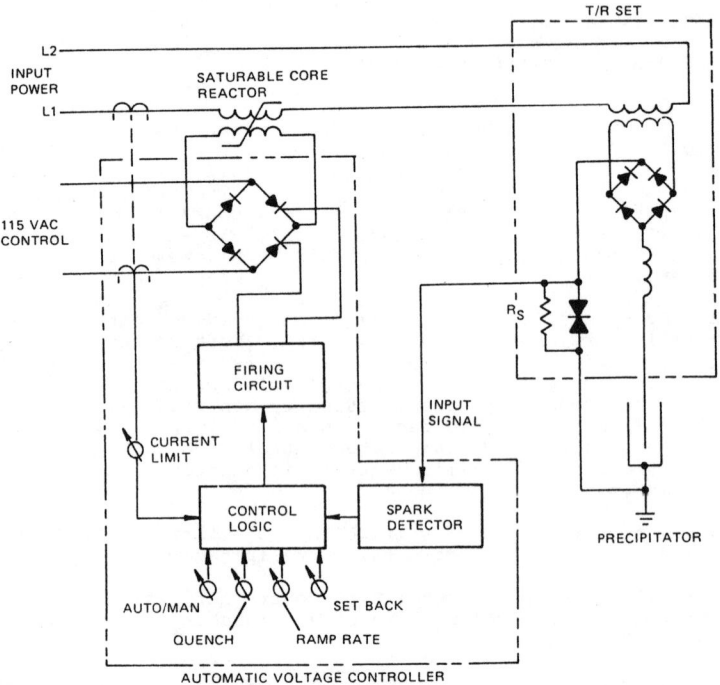

Figure 297. Block diagram saturable core reactor type system.[28]

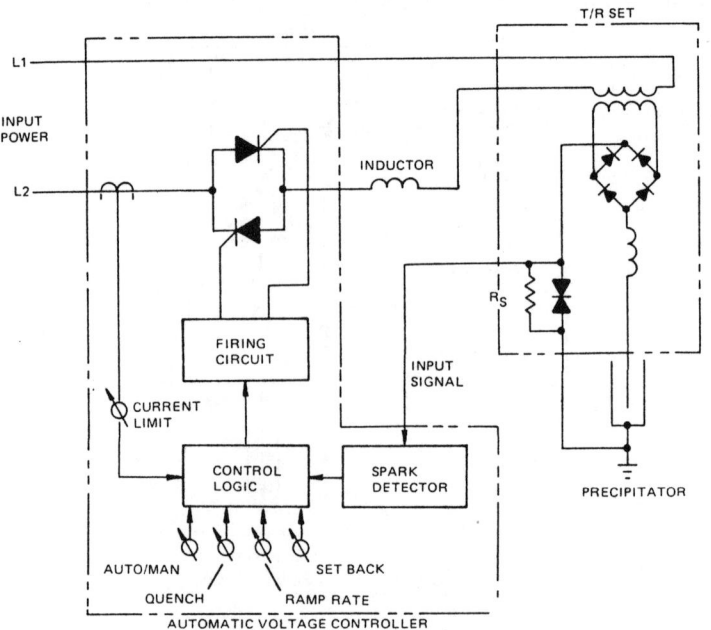

Figure 298. Block diagram Thyristor-type system.[283]

Secondary Voltage And Current Meters

Most precipitator control rooms have panel meters for each TR set which show the primary and secondary voltage and current and the sparking rate. Secondary voltage-current relationships can be obtained for both clean and dirty plate conditions and interpretations can be made of precipitator behavior based on the V-I data. The secondary voltage-current meters operate on the same principle as voltage diviers which were discussed in a previous section. Secondary voltage-current meters are supplied by the precipitator vendor and are not considered specialty items. Usually a major manufacturer such as General Electric sells the meters off-the-shelf, and a meter company such as Meter Master, Simpson, Triplett, etc. makes and calibrates the meter scale to specifications.

If meters are not installed on the transformer secondary, a quick, temporary voltage divider network can be installed on the precipitator side of the rectifier network as discussed previously. Many companies sell voltage dividers and a few of these are given below:

Beckman Instruments-Helipot Division
2500 Harbor Boulevard
Fullerton, California 92634
(714) 871-4848

CPS Inc.
110 Wolfe Road
Sunnyvale, California 94086
(408) 738-0530

Del Electronics Corporation
250 East Sandford Boulevard
Mt. Vernon, New York 10550
(914) 699-2000

EECO
1441 East Chesnut Avenue
Santa Ana, California 92701
(714) 835-6000

Electro Scientific Industries
13900 N.W. Science Park Drive
Portland, Oregon 97229
(503) 641-4141

Genrad
300 Baker Avenue
Concord, Massachusetts
(617) 369-8770

Guideline Instruments, Inc.
2 Westchester Plaza
Elmsford, New York 10523
(914) 592-9101

Heath Company
Benton Harbor, Michigan 49022
(616) 982-3200

Hipotronics Inc.
Route 22
Brewster, New York 10509
(914) 279-8031

ILC Data Device Corporation
105 Wilbur Place
Prpt. Intl. Plaza
Bohemia, New York 11716
(516) 567-5600

Kepco Inc.
131-38 Sanford Avenue
Flushing, New York 11352
(212) 461-7000

Pearson Electronics Inc.
4007 Transport Street
Palo Alto, California 94303
(415) 494-6444

Sensitive Research Instruments
25 Dock Street
Mr. Vernon, New York 10550
(914) 699-9717

A representative example of a voltage divider made by Hipotronics has a guaranteed accuracy of 0.5% DC and 1.0% AC. There are three stock models available, 50 KV, 100 KV, and 200 KV with other models with ratings to one megavolt available on request. Some of the specifications for the standard models are given below:[283]

	Model KV50A	Model KV100A	Model KV200A
Accuracy:			
DC	0.5%	0.5%	0.5%
AC	1.0%	1.0%	1.0%
Tracking	0.5%	0.5%	0.5%
Movement	Taut band	Taut band	Taut band
Meter:			
Scale	100 divisions mirror scale	100 divisions mirror scale	100 divisions mirror scale
Size	5½"	5½"	5½"
Voltage Coefficient:			
DC	0.025%/C°	0.025%/C°	0.025%/C°
AC	0.1%/C°	0.1%/C°	0.1%/C°
Frequency response	DC and 40 to 1000 Hz	DC and 40 to 1000 Hz	DC and 40 to 1000 Hz
Connecting cable	25 feet	25 feet	25 feet
Meter ranges (KV) Volts Division	0-10/25/50 100/250/500	0-20/50/100 200/500/1000	0-40/100/200 400/1000/2000
Impedance	190 megohms @ 200 pfd.	380 megohms @ 100 pfd.	760 megohms @ 50 pfd.
Size	8½" W x 10½" D x 15½" H	8½" W x 10½" D x 15½" H	9½" W x 10½" D x 40" H

Opacity Meters

Opacity meters can be used effectively in monitoring the performance of emission control equipment continuously. In addition, optical density output can be correlated with particulate grain loading to allow determination of mass emissions on a continuous basis. Opacity meters are invaluable in gauging precipitator performance quickly when small changes are made in coal, precipitator controls, or boiler conditions. Some of the more important variables which affect performance the most are boiler load, boiler outlet gas temperature, boiler excess air level, precipitator operating voltage, precipitator rapping intensity and direction, and precipitator internal condition.

A number of techniques are used to determine relative stack emission levels. These techniques and corresponding instrumentation were discussed in detail in Section 7 of this report.

Hopper Level Meters

Preventing precipitator hoppers from completely filling with fly ash is extremely important. Overflow can lead to shorted electrical systems or fly ash reentrainment, either of which would adversely affect precipitator performance. A number of hopper level detectors have been developed to help eliminate the overflow problem. These detectors have been previously discussed in Section 4. Some of the principles of operation used in detection are:

Non-contacting radiation principle - a narrow beam of gamma rays is directed across the hopper to a radiation detector located on the opposite wall. The rays are absorbed when ash builds up causing a relay to activate an alarm.

Rod oscillation dampening - a rod is installed at the desired ash level. A drive coil drives the rod into self-sustained mechanical oscillations and a signal is produced by a pick-up coil located opposite the drive coil. When fly ash reaches the level of the rod, a dampening of the oscillations occurs and the signal from the pick-up coil is reduced.

Capacitance sensor assembly - the detector assembly senses a change in ash level as a function of the capacitance change between the detector and the vessel wall. This change is then transmitted to a control instrument.

Radio frequency - a low power RF signal is radiated from a sensing probe and changes in the impedance of the probe caused by a change in ash level are monitored.

After alarms are given indicating dangerous accumulations of fly ash, systems for removal of the ash are activated. These systems are discussed in detail in Section 4.

Section 12

An Electrostatic Precipitator Computer Model

INTRODUCTION

In recent years, increasing emphasis has been placed on developing theoretical relationships which accurately describe the individual physical mechanisms involved in the precipitation process and on incorporating these relationships into a complete mathematical model for electrostatic precipitation. From a practical standpoint, a reliable theoretical model for electrostatic precipitation would offer several valuable applications:

(1) precipitator design could be easily and completely performed by calculation from fundamental principles;

(2) a theoretical model could be used in conjunction with a pilot-plant study in order to design a full-scale precipitator;

(3) precipitator bids submitted by various manufacturers could be evaluated by a purchaser with respect to meeting the design efficiency and the costs necessary to obtain the design efficiency;

(4) the optimum operating efficiency of an existing precipitator could be established and the capability to meet particulate emissions standards could be ascertained; and

(5) an existing precipitator performing below its optimum efficiency could be analyzed with respect to the different operating variables in a procedure to troubleshoot and diagnose problem areas.

In addition to its many applications, a mathematical model can be a valuable tool for analyzing precipitator performance due to its cost- and time-savings capability. The approach is cost-effective because it (1) allows for the analysis and projection of precipitator operation based on a limited amount of data (extensive field testing is not necessary), (2) can predict trends caused by changing certain precipitator parameters and thus, in many cases, can prevent costly modifications to a precipitator which will not significantly improve the performance, (3) can be used as a tool in sizing precipitators and prevent excessive costs due to undersizing or significant oversizing, and (4) can be used to obtain large amounts of information without extensive use of manpower but, instead, with reasonable use of a computer.

The approach is time-effective because (1) large amounts of

information can be generated quickly, (2) it does not necessarily depend on time-consuming field tests which involve travel, extensive analysis, and plant and precipitator shut-downs, (3) it can prevent losses in time due to unnecessary or insufficient modifications to a precipitator, and (4) it can prevent losses in time due to the construction of an undersized precipitator.

In this section, the latest version[137,152] of a mathematical model of electrostatic precipitation developed under the sponsorship of the U.S. Environmental Protection Agency is briefly described. Since the model is described in great detail elsewhere, the capabilities and applications of the model will be stressed here, rather than mathematical details. In the latest version, earlier work[153] has been improved and extended. Major improvements to the fundamental basis of the model include the capability of generating theoretical voltage-current characteristics for wire-plate geometries, a new method for describing the effects of rapping reentrainment, a new procedure for accounting for the effects of particles on the electrical conditions, and the incorporation of experimentally determined correction factors to account for unmodeled effects. The computer program which performs the calculations in the model has been made more user-oriented by making the input data less cumbersome, by making the output data more complete, by making modifications which save computer time, and by providing for the construction of log-normal particle size distributions.

CAPABILITIES OF THE MODEL

The present version of the model has the following capabilities

(1) it predicts collection efficiency as a function of particle diameter, electrical operating conditions, and gas properties;

(2) it can calculate clean-plate, clean-air voltage-current characteristics for wire-plate geometries;

(3) it determines particle charging by unipolar ions as a function of particle diameter, electrical conditions, and residence time;

(4) it can estimate the effects of particles on the electrical conditions under the assumption that effects due to the particulate layer can be ignored;

(5) it accounts for electrical sectionalization;

(6) it predicts particle capture at the collection electrode based on the assumptions of completely random, turbulent flow, uniform gas velocity, and particle migration velocities which are small compared to the gas velocity;

(7) it employs empirical correction factors which adjust the particle migration velocities obtained without rapping losses in order to account for unmodeled effects;

(8) it accounts for the nonideal effects of nonuniform gas velocity distribution, gas bypassage of electrified regions, and particle reentrainment from causes other than rapping by using empirical correction factors to scale down the ideally calculated particle migration velocities; and

(9) it accounts for rapping reentrainment by using empirical

relationships for the quantity and size distribution of the re-entrained mass.

In its present form, the model has the capability of predicting trends caused by changes in specific collection area, applied voltage, current density, mass loading, and particle size distribution. Comparisons of the predictions of the model with laboratory-scale precipitators[138,284,285] and full-scale precipitators collecting fly ash from coal-fired boilers[19,285] indicate that the model can be used successfully to predict precipitator performance.

BASIC FRAMEWORK OF THE MODEL

The mathematical model is based on an exponential-type relationship given by equation (2). Although the previously discussed assumptions upon which equation (2) is derived are never completely satisfied in an industrial precipitator, they can be closely approached with respect to the treatment of fine particles.

The assumption that the particle migration velocity near the collection surface is constant for all particles has the most significant effect on the structure of the model. This assumption implies two things: (1) all particles are of the same diameter and (2) the electrical conditions are constant.

Because all particles entering a precipitator are not of the same diameter, the assumption of uniform particle diameters creates a problem. This problem is dealt with in the model by performing all calculations for single-diameter particles and then summing the results to determine the effect of the electrostatic precipitation process on the entire particle size distribution.

Because the electrical conditions change along the length of a precipitator, the assumption of constant electrical conditions creates a problem. This problem is dealt with in the model by dividing the precipitator into small length increments. These length increments can be made small enough that the electrical conditions remain essentially constant over the increment. The number of particles of a given diameter which are collected in the different length increments are summed to determine the collection efficiency of particles of a single diameter over the entire length of the precipitator.

In summary, a precipitator is divided into essentially many small precipitators in series. Equation (2) is valid in each of these small precipitators for fine particles of a given diameter.

The collection fraction, $n_{i,j}$, for the ith particle size in the jth increment of length of the precipitator is mathematically represented in the form

$$n_{i,j} = 1 - \exp(-w_{i,j} A_j/Q), \qquad (70)$$

where $w_{i,j}$ (m/sec) is the migration velocity near the collection electrode of the ith particle size in the jth increment of length and A_j (m^2) is the collection plate area in the jth increment of length.

The collection fraction (fractional efficiency) n_i for a given particle size over the entire length of the precipitator is determined from

$$n_i = \frac{\sum_j n_{i,j} N_{i,j}}{N_{i,1}} , \qquad (71)$$

where $N_{i,j}$ is the number of particles of the ith particle size per cubic meter of gas entering the jth increment.

Effective or length-averaged migration velocities (w_i^e) are calculated for the different particle diameters from

$$w_i^e = \frac{Q}{A_T} \ln\left(\frac{1}{1-n_i}\right) , \qquad (72)$$

where A_T (m²) is the total collecting area.

The overall mass collection efficiency η for the entire polydisperse aerosol is obtained from

$$\eta = \sum_i n_i P_i , \qquad (73)$$

where P_i is the percentage by mass of the ith particle size in the inlet size distribution.

In order to determine the migration velocities for use in equation (70), the electrical conditions and the particle charging process in a precipitator must be modeled. If the operating voltage and current density are known, then the electric potential and electric field distributions are determined by using a relaxation technique.[286,287] In this numerical technique, the appropriate partial differential equations which describe the electrodynamic field are solved simultaneously under boundary conditions existing in a wire-plate geometry. In order to find the solutions for the electric potential and space charge density distributions, the known boundary conditions on applied voltage and current density are held fixed while the space charge density at the wire is adjusted until all the boundary conditions are satisfied. For each choice of space charge density at the wire, the procedure iterates on a grid of electric potential and space charge density until convergence is obtained and then checks to see if the boundary condition on the current density is met. If the boundary condition on the current density is not met, then the space charge density at the wire is adjusted and the iteration procedure is repeated.

Particle charge is calculated by using a unipolar, ionic-charging theory.[8,288] Particle charge is predicted as a function of particle diameter, exposure time, and electrical conditions. The charging equation is derived based on concepts from kinetic theory and determines the charging rate in terms of the probability of collisions between particles and ions. The theory accounts simultaneously for the effects of field and thermal charging and accounts for the effect of the applied electric field on the thermal charging process.

The nonideal effects of major importance in a precipitator are (1) nonuniform gas velocity distribution, (2) gas bypassage of electrified regions, and (3) particle reentrainment. These nonideal effects will reduce the ideal collection efficiency that may be achieved by a precipitator operating with a given specific collection area. Since the model is structured around an exponential-type equation for individual particle diameters, it is convenient to represent certain nonideal effects in the form of correction factors which apply to the exponential argument. The model employs correction factors which are used as divisors for the ideally calculated effective migration velocities in order to account for nonuniform gas velocity distribution, gas bypassage, and particle reentrainment without rapping.[289,290] The resulting apparent effective migration velocities are empirical quantities.

LATEST IMPROVEMENTS TO THE MODEL

Calculation Of Voltage-Current Characteristics

A new technique[4] has been developed for theoretically calculating electrical conditions in wire-plate geometries and has been incorporated into the model. In this numerical technique, the appropriate partial differential equations which describe the electrodynamic field are solved simultaneously, subject to a suitable choice of boundary conditions. The procedure yields the voltage-current curve for a given wire-plate geometry and determines the electric potential, electric field, and charge density distributions for each point on the curve.

The key element in this technique is the theoretical calculation of the space charge density near the corona wire for a specified current density at the plate. In order to find the solutions for the electric potential and space charge density distributions, the known boundary conditions on space charge density near the wire and current density are held fixed while the electric potential at the wire is adjusted until all boundary conditions are satisfied. For each choice of electric potential at the wire, the procedure iterates on a grid of electric potential and space charge density until convergence is obtained and then checks to see if the boundary condition on the current density is met. If the boundary condition on the current density is not met, then the electric potential at the wire is adjusted and the iteration procedure is repeated. The entire procedure is repeated for increasing values of current density in order to generate a voltage-current curve. Comparisons[4,291] of the predictions of this technique with experimental data show that the agreement between theory and experiment is within 15%.

Method For Predicting Trends Due To Particulate Space Charge

A new method has been incorporated into the model in order to provide a more comprehensive representation of the effects of particulate space charge on the electrical operating conditions in a precipitator. In this method, the precipitator is divided into successive length increments which are equal to the wire-to-wire spacing. Each of these increments is divided into several subincrements. The first calculation in the procedure involves the determination of a clean-gas, voltage-current curve which terminates at some specified value of applied voltage. At the specified applied voltage, the average electric field and ion density are calculated in each subincrement. This allows for the nonuniformity of the electric field and current density distributions to be taken into account.

As initially uncharged particles enter and proceed through the precipitator, the mechanisms of particle charging and particle collection are considered in each subincrement. In each subincrement, the average ion density, average particulate density, weighted particulate mobility, and effective mobility due to both ions and particles are determined. At the end of each increment, the effective mobilities for the subincrements are averaged in order to obtain an average effective mobility for the increment. Then, for the specified value of applied voltage, the average effective mobility is used to determine the reduced current for the increment by either calculating a new voltage-current curve or using an approximation procedure. Although it is not presently utilized, the method allows for iterations over each length increment so that schemes which ensure self-consistency can be implemented at a future date.

In its present state of development, this method provides good estimates of reduced current due to the presence of particles. The reduced current is a function of mass loading, particle size distribution, gas volume flow, and position along the length of the precipitator. However, this method does not have the capability of predicting the redistribution of the electric field due to the presence of particles. Work is going on at the present time to improve the model in this respect.

Method For Estimating Effects Due To Rapping Reentrainment

As part of a program sponsored by the Electric Power Research Institute, an approach to representing losses in collection efficiency due to rapping reentrainment has been developed based on studies performed on six different full-scale precipitators collecting fly ash.[19] These studies have been discussed earlier in this text. In these studies, outlet mass loadings and particle size distributions were measured both with rapping losses and without rapping losses. Outlet mass loadings and particle size distributions which can be attributed to rapping were obtained based on the data acquired in these studies. The results of these studies have been incorporated into the model.

The rapping emissions obtained from the measurements are graphed in Figure 278 as a function of the amount of dust calculated to have been removed by the last electrical section. The dust removal in the last electrical section was approximated by using an exponential relationship for the collection process and the overall mass collection fraction determined from mass train measurements under normal operating conditions, as described earlier. These data suggest a correlation between rapping losses and particulate collection rate in the last electrical section. Data for the two hot-side installations (4 and 6) which were tested show higher rapping losses than for the cold-side units, and, thus, hot- and cold-side units are treated differently in the model with respect to rapping reentrainment.

The apparent particle size distribution of emissions attributable to rapping at each installation was obtained by subtracting the cumulative distributions during nonrapping periods from those with rappers in operation and dividing by the total emissions (based on impactor measurements) resulting from rapping in order to obtain a cumulative percent distribution. Although the data indicated considerable scatter, the average particle size distribution shown in Figure 280 has been constructed for use in modeling rapping puffs. In the model, the data are approximated by a log-normal distribution with a mass median diameter of 6.0 µm and a geometric standard deviation of 2.5 as shown in Figure 299.

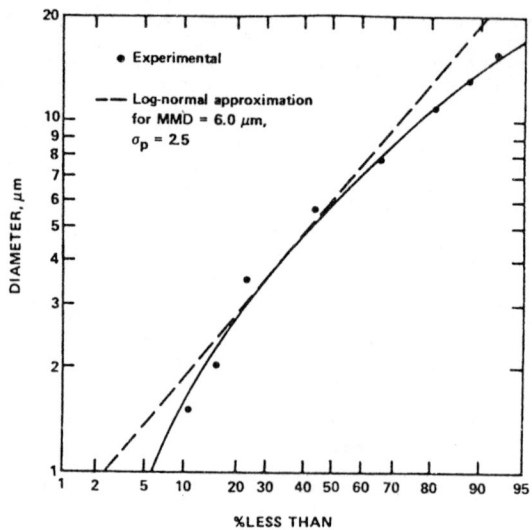

Figure 299. Average rapping puff size distribution and log-normal approximation for six full-scale precipitators. These data are a result of work sponsored by the Electric Power Research Institute.[19]

In summary, the model determines a rapping puff by using the information in Figure 278 to obtain the outlet mass loading due to rapping and by using a log-normal approximation to the data in Figure 280 to represent the particle size distribution of the outlet mass loading due to rapping. This "rapping puff" is added to the "no-rap" outlet emissions to obtain the total outlet emissions as a function of mass loading and particle size distribution.

Empirical Corrections To No-Rap Migration Velocities

Comparisons of measured apparent effective migration velocities for full-scale precipitators under "no-rap" conditions with those predicted by the model indicate that the field-measured values exceed the theoretically projected values (in the absence of back corona, excessive sparking, or severe mechanical problems) in the smaller size range. Based on these comparisons, a size-dependent correction factor has been constructed and incorporated into the model.[19] This correction factor is shown in Figure 300.

The empirical correction factor accounts for those effects which enhance particle collection efficiency but are not included in the present model. These effects might include particle charging near corona wires, particle charging by free electrons, particle concentration gradients, the electric wind, and flow field phenomena.

In future work which is planned, efforts will be made to develop appropriate theoretical relationships to describe the above effects and to incorporate them into a more comprehensive model for electrostatic precipitation.

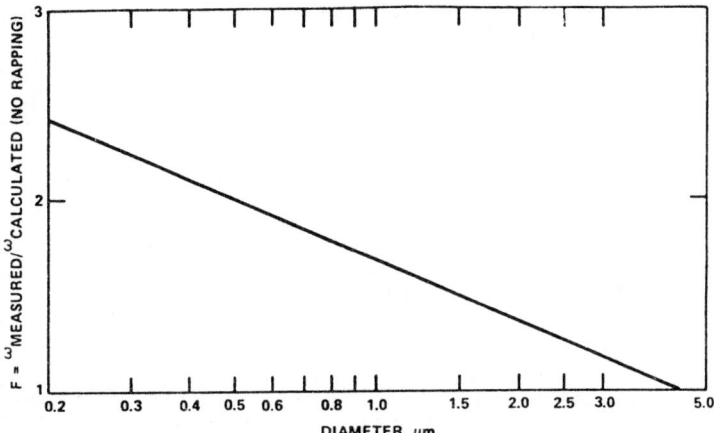

Figure 300. Empirical correction factors for the "no-rap" migration velocities calculated from the mathematical model. This work was sponsored by the Electric Power Research Institute.[19]

User-Oriented Improvements

The computer program which performs the calculations in the model has been modified to make the input data less cumbersome and the output data more complete. The performance of a precipitator can be analyzed as a function of particle size distribution, current density, specific collection area, and nonideal conditions without repetition of input data which remain fixed. All input data are now printed out in a format which is easily utilized. A summary table of precipitator operating conditions and performance is printed out as the last section of data for a given set of conditions.

Several modifications have been made in order to save computer time. The particle charging algorithm has been modified, and this has decreased the computer time required for particle charging calculations by approximately 40%. In addition, particle charge calculations for a given diameter will terminate in a given electrical section whenever the charging rate becomes negligible. This can reduce the time required to perform particle charging calculations by up to a factor of two or more in some cases. The computer program has been modified so that several sets of nonideal conditions can be analyzed in conjunction with the results of one ideal calculation. This allows for the analysis of an extended range of nonideal conditions with only a small increase in computer time. As another means of saving computer time, the computer program now contains an estimation procedure for use in analyzing precipitator performance. This procedure results in considerable savings in computer time since involved numerical techniques are not employed. The estimation procedure runs approximately 20 times faster than the rigorous calculation. This procedure can be used to good advantage to determine gross trends or to establish a limited range of interest in which to apply the more rigorous calculation. The procedure can also be used to good advantage for checking the validity of input data before making extensive rigorous calculations.

The computer program now has the capability of constructing log-normal particle size distributions based on specified values of the mass median diameter and geometric standard deviation. This capability can be used to construct inlet and rapping puff particle size distributions. Thus, the effects of different log-normal particle size distributions can be readily obtained. Also, the program can fit any specified particle size distribution to a log-normal distribution.

APPLICATIONS AND USEFULNESS OF THE MODEL

The different practical applications of the model have been discussed elsewhere.[3,137,153] These include the examination of the effects of particle size distribution, electrical conditions, specific collection area, dust resistivity, and nonideal conditions on the performance of a precipitator. These applications have now been incorporated into procedures for troubleshooting and sizing precipitators.[137] These procedures, which provide specific guidelines for applying the model to troubleshooting and sizing applications, are discussed next in order to demonstrate the usefulness of the model.

Use Of The Model For Troubleshooting

The mathematical model of electrostatic precipitation can be used as a tool in troubleshooting precipitators that are not meeting the overall mass collection efficiency which is expected or anticipated. When using the model for troubleshooting, certain experimental data should be obtained in order to properly utilize the model. These data include operating voltages and currents in the different electrical sections, inlet mass loading and particle size distribution, ash resistivity, average gas flow rate and velocity, and average gas temperature and pressure. By using these limited experimental data, the geometry of the precipitator, and the mathematical model, certain steps which are given below can be taken in an attempt to diagnose the possible reason or reasons for the level of performance of the precipitator.

Step 1: Determine optimum collection efficiency.

The model is used to simulate the operation of the precipitator under ideal, no-rap conditions ($\sigma_g = 0$ and $S = 0$) with the actual operating parameters, where σ_g is the normalized standard deviation of the gas velocity distribution and S is the fraction of the gas volume bypassing each electrical section. This calculation establishes the optimum overall mass collection efficiency that can be expected under the given operating conditions. It should be noted that this optimum efficiency may not always represent the best performance of the precipitator since accumulation of material on the discharge and collection electrodes, broken discharge electrodes, electrode misalignment, or operation of the precipitator at lower than permissible voltages and currents would result in less than optimum electrical operating conditions. If possible, measures should be taken to ensure that the electrical conditions in the precipitator are at their best when obtaining data for use in the troubleshooting procedure. In any event, the starting point in the troubleshooting procedure can be taken to be the calculated optimum efficiency under the actual operating conditions.

Step 2: Check to see if the calculated optimum efficiency is equal to or less than the measured value.

If the calculated optimum value of efficiency is equal to or less than the measured value, then the precipitator can be assumed to be performing as well as possible for the given set of operating conditions. Changes in the inlet particle size distribution, the electrical operating conditions, or the gas volume flow can result in a reduction of collection efficiency for a given precipitator even though the precipitator is performing at its best. Thus, in certain cases, a precipitator may not be able to attain the overall mass collection efficiency it once achieved or was designed to achieve solely due to a change in the process variables. As a consequence, the precipitator may no longer be sized properly for the operating conditions encountered. The options that are available for improving the performance of the precipitator are limited to the possible improvement of the electrical operating conditions or a reduction in the gas flow rate through the precipitator.

Step 3: Check to see if the calculated optimum efficiency is only a little larger than the measured value.

If the calculated optimum value of efficiency is only a little larger than the measured value, then the precipitator is probably functioning well but nonideal conditions are having some effect on the performance. In this case, calculations should be made with the model in order to obtain NO-RAP + RAP overall mass collection efficiencies for various small values of σ_g and S and the rapping reentrainment parameters which are built into the computer program. If the measured efficiency can be predicted by the model with values of $\sigma_g \leq 0.25$ and $S \leq 0.1$, then it is questionable whether or not improvements in the gas flow properties and mechanical design will result in an appreciable improvement in precipitator performance. A less costly and possibly more profitable exercise would be to vary the rapping intensities and frequencies in an attempt to minimize losses in collection efficiency due to rapping reentrainment. If $\sigma_g > 0.25$ or $S > 0.1$, then these quantities should be measured. If the measured values of σ_g and S are consistent with those predicted by the model, then the gas flow properties and mechanical design should be improved.

Step 4: Check to see if the calculated optimum efficiency is significantly larger than the measured value.

If the calculated optimum value of efficiency is significantly larger than the measured value, then the precipitator is functioning poorly. Poor performance of a precipitator may be due to either one or a combination of several factors that can be analyzed with the model. These factors include the electrical operating conditions, nonuniform gas velocity distribution, gas bypassage of electrified regions, particle reentrainment without rapping, and rapping reentrainment. In the following steps, procedures are outlined that can be taken in an attempt to pinpoint the problem areas.

Step 5: Determine whether or not the operating currents are completely useful in the precipitation process.

At this point, the electrical operating conditions should be examined in order to determine whether or not the operating currents are completely useful in the precipitation process. If excessive sparking or back corona is occurring in the precipitator, then the measured currents will not be totally useful in the precipitation process and, in fact, the nature of the currents may be very detrimental to precipitator performance. Use in the model of currents measured under these conditions will result in the prediction of much higher collection efficiencies than will be attained by the precipitator.

Step 5a: Check for excessive sparking.

Sparking results in localized currents that are not very effective in charging particles. In addition, excessive sparking can lead to increased particle reentrainment by producing disruptions at the surface of the collected particulate layer and by producing reduced holding forces over large regions of the collected layer due to reduced currents to these regions.

If sparking is occurring, then the extent of the sparking should be determined by using spark rate meters or other appropriate instrumentation. If excessive sparking is occurring, then the applied voltage should be lowered until the spark rate is at a level which is not detrimental to the performance of the precipitator. Although the operating voltages and currents will be reduced, the performance of the precipitator will improve and the use of these operating electrical conditions in the model will give better agreement between predicted and measured collection efficiencies.

Step 5b: Check for the existence of back corona.

If excessive sparking is not occurring, then a check should be made to determine whether or not a condition of back corona exists in the precipitator. When back corona exists, both positive and negative ions move in the interelectrode space and this results in a reduction in the negative charge that can be acquired by a particle.

Two methods can be used to check for the existence of back corona. First, the measured value of ash resistivity and Figure 208 can be used to estimate the maximum allowable current density. If the current density in the precipitator greatly exceeds this value, then the precipitator is probably operating in back corona. As a second method of checking for the existence of back corona, the voltage-current curves for the different electrical sections can be checked to see if at some point on the curve increased current is obtained at a reduced applied voltage. If this is the case and the precipitator is operating in this region of the voltage-current curve, then back corona is occurring in the precipitator.

If back corona is occurring, then the applied voltage should be lowered in order to obtain a current density which will not lead to the formation of back corona. The reduced voltages and currents will result in improved performance of the precipitator and the use of these operating electrical conditions in the model will give better agreement between predicted and measured collection efficiencies.

Step 5c: Consider electrode misalignment.

As a further consideration concerning the electrical conditions, the electrode alignment should be taken into account. Consideration of electrode alignment is especially important when troubleshooting hot precipitators. In hot precipitators, the collection plates may buckle if proper precautions have not been taken to allow for the expansion of the plates at the elevated temperatures. If buckling of the plates occurs, then higher currents will be measured but they will be localized. Currents of this type are not desirable for treating particles. The existence of this type of misalignment should be evidenced by steep voltage-current curves with a narrow voltage range from corona initiation to sparkover.

Use in the model of measured currents obtained from this type of situation will result in predicted collection efficiencies that are well above those which are attained.

Step 6: Estimate the effect that various nonideal conditions could have on the performance of the precipitator.

If the poor performance of the precipitator cannot be traced to the electrical operating conditions, then the nonideal effects of nonuniform gas velocity distribution, gas bypassage of electrified regions, and particle reentrainment should be considered next. The effect of σ_g and S on the NO-RAP + RAP overall mass collection efficiency of the precipitator should be analyzed in a systematic fashion with the model.

Step 6a: Estimate the possible effect of nonuniform velocity distribution on the performance of the precipitator.

In order to determine whether or not a nonuniform gas velocity distribution could be responsible for the poor performance of the precipitator, calculations should be made for S = 0 and values of σ_g ranging from 0 to at least 2.0. If a certain value of σ_g in the chosen range produces the necessary reduction in collection efficiency and this value is not completely out of line with available information concerning the gas flow, interfacing of the precipitator with the duct work, existence of gas diffusion plates, etc., then the actual value of σ_g should be determined experimentally by making a velocity traverse in a plane at the inlet of the precipitator. If the measured value of σ_g is greater than 0.25, then measures should be taken to improve the gas flow distribution.

Step 6b: Estimate the possible effect of gas sneakage and/or particle reentrainment without rapping on the performance of the precipitator.

In order to determine the extent of gas bypassage of the electrified regions and/or particle reentrainment without rapping that would be necessary to cause the poor performance of the precipitator, calculations should be made for σ_g = 0 and values of S ranging from 0 to 0.9. There will be a value of S in this range that will result in the necessary reduction in collection efficiency. Depending on the value of S, different interpretations can be made. If S is not too large (S $\leq$ 0.2), then the poor performance might be attributed to either excessive gas bypassage of the electrified regions or excessive particle reentrainment without rapping or very poor gas velocity distribution or a combination of all three of these effects where neither effect alone is very detrimental to the performance of the precipitator. In this case, measurements should be made under air-load conditions to determine σ_g and the fraction of the gas volume flow passing through nonelectrified regions in each baffled section. If the measured values of these quantities are such that they can account for a major part of the reduction in collection efficiency, then the appropriate corrective measures can be made to the mechanical design of the precipitator. If the measured values of these quantities are such that they can not account for a major part of the reduction in collection efficiency, then it is possible that particle reentrainment without rapping is having an adverse effect on the performance of the precipitator. This could be due to factors which include a high average gas velocity, a very nonuniform gas velocity distribution, a low value of ash resistivity, excessive sparking, low operating current densities, and hopper problems. All of these factors can lead to particle reentrainment from causes other than rapping and should be taken into account in the troubleshooting analysis.

If S is large (S > 0.2), then the poor performance of the precipitator is probably due primarily to extremely excessive particle reentrainment. This could be a result of one or more of the same factors mentioned above. In this case, reentrainment of particles from the hoppers, caused by poor gas flow qualities or by hopper malfunctions, should receive more serious attention as a possible cause of the poor performance. If very large values of S are needed to predict the reduction in collection efficiency, then it is also possible that rapping reentrainment is occurring to a much greater extent than that predicted by the rapping reentrainment calculation and that this is reflected in the value of S. If the value of S is large, then hopper operation should be checked, outlet mass loadings should be obtained with and without rapping, and real-time measurements of the outlet mass loading should be made. These measures should indicate whether the problem is due to hopper operation or rapping reentrainment or reentrainment without rapping or some combination of the three.

The troubleshooting procedure described above can be a valuable tool in helping to diagnose the causes of poor performance of a precipitator. Since the procedure involves only limited experimental data, it is not costly to perform. Use of the procedure can also result in time and cost savings by giving direction and helping to focus on those quantities which actually need to be measured. A further benefit of using the procedure is the possibility that costly modifications to the precipitator that will not result in significant improvement in the performance can be avoided.

Use Of The Model For Sizing Of Precipitators

The mathematical model of electrostatic precipitation can be used as a guide in sizing precipitators. Although this method of sizing precipitators can be very successful, care must be taken to ensure proper usage of the model and to prevent the use of erroneous input data. Misuse of the model could result in a large error in sizing a precipitator.

When using the model for the purpose of sizing a precipitator, certain data which are used as input to the model should be obtained from measurements made using the actual gas stream or one which will be very similar to the actual gas stream. If a gas stream other than the actual one is used to obtain representative data, then steps should be taken to assure that the process variables producing the effluent gas stream and particles are not too different. Also, it is very important that the temperature and composition of the gas stream be close to that which will be experienced in the precipitator to be sized.

The following is a list and discussion of those quantities whose values should be determined from measurements under conditions similar to those which will be experienced in the precipitator to be sized:

The temperature, pressure, and composition of the gas stream should be measured.

The particle size distribution and mass loading in the gas

stream should be measured at a location from the source that would be representative of where the gas stream would enter the precipitator.

The bulk resistivity of the particles should be measured both in situ and in the laboratory. In making these measurements, the gaseous environment must not only be preserved but, in addition, the electric field strength at which the measurements are made must be close to that which will be experienced in the precipitator in order to obtain the appropriate measurement. If agreement can not be obtained between the in situ and laboratory measurement, then the higher of the two values should be used in order to size the precipitator.

The effective mobility of the negative ions which would be produced during negative corona discharge in the gas stream should be measured.

If any or all of the above quantities are not measured or can not be measured, then their values can only be estimated by using the best data available and prior experience for similar sets of conditions. Using values of these quantities that are not obtained from measurements with the actual or a similar gas stream is risky and these values should be estimated in a conservative manner.

Once the values of the quantities discussed above are determined, the model can be used in a procedure to predict what precipitator sizes are needed to attain various levels of overall mass collection efficiency. The steps which should be taken in this procedure are discussed next.

Step 1: Establish an estimate of the electrical conditions under which the precipitator should operate.

In establishing an estimate of the electrical operating conditions, a determination of the maximum allowable current density should be made first. The maximum allowable current density can be estimated by using the determined value of ash resistivity and the curve given in Figure 208. If voltage-current data are available for similar conditions, then these should also be used in helping to determine the maximum allowable current density.

Once the maximum allowable current density is estimated, then the applied voltages which will produce this current density in the different electrical sections must be estimated. These voltages may be obtained from voltage-current data which are available for similar conditions except it is not necessary that the ash resistivity be duplicated. Alternatively, the model can be used with the option which calculates voltage-current curves for a wire-plate geometry in order to determine voltage-current characteristics with the effect of resistivity being ignored. Then, the applied voltages necessary to produce the maximum allowable current density can be estimated. In utilizing the voltage-current calculation, a value for the roughness factor of the discharge electrodes must be specified. The value of this parameter normally lies between 0.5 and 1.0 and small changes in the value lead to significantly different results. Since the value of this parameter is difficult to project in advance and the value changes during the operation of the precipitator, care must be taken in specifying this value and in analyzing the results obtained. Calculations used to size the precipitator should be made for several values of the roughness factor between 0.5 and 1.0 and the most conservative prediction of precipitator performance should be used as the basis for sizing the precipitator. Also, if values of the roughness factor in a particular range yield results that are obviously out of line with similar applications, then this range should be eliminated from consideration.

Since the ash resistivity is difficult to determine precisely and environmental changes can produce significant changes in its value, the size of a precipitator should be determined based on a maximum allowable current density which is estimated based on a somewhat higher value of resistivity than anticipated. A reasonable and conservative approach might be to base the estimated maximum allowable current density on a value of resistivity that is one-half an order of magnitude greater than the anticipated value.

Step 2: Determine the geometrical parameters to be used.

At this point, the geometrical characteristics of the precipitator should be established since these data are necessary as input to the model. The values of the plate spacing, discharge electrode spacing, and diameter of the discharge electrodes which are used in the model must be the actual values. In order to size the precipitator, it is not necessary to know the actual values of the cross-sectional area, height, area, and number of the plates, length of the electrical sections, or total electrified length. Although the values of these quantities can be chosen arbitrarily, they should be as representative as possible.

In the model, different overall mass collection efficiencies can be determined for different specific collection areas and then, based on the actual gas volume flow through the precipitator, the total collection plate area necessary to achieve a given efficiency can be determined. Knowing the required collection plate area, the precipitator can be designed with respect to cross-sectional area, plate height, and length. In designing the precipitator so that it will have the required collection plate area, certain considerations should be made. First, the height of the collection plates should not be too high since this can lead to increased reentrainment from rapping and to greater difficulty in providing sufficient rapping force to the entire area of the plate. In practice, the height of collection plates ranges from approximately 3.05 (10) to 12.2 (40) meters (feet). Second, the precipitator should be long enough so that it can contain several baffled, independent electrical sections. Increasing the number of baffled electrical sections leads to better operating electrical conditions and reduced losses in collection efficiency due to gas sneakage and hopper boil-up. Third, the gas velocity through the precipitator should be 1.53 m/sec (5 ft/sec) or less in order to help prevent reentrainment without rapping and to allow sufficient residence time to recollect material reentrained due to rapping.

Step 3: Determine the nonideal conditions for which the precipitator will be sized.

Since a certain degree of a gas flow nonuniformity and gas bypassage of electrified regions and/or particle reentrainment without rapping can be expected to exist in a precipitator, these factors must be considered in sizing the precipitator. Experience in simulating the operation of full-scale, industrial precipitators indicates that values of σ_g = 0.25 and S = 0.1 are appropriate for modeling precipitators which are in good working condition. Losses in overall mass collection efficiency due to rapping reentrainment are built into the model and cannot be varied without changing the computer program itself. Since the procedure which determines the effect of rapping reentrainment on precipitator performance is based on average data acquired from six different full-scale precipitators, the effects of rapping reentrainment might not be estimated in a conservative manner.

If a conservative approach is taken in sizing the precipitator, then the values of σ_g and S should be taken to be somewhat higher than 0.25 and 0.1, respectively. Values of σ_g = 0.4 and S = 0.2 should be conservative. This value of S should also allow for above average losses in collection efficiency due to rapping reentrainment. If the precipitator is sized in a conservative manner, then the chances that the precipitator will be able to meet the particulate emissions standards once it is built are improved even though undesirable nonideal conditions exist. As a consequence, the process producing the emissions does not have to be shut down until the problems with the precipitator are diagnosed and corrected. The problems with the precipitator can be diagnosed with the troubleshooting procedure while the precipitator is in operation and appropriate corrective measures can be made during a scheduled shut down. Thus, in many cases, the added cost of a conservative design can be partially or fully recovered.

Step 4: Consider the effect of adverse changes in particle size distribution in sizing the precipitator.

Since any decrease in the mass median diameter or increase in the dispersiveness of the inlet particle size distribution will result in a fundamental reduction in precipitator performance, this factor should be considered in sizing a precipitator. Any changes in the process variables controlling the source of the emissions can result in significant changes in particle size distribution. Thus, the possibility of a change from the anticipated particle size distribution to a less favorable one should be incorporated into the sizing procedure. In a conservative approach, the measured or anticipated inlet particle size distribution can be fit to a log-normal distribution and the fitted mass median diameter and geometric standard deviation can be decreased and increased by 25%, respectively. These new values should then be used in the model in order to obtain the inlet particle size distribution for use in sizing the precipitator.

Step 5: Generate a curve of overall mass collection efficiency versus specific collection area.

At this point, since all appropriate input data have been or can be determined, the computer program for the mathematical model can be executed in order to size the precipitator. The precipitator can be sized by generating a curve of overall mass collection efficiency versus specific collection area.

Based on the curve of overall mass collection efficiency versus specific collection area and the particulate emissions standard, the precipitator size needed to attain the required efficiency can be determined. In sizing the precipitator in a conservative manner, the precipitator should be sized to attain an efficiency which is somewhat higher than that which is required. This is necessary in order to provide a margin of safety in design irrespective of any uncertainties in operating parameters and of any nonidealities which might exist. In order to provide this margin of safety, the projected collection plate area needed to attain the required efficiency should be increased by a certain percentage, possibly 10-15%. This added collection plate area is also an advantage in that it offers the possibility that the precipitator will be able to adequately treat gas flows which are somewhat higher than the design gas flow.

Step 6: Allow for the outage of electrical sections.

In designing the precipitator, a high degree of electrical sectionalization should be provided. As stated previously, this leads to improved electrical operating conditions. In addition, if certain electrical sections are not working, this condition does not disable a large portion of the precipitator.

In sizing a precipitator, proper allowance should be made for the possibility that from time to time certain electrical sections will not be functioning. This can be done by increasing the collection plate area obtained in Step 5. The additional collection plate area should be provided in the form of added electrical sections. If reliable data or past experiences are not sufficinet for estimating the number of electrical sections that might be inoperable at any given time, than a reasonable approach might be to add an extra electrical section for approximately every four electrical sections that are required in Step 5.

The above guidelines and procedure cover the important considerations which must be made in sizing an electrostatic precipitator. If the guidelines and procedure are followed correctly, then the mathematical model of electrostatic precipitation can be a valuable tool for sizing electrostatic precipitators. Since the procedure includes reasonable conservative measures to account for several different uncertainties, the cumulative effect should lead to a precipitator which is sized conservatively but not excessively oversized.

The procedure for sizing a precipitator can be utilized by manufacturers to assist in designing a precipitator and by purchasers to assess bids submitted by the various manufacturers. It can also be used by government regulatory agencies in helping to establish particulate emissions standards which are economically feasible and consistent with the best available control technology.

The troubleshooting and sizing procedures can both be utilized in conjunction with pilot precipitator studies. The troubleshooting procedure can be used to characterize the performance of the pilot precipitator and to establish the values of the parameters characterizing the operation of the precipitator. This will establish baseline information for which the model predictions and experimental data are in agreement. The sizing procedure can then be used to project full-scale precipitator performance under various operating conditions in order to obtain the size necessary to give the required collection efficiency.

It should be noted that care should be taken in projecting full-scale performance based on pilot data. Normally, better electrical conditions can be obtained in a pilot unit than a full-scale unit because of the reduced collection electrode area. In addition, particle reentrainment characteristics, gas velocity distribution, and gas bypassage of electrified regions in the pilot unit and the constructed full-scale unit may differ significantly.

Section 13

Features of a Well-Equipped Electrostatic Precipitator

There are several, important features that a well-equipped electrostatic precipitator should possess. These features are necessary in order to achieve high collection efficiency, operational and mechanical reliability, and ease in locating potential problems and in troubleshooting existing problems. In this section, these features are listed and discussed. Most of these features have been pointed out or discussed earlier in the text. Thus, the following list serves to bring these features together in a single location for easy reference.

- Adjustable gas flow distribution screens (or other devices) should be located at the inlet of an electrostatic precipitator in order to reduce the turbulence in the gas stream and to improve the gas velocity distribution. Adjustable devices are needed because flow model studies or other methods of prediction may not prove to be reliable. In some cases two or more devices may be necessary in order to achieve good gas flow qualities. (It has been demonstrated that this can be done without incurring excessive pressure drops). The average gas velocity entering the electrostatic precipitator should be no higher than 1.22 m/sec (4 ft/sec). The uniformity of the gas velocity distribution at the inlet of the electrostatic precipitator should, as a minimum, meet existing IGCI requirements.

- The electrostatic precipitator should have chambers which can be isolated for on-line maintenance and repair. It should have an adequate number of inlet and outlet sampling ports for each chamber. A minimum of six is necessary at each location in order to provide proper sampling access. The sampling ports should be of 6 in. diameter pipe instead of the commonly used 4 in. diameter pipe. This would facilitate the design and use of sampling instrumentation. Thermocouples should be located at the inlet and outlet of each chamber for proper monitoring of temperature. The use of induced draft fans will make gas and particulate sampling less difficult and less hazardous. The electrostatic precipitator should have a totally enclosed roof penthouse.

- The electrostatic precipitator should have hopper baffles and baffles above the electrodes to minimize gas bypass of electrified regions and to prevent significant gas flow from occurring in the hoppers.

- The electrostatic precipitator should have at least four, and preferably six, electrical sections in the direction of gas flow. There should be adequate electrical sectionalization with no more than between 1,861 - 2,791 m² (20,000 - 30,000 ft²) of collection plate area per transformer/rectifier (TR) set with two bushings per TR set. A rigid discharge electrode system is desirable because of its stability and reliability. The collection electrodes should be mounted in guides for proper alignment and stability. A dried, heated purge air system should be provided for keeping insulator feed-thrus free of particles and condensed gases. Secondary current and voltage panel meters are needed for monitoring actual precipitator electrical operating conditions and for troubleshooting. The power supplies should have controllers which can operate in either a spark rate or current limit mode to produce the maximum useful voltages and currents. Each electrical section should be provided with access from the inlet, outlet, top, and bottom for ease of inspection, wire replacement, alignment, and collection of representative ash samples, if necessary.

- The electrostatic precipitator should have independent discharge and collection electrode rappers. The rapping systems should be programmable with frequency and intensity adjustment capability so that precipitator performance can be optimized with respect to the rapping process. The rapping system for the collection electrodes should be capable of producing accelerations in all parts of the plate of over 50 times that of the gravitational acceleration. The discharge electrode system should be cleaned by impulse rappers rather than vibrators. The hoppers should be sufficiently heated or insulated to prevent condensation and resultant pluggage. Hopper level indicators should be installed to monitor hopper performance. Ash collected in the hoppers should be removed with a system which minimizes air flow into or out of the hoppers and should be conveyed away with an air transport system.

- The outlet of the precipitator should be instrumented with an opacity meter for continuous monitoring of precipitator performance. This will provide continuous information which will indicate changes in precipitator operation which could be caused by changes in the process variables or precipitator malfunctions. The opacity information is also useful in troubleshooting.

References

1a. Engelbrecht, H. L. Air Flow Model Studies for Electrostatic Precipitation, p. 72-73. From: Symposium on the Transfer and Utilization of Particulate Control Technology: Volume 1. Electrostatic Precipitators. EPA-600/7-79-044a, Environmental Protection Agency, Research Triangle Park, North Carolina, February 1979.

b. Szabo, M. and R. Gerstle. Electrostatic Precipitator Malfunctions in the Electric Utility Industry, section 2, p. 16. EPA-600/2-77-006, prepared by PEDCo for the Environmental Protection Agency, Research Triangle Park, North Carolina, January 1977.

2. Smith, W., K. Cushing, and J. McCain. Procedures Manual for Electrostatic Precipitator Evaluation, p.18. EPA-600/7-77-059, prepared by Southern Research Institute for the Environmental Protection Agency, Research Triangle Park, North Carolina, June 1977.

3. McDonald, J. and L. Sparks. A Precipitator Performance Model: Application to the Nonferrous Metals Industry. Proceedings: Particulate Collection Problems Using ESPs in the Metallurgical Industry. EPA-600/2-77-208, U.S. Environmental Protection Agency, Raleigh Durham, North Carolina, 1977. 72 pp.

4. McDonald, J., W. Smith, H. Spencer, and L. Sparks. A Mathematical Model for Calculating Electrical Conditions in Wire-Duct Electrostatic Precipitation Devices. J. Apply. Phys., 48(6):2231-2246, 1977.

5. Pauthenier, M. and M. Moreau-Hanot. Charging of Spherical Particles in an Ionizing Field. J. Phys. Radium, 3(7):590-613, 1932.

6. White, H. Particle Charging in Electrostatic Precipitation. Trans. Amer. Inst. Elec. Eng. Part 1, 70:1186-1191, 1951.

7. Murphy, A., F. Adler, and G. Penney. A Theoretical Analysis of the Effects of an Electric Field on the Charging of Fine Particles. Trans. Amer. Inst. Elec. Eng., 78:318-326, 1959.

8. Pontius, D., L. Felix, J. McDonald, and W. Smith. Fine Particle Charging Development. EPA-600/2-77-173, U.S. Environmental Protection Agency, Raleigh Durham, North Carolina, 1977.

9. Smith, W., L. Felix, D. Hussey, and D. Pontius. Experimental Investigations of Fine Particle Charging by Unipolar Ions - A Review, J. Aerosol Sci., 9:101-124 (1978).

10. White, H. Industrial Electrostatic Precipitation. Addison-Wesley, Reading, Massachusetts, 1963. p. 157.

11. Fuchs, N. The Mechanics of Aerosols. Chapter 2. Macmillan, New York, 1964.

12. White, H. Reference 10, pp. 166-170.

13. White, H. Reference 10, pp. 185-190.

14. Penney, G., and S. Craig. Pulsed Discharges Preceding Sparkover at Low Voltage Gradients. AIEE Winter General Meeting, New York, 1961.

15. Pottinger, J. The Collection of Difficult Materials by Electrostatic Precipitation. Australian Chem. Process Eng., 20(2): 17-23, 1967.

16. Spencer, H. Electrostatic Precipitators: Relationship Between Resistivity, Particle Size, and Sparkover. EPA-600/2-76-144, U.S. Environmental Protection Agency, Raleigh Durham, North Carolina, 1976.

17. White, H. Reference 10, pp. 238-293.

18. Preszler, L. and T. Lajos. Uniformity of the Velocity Distribution Upon Entry into an Electrostatic Precipitator of a Flowing Gas. Staub Reinhalt. Luft (In English), 32(11):1-7, 1972.

19. Gooch, J. P., and G. H. Marchant, Jr. Electrostatic Precipitator Rapping Reentrainment and Computer Model Studies. EPRI RP-792, Vol. 3, August 1978.

20. Spencer, H. A Study of Rapping Reentrainment in a Nearly Full Scale Pilot Electrostatic Precipitator. EPA-600/2-76-140, U.S. Environmental Protection Agency, Raleigh Durham, North Carolina, 1976.

21. White, H. Electrostatic Precipitation of Fly Ash, Journal of the Air Pollution Control Association, 27(1):15-21, January 1977.

22. Oglesby, S. and G. Nichols. Electrostatic Precipitation. Marcel-Dekker, Inc., New York, 1978.

23. Oglesby, S. and G. Nichols. A Manual of Electrostatic Precipitator Technology, Part 1 - Fundamentals. NTIS PB-196 380, U.S. Environmental Protection Agency, Research Triangle Park, North Carolina, August 25, 1970.

24. Oglesby, S. and G. Nichols. Comparison of Precipitator Design Methods. Paper presented at the Conference on European Electrostatic Precipitators for Controlling Particle Emissions from Pulp Mills, University of Washington, March 5, 1974.

25. White, H. Electrostatic Precipitation of Fly Ash. Journal of the Air Pollution Control Association, 27(4):308-312, April 1977.

26. Hall, H. Design and Application of High-Voltage Power Supplies in Electrostatic Precipitation. H. J. Hall Associates, Inc., Princeton, New Jersey.

27. Smith, W., K. Cushing, and J. McCain. Procedures Manual for Electrostatic Precipitator Evaluation. EPA-600/7-77-059, Environmental Protection Agency, Research Triangle Park, North Carolina, June 1977.

28. Schummer, H. and W. Steinbauer. Siemens Rev., 34(12):458-463, 1967.

29. Piulle, W. Precipitator Performance Hinges on Control. Power, 119(1):23-26, January 1975.

30. Gelfand, P. Electrostatic Precipitator Voltage Control Using Silicon-Controlled Rectifiers. IEEE Transactions on Industry Applications, 10(5):662-665, September/October 1974.

31. Engelbrecht, H. Rigid Frame Precipitators. Proceedings: Operation and Maintenance of Electrostatic Precipitators, Air Pollution Control Association, April 1978.

32. Oglesby, S. and G. Nichols. Reference 15, p. 272.

33. Oglesby, S. and G. Nichols. Reference 15, p. 273.

34. Electric Light and Power, 55(6):31, June 1978.

35. Oglesby, S. and G. Nichols. Reference 15, p. 126.

36. Written communication between SoRI and Wheelabrator-Frye, Inc.

37. Lynch, J. A Review of Rapper System Problems Associated with Industrial Electrostatic Precipitators. Proceedings: Operation and Maintenance of Electrostatic Precipitators, Dearborn, Michigan, April 10-12, 1978.

38. Oglesby, S. and G. Nichols. Reference 15, p. 280.

39. Oglesby, S. and G. Nichols. Reference 15, p. 282.

40. Smith, W., K. Cushing, and J. McCain. Reference 20, p. 33.

41. Smith, W., K. Cushing, and J. McCain. Reference 20, p. 32.

42. Information obtained from industry survey by SoRI personnel.

43. Dumbauld, J. Electrostatic Precipitator Hopper Evaluation Problems and Their Solutions. Proceedings: Operation and Maintenance of Electrostatic Precipitators, Dearborn, Michigan, April 10-12, 1978.

44. Communication from Environmental Elements Corporation.

45. AMCA Bulletin 210. Standard Test Code for Air Moving Devices. Air Moving and Conditioning Association, Detroit, Michigan, 1960.

46. Baines, W. and E. Peterson. An Investigation of Flow Through Screens. ASME Trans., July 1961.

47. Dryden and Schubauer. The Use of Damping Screens for the Reduction of Turbulence. Journal of Aero. Science, 14(4), 1947.

48. Communication from WAHLCO, Inc., 3600 West Segerstrom Avenue, Santa Ana, California 92704, phone: (714) 979-7300.

49. Southern Research Institute. A Review of Technology for Control of Industrial Particulate Emissions. Report to Argonne National Laboratory, Energy Research and Development Administration, Argonne, Illinois, Mary 1977.

50. Smith, W., K. Cushing, and J. McCain. Reference 20, p. 102.

51. Smith, W. and J. McCain. Particle Size Measurements in Industrial Flue Gases, Air Pollution Control, Part III. Edited by Werner Strauss, published by John Wiley and Sons, Inc., 1978.

52. Smith, W., K. Cushing, and J. McCain. Reference 20, p. 106.

53. Smith, W., K. Cushing, and J. McCain. Reference 20, p. 107.

54. Smith, W., P. Cavanaugh, and R. Wilson. Technical Manual: A Survey of Equipment and Methods for Particulate Sampling in Industrial Process Streams. EPA-600/2-77-173, U.S. Environmental Protection Agency, Research Triangle Park, North Carolina, March 1978.

55. Wilson, R., Jr., P. Cavanaugh, K. Cushing, W. Farthing, and W. Smith. Guidelines for Particulate Sampling in Gaseous Effluents from Industrial Processes. EPA-600/7-79-028, U.S. Environmental Protection Agency, Research Triangle Park, North Carolina, January 1979.

56. Smith, W., P. Cavanaugh, and R. Wilson. Reference 54, p. 95.

57. Cohen, J. and D. Montan. Theoretical Considerations, Design, and Evaluation of a Cascade Impactor. Amer. Ind. Hyg. Assoc. Journal, 95-104, 1976.

58. Marple, V. and K. Willeke. Impactor Design. Atmos. Environ., 10:891-896, 1976.

59. Mercer, T. On the Calibration of Cascade Impactors. Ann. Occup. Hyg., 6:1-17, 1963.

60. Newton, G., O. Raabe, and B. Mokler. Cascade Impactor Design and Performance. J. Aerosol Sci., 8:339-347, 1977.

61. Marple, V. and B. Y. H. Liu. Characteristics of Laminar Jet Impactors. Environ. Sci. and Tech., 8(7):648-654, 1974.

62. Rao, A. and K. Whitby. Nonideal Collection Characteristics of Single Stage and Cascade Impactors. Amer. Ind. Hyg. Assoc. J., 38:174-179, 1977.

63. Cushing, K., G. Lacey, J. McCain, and W. Smith. Particulate Sizing Techniques for Control Device Evaluation: Cascade Impactor Calibrations. EPA-600/2-76-280, U.S. Environmental Protection Agency, Research Triangle Park, North Carolina, 1976.

64. Lundgren, D. An Aerosol Sampler for Determination of Particle Concentration as a Function of Size and Time. J. Air Pollut. Contr. Assoc., 17(4):225-259, 1967.

65. Ranz, W. and J. Wong. Impaction of Dust and Smoke Particles. Ind. Eng. Chem., 44(6):1371-1381, 1952.

66. Davies, C. and M. Aylward. The Trajectories of Heavy, Solid Particles in a Two-Dimensional Jet of Ideal Fluid Impinging Normally Upon a Plate. Proc. Phys. Soc., 64:889-991, 1951.

67. Marple, V. A Fundamental Study of Inertial Impactors. University Microfilms, Ann Arbor, Michigan, 1970.

68. Mercer, T. and R. Stafford. Impaction from Round Jets. Ann. Occup. Hyg., 12:41-48, 1969.

69. Smith, W., P. Cavanaugh, and R. Wilson. Reference 54, pp. 99-104.

70. Smith, W., P. Cavanaugh, and R. Wilson. Reference 54, pp. 105-106.

71. Calvert, S., C Lake, and R. Parker. Cascade Impactor Calibration Guidelines. EPA-600/2-76-118. U.S. Environmental Protection Agency, Research Triangle Park, North Carolina, 1976.

72. McCain, J., K. Cushing and A. Bird, Jr. Field Measurements of Particle Size Distribution with Inertial Sizing Devices. EPA-650/2-73-035. U.S. Environmental Protection Agency, Research Triangle Park, North Carolina, 1973.

73. Felix, L., G. Clinard, G. Lacey, and J. McCain. Inertial Cascade Impactor Substrate Media for Flue Gas Sampling. EPA-600/7-77-060. U.S. Environmental Protection Agency, Research Triangle Park, North Carolina, 1977.

74. Brink, J., Jr., E. Kennedy, and H. Yu. Particle Size Measurements with Cascade Impactors. 65th Annual Meeting, AIChE, New York, New York, 1972.

75. Ragland, J., K. Cushing, J. McCain, and W. Smith. HP-25 Programmable Pocket Calculator Applied to Air Pollution Measurement Studies: Stationary Sources. Interagency Energy-Environment Research and Development Program Report, EPA-600/2-77-05, June 1977.

76. Ragland, J., K. Cushing, J. McCain, and W. Smith. HP-65 Programmable Pocket Calculator Applied to Air Pollution Measurement Studies: Stationary Sources. U.S. Environmental Protection Agency Report, EPA-600/2-76-002, October 1976.

77. Smith, W., P. Cavanaugh, and R. Wilson. Reference 54, p. 110.

78. Smith, W., P. Cavanaugh, and R. Wilson. Reference 54, p. 112.

79. Chan, T. and M. Lippmann. Particle Collection Efficiencies of Air Sampling Cyclones: An Empirical Theory. Environ. Sci. Technol., 11(4):377-382, 1977.

80. Smith, W., and R. Wilson. Development and Laboratory Evaluation of a Five-Stage Cyclone System. EPA-600/7-78-008, U.S. Environmental Protection Agency, Research Triangle Park, North Carolina, 1978.

81. Rusanov, A. Determination of the Basic Properties of Dusts and Gases in "Ochistka Dymovykl Gasov V Promyshlennoy Energtike". 405-440, 1969.

82. Smith, W., K. Cushing, G. Lacey, and J. McCain. Particulate Sizing Techniques for Control Device Evaluation. EPA-650/2-74-102A, U.S. Environmental Protection Agency, Research Triangle Park, North Carolina, 1975.

83. Hamersma, J., S. Reynolds, and R. Maddalone. Procedures Manual for Level 1 Environmental Assessment. EPA-600/2-76-160A, U.S. Environmental Protection Agency, Research Triangle Park, North Carolina, 1976.

84. Smith, W., P. Cavanaugh, and R. Wilson. Reference 54, p. 120.

85. Smith, W., P. Cavanaugh, and R. Wilson. Reference 54, p. 121.

86. Whitby, K., and B. Y. H. Liu. J. Colloid Interface Science, 25:537, 1967.

87. Willeke, K., and B. Y. H. Liu. Single Particle Optical Counter: Principle and Application. In: Fine Particles, Aerosol Generation, Measurement, Sampling, and Analysis. Academic Press, B. Y. H. Liu, ed., 1976. pp. 698-725.

88. Smith, W., P. Cavanaugh, and R. Wilson. Reference 54, p. 124.

89. Smith, W., P. Cavanaugh, and R. Wilson. Reference 54, p. 125.

90. Marple, V. The Aerodynamic Size Calibration of Optical Particle Counters by Inertial Impactors. Particle Tech. Lab. Pub. 306, presented at Aerosol Measurement Workshop, University of Florida, Gainesville, Florida, 1976.

91. McCain, J., K. Cushing, and W. Smith. Methods for Determining Particulate Mass and Size Properties: Laboratory and Field Measurements. J. Air Pollut. Contr. Assoc., 24(12):1172-1176, 1974.

92. Smith, W., P. Cavanaugh, and R. Wilson. Reference 54, p. 124.

93. Breslin, A., S. Guggenheim, and A. George. Staub (English Translation), 31(8):1-5, 1971.

94. Sinclair, D., and G. Hoopes. A Novel Form of Diffusion Battery. Amer. Ind. Hyg. Assoc. J., 36(1):39-42, 1975.

95. Junge, C., and E. McLaren. Relationship of Cloud Nuclei Spectra to Aerosol Size Distribution and Composition. J. of Atmos. Sci., 28(3):382-390, 1971.

96. Haberl, Jr., and S. Fusco. Condensation Nuclei Counters: Theory and Principles of Operation. Prepared for presentation at the 11th Conference on Methods in Air Pollution and Industrial Hygiene Studies at the University of California, Berkeley, California, sponsored by California Air Resources Board and California Department of Public Health, 1970.

97. Sinclair, D. A Portable Diffusion Battery: Its Application to Measuring Aerosol Size Characteristics. Amer. Ind. Hyg. Assoc. J., 33(11):729-735, 1972.

98. Ragland, J., W. Smith, and J. McCain. Design, Construct, and Test a Field Usable Prototype System for Sizing Particles Smaller than 0.5 μm Diameter. EPA Contract Number 68-02-2114, U.S. Environmental Protection Agency, Research Triangle Park, North Carolina, 1978.

99. Soderholm, S. Modification of a Commercial Condensation Nuclei Counter for Steady Flow. Atmos. Environ., 10:659-660, 1976.

100. Fuchs, N., I. Stechkina, and V. Starosselskii. On the Determination of Particle Size Distribution in Polydisperse Aerosols by the Diffusion Method. Brit. J. Appl. Phys., 16:280-281, 1962.

101. Sinclair, D., R. Countese, B. Y. H. Liu, and D. Y. H. Pui. Experimental Verification of Diffusion Battery Theory. J. Air Pollut. Contr. Assoc., 26(7):661-663, 1976.

102. Sinclair, D. and G. Hoopes. A Novel Form of Diffusion Battery. Amer. Ind. Hyg. Assoc. J., 36(1):39-42, 1975.

103. Breslin, A., S. Guggenheim, and A. George. Compact High-Efficiency Diffusion Batteries. Staub Reinhaltung der Luft, 33(4):187-190, 1973.

104. Twomey, S. The Determination of Aerosol Size Distributions from Diffusional Decay Measurements. J. of Franklin Inst., 275:121-138, 1963.

105. Sansone, E., and D. Weyel. A Note on the Penetration of a Circular Tube by an Aerosol with a Log-Normal Size Distribution. J. Aerosol Sci., 2:413-415, 1971.

106. Smith, W., P. Cavanaugh, and R. Wilson. Reference 54, pp. 132-133.

107. Smith, W., P. Cavanaugh, and R. Wilson. Reference 54, p. 135.

108. Megaw, W., and A. Wells. A High Resolution Charge and Mobility Spectrometer for Radioactive Submicrometer Aerosols. J. Physics E., 1013-1016, 1969.

109. Maltoni, G., C. Melandri, V. Prodi, G. Tarroni, A. DeZaiacomo, G. Bompane, and M. Formignani. An Improved Parallel Plate Mobility Analyzer for Aerosol Particles. J. Aerosol Sci., 4:447-455, 1973.

110. Krutson, E. Extended Electric Mobility Method. In: Proceedings of Symposium on Fine Particles, Minneapolis, Minnesota, 1975.

111. Markowski, G. and D. Ensor. Development of an In-Stack Impactor/Precipitator for Sizing Submicron Particles. EPRI FP-501, Electric Power Research Institute, Palo Alto, California.

112. Smith, W., P. Cavanaugh, and R. Wilson. Reference 54, p. 137.

113. Whitby, K., and W. Clark. Electric Aerosol Particle Counting and Size Distribution Measuring System for the 0.015 to 1 Micron Size Range. Tellus, 18:573-586, 1966.

114. Liu, B. Y. H., K. Whitby, and D. Y. H. Pui. A Portable Electrical Analyzer for Size Distribution Measurement of Sub-Micron Aerosols. J. Air Pollut. Contr. Assoc., 24(11):1067-1072, 1974.

115. Sem, G. Submicron Particle Sizing Experience on a Smoke Stack Using the Electrical Aerosol Size Analyzer. EPA-600/2-77-060, U.S. Environmental Protection Agency, Research Triangle Park, North Carolina, 1975.

116. Lacey, G., K. Cushing, and W. Smith. Compact, In-Stack, Three-Size-Cut Particle Classifier. Report prepared by Southern Research Institute, Contract No. 68-02-1736, for the Environmental Protection Agency, Research Triangle Park, North Carolina, October 5, 1976.

117. Gooch, J., and G. Marchant, Jr. Reference 19, pp. 3-13 through 3-18.

118. Cadle, R. Particle Size Measurement. Interscience Publishers, Inc., New York, New York, 1955.

119. Allen, T. Particle Size Measurement. Chapman and Hall Ltd., London, England, 1975.

120. Smith, W., P. Cavanaugh, and R. Wilson. Reference 54, p. 144.

121. Godridge, A., S. Badzioch, and P. Hawksley. A Particle Size Classifier for Preparing Graded Sub-Sieve Fractions. J. Sci. Instrum., 39:611-613, 1962.

122. Göetz, A. and T. Kallai. Instrumentation for Determining Size and Mass Distribution of Submicron Aerosols. APCA J., 12:479-486, 1962.

123. Göetz, A., H. Stevenson, and O. Preining. The Design and Performance of the Aerosol Spectrometer. APCA J., 10:378-838, 1960.

124. Gerber, H. On the Performance of the Göetz Aerosol Spectrometer. Atmos. Environ., 5:1009-1031, 1971.

125. Stöber, W., and H. Flachsbart. Size-Separating Precipitation of Aerosols in a Spinning Spiral Duct. Environ. Sci. Technol., 3(12):1280-1296, 1969.

126. Swayer, K. F., and W. Walton. The "Conifuge" - A Size-Separating Sampling Device for Airborne Particles. J. Sci. Instrum., 27:272-276, 1950.

127. Keith, C., and J. Derrick. Measurement of the Particle Size Distribution and Concentration of Cigarette Smoke by the "Conifuge". J. Colloid. Sci., 14:340-356, 1960.

128. Tillery, M. Design and Calibration of a Modified Conifuge. Assessment of Airborne Radioactivity, IAEA, Vienna, 1967.

129. Smith, W., P. Cavanaugh, and R. Wilson. Reference 54, p. 148.

130. McCrone, W., and J. Delly. The Particle Atlas, Edition Two. Ann Arbor Science, Ann Arbor, Michigan, 1973.

131. Smith, W., P. Cavanaugh, and R. Wilson. Reference 54, p. 150.

132. Smith, W., P. Cavanaugh, and R. Wilson. Reference 54, p. 152.

133. Kaye, B. Symposium on Particle Size Analysis Society for Analytical Chemistry, Loughborough, England, 1966.

134. Allen-Bradley Sonic Sifter. U.S. Patent 3,045,817.

135. Smith, W., P. Cavanaugh, and R. Wilson. Reference 54, p. 154.

136. Nichols, G., and J. McCain. Particulate Collection Efficiency Measurements on Three Electrostatic Precipitators. EPA-600/2-75-056, U.S. Environmental Protection Agency, Research Triangle Park, North Carolina, October 1975.

137. McDonald, J. A Mathematical Model of Electrostatic Precipitation (Revision 1): Volume II. User Manual. EPA-600/7-78-111b, U.S. Environmental Protection Agency, Research Triangle Park, North Carolina, June 1978. p. 60.

138. Electrostatic Precipitators for Control of Fine Particle Emissions. Final report prepared by Southern Research Institute for the Environmental Protection Agency, Research Triangle Park, North Carolina under Contract No. 68-02-2114.

139. Oglesby, S. and G. B. Nichols. Reference 23, p. 251.

140. Oglesby, S. and G. B. Nichols. Reference 23, p. 254.

141. Banks, S. M., J. R. McDonald, and L. E. Sparks. Voltage-Current Data From Electrostatic Precipitators Under Normal and Abnormal Conditions. Proceedings: Particulate Collection Problems Using ESPs in the Metallurgical Industry. EPA-600/2-77-208, U.S. Environmental Protection Agency, Research Triangle Park, North Carolina, 1977. 129 pp.

142. McDonald, J. R. Mathematical Modelling of Electrical Conditions, Particle Charging, and the Electrostatic Precipitation Process. Ph.D Dissertation, Auburn University, Auburn, AL, 1977. 186 pp.

143. White, H. Reference 10, p. 222.

144. White, H. Reference 10, p. 92.

145. Peek, F. W., Jr. Dielectric Phenomena in High Voltage Engineering. 3rd ed., McGraw-Hill, New York. p. 64, 1929.

146. White, H. Reference 10, pp. 105-106.

147. White, H. Reference 10, p. 89 and p. 107.

148. Voshall, R. E., J. L. Packs, and A. V. Phelps. Mobility of Negative Ions in O_2 at Low E/N. J. Chem. Phys. 43:1990, 1965.

149. Tassicker, O. J. Experiences With an Electrostatic Precipitation Analyzer in the Evaluation of Difficult Dusts. Proceedings International Clean Air Conference, Melbourne, Australia, May, 1972.

150. Spencer, H. W. Experimental Determination of the Effective Ion Mobility of Simulated Flue Gas. In: Proceedings of 1975 IEEE-IAS Conference, Atlanta, Georgia, 1975.

151. McDonald, J. R., S. M. Banks, and L. E. Sparks. Measurement of Effective Ion Mobilities in a Corona Discharge in Industrial Flue Gases. Proceedings: Symposium on the Transfer and Utilization of Particulate Control Technology. Volume 1, Electrostatic Precipitators, EPA-600/7-79-044a, U.S. Environmental Protection Agency, Research Triangle Park, North Carolina, July 1978.

152. McDonald, J. R. A Mathematical Model of Electrostatic Precipitation (Revision 1): Modeling and Programing. EPA-600/7-78-111a, U.S. Environmental Protection Agency, Research Triangle Park, North Carolina, June 1978. pp. 175.

153. Gooch, J. P., J. R. McDonald, and S. Oglesby, Jr. A Mathematical Model of Electrostatic Precipitation. EPA-650/2-75-037, U.S. Environmental Protection Agency, Research Triangle Park, North Carolina, 1975. pp. 77-79.

154. Dismukes, E. B. and J. P. Gooch, Fly Ash Conditioning with Sulfur Trioxide. EPA-600/2-77-242, U.S. Environmental Protection Agency, Research Triangle Park, North Carolina, 1977.

155. A Field Demonstration Study to Evaluate Sodium Injection for Reducing Fly Ash Resistivity. Contract No. 68-02-2656, U.S. Environmental Protection Agency, Research Triangle Park, North Carolina.

156. Dismukes, E. B. Techniques for Conditioning Fly Ash. Proceedings: Conference on Particulate Collection Problems in Converting to Low Sulfur Coals. EPA-600/7-76-016, U.S. Environmental Protection Agency, Research Triangle Park, pp. 107, 1976.

157. Cragle, S. H. Operating Experience with ESP Conditioning in Relation to an Electrostatic Precipitator Upgrading Program. Proceedings: Conference on Particulate Collection Problems in Converting to Low Sulfur Coals. EPA-600/7-76-016, U.S. Environmental Protection AGency, Research Triangle Park, pp. 3, 1976.

158. Borsheim, R. and R. P. Bennett. Chemical Conditioning of Low-Sulfur Western Coal. Presented at 39th Annual Meeting, American Power Conference, Chicago, Illinois, April, 1977.

159. Effects of Conditioning Agents on Emissions from Coal-Fired Boilers. Contract No. 68-02-2628, U.S. Environmental Protection Agency, Research Triangle Park, North Carolina.

160. Flue Gas Conditioning for Enhanced Precipitation of Difficult Ashes. Contract No. RP724-2, Electric Power Research Institute, Chattanooga, Tennessee.

161. Selle, S. J., P. H. Tufte, and G. H. Gronhovd. A Study of the Electrical Resistivity of Fly Ashes from Low-Sulfur Western Coals Using Various Methods. Paper 72-107 presented at the 65th Annual Meeting of the Air Pollution Control Association, Miami Beach, Florida, 1972.

162. Bickelhaupt, R. E. Electrical Volume Conduction in Fly Ash. APCA Journ., 24(3):251-255, 1974.

163. Lederman, P. B., P. P. Bibbo, and J. Bush. Chemical Conditioning of Fly Ash for Hot-Side Precipitation. Proceedings: Symposium on the Transfer and Utilization of Particulate Control Technology. Volume 1, Electrostatic Precipitators, EPA-600/7-79-044a, U.S. Environmental Protection Agency, Research Triangle Park, North Carolina, 1978, pp. 79-98.

164. Dismukes, E. B. Conditioning of Fly Ash with Sulfur Trioxide and Ammonia. TVA No. F75 PRS-5, Tennessee Valley Authority, Chattanooga, Tennessee and EPA No. 600/2-75-015, U.S. Environmental Protection Agency, Washington, D.C., 1975.

165. Dalmon, J. and D. Tidy. The Cohesive Properties of Fly Ash in Electrostatic Precipitation. Atmos. Environ. (Oxford, England), 6(2):81-92, 1972.

166. Dismukes, E. Conditioning of Fly Ash with Ammonia. JAPCA, 25(2):152-156, 1975.

167. McDonald, J. R. Reference 137, pp. 64-68.

168. Hall, H. J. Trends in Electrical Energization of Electrostatic Precipitators. Presented at Electrostatic Precipitator Symposium, Birmingham, Alabama, Paper I-C, February 23-25, 1971.

169. Penney, G. W. and E. H. Klingler. Contact Potentials and Adhesion of Dust. Trans. Amer. Inst. Elec. Eng. Part I, 81:200-204, 1962.

170. Nichols, G. B. Techniques for Measuring Fly Ash Resistivity. EPA-650/2-74-079, NTIS PB244140, U.S. Environmental Protection Agency, Research Triangle Park, 1974. pp. 5.

171. Bickelhaupt, R. E. Surface Resistivity and the Chemical Composition of Fly Ash. APCA Journal, 25(2):148-152, 1975.

172. Bickelhaupt, R. E. A Technique for Predicting Fly Ash Resistivity. Proceedings: Symposium on the Transfer and Utilization of Particulate Control Technology, U.S. Environmental Protection Agency, Research Triangle Park, North Carolina, July, 1978.

173. Nichols, G. B. Reference 169, p. 8.

174. Nichols, G. B. Reference 169, p. 13.

175. Baker, J. W. and K. M. Sullivan. Reproducibility of Ash Resistivity Determinations. Presentated at the Joint Power Generation Conference, Long Beach, California, September 18-21, 1977.

176. Personal communications with Dr. R. E. Bickelhaupt.

177. Babcock & Wilcox. Steam/its generation and use. Chapter 6. Babcock & Wilcos, New York, New York, 1975.

178. ASME PTC-28. Determining the Properties of Fine Particulate Matter. Section 4.05, Method for Determination of Bulk Electrical Resistivity, pp. 15-17, 1965.

179. Nichols, G. B. Reference 169, p. 18.

180. Nichols, G. B. Reference 169, p. 19.

181. Bickelhaupt, R. E. Measurement of Fly Ash Resistivity Using Simulated Flue Gas Environments. EPA-600/7-78-035, U.S. Environmental Protection Agency, Research Triangle Park, North Carolina, March 1978.

182. Bickelhaupt, R. E. Reference 180, p. 7.

183. Bickelhaupt, R. E. Reference 180, p. 12.

184. Nevens, T. D., et al. A Comparative Evaluation of Cells for Ash Resistivity Measurement. Presented at IEEE-ASME Joint Power Generation Conference, Long Beach, California, September 18-21, 1977.

185. Kanowski, S. and R. W. Coughlin. Catalytic Conditioning of Fly Ash Without Addition of SO_3 from External Sources. Environmental Science and Technology, 11(1):67-70, 1977.

186. Bickelhaupt, R. E. Reference 180, p. 15.

187. Bickelhaupt, R. E. Reference 180, p. 17.

188. Nichols, G. B. and S. M. Banks. Test Methods and Apparatus for Conducting Resistivity Measurements. Final Report, Contract No. 68-02-1083, U.S. Environmental Protection Agency, Research Triangle Park, North Carolina, September, 1977.

189. Nichols, G. B. Reference 169, p. 24.

190. Nichols, G. B. Reference 169, p. 26.

191. Nichols, G. B. and S. M. Banks. Reference 187, p. 10.

192. Cohen, L. and R. W. Dickinson. The Measurement of the Resistivity of Power Station Fine Dust. J. Sci. Instrum. (London), 40:72-75, 1963.

193. Nichols, G. B. Reference 169, p. 31.

194. Tassicker, O. J., Z. Herceg, and K. J. McLean. A New Method and Apparatus to Assist the Prediction of Electrostatic Precipitator Performance. Institution of Engineers, Australia. Electrical Engineering Transactions (Sydney). EE5(2):277-278, September 1969.

195. Eishold, H. G. A Measuring Device for Determining the Specific Electrical Resistance of Dust. Staub Reinhaltung der Luft in English (Düsseldorf). 26(1):14-18, January 1966.

196. Nichols, G. B. and J. P. Gooch. An Electrostatic Precipitator Performance Model. Report to Environmental Protection Agency on Contract No. CPA 70-166 by Southern Research Institute, Birmingham, Alabama. July 1972. 171 p.

197. White, H. Reference 10, pp. 238-293.

198. Burton, C. L., and D. A. Smith. Precipitator Gas Flow Distribution. JAPCA, 25(2):139-143, February, 1975.

199. Engelbrecht, H. L. Air Flow Model Studies for Electrostatic Precipitators. Proceedings: Symposium on the Transfer and Utilization of Particulate Control Technology. Volume 1, Electrostatic Precipitators, EPA-600/7-79-044a, U.S. Environmental Protection Agency, Research Triangle Park, North Carolina, February, 1979, pp. 57.

200. Industrial Gas Cleaning Institute, Inc. Criteria for Performance Guarantee Determinations, Publication No. EP-3. August 1965.

201. Gooch, J., and G. Marchant. Reference 19, pp. 5-71 to 5-72.

202. Gooch, J., and G. Marchant. Reference 19, pp. 5-87 to 5-97.

203. Gooch, J. P., J. R. McDonald, and S. Oglesby, Jr. Reference 153, pp. 48-53.

204. Gooch, J. P., J. R. McDonald, and S. Oglesgy, Jr. Reference 153, pp. 54-62.

205. Gilbert, Gerald B. Experimental Flow Modeling for Power Plant Equipment. Power Engineering Magazine. May 1974.

206. Tassicker, O. J. Some Aspects of Electrostatic Precipitator Research in Australia. J. Air Pollution Control Assoc., 25(2):122-128, 1975.

207. Tassicker, O. J. Aspects of Forces on Charged Particles in Electrostatic Precipitators. Dissertation, Wollongong University College, University of New South Wales, Australia, 1972.

208. Sproull, W. T. Fundamentals of Electrode Rapping in Industrial Electrical Precipitators. J. Air Pollution Control Assoc., 15(2):50-55, 1965.

209. White, H. J. Reference 10, pp. 331-354.

210. Sproull, W. T. Minimizing Rapping Losses in Precipitators at a 2000 Megawatt Coal-Fired Power Station. J. Air Pollut. Contr. Assoc., 22:181-186, 1972.

211. Juricic, D. and G. Herrmann. Response of Collecting Plates in Electrostatic Precipitators Due to Shear Rapping. Journal of Mechanical Design, 100:105-112, January, 1978.

212. Juricic, D. and G. Herrmann. On the Dynamics of Electrostatically Precipitated Fly Ash. Paper No. 78-WA/FU-3, presented at the Winter Annual Meeting of the American Society of Mechanical Engineers, San Francisco, Dec. 10-15, 1978.

213. Plato, H. Rapping of Collecting Plates in Electrostatic Precipitators. Staub-Reinhalt, Luft (in English), 29(8): 22-30, 1969.

214. Sanayev, Yu. I., and I. K. Reshidov. Study of Dust Reentrainment Phenomena and Their Influence on Efficiency of Industrial Electrostatic Precipitators. Promyshlennaya i Sanitarnaya Ochistka Gazov, (Moscow), (1):1-5, 1974.

215. Schwartz, L. B., and M. Lieberstein. Effect of Rapping Frequency on the Efficiency of an Electrostatic Precipitator at a Municipal Incinerator. Proceedings of the Fourth Annual Environmental Engineering and Science Conference, Louisville, Kentucky, March 4-5, 1975.

216. Nichols, G. B., H. W. Spencer, and J. D. McCain. Rapping Reentrainment Study. Report SoRI-EAS-75-307 to Tennessee Valley Authority, TVA Agreement TV36921A, November 1975.

217. Gooch, J. P. Electrostatic Precipitator Performance Proceedings: Symposium on the Transfer and Utilization of Particulate Control Technology. Volume 1, Electrostatic Precipitators, EPA-600/7-79-044a, U.S. Environmental Protection Agency, Research Triangle Park, North Carolina, February 1979, pp. 1-18.

218. Gooch, J. P., J. R. McDonald, and S. Oglesby, Jr. Reference 153, pp. 58-61.

219. U.S. Environmental Protection Agency. Standards of Performance for New Stationary Sources. Federal Register, 43(160):41776-41782, 1977.

220. U.S. Environmental Protection Agency. Standards of Performance for New Stationary Sources. Federal Register, 42(187):42020-42028, 1976.

221. American Society of Mechanical Engineers. Determining Dust Concentrations in a Gas Stream, Power Test Code 27. New York, New York, 1957.

222. Smith, W., P. Cavanaugh, and R. Wilson. Reference 54, p. 5.

223. Hemeon, W. and A. Black. Stack Dust Sampling: In-Stack Filter or EPA Train. Journal of the Air Pollution Control Association, 22(7):516, July 1972.

224. Brenchley, D., C. Turley, and R. Yarmac. Industrial Source Sampling. Ann Arbor Science Publishers, Inc., Ann Arbor, Michigan, 1973.

225. Rom, J. Maintenance, Calibration, and Operation of Isokinetic Source Sampling Equipment. U.S. Environmental Protection AGency, Research Triangle Park, North Carolina, 1972. APTD-0576.

226. Smith, W., P. Cavanaugh, and R. Wilson. Reference 54, p. 6.

227. Smith, W., P. Cavanaugh, and R. Wilson. Reference 54, p. 16.

228. U.S. Environmental Protection Agency. State Implementation Plan Emission Regulations for Particulate Matter: Fuel Combustion, Strategies and Air Standards Division, August 1976. EPA-450/2-76-010.

229. Code of Federal Regulations 40, Part 60, Subpart D, Paragraph 60.42-60.44, July 1, 1977.

230. Discussions with EPA.

231. EPA NSPS Proposal Eyes "Full Scrubbing". Electric Light and Power, 56(10):1 and 7, October 1978.

232. EPA Sets New Sulfur Limits. Electric Light and Power, 57(7):1 and 4, July 1979.

233. Farthing, W. E. and A. H. Dean. Summary Document on Control Stategies for Visible Emissions. Final Report prepared by SoRI for the FLAKT, INC., May 5, 1978.

234. Peterson, C. M. In-Stack Transmissometer Techniques for Measuring Opacities of Particulate Emissions from Stationary Sources. U.S. Environmental Protection Agency, Research Triangle Park, North Carolina, 1972. EPA-R2-72-099.

235. Ensor, D. S. and M. J. Pilat. The Effect of Particle Size Distribution on Light Transmittance Measurement. American Industrial Hygiene Association Journal, 32(5): 287-292, 1971.

236. U.S. Environmental Protection Agency. Appendix B, Performance Specification 1 - Performance Specifications and Specification Test Procedures for Transmissometer Systems.

237. Nader, J. S., F. Jaye, and W. Conner. Performance Specifications for Stationary-Source Monitoring Systems for Gases and Visible Emissions. U.S. Environmental Protection Agency, Research Triangle Park, N.C., 1974. EPA-650/2-74-013.

238. Ensor, D. S. Plume Opacity Measurements. In: Proceedings of the Symposium on the Control of Fine-particulate Emissions from Industrial Sources, Particulate Technical Sub-Group of the U.S.-U.S.S.R. Working Group on Stationary Source Air Pollution Control Technology, San Francisco, California, 1974.

239. Ensor, D. S. and M. J. Pilat. Calculation of Smoke Plume Opacity from Particulate Air Pollutant Properties. Jounral of the Air Pollution Control Association, 21(8):496-501, 1971.

239a. Sparks, L. E. In-Stack Plume Opacity from the Electrostatic/Scrubber System at Harrington Unit 1, May 1979. EPA 600/7-79-118.

240. Schutz, A. Technical Dust Control Principles and Practice. Staub-Reinhalt, Luft, 26(10):1-8, 1966.

241. Sem, G. J., et al. State of the Art, 1971 Instrumentation for Measurement of Particulate Emissions from Combustion Sources. Vol. II: Particulate Mass - Detail Report. Environmental Protection AGency, Research Triangle Park, North Carolina, 1971. EPA APTD-0734.

242. Schneider, W. A. Opacity Monitoring of Stack Emissions: A Design Tool with Promising Results. In: The 1974 Electric Utility-Generation Planbook, McGraw-Hill, New York, N. Y., 1974.

243. Duwel, L. Latest State of Development of Control Instruments for the Continuous Monitoring of Dust Emissions. Staub-Reinhalt, Luft, 28(3):42-53, 1968.

244. Bühne, W. K., and L. Duwel. Recording Dust Emission Measurements in the Cement Industry with the RM4 Smoke Density Meter made by Messrs. Sick. Staub-Reinhalt, Luft, 32(8): 19-26, 1972.

245. Larssen, S., D. S. Ensor, and M. J. Pilat. Relationship of Plume Opacity to the Properties of Particulates Emitted from Kraft Recovery Furnaces. Tappi, 55(1):88-92, 1972.

246. Reisman, E. R., W. B. Gerber, and N. D. Potter. In Stack Transmissometer Measurement of Particulate Opacity and Mass Concentration. EPA-650/2-74-120, U.S. Environmental Protection Agency, Research Triangle Park, N.C., 1975.

247. Nader, J. S. Source Monitoring. In: Air Pollution, 3rd Edition, Vol. III, Measuring, Monitoring, and Surveillance of Air Pollution, A. C. Stern, Ed. Academic Press, New York, N. Y., 1976.

248. Ensor, D. S. and L. D. Bevan. Application of Nephelometry to the Monitoring of Air Pollution Sources. Paper 73-AP-14, presented at the 1977 Annual Meeting of the Air Pollution Control Association, Pacific Northwest International Section, Seattle, Washington, 1973.

249. Ensor, D. S. Plume Opacity Measurements. In: Proceedings of the Symposium on Control of Fine-Particulate Emissions from Industrial Sources, Particulate Technical Sub-Group of the U.S.-U.S.S.R. Working Group on Stationary Source Air Pollution Control Technology, San Francisco, California, 1974.

250. Ensor, D. S., L. D. Bevan, and G. Markowski. Application of Nephelometry to the Monitoring of Air Pollution Sources. In: Proceedings of the Sixty-Seventh Annual Meeting, Air Pollution Control Association, Denver, Colorado, 1974.

251. Shofner, F., G. Kreikebaum, and H. Schmitt. *In situ* Continuous Measurement of Particulate Mass Concentration. Presented at the 68th Annual Meeting and Exhibition of the Air Pollution Control Association, Boston, Massachusetts, 1975.

252. Schmitt, H., R. Nuspliger, and G. Kreikebaum. Continuous *In Situ* Particulate Mass Concentration Measurement of Industrial Discharges. Presented at the 70th Annual Meeting of the Air Pollution Control Association, Toronto, Ontario, Canada, 1977.

253. Tipton, D. A Particle Analyzer for Stack Emissions. Powder Tech., 14:245-252, 1976.

254. Gooch, J. P., and G. H. Marchant. Reference 19, pp. 5-75 to 5-98.

255. Gooch, J. P., and G. H. Marchant. Reference 19, pp. 5-1 to 5-35.

256. Marchant, G. H., Jr. and J. P. Gooch. Performance and Economic Evaluation of a Hot-Side Electrostatic Precipitator. EPA-600/7-78-214, U.S. Environmental Protection Agency, Research Triangle Park, North Carolina (1978).

257. Gooch, J. P., and G. H. Marchant. Reference 19, pp. 5-98 to 5-141.

258. Gooch, J. P., and G. H. Marchant. Reference 19, pp. 5-165 to 5-212.

259. Breish, E. W. Method and Cost Analysis of Alternative Collectors for Low Sulfur Coal Fly Ash. Proceedings: Symposium on the Transfer and Utilization of Particulate Control Technology. Volume 1, Electrostatic Precipitators. EPA-600/7-79-044a, U.S. Environmental Protection Agency, Research Triangle Park, North Carolina, February 1979, pp. 121-130.

260. Bennett, R. P., and A. E. Kober. Chemical Enhancement of Electrostatic Precipitator Efficiency. Proceedings: Symposium on the Transfer and Utilization of Particulate Control Technology. Volume 1, Electrostatic Precipitators, EPA-600/7-79-044a, U.S. Environmental Protection Agency, Research Triangle Park, North Carolina, February 1979, pp. 113-120.

261. Potter, E. C., and C. A. J. Paulson. Improvement of Electrostatic Precipitator Performance by Carrier Gas Additives. Chem. Ind. (London) 1974:532-533, July 6, 1974.

262. Flue Gas Conditioning. Environmental Science and Technology, 12(13):1362-1365, December 1978.

263. Dismukes, E. Gas Conditioning for Electrostatic Precipitators. Paper presented at the Western Precipitator Symposium, April 1977.

264. Dismukes, E. B. Conditioning of Fly Ash with Sulfamic Acid, Ammonium Sulfate, and Ammonium Bisulfate. EPA-650/2-74-114, U.S. Environmental Protection Agency, 1974.

265. Verhoff, F. and J. Banchero. The Equilibrium Partial Pressures above Sulfuric Acid Solutions. AIChE J., 18(8): 1265-1268 (1972).

266. Bickelhaupt, R. E. Sodium Conditioning to Reduce Fly Ash Resistivity. EPA-650/2-74-092, Environmental Protection Agency, 1974.

267. Selle, S. J., and L. L. Hess. Factors Affecting ESP Performance on Western Coals and Experience with North Dakota Lignites. Symposium on Particulate Control in Energy Processes, San Francisco, May 11-13, 1976.

268. Telephone conversation with the Occupational Safety and Health Administration, Wash. D. C.

269. Engineering and Safety Service. Special Hazards Bulletin. American Insurance Association, New York, New York, August 1975.

270. Communication from Research-Cottrell.

271. Communication from Lodge-Cottrell.

272. Babcock & Wilcox. Steam/Its Generation and Use. 38th Edition, New York, New York, 1975.

273. Ross, R. D., Editor. Air Pollution and Industry. Van Nostrand Reinhold Company, New York, New York, 1972.

274. Communication from major electrostatic precipitator manufacturers.

275. Oglesby, S. and G. Nichols. A Manual of Electrostatic Precipitator Technology, Part I - Fundamentals. Prepared by Southern Research Institute under Contract CPA 22-69-73 for the National Air Pollution Control Administration, Cincinnati, Ohio, August 25, 1970.

276. Power, 119:56-58, August 1975.

277. Szabo, M. and R. Gerstle. Electrostatic Precipitator Malfunctions in the Electric Utility Industry. Prepared by PEDCo-Environmental Specialists, Inc., Cincinnati, Ohio, under Contract No. 68-02-2105 for the Industrial Environmental Research Laboratory, Research Triangle Park, North Carolina, January 1977. EPA-600/2-77-006 or NTIS No. PB 263 504.

278. Engelbrecht, H. Plant Engineer's Guide to Electrostatic Precipitator Inspection and Maintenance. Plant Engineering, pp. 193-196, April 29, 1976.

279. Bump, R. Electrostatic Precipitator Maintenance Survey. Journal of the Air Pollution Control Association, 26(11): 1061-1064, 1976.

280. A Review of Technology for Control of Fly Ash Emissions from Coal in Electric Power Generation. Prepared by Southern Research Institute for Argonne National Laboratory under Contract 31-109-38-3550, July 1, 1977.

281. Proceedings: Operation & Maintenance of Electrostatic Precipitators. Michigan Chapter - East Central Section Air Pollution Control Association, Dearborn, Michigan, April 10-12, 1978.

282. Scheider, G., T. Horzeila, J. Cooper, and P. Striegl. Selecting and Specifying Electrostatic Precipitators. Chemical Engineering, pp. 94-108, May 26, 1975.

283. Communication from vendor.

284. Gooch, J. P., and J. R. McDonald. Mathematical Modelling of Fine Particle Collection by Electrostatic Precipitation. Atmospheric Emissions and Energy-Source Pollution, AIChE Symposium Series, 73(165):146, 1977.

285. Gooch, J. P., and J. R. McDonald. Mathematical Modelling of Fine Particle Collection by Electrostatic Precipitation. Conference on Particulate Collection Problems in Converting to Low Sulfur Coals, Interagency Energy-Environment Research and Development Series. EPA-600/7-76-016, U.S. Environmental Protection Agency, 1976. 68 pp.

286. Leutert, G., and B. Böhlen. The Spatial Trend of Electric Field Strength and Space Charge Density in Plate-Type Electrostatic Precipitators. Staub, 32(7):27, 1972.

287. Gooch, J. P., J. R. McDonald, and S. Oglesby, Jr. Reference 153, pp. 12-19.

288. Smith, W. B., and J. R. McDonald. Development of a Theory for the Charging of Particles by Unipolar Ions. J. Aerosol Sci., 7:151-166, 1976.

289. Gooch, J. P., J. R. McDonald, and S. Oglesby, Jr. Reference 153, pp. 48-62.

290. McDonald, J. R. Reference 152, pp. 29-33.

291. McDonald, J. R., and D. H. Pontius. Electrostatic Precipitators. AIChE Conference on Theory, Practice and Process Principles for Physical Separations, Pacific Grove, California, November, 1977. (To be published in December, 1979).

292. Nichols, G. B., and J. P. Gooch. An Electrostatic Precipitator Performance Model. Final Report, Contract No. CPA 70-166, U.S. Environmental Protection Agency, Research Triangle Park, North Carolina, 1972. pp. 112-160.

293. Hedley, A. B., in: The Mechanism of Corrosion by Fuel Impurities (H. R. Johnson and D. L. Littler, editors), Butterworth, London, p. 204, 1963.

294. Cuffe, S. T., Gerstle, R. W., Orning, A. A., and Schwartz, C. H., J. Air Poll. Control Assoc., 14:353, 1964.

295. Snowden, P. N., and Ryan, M. H. Sulfuric Acid Condensation from Flue Gases Containing Sulfur Oxides. J. Inst. Fuel, 42:188, 1969.

296. Mueller, Peter. Study of the Influence of Sulfuric Acid on the Dew Point Temperature of the Flue Gas. Chemie - Ing. - Tech. 31:345, 1959.

297. Abel, Emil. The Vapor Phase Above the System Sulfuric Acid - Water. J. Phys. Chem. 50:260, 1946.

298. Gmitro, J. I., and Vermuelen, T. Vapor-Liquid Equilibria for Aqueous Sulfuric Acid. Univ. of California Radiation Laboratory Report 10866, Berkeley, California, June 24, 1963.

299. Greenewalt, C. H. Partial Pressure of Water Out of Aqueous Solutions of Sulfuric Acid. Ind. and Eng. Chem., 17:522-523, May 1925.

300. Johnstone, H. F. An Electrical Method for the Determination of the Dew Point of Flue Gases. Univ. of Illinois Eng. Exp. Station, Circular 20, 1929.

301. Flint, D. The Investigation of Dew Point and Related Condensation Phenomena in Flue Gases. J. Inst. Fuel, 21:248, 1948.

302. Burnside, W., W. G. Marshall, and J. M. Miller. The Influence of Superheater Metal Temperature on the Acid Dew Point of Flue Gases. J. Inst. Fuel, 29:261, 1956.

303. Corbett, P. F., and D. Flint. The Influence of Certain Smokes and Dusts on the SO_3 Content of the Flue Gases in Power Station Boilers. J. Inst. Fuel, 25:410, 1953.

304. Dooley, A., and G. Whittingham. The Oxidation of Sulfur Dioxide in Gas Flames. Trans. Faraday Soc., 42:354, 1946.

305. Whittingham, G. The Influence of Carbon Smokes on the Dew Point and Sulfur Trioxide Content of Flame Gases. J. Appl. Chem., 1:382, September 1951.

306. Flint, D., and R. W. Kear. The Corrosion of a Steel Surface by Condensed Films of Sulfuric Acid. J. Appl. Chem., 1:388, 1951.

307. Lee, G. K., F. D. Friedrich, and E. R. Mitchell. Effect of Fuel Characteristics and Excess Combustion Air on Sulfuric Acid Formation in a Pulverized-Coal-Fired Boiler. Department of Energy, Mines, and Resources, Mines Branch (Canada), 9p., 1967.

308. Friedrich, F. D., G. K. Lee, and E. R. Mitchell. Combustion and Fouling Characteristics of Two Canadian Lignites. Department of Energy, Mines, and Resources, Mines Branch (Canada), Research Report R208, 31p., August 1969.

309. Kear, R. W. The Influence of Carbon Smokes on the Corrosion of Metal Surfaces Exposed to Flue Gases. J. Appl. Chem., 1:393, September 1951.

310. Black, A. W., C. F. Stark, and W. H. Underwood. Dew Point Meter Measurements in Boiler Flue Gases. ASME Paper No. 60-WA-285, December 1960.

311. Clark, N. D., and G. D. Childs. Boiler Flue Gas Measurements Using a Dew Point Meter. Trans. ASME 87(A-1), p. 8, 1965.

312. Taylor, A. A. Relation Between Dew Point and the Concentration of Sulfuric Acid in Flue Gases. J. Inst. Fuel 16:25, 1942.

313. Lisle, E. S. and J. D. Sensenbaugh. The Determination of Sulfur Trioxide and Acid Dew Point in Flue Gases. Combustion, 36(1):12, 1965.

314. Taylor, H. D. The Condensation of Sulfuric Acid on Cooled Surfaces Exposed to Hot Gases Containing Sulfur Trioxide. Trans. Faraday Soc., 47:1114, 1951.

315. Piper, John D., and H. Van Vliet. The Effect of Temperature Variation on Composition, Fouling Tendency, and Corrosiveness of Combustion Gas from Pulverized-Fuel-Fired Steam Generators. Trans. ASME, 80:1251, August 1958.

316. Fontana, M. G. Corrosion: A Compilation, The Press of Hollenback, 1957.

317. Thurlow, G. G. An Air Cooled Metal Probe for the Investigation of the Corrosive Nature of Boiler Flue Gases. J. Inst. Fuel, 25:252-255 and 260, 1952.

318. The Boiler Availability Committee (London). Testing Techniques for Determining the Corrosive and Fouling Tendencies of Boiler Flue Gases. (Bulletin No. MC/316), p. 18, March 1961.

319. Southern Research Institute, Final Report on Contract CPA 70-149. A Study of Resistivity and Conditioning of Fly Ash, to Division of Control Systems, Office of Air Programs, Environmental Protection Agency.

320. Halstead, W. D. The Behavior of Sulfur and Chlorine Compounds in Pulverized-Coal-Fired Boilers. J. Inst. Fuel, 42:344, September 1969.

321. Kear, R. W. The Effect of Hydrochloric Acid on the Corrosive Nature of Combustion Gases Containing Sulfur Trioxide. J. Appl. Chem., 5:237, May 1955.

322. Canady, B. L. High Pressure Jetting of Regenerative Air Preheaters. Combustion, p. 55, February 1955.

323. Roddy, Charles P. Sulfur and Air Heater Corrosion. Power Engineering, p. 40, January 1968.

324. Barkley, J. F., et al. Corrosion and Deposits in Regenerative Air Heaters. U.S. Bureau of Mines Report of Investigations 4996, 23 pp., August 1953.

325. Brownell, Wayne E. Analysis of Fly Ash Deposits from Hoot Lake Station. Report to The Air Preheater Corp., Wellesville, New York, 12 pp., December 1961.

326. IGCI/ABMA Joint Technical Committee Survey. Criteria for the Application of Dust Collectors to Coal-Fired Boilers. April 1965.

326a. Dismukes. E. B. The Study of Resistivity and Conditioning of Fly Ash. U.S. Environmental Protection Agency, Research Triangle Park, North Carolina, February 1972. EPA-R2-72-087. NTIS PB-212 607.

327. Clark, Norman D. Higher Efficiency Through Lower Stack Temperature. The Air Preheater Corp., Wellesville, New York.

328. Kear, R. W. A Constant Temperature Corrosion Probe. J. Inst. Fuel, 32:267, 1959.

329. Alexander, P. A., R. S. Fielder, P. J. Jackson, and E. Raask. An Air-Cooled Probe for Measuring Acid Deposition in Boiler Flue Gases. J. Inst. Fuel, 33:31, 1960.

330. CERL (private communication).

Appendix A

Power Plant and Air Quality Data for Plants with ESP

TABLE 44. POWER PLANT AND AIR QUALITY DATA FOR THOSE PLANTS WITH ELECTROSTATIC PRECIPITATORS

	Company Name*	Plant Name	Boiler Number	Average Heat Content of Coal, Btu/lb	Average Sulfur Content, %	Average Ash Content, %
1	Alabama Power	Barry	5	11,995	2.34	11.67
2	Alabama Power	Gorgas	8	11,591	1.22	14.53
3	Alabama Power	Gorgas	9	11,591	1.22	14.53
4	Alabama Power	Gorgas	10	11,591	1.22	14.53
5	Alabama Power	Gadsden	1	11,905	1.99	13.14
6	Alabama Power	Gadsden	2	11,905	1.99	13.14
7	Allegheny Power (Monongahela)	Albright	3	11,757	2.09	15.84
8	Allegheny Power (Monongahela)	Fort Martin	1	12,100	2.41	13.30
9	Allegheny Power (Monongahela)	Fort Martin	2	12,100	2.41	13.30
10	Allegheny Power (Monongahela)	Harrison	1	12,246	4.05	15.31
11	Allegheny Power (Monongahela)	Harrison	2	12,246	4.05	15.31
12	Allegheny Power (Monongahela)	Harrison	3	12,246	4.05	15.31
13	Allegheny Power (Monongahela)	Willow Island	2	11,238	3.82	17.36
14	Allegheny Power (West Pa.)	Armstrong	1	11,327	2.67	17.49
15	Allegheny Power (West Pa.)	Armstrong	2	11,327	2.67	17.49
16	Allegheny Power (West Pa.)	Hatfield	1	12,007	2.46	15.39
17	Allegheny Power (West Pa.)	Hatfield	2	12,007	2.46	15.39
18	Allegheny Power (West Pa.)	Hatfield	3	12,007	2.46	15.39
19	Allegheny Power (West Pa.)	Mitchell	33	12,700	2.17	9.85
20	Allegheny Power (West Pa.)	Springdale	88	13,266	1.56	7.08
21	Appalachian Power	Cabin Creek	81,82	13,007	1.20	8.69
22	Appalachian Power	Cabin Creek	91,92	13,007	1.20	8.69
23	Appalachian Power	Clinch River	1	12,012	0.86	15.13
24	Appalachian Power	Clinch River	2	12,012	0.86	15.13
25	Appalachian Power	Clinch River	3	12,012	0.86	15.13
26	Arizona Public Service	Four Corners	4	8,924	0.63	21.76
27	Arizona Public Service	Four Corners	5	8,924	0.63	21.76
28	Big Rivers Electric	Kenneth Coleman	1	10,890	3.76	12.54
29	Big Rivers Electric	Kenneth Coleman	2	10,890	3.76	12.54
30	Big Rivers Electric	Kenneth Coleman	3	10,890	3.76	12.54
31	Big Rivers Electric	Robert Reid	1	10,344	3.72	15.40
32	Cardinal Operating Co.	Cardinal	1	11,338	2.97	16.01
33	Cardinal Operating Co.	Cardinal	2	11,338	2.97	16.01
34	Carolina Power & Light	Asheville	1	11,936	1.38	11.03
35	Carolina Power & Light	Asheville	2	11,936	1.38	11.03
36	Carolina Power & Light	Cape Fear	9	12,340	1.25	11.80
37	Carolina Power & Light	Cape Fear	10	12,340	1.25	11.80
38	Carolina Power & Light	H. B. Robinson	1	12,170	1.05	10.98
39	Carolina Power & Light	H. F. Lee	1	12,702	1.10	9.90
40	Carolina Power & Light	H. F. Lee	2	12,702	1.10	9.90
41	Carolina Power & Light	Louis Sutton	1	11,832	1.26	14.63
42	Carolina Power & Light	Louis Sutton	2	11,832	1.26	14.63
43	Carolina Power & Light	Louis Sutton	3	11,832	1.26	14.63
44	Carolina Power & Light	Roxboro	1	12,488	1.10	9.90
45	Carolina Power & Light	Roxboro	2	12,488	1.10	9.90
46	Carolina Power & Light	Roxboro	3A	12,488	1.10	9.90
47	Carolina Power & Light	Roxboro	3B	12,488	1.10	9.90
48	Carolina Power & Light	W. H. Weatherspoon	1	12,668	1.13	9.15
49	Carolina Power & Light	W. H. Weatherspoon	2	12,668	1.13	9.15
50	Carolina Power & Light	W. H. Weatherspoon	3	12,668	1.13	9.15
51	Cedar Falls Utilities	Streeter	7	12,085	2.73	6.49
52	Central Illinois Light	E. D. Edwards	1	10,376	2.83	10.30
53	Central Illinois Light	E. D. Edwards	2	10,376	2.83	10.30
54	Central Illinois Light	E. D. Edwards	3	10,376	2.83	10.30
55	Central Illinois Light	R. S. Wallace	7	10,338	2.59	9.17
56	Central Illinois Light	R. S. Wallace	8	10,338	2.59	9.17
57	Central Illinois Light	R. S. Wallace	9	10,338	2.59	9.17
58	Central Illinois Light	R. S. Wallace	10	10,338	2.59	9.17
59	Central Illinois Pub. Service	Coffeen	1	9,367	4.43	20.33
60	Central Illinois Pub. Service	Coffeen	2	9,367	4.43	20.33
61	Central Illinois Pub. Service	Grand Tower	7	11,252	3.33	11.88
62	Central Illinois Pub. Service	Grand Tower	8	11,252	3.33	11.88
63	Central Illinois Pub. Service	Grand Tower	9	11,252	3.33	11.88
64	Central Illinois Pub. Service	Meredosia	1	10,826	3.50	9.34
65	Central Illinois Pub. Service	Meredosia	2	10,826	3.50	9.34
66	Central Illinois Pub. Service	Meredosia	3	10,826	3.50	9.34
67	Central Illinois Pub. Service	Meredosia	4	10,826	3.50	9.34
68	Central Illinois Pub. Service	Meredosia	5	10,826	3.50	9.34
69	Central Operating	Philip Sporn	5	11,453	1.26	15.10
70	Charleston Bottoms REC	H. L. Spurlock	1	---	--	---

*The numbers in the first column correspond to the same plant names in Tables 45 and 46 as they do in Table 44.

(Continued)

Appendix A—Power Plant and Air Quality Data 427

TABLE 44. (Continued)

	Company Name*	Plant Name	Boiler Number	Average Heat Content of Coal, Btu/lb	Average Sulfur Content, %	Average Ash Content, %
71	Cincinnati Gas & Electric	Miami Fort	6-1	10,918	3.21	14.54
72	Cincinnati Gas & Electric	W. C. Beckjord	1	10,561	2.63	18.37
73	Cincinnati Gas & Electric	W. C. Beckjord	2	10,561	2.63	18.37
74	Cincinnati Gas & Electric	W. C. Beckjord	3	10,561	2.63	18.37
75	Cincinnati Gas & Electric	W. C. Beckjord	4	10,561	2.63	18.37
76	Cincinnati Gas & Electric	W. C. Beckjord	5	10,561	2.63	18.37
77	Cincinnati Gas & Electric	W. C. Beckjord	6	10,561	2.63	18.37
78	City of Colorado Springs DPU	Martin Drake	5	---	--	---
79	City of Colorado Springs DPU	Martin Drake	6	---	--	---
80	City of Colorado Springs DPU	Martin Drake	7	---	--	---
81	City of Peru	Peru	2	11,501	2.87	9.96
82	City of Springfield Lt. & Pr.	Lakeside	5	10,578	3.91	12.39
83	City of Springfield Lt. & Pr.	Lakeside	6	10,578	3.91	12.39
84	City of Springfield Lt. & Pr.	Lakeside	7	10,578	3.91	12.39
85	City of Springfield Lt. & Pr.	Lakeside	8	10,578	3.91	12.39
86	City of Springfield Lt. & Pr.	V. Y. Dallman	31	10,791	3.83	11.59
87	City of Springfield Lt. & Pr.	V. Y. Dallman	32	10,791	3.83	11.59
88	City Util. of Springfield, Mo.	James River	5	11,688	3.74	17.97
89	Cleveland Electric Illumtg.	Ashtabula	7	11,589	3.20	14.31
90	Cleveland Electric Illumtg.	Ashtabula	8	11,589	3.20	14.31
91	Cleveland Electric Illumtg.	Ashtabula	9	11,589	3.20	14.31
92	Cleveland Electric Illumtg.	Ashtabula	10	11,589	3.20	14.31
93	Cleveland Electric Illumtg.	Ashtabula	11	11,589	3.20	14.31
94	Cleveland Electric Illumtg.	Avon Lake	9	11,684	2.96	12.02
95	Cleveland Electric Illumtg.	Avon Lake	10	11,684	2.96	12.02
96	Cleveland Electric Illumtg.	Avon Lake	11	11,684	2.96	12.02
97	Cleveland Electric Illumtg.	Avon Lake	12	11,684	2.96	12.02
98	Cleveland Electric Illumtg.	East Lake	5	11,845	3.50	11.20
99	Cleveland Electric Illumtg.	Lake Shore	91	12,059	3.32	11.82
100	Cleveland Electric Illumtg.	Lake Shore	92	12,059	3.32	11.82
101	Cleveland Electric Illumtg.	Lake Shore	93	12,059	3.32	11.82
102	Cleveland Electric Illumtg.	Lake Shore	94	12,059	3.32	11.82
103	Cleveland Electric Illumtg.	Lake Shore	18	12,059	3.32	11.82
104	Columbus & Southern Ohio Elec.	Conesville	4	10,455	4.91	18.35
105	Commonwealth Edison	Crawford	7	9,239	0.42	4.98
106	Commonwealth Edison	Crawford	8	9,239	0.42	4.98
107	Commonwealth Edison	Dixon	4	10,539	2.89	10.94
108	Commonwealth Edison	Dixon	5	10,539	2.89	10.94
109	Commonwealth Edison	Fisk	18-1	9,261	0.40	4.61
110	Commonwealth Edison	Fisk	18-2	9,261	0.40	4.61
111	Commonwealth Edison	Fisk	19	9,261	0.40	4.61
112	Commonwealth Edison	Joliet	3	10,033	2.89	13.39
113	Commonwealth Edison	Joliet	4	10,033	2.89	13.39
114	Commonwealth Edison	Joliet	5	10,033	2.89	13.39
115	Commonwealth Edison	Joliet	71	10,033	2.89	13.39
116	Commonwealth Edison	Joliet	72	10,033	2.89	13.39
117	Commonwealth Edison	Joliet	81	10,033	2.89	13.39
118	Commonwealth Edison	Joliet	82	10,033	2.89	13.39
119	Commonwealth Edison	Kincaid	1	9,718	3.99	15.16
120	Commonwealth Edison	Kincaid	2	9,718	3.99	15.16
121	Commonwealth Edison	Powerton	51	10,699	3.63	8.44
122	Commonwealth Edison	Powerton	52	10,699	3.63	8.44
123	Commonwealth Edison	Sabrooke	4	10,722	0.92	15.90
124	Commonwealth Edison	Waukegan	14	10,045	1.21	9.40
125	Commonwealth Edison	Waukegan	15	10,045	1.21	9.40
126	Commonwealth Edison	Waukegan	16	10,045	1.21	9.40
127	Commonwealth Edison	Waukegan	17	10,045	1.21	9.40
128	Commonwealth Edison	Waukegan	7	10,045	1.21	9.40
129	Commonwealth Edison	Waukegan	8	10,045	1.21	9.40
130	Commonwealth Edison	Will County	1	9,377	1.58	8.35
131	Commonwealth Edison	Will County	2	9,377	1.58	8.35
132	Commonwealth Edison	Will County	3	9,377	1.58	8.35
133	Commonwealth Edison	Will County	4	9,377	1.58	8.35
134	Commonwealth Edison/Indiana	State Line	1-1	9,730	1.53	11.00
135	Commonwealth Edison/Indiana	State Line	2-1	9,730	1.53	11.00
136	Commonwealth Edison/Indiana	State Line	3-1	9,730	1.53	11.00
137	Commonwealth Edison/Indiana	State Line	4-1	9,730	1.53	11.00
138	Commonwealth Edison/Indiana	State Line	5-1	9,730	1.53	11.00
139	Commonwealth Edison/Indiana	State Line	6-1	9,730	1.53	11.00
140	Commonwealth Edison/Indiana	State Line	1-2	9,730	1.53	11.00
141	Commonwealth Edison/Indiana	State Line	2-2	9,730	1.53	11.00
142	Commonwealth Edison/Indiana	State Line	3-2	9,730	1.53	11.00
143	Commonwealth Edison/Indiana	State Line	1-3	9,730	1.53	11.00
144	Commonwealth Edison/Indiana	State Line	1-4	9,730	1.53	11.00
145	Consolidated Edison/New York	Astoria	10	---	--	---
146	Consolidated Edison/New York	Astoria	20	---	--	---
147	Consolidated Edison/New York	Astoria	30	---	--	---
148	Consolidated Edison/New York	Astoria	40	---	--	---
149	Consolidated Edison/New York	Astoria	50	---	--	---
150	Consolidated Edison/New York	Ravenswood	30	---	--	---
151	Consumers Power	B. C. Cobb	1	11,462	3.27	11.34
152	Consumers Power	B. C. Cobb	2	11,462	3.27	11.34
153	Consumers Power	B. C. Cobb	3	11,462	3.27	11.34
154	Consumers Power	B. C. Cobb	4	11,462	3.27	11.34
155	Consumers Power	B. C. Cobb	5	11,462	3.27	11.34
156	Consumers Power	D. E. Karn	1	11,138	3.21	14.55
157	Consumers Power	D. E. Karn	2	11,138	3.21	14.55
158	Consumers Power	J. C. Weadock	7	11,240	2.73	13.27
159	Consumers Power	J. C. Weadock	8	11,240	2.73	13.27
160	Consumers Power	J. H. Campbell	1	11,187	3.61	16.12
161	Consumers Power	J. H. Campbell	2	11,187	3.61	16.12
162	Dairyland Power Cooperative	Alma	1	11,666	2.97	17.89
163	Dairyland Power Cooperative	Alma	2	11,666	2.97	17.89
164	Dairyland Power Cooperative	Alma	3	11,666	2.97	17.89
165	Dairyland Power Cooperative	Alma	4	11,666	2.97	17.89
166	Dairyland Power Cooperative	Alma	5	11,666	2.97	17.89
167	Dairyland Power Cooperative	Genoa #3	1	10,600	4.10	24.59
168	Dairyland Power Cooperative	Stoneman	1	11,658	3.60	18.79
169	Dairyland Power Cooperative	Stoneman	2	11,658	3.60	18.79
170	Dallas Power & Light	Big Brown	1	7,000	0.60	10.40
171	Dallas Power & Light	Big Brown	2	7,000	0.60	10.40
172	Dallas Power & Light	Monticello	1	---	--	---

(Continued)

TABLE 44. (Continued)

	Company Name*	Plant Name	Boiler Number	Average Heat Content of Coal, Btu/lb	Average Sulfur Content, %	Average Ash Content, %
173	Dayton Power & Light	Frank M. Tait	4	11,465	0.97	13.67
174	Dayton Power & Light	Frank M. Tait	5	11,465	0.97	13.67
175	Dayton Power & Light	Frank M. Tait	7-1	11,465	0.97	13.67
176	Dayton Power & Light	Frank M. Tait	7-2	11,465	0.97	13.67
177	Dayton Power & Light	Frank M. Tait	8-1	11,465	0.97	13.67
178	Dayton Power & Light	Frank M. Tait	8-2	11,465	0.97	13.67
179	Dayton Power & Light	J. M. Stuart	1	11,053	1.68	15.88
180	Dayton Power & Light	J. M. Stuart	2	11,053	1.68	15.88
181	Dayton Power & Light	J. M. Stuart	3	11,053	1.68	15.88
182	Dayton Power & Light	J. M. Stuart	4	11,053	1.68	15.88
183	Dayton Power & Light	O. M. Hutchings	1	12,186	0.86	10.71
184	Dayton Power & Light	O. M. Hutchings	2	12,186	0.86	10.71
185	Dayton Power & Light	O. M. Hutchings	3	12,186	0.86	10.71
186	Dayton Power & Light	O. M. Hutchings	4	12,186	0.86	10.71
187	Dayton Power & Light	O. M. Hutchings	5	12,186	0.86	10.71
188	Dayton Power & Light	O. M. Hutchings	6	12,186	0.86	10.71
189	Delmarva Power & Light	Delaware City	4	14,170	6.70	0.30
190	Delmarva Power & Light	Indian River	1	12,130	1.63	11.76
191	Delmarva Power & Light	Indian River	2	12,130	1.63	11.76
192	Delmarva Power & Light	Indian River	3	12,130	1.63	11.76
193	Detroit Edison	Conners Creek	15	11,645	1.81	13.75
194	Detroit Edison	Conners Creek	16	11,645	1.81	13.75
195	Detroit Edison	Conners Creek	17	11,645	1.81	13.75
196	Detroit Edison	Conners Creek	18	11,645	1.81	13.75
197	Detroit Edison	Harbor Beach	1	11,500	3.03	12.38
198	Detroit Edison	Marysville	9	11,698	2.87	13.46
199	Detroit Edison	Marysville	10	11,698	2.87	13.46
200	Detroit Edison	Marysville	11	11,698	2.87	13.46
201	Detroit Edison	Marysville	12	11,698	2.87	13.46
202	Detroit Edison	Monroe	1	12,475	2.77	12.10
203	Detroit Edison	Monroe	2	12,475	2.77	12.10
204	Detroit Edison	Monroe	3	12,475	2.77	12.10
205	Detroit Edison	Monroe	4	12,475	2.77	12.10
206	Detroit Edison	Pennsalt	23	11,635	1.44	13.38
207	Detroit Edison	Pennsalt	24	11,635	1.44	13.38
208	Detroit Edison	River Rouge	2	11,999	3.37	11.75
209	Detroit Edison	St. Clair	1	11,790	3.01	13.48
210	Detroit Edison	St. Clair	2	11,790	3.01	13.48
211	Detroit Edison	St. Clair	3	11,790	3.01	13.48
212	Detroit Edison	St. Clair	4	11,790	3.01	13.48
213	Detroit Edison	St. Clair	6	11,790	3.01	13.48
214	Detroit Edison	St. Clair	7	11,790	3.01	13.48
215	Detroit Edison	Wyandotte	9	11,777	1.13	12.34
216	Detroit Edison	Wyandotte	10	11,777	1.13	12.34
217	Detroit Edison	Wyandotte	11	11,777	1.13	12.34
218	Detroit Edison	Wyandotte	12	11,777	1.13	12.34
219	Duke Power	Allen	1	11,965	0.89	12.53
220	Duke Power	Allen	2	11,965	0.89	12.53
221	Duke Power	Allen	3	11,965	0.89	12.53
222	Duke Power	Allen	4	11,965	0.89	12.53
223	Duke Power	Allen	5	11,965	0.89	12.53
224	Duke Power	Belews Creek	1	---	--	---
225	Duke Power	Buck	5	12,125	0.88	11.54
226	Duke Power	Buck	6	12,125	0.88	11.54
227	Duke Power	Buck	7	12,125	0.88	11.54
228	Duke Power	Buck	8	12,125	0.88	11.54
229	Duke Power	Buck	9	12,125	0.88	11.54
230	Duke Power	Cliffside	1	12,368	1.30	13.57
231	Duke Power	Cliffside	2	12,368	1.30	13.57
232	Duke Power	Cliffside	3	12,368	1.30	13.57
233	Duke Power	Cliffside	4	12,368	1.30	13.57
234	Duke Power	Dan River	1	11,963	0.92	12.69
235	Duke Power	Dan River	2	11,963	0.92	12.69
236	Duke Power	Dan River	3	11,963	0.92	12.69
237	Duke Power	Lee	1	11,545	1.17	14.21
238	Duke Power	Lee	2	11,545	1.17	14.21
239	Duke Power	Lee	3	11,545	1.17	14.21
240	Duke Power	Marshall	1	11,737	0.96	13.55
241	Duke Power	Marshall	2	11,737	0.96	13.55
242	Duke Power	Marshall	3	11,737	0.96	13.55
243	Duke Power	Marshall	4	11,737	0.96	13.55
244	Duke Power	Riverbend	7	11,834	0.89	13.64
245	Duke Power	Riverbend	8	11,834	0.89	13.64
246	Duke Power	Riverbend	9	11,834	0.89	13.64
247	Duke Power	Riverbend	10	11,834	0.89	13.64
248	Duquesne Light Co.	Cheswick	1	11,038	2.16	20.33
249	Duquesne Light Co.	Elrama	1	10,996	2.13	20.07
250	Duquesne Light Co.	Elrama	2	10,996	2.13	20.07
251	Duquesne Light Co.	Elrama	3	10,996	2.13	20.07
252	Duquesne Light Co.	Elrama	4	10,996	2.13	20.07
253	Duquesne Light Co.	Phillips	1	11,342	1.89	16.74
254	Duquesne Light Co.	Phillips	2	11,342	1.89	16.74
255	Duquesne Light Co.	Phillips	3	11,342	1.89	16.74
256	Duquesne Light Co.	Phillips	4	11,342	1.89	16.74
257	Duquesne Light Co.	Phillips	5	11,342	1.89	16.74
258	Duquesne Light Co.	Phillips	6	11,342	1.89	16.74
259	East Kentucky Power Coop.	John S. Cooper	1	11,435	2.35	15.32
260	East Kentucky Power Coop.	John S. Cooper	2	11,435	2.35	15.32
261	East Kentucky Power Coop.	William Dale	3	11,380	1.62	14.00
262	East Kentucky Power Coop.	William Dale	4	11,380	1.62	14.00
263	Electric Energy, Inc.	Joppa	1-2	11,439	2.38	10.21
264	Electric Energy, Inc.	Joppa	3-4	11,439	2.38	10.21
265	Electric Energy, Inc.	Joppa	5-6	11,439	2.38	10.21
266	Empire District Electric	Asbury	1	10,238	4.43	24.13
267	Georgia Power	Arkwright	1	11,904	2.00	12.77
268	Georgia Power	Arkwright	2	11,904	2.00	12.77
269	Georgia Power	Arkwright	3	11,904	2.00	12.77
270	Georgia Power	Arkwright	4	11,904	2.00	12.77
271	Georgia Power	Hammond	1	11,329	3.25	9.49
272	Georgia Power	Hammond	2	11,329	3.25	9.49
273	Georgia Power	Hammond	3	11,329	3.25	9.49

(Continued)

TABLE 44. (Continued)

	Company Name*	Plant Name	Boiler Number	Average Heat Content of Coal, Btu/lb	Average Sulfur Content, %	Average Ash Content, %
274	Georgia Power	Hammond	4	11,329	3.25	9.49
275	Georgia Power	H. L. Bowen	1	11,444	3.13	10.73
276	Georgia Power	H. L. Bowen	2	11,444	3.13	10.73
277	Georgia Power	H. L. Bowen	3	11,444	3.13	10.73
278	Georgia Power	Jack McDonough	1	11,887	1.05	12.99
279	Georgia Power	Jack McDonough	2	11,887	1.05	12.99
280	Georgia Power	Plant Harllee	1	12,156	0.94	10.53
281	Georgia Power	Plant Harllee	2	12,156	0.94	10.53
282	Georgia Power	Plant Harllee	3	12,156	0.94	10.53
283	Georgia Power	Plant Harllee	4	12,156	0.94	10.53
284	Georgia Power	Mitchell	1	11,519	1.42	15.01
285	Georgia Power	Mitchell	2	11,519	1.42	15.01
286	Georgia Power	Mitchell	3	11,519	1.42	15.01
287	Georgia Power	Yates	1	12,284	2.22	9.25
288	Georgia Power	Yates	2	12,284	2.22	9.25
289	Georgia Power	Yates	3	12,284	2.22	9.25
290	Georgia Power	Yates	4	12,284	2.22	9.25
291	Georgia Power	Yates	5	12,284	2.22	9.25
292	Georgia Power	Yates	6	12,284	2.22	9.25
293	Georgia Power	Yates	7	12,284	2.22	9.25
294	Gulf Power	Lansing Smith	1	11,510	2.84	11.18
295	Gulf Power	Lansing Smith	2	11,510	2.84	11.18
296	Gulf Power	Crist	4	11,883	3.11	10.92
297	Gulf Power	Crist	5	11,883	3.11	10.92
298	Gulf Power	Crist	6	11,883	3.11	10.92
299	Gulf Power	Crist	7	11,883	3.11	10.92
300	Gulf Power	Scholz	1	12,455	1.41	12.55
301	Gulf Power	Scholz	2	12,455	1.41	12.55
302	Hartford Electric	Middletown	1	11,746	2.25	15.00
303	Hartford Electric	Middletown	2	11,746	2.25	15.00
304	Henderson Municipal	Station 2	1	10,347	3.80	15.48
305	Henderson Municipal	Station 2	2	10,347	3.80	15.48
306	Holland Board of Public Works	James De Young	5	12,404	3.22	7.97
307	Illinois Power Company	Baldwin	1	10,285	3.27	12.79
308	Illinois Power Company	Baldwin	2	10,285	3.27	12.79
309	Illinois Power Company	Hennepin	1	10,890	3.00	10.00
310	Illinois Power Company	Hennepin	2	10,890	3.00	10.00
311	Illinois Power Company	Vermilion	1	10,858	2.90	11.33
312	Illinois Power Company	Vermilion	2	10,858	2.90	11.33
313	Illinois Power Company	Wood River	4	10,991	2.97	10.30
314	Illinois Power Company	Wood River	5	10,991	2.97	10.30
315	Indiana-Kentucky Elec. Corp.	Clifty Creek	1	10,852	3.64	11.69
316	Indiana-Kentucky Elec. Corp.	Clifty Creek	2	10,852	3.64	11.69
317	Indiana-Kentucky Elec. Corp.	Clifty Creek	3	10,852	3.64	11.69
318	Indiana-Kentucky Elec. Corp.	Clifty Creek	4	10,852	3.64	11.69
319	Indiana-Kentucky Elec. Corp.	Clifty Creek	5	10,852	3.64	11.69
320	Indiana-Kentucky Elec. Corp.	Clifty Creek	6	10,852	3.64	11.69
321	Indiana & Michigan Elec. Co.	Tanners Creek	4	10,995	3.43	13.03
322	Indianapolis Power & Light Co.	C. C. Perry K	11	11,299	2.29	9.35
323	Indianapolis Power & Light Co.	C. C. Perry K	12	11,299	2.29	9.35
324	Indianapolis Power & Light Co.	C. C. Perry K	13	11,299	2.29	9.35
325	Indianapolis Power & Light Co.	C. C. Perry K	14	11,299	2.29	9.35
326	Indianapolis Power & Light Co.	C. C. Perry K	15	11,299	2.29	9.35
327	Indianapolis Power & Light Co.	C. C. Perry K	16	11,299	2.29	9.35
328	Indianapolis Power & Light Co.	E. W. Stout	50	11,076	2.64	9.30
329	Indianapolis Power & Light Co.	E. W. Stout	60	11,076	2.64	9.30
330	Indianapolis Power & Light Co.	H. T. Pritchard	3	11,112	2.39	9.70
331	Indianapolis Power & Light Co.	H. T. Pritchard	4	11,112	2.39	9.70
332	Indianapolis Power & Light Co.	H. T. Pritchard	5	11,112	2.39	9.70
333	Indianapolis Power & Light Co.	H. T. Pritchard	6	11,112	2.39	9.70
334	Indianapolis Power & Light Co.	Petersburg	1	10,954	2.98	9.77
335	Indianapolis Power & Light Co.	Petersburg	2	10,954	2.98	9.77
336	Interstate Power Company	Dubuque	1	11,169	2.86	13.19
337	Interstate Power Company	M. L. Kapp	2	11,211	2.92	10.85
338	Iowa Electric Light & Power	Prairie Creek Station 1-2-3	3	10,941	2.48	9.10
339	Iowa Electric Light & Power	Sixth Creek Station	3-4	10,285	2.34	8.04
340	Iowa Electric Light & Power	Sixth Creek Station	5-6	10,285	2.34	8.04
341	Iowa Electric Light & Power	Sixth Creek Station	7-8	10,285	2.34	8.04
342	Iowa Electric Light & Power	Sixth Creek Station	9-10	10,285	2.34	8.04
343	Iowa-Illinois Gas & Electric	Riverside	5	10,805	2.48	8.68
344	Iowa-Illinois Gas & Electric	Riverside	6	10,805	2.48	8.68
345	Iowa-Illinois Gas & Electric	Riverside	7	10,805	2.48	8.68
346	Iowa-Illinois Gas & Electric	Riverside	8	10,805	2.48	8.68
347	Iowa-Illinois Gas & Electric	Riverside	9	10,805	2.48	8.68
348	Iowa Power & Light Company	Council Bluffs	1	10,143	1.09	8.96
349	Iowa Power & Light Company	Council Bluffs	2	10,143	1.09	8.96
350	Iowa Power & Light Company	Des Moines	10	9,549	2.94	13.65
351	Iowa Power & Light Company	Des Moines	11	9,549	2.94	13.65
352	Iowa Public Service Company	Maynard	14	10,960	2.86	10.21
353	Iowa Public Service Company	Neal	1	9,981	0.60	11.22
354	Iowa Public Service Company	Neal	2	9,981	0.60	11.22
355	Iowa Southern Utilities	Burlington	1	10,183	2.58	13.70
356	Kansas City Bd. of Pub. Util.	Kaw	3	11,784	3.90	13.75
357	Kansas City Bd. of Pub. Util.	Quindaro No. 3	1	11,492	1.61	11.14
358	Kansas City Bd. of Pub. Util.	Quindaro No. 3	2	11,492	1.61	11.14
359	Kansas City Power & Light	Grand Avenue	7	12,336	3.71	11.08
360	Kansas City Power & Light	Hawthorn	5	10,566	1.40	9.53
361	Kansas City Power & Light	Montrose	1	9,413	5.51	23.19
362	Kansas City Power & Light	Montrose	2	9,413	5.51	23.19
363	Kansas City Power & Light	Montrose	3	9,413	5.51	23.19
364	Kentucky Power Company	Big Sandy	1	11,835	0.97	12.49
365	Kentucky Power Company	Big Sandy	2	11,835	0.97	12.49
366	Kentucky Utilities Company	E. W. Brown	1	11,804	1.72	13.25
367	Kentucky Utilities Company	E. W. Brown	3	11,804	1.72	13.25
368	Kentucky Utilities Company	Ghent	1	10,917	2.76	10.19
369	Kentucky Utilities Company	Green River	4	11,364	2.58	10.15
370	Kentucky Utilities Company	Tyrone	5	11,570	0.90	12.39
371	Lansing Bd. of Water & Light	Eckert	1	12,319	2.98	10.74
372	Lansing Bd. of Water & Light	Eckert	2	12,319	2.98	10.74
373	Lansing Bd. of Water & Light	Eckert	3	12,319	2.98	10.74
374	Lansing Bd. of Water & Light	Eckert	4	12,319	2.98	10.74
375	Lansing Bd. of Water & Light	Eckert	5	12,319	2.98	10.74

(Continued)

TABLE 44. (Continued)

	Company Name*	Plant Name	Boiler Number	Average Heat Content of Coal, Btu/lb	Average Sulfur Content, %	Average Ash Content, %
376	Lansing Bd. of Water & Light	Eckert	6	12,319	2.98	10.74
377	Lansing Bd. of Water & Light	Erickson	1	12,270	2.92	11.61
378	Lansing Bd. of Water & Light	Ottawa	1	12,437	2.74	7.96
379	Lansing Bd. of Water & Light	Ottawa	2	12,437	2.74	7.96
380	Lansing Bd. of Water & Light	Ottawa	3	12,437	2.74	7.96
381	Lansing Bd. of Water & Light	Ottawa	4	12,437	2.74	7.96
382	Lansing Bd. of Water & Light	Ottawa	5	12,437	2.74	7.96
383	Louisville Gas & Elec. Co.	Cane Run	1	11,075	3.76	14.02
384	Louisville Gas & Elec. Co.	Cane Run	2	11,075	3.76	14.02
385	Louisville Gas & Elec. Co.	Cane Run	3	11,075	3.76	14.02
386	Louisville Gas & Elec. Co.	Cane Run	4	11,075	3.76	14.02
387	Louisville Gas & Elec. Co.	Cane Run	5	11,075	3.76	14.02
388	Louisville Gas & Elec. Co.	Cane Run	6	11,075	3.76	14.02
389	Louisville Gas & Elec. Co.	Mill Creek	1	11,152	3.80	13.76
390	Louisville Gas & Elec. Co.	Mill Creek	2	11,152	3.80	13.76
391	Louisville Gas & Elec. Co.	Paddy's Run	1	11,368	3.42	12.57
392	Louisville Gas & Elec. Co.	Paddy's Run	2	11,368	3.42	12.57
393	Louisville Gas & Elec. Co.	Paddy's Run	3	11,368	3.42	12.57
394	Louisville Gas & Elec. Co.	Paddy's Run	4	11,368	3.42	12.57
395	Louisville Gas & Elec. Co.	Paddy's Run	5	11,368	3.42	12.57
396	Louisville Gas & Elec. Co.	Paddy's Run	6	11,368	3.42	12.57
397	Madison Gas & Elec. Co.	Blount Street	9	11,535	3.06	8.79
398	Metropolitan Edison Co.	Crawford	7	12,660	1.21	10.70
399	Metropolitan Edison Co.	Crawford	8	12,660	1.21	10.70
400	Metropolitan Edison Co.	Portland	1	12,473	1.53	11.35
401	Metropolitan Edison Co.	Portland	2	12,473	1.53	11.35
402	Metropolitan Edison Co.	Titus	1	12,224	0.96	11.97
403	Metropolitan Edison Co.	Titus	2	12,224	0.96	11.97
404	Metropolitan Edison Co.	Titus	3	12,224	0.96	11.97
405	Michigan State University	Power Plant '65	1	12,639	0.98	9.70
406	Michigan State University	Power Plant '65	2	12,639	0.98	9.70
407	Michigan State University	Power Plant '65	3	12,639	0.98	9.70
408	Mississippi Power Company	Jack Watson	4	11,885	2.70	10.88
409	Mississippi Power Company	Jack Watson	5	11,885	2.70	10.88
410	Montana Power Company	J. E. Corette	1	8,582	0.67	8.22
411	Municipal Power & Light	Station One	6	---	--	---
412	Muscatine Power & Light	Muscatine Municipal	8	10,712	3.09	10.22
413	N. Y. State Elec. & Gas	Goudey	11	11,341	2.20	18.45
414	N. Y. State Elec. & Gas	Goudey	12	11,341	2.20	18.45
415	N. Y. State Elec. & Gas	Goudey	13	11,341	2.20	18.45
416	N. Y. State Elec. & Gas	Greenidge	4	11,638	1.98	15.20
417	N. Y. State Elec. & Gas	Greenidge	5	11,638	1.98	15.20
418	N. Y. State Elec. & Gas	Greenidge	6	11,638	1.98	15.20
419	N. Y. State Elec. & Gas	Hickling	1	10,917	1.98	15.20
420	N. Y. State Elec. & Gas	Hickling	2	10,917	1.98	15.20
421	N. Y. State Elec. & Gas	Hickling	3	10,917	1.98	15.20
422	N. Y. State Elec. & Gas	Hickling	4	10,917	1.98	15.20
423	N. Y. State Elec. & Gas	Milliken	1	11,317	2.08	16.37
424	N. Y. State Elec. & Gas	Milliken	2	11,317	2.08	16.37
425	No. Indiana Pub. Service Co.	Bailly	7	11,109	3.62	10.00
426	No. Indiana Pub. Service Co.	Bailly	8	11,109	3.62	10.00
427	No. Indiana Pub. Service Co.	Dean H. Mitchell	4	11,146	3.18	9.32
428	No. Indiana Pub. Service Co.	Dean H. Mitchell	5	11,146	3.18	9.32
429	No. Indiana Pub. Service Co.	Dean H. Mitchell	6	11,146	3.18	9.32
430	No. Indiana Pub. Service Co.	Dean H. Mitchell	11	11,146	3.18	9.32
431	No. Indiana Pub. Service Co.	Michigan City	4	10,558	3.36	11.14
432	No. Indiana Pub. Service Co.	Michigan City	5	10,558	3.36	11.14
433	No. Indiana Pub. Service Co.	Michigan City	6	10,558	3.36	11.14
434	No. Indiana Pub. Service Co.	Michigan City	12	10,558	3.36	11.14
435	No. Indiana Pub. Service Co.	ADVANCE	1	12,341	2.38	8.95
436	No. Indiana Pub. Service Co.	ADVANCE	2	12,341	2.38	8.95
437	No. Indiana Pub. Service Co.	ADVANCE	3	12,341	2.38	8.95
438	Northern States Power Co.	A. S. King	1	10,567	3.32	15.17
439	Northern States Power Co.	Black Dog	1	10,108	2.27	11.73
440	Northern States Power Co.	Black Dog	2	10,108	2.27	11.73
441	Northern States Power Co.	Black Dog	3	10,108	2.27	11.73
442	Northern States Power Co.	Black Dog	4	10,108	2.27	11.73
443	Northern States Power Co.	High Bridge	9	9,666	1.82	9.71
444	Northern States Power Co.	High Bridge	10	9,666	1.82	9.71
445	Northern States Power Co.	High Bridge	11	9,666	1.82	9.71
446	Northern States Power Co.	High Bridge	12	9,666	1.82	9.71
447	Northern States Power Co.	Minnesota Valley	4	10,044	1.28	9.20
448	Ohio Edison Company	Edgewater	11	12,267	2.68	10.13
449	Ohio Edison Company	Edgewater	12	12,267	2.68	10.13
450	Ohio Edison Company	Edgewater	13	12,267	2.68	10.13
451	Ohio Edison Company	Gorge	25	10,792	3.22	15.25
452	Ohio Edison Company	Gorge	26	10,792	3.22	15.25
453	Ohio Edison Company	Norwalk	5	11,322	3.55	13.10
454	Ohio Edison Company	R. E. Burger	1	11,457	3.25	13.93
455	Ohio Edison Company	R. E. Burger	2	11,457	3.25	13.93
456	Ohio Edison Company	R. E. Burger	3	11,457	3.25	13.93
457	Ohio Edison Company	R. E. Burger	4	11,457	3.25	13.93
458	Ohio Edison Company	R. E. Burger	5	11,457	3.25	13.93
459	Ohio Edison Company	R. E. Burger	6	11,457	3.25	13.93
460	Ohio Edison Company	R. E. Burger	7	11,457	3.25	13.93
461	Ohio Edison Company	R. E. Burger	8	11,457	3.25	13.93
462	Ohio Edison Company	W. H. Sammis	1	11,367	2.99	15.79
463	Ohio Edison Company	W. H. Sammis	2	11,367	2.99	15.79
464	Ohio Edison Company	W. H. Sammis	3	11,367	2.99	15.79
465	Ohio Edison Company	W. H. Sammis	4	11,367	2.99	15.79
466	Ohio Edison Company	W. H. Sammis	5	11,367	2.99	15.79
467	Ohio Edison Company	W. H. Sammis	6	11,367	2.99	15.79
468	Ohio Edison Company	W. H. Sammis	7	11,367	2.99	15.79
469	Ohio Electric Company	Gavin	1	---	--	---
470	Ohio Power Company	Mitchell	1	11,601	3.35	15.20
471	Ohio Power Company	Mitchell	2	11,601	3.35	15.20
472	Ohio Power Company	Muskingum River	1	10,448	4.64	19.35
473	Ohio Power Company	Muskingum River	2	10,448	4.64	19.35
474	Ohio Power Company	Muskingum River	3	10,448	4.64	19.35
475	Ohio Power Company	Muskingum River	4	10,448	4.64	19.35
476	Ohio Power Company	Muskingum River	5	10,448	4.64	19.35
477	Ohio Valley Elec. Corp.	Kyger Creek	1	11,586	3.89	14.52

(Continued)

Appendix A—Power Plant and Air Quality Data

TABLE 44. (Continued)

	Company Name*	Plant Name	Boiler Number	Average Heat Content of Coal, Btu/lb	Average Sulfur Content, %	Average Ash Content, %
478	Ohio Valley Elec. Corp.	Kyger Creek	2	11,586	3.89	14.52
479	Ohio Valley Elec. Corp.	Kyger Creek	3	11,586	3.89	14.52
480	Ohio Valley Elec. Corp.	Kyger Creek	4	11,586	3.89	14.52
481	Ohio Valley Elec. Corp.	Kyger Creek	5	11,586	3.89	14.52
482	Omaha Public Power Dist.	North Omaha	1	10,953	1.48	9.12
483	Omaha Public Power Dist.	North Omaha	2	10,953	1.48	9.12
484	Omaha Public Power Dist.	North Omaha	3	10,953	1.48	9.12
485	Omaha Public Power Dist.	North Omaha	4	10,953	1.48	9.12
486	Omaha Public Power Dist.	North Omaha	5	10,953	1.48	9.12
487	Otter Tail Power Company	Hoot Lake	2	7,093	0.72	6.16
488	Otter Tail Power Company	Hoot Lake	3	7,093	0.72	6.16
489	Owensboro Munic. Utilities	Elmer Smith	1	10,993	3.11	10.48
490	Owensboro Munic. Utilities	Elmer Smith	2	10,993	3.11	10.48
491	Owensboro Munic. Utilities	Owensboro Plant 1	1	11,027	3.12	10.35
492	Owensboro Munic. Utilities	Owensboro Plant 1	2	11,027	3.12	10.35
493	Owensboro Munic. Utilities	Owensboro Plant 1	3	11,027	3.12	10.35
494	Owensboro Munic. Utilities	Owensboro Plant 1	4	11,027	3.12	10.35
495	Pacific Power & Light Co.	Centralia	1	7,552	0.49	14.88
496	Pacific Power & Light Co.	Centralia	2	7,552	0.49	14.88
497	Pacific Power & Light Co.	Jim Bridger	1	---	--	---
498	Pella Munic. Power & Light	Pella	6	9,410	6.43	17.24
499	Pella Munic. Power & Light	Pella	7	9,410	6.43	17.24
500	Pennsylvania Electric Co.	Homer City	1	11,766	2.40	19.30
501	Pennsylvania Electric Co.	Homer City	2	11,766	2.40	19.30
502	Pennsylvania Electric Co.	Conemaugh	1	11,437	2.29	18.68
503	Pennsylvania Electric Co.	Conemaugh	2	11,437	2.29	18.68
504	Pennsylvania Electric Co.	Front Street	7	12,101	2.12	13.17
505	Pennsylvania Electric Co.	Front Street	8	12,101	2.12	13.17
506	Pennsylvania Electric Co.	Front Street	9	12,101	2.12	13.17
507	Pennsylvania Electric Co.	Front Street	10	12,101	2.12	13.17
508	Pennsylvania Electric Co.	Keystone	1	11,640	2.24	20.36
509	Pennsylvania Electric Co.	Keystone	2	11,640	2.24	20.36
510	Pennsylvania Electric Co.	Seward	12	12,076	2.97	18.16
511	Pennsylvania Electric Co.	Seward	14	12,076	2.97	18.16
512	Pennsylvania Electric Co.	Seward	15	12,076	2.97	18.16
513	Pennsylvania Electric Co.	Shawville	1	12,461	2.06	12.53
514	Pennsylvania Electric Co.	Shawville	2	12,461	2.06	12.53
515	Pennsylvania Electric Co.	Shawville	3	12,461	2.06	12.53
516	Pennsylvania Electric Co.	Shawville	4	12,461	2.06	12.53
517	Pennsylvania Electric Co.	Warren	1	12,196	2.12	11.92
518	Pennsylvania Electric Co.	Warren	2	12,196	2.12	11.92
519	Pennsylvania Electric Co.	Warren	3	12,196	2.12	11.92
520	Pennsylvania Electric Co.	Warren	4	12,196	2.12	11.92
521	Pennsylvania Power Company	New Castle	1	12,462	3.24	10.70
522	Pennsylvania Power Company	New Castle	2	12,462	3.24	10.70
523	Pennsylvania Power Company	New Castle	3	12,462	3.24	10.70
524	Pennsylvania Power Company	New Castle	4	12,462	3.24	10.70
525	Pennsylvania Power Company	New Castle	5	12,462	3.24	10.70
526	Penn. Power & Light Co.	Brunner Island	1	12,460	1.99	13.74
527	Penn. Power & Light Co.	Brunner Island	2	12,460	1.99	13.74
528	Penn. Power & Light Co.	Brunner Island	3	12,460	1.99	13.74
529	Penn. Power & Light Co.	Holtwood	17	10,205	0.70	19.20
530	Penn. Power & Light Co.	Martins Creek	1	12,639	2.07	11.46
531	Penn. Power & Light Co.	Martins Creek	2	12,639	2.07	11.46
532	Penn. Power & Light Co.	Montour	1	12,565	1.79	13.17
533	Penn. Power & Light Co.	Montour	2	12,565	1.79	13.17
534	Penn. Power & Light Co.	Sunbury	1A	11,407	1.99	15.52
535	Penn. Power & Light Co.	Sunbury	1B	11,407	1.99	15.52
536	Penn. Power & Light Co.	Sunbury	2A	11,407	1.99	15.52
537	Penn. Power & Light Co.	Sunbury	2B	11,407	1.99	15.52
538	Penn. Power & Light Co.	Sunbury	3	11,407	1.99	15.52
539	Penn. Power & Light Co.	Sunbury	4	11,407	1.99	15.52
540	Philadelphia Electric Co.	Eddystone	1	13,026	2.37	8.62
541	Philadelphia Electric Co.	Eddystone	2	13,026	2.37	8.62
542	Potomac Electric	Benning	25	13,106	0.90	9.01
543	Potomac Electric	Benning	26	13,106	0.90	9.01
544	Potomac Electric	Chalk Point	1	12,341	1.70	12.16
545	Potomac Electric	Chalk Point	2	12,341	1.70	12.16
546	Potomac Electric	Dickerson	1	12,209	1.64	12.86
547	Potomac Electric	Dickerson	2	12,209	1.64	12.86
548	Potomac Electric	Dickerson	3	12,209	1.64	12.86
549	Potomac Electric	Morgantown	1	12,693	1.78	13.34
550	Potomac Electric	Morgantown	2	12,693	1.78	13.34
551	Potomac Electric	Potomac River	1	12,683	0.84	10.69
552	Potomac Electric	Potomac River	2	12,683	0.84	10.69
553	Potomac Electric	Potomac River	3	12,683	0.84	10.69
554	Potomac Electric	Potomac River	4	12,683	0.84	10.69
555	Public Serv. Co. of Colorado	Arapahoe	2	10,234	0.73	7.48
556	Public Serv. Co. of Colorado	Arapahoe	3	10,234	0.73	7.48
557	Public Serv. Co. of Colorado	Arapahoe	4	10,234	0.73	7.48
558	Public Serv. Co. of Colorado	Cameo	2	11,008	0.52	10.08
559	Public Serv. Co. of Colorado	Cherokee	1	10,768	0.51	8.47
560	Public Serv. Co. of Colorado	Cherokee	2	10,768	0.51	8.47
561	Public Serv. Co. of Colorado	Cherokee	3	10,768	0.51	8.47
562	Public Serv. Co. of Colorado	Cherokee	4	10,768	0.51	8.47
563	Public Serv. Co. of Colorado	Comanche	1	8,620	0.30	5.10
564	Public Serv. Co. of Colorado	Valmont	5	10,400	0.82	7.53
565	Public Serv. Co. of Colorado	Zuni	3	---	--	---
566	Public Serv. Co. of Indiana	Cayuga	1	10,363	2.17	13.38
567	Public Serv. Co. of Indiana	Cayuga	2	10,363	2.17	13.38
568	Public Serv. Co. of Indiana	Edwardsport	7-1	10,231	2.79	12.56
569	Public Serv. Co. of Indiana	Edwardsport	7-2	10,231	2.79	12.56
570	Public Serv. Co. of Indiana	Edwardsport	8-1	10,231	2.79	12.56
571	Public Serv. Co. of Indiana	Gallagher	1	11,149	3.40	10.73
572	Public Serv. Co. of Indiana	Gallagher	2	11,149	3.40	10.73
573	Public Serv. Co. of Indiana	Gallagher	3	11,149	3.40	10.73
574	Public Serv. Co. of Indiana	Gallagher	4	11,149	3.40	10.73
575	Public Serv. Co. of Indiana	Noblesville	1	10,742	2.74	9.87
576	Public Serv. Co. of Indiana	Noblesville	2	10,742	2.74	9.87
577	Public Serv. Co. of Indiana	Noblesville	3	10,742	2.74	9.87
578	Public Serv. Co. of Indiana	Wabash River	5	10,907	2.54	11.58
579	Public Serv. Co. of Indiana	Wabash River	6	10,907	2.54	11.58
580	Public Serv. Co. of Indiana	Wabash River	1	10,907	2.54	11.58

(Continued)

TABLE 44. (Continued)

	Company Name*	Plant Name	Boiler Number	Average Heat Content of Coal, Btu/lb	Average Sulfur Content, %	Average Ash Content, %
581	Public Serv. Co. of Indiana	Wabash River	2	10,907	2.54	11.58
582	Public Serv. Co. of Indiana	Wabash River	3	10,907	2.54	11.58
583	Public Serv. Co. of Indiana	Wabash River	4	10,907	2.54	11.58
584	Pub. Serv. Co. of N. Hamp.	Merrimack	1	13,443	2.08	7.09
585	Pub. Serv. Co. of N. Hamp.	Merrimack	2	13,443	2.08	7.09
586	Pub. Serv. Co. of N. Mexico	San Juan	2	8,838	0.80	21.20
587	Richmond Power & Light	Whitewater Valley	1	11,506	3.00	10.00
588	Richmond Power & Light	Whitewater Valley	2	11,506	3.00	10.00
589	Rochester Dept. of Pub. Utl.	Silver Lake	4	12,400	1.95	7.20
590	Rochester Gas & Elec. Corp.	Rochester 3	12	12,680	1.98	9.75
591	Rochester Gas & Elec. Corp.	Rochester 7	1	12,706	2.06	9.78
592	Rochester Gas & Elec. Corp.	Rochester 7	2	12,706	2.06	9.78
593	Rochester Gas & Elec. Corp.	Rochester 7	3	12,706	2.06	9.78
594	Rochester Gas & Elec. Corp.	Rochester 7	4	12,706	2.06	9.78
595	Salt River Project	Navajo	1	---	--	---
596	Salt River Project	Navajo	2	---	--	---
597	S. Carolina Elec. & Gas	Canadys	1	12,407	1.20	12.68
598	S. Carolina Elec. & Gas	Canadys	2	12,407	1.20	12.68
599	S. Carolina Elec. & Gas	Canadys	3	12,407	1.20	12.68
600	S. Carolina Elec. & Gas	McMeekin	1	12,304	1.55	12.44
601	S. Carolina Elec. & Gas	McMeekin	2	12,304	1.55	12.44
602	S. Carolina Elec. & Gas	Urquhart	1	12,378	1.69	12.79
603	S. Carolina Elec. & Gas	Urquhart	2	12,378	1.69	12.79
604	S. Carolina Elec. & Gas	Urquhart	3	12,378	1.69	12.79
605	S. Carolina Elec. & Gas	Wateree	1	12,179	1.48	12.29
606	S. Carolina Elec. & Gas	Wateree	2	12,179	1.48	12.29
607	S. Carolina Elec. & Gas	Winyah	1	---	--	---
608	S. Carolina Elec. & Gas	Grainger	1	11,655	1.33	13.51
609	S. Carolina Elec. & Gas	Grainger	2	11,655	1.33	13.51
610	S. Carolina Pub. Serv. Auth.	Jefferies	3	11,771	0.96	13.39
611	S. Carolina Pub. Serv. Auth.	Jefferies	4	11,771	0.96	13.39
612	S. Indiana Gas & Elec. Co.	F. B. Culley	1	10,756	3.72	11.46
613	S. Indiana Gas & Elec. Co.	F. B. Culley	2	10,756	3.72	11.46
614	S. Indiana Gas & Elec. Co.	F. B. Culley	3	10,756	3.72	11.46
615	Southern California Edison	Mohave	1	12,288	0.40	9.86
616	Southern California Edison	Mohave	2	12,288	0.40	9.86
617	Southern Elec. Gen. Co.	Gaston	1	11,744	1.17	14.40
618	Southern Elec. Gen. Co.	Gaston	2	11,744	1.17	14.40
619	Southern Elec. Gen. Co.	Gaston	3	11,744	1.17	14.40
620	Southern Elec. Gen. Co.	Gaston	4	11,744	1.17	14.40
621	Southern Ill. Power Coop.	Marion	1	10,770	4.17	14.81
622	Southern Ill. Power Coop.	Marion	2	10,770	4.17	14.81
623	Southern Ill. Power Coop.	Marion	3	10,770	4.17	14.81
624	Tampa Electric Company	Big Bend	1	11,131	3.46	11.41
625	Tampa Electric Company	Big Bend	2	11,131	3.46	11.41
626	Tampa Electric Company	F. J. Gannon	1	11,235	3.12	11.22
627	Tampa Electric Company	F. J. Gannon	2	11,235	3.12	11.22
628	Tampa Electric Company	F. J. Gannon	3	11,235	3.12	11.22
629	Tampa Electric Company	F. J. Gannon	4	11,235	3.12	11.22
630	Tampa Electric Company	F. J. Gannon	5	11,235	3.12	11.22
631	Tampa Electric Company	F. J. Gannon	6	11,235	3.12	11.22
632	Tennessee Valley Authority	Allen	1	11,058	3.12	11.48
633	Tennessee Valley Authority	Allen	2	11,058	3.12	11.48
634	Tennessee Valley Authority	Allen	3	11,058	3.12	11.48
635	Tennessee Valley Authority	Bull Run	1	11,171	0.85	15.31
636	Tennessee Valley Authority	Colbert A	1	11,116	3.98	15.03
637	Tennessee Valley Authority	Colbert A	2	11,116	3.98	15.03
638	Tennessee Valley Authority	Colbert A	3	11,116	3.98	15.03
639	Tennessee Valley Authority	Colbert A	4	11,116	3.98	15.03
640	Tennessee Valley Authority	Colbert B	5	11,254	4.11	15.02
641	Tennessee Valley Authority	Cumberland	1	10,536	3.65	16.27
642	Tennessee Valley Authority	Cumberland	2	10,536	3.65	16.27
643	Tennessee Valley Authority	Gallatin	1	10,749	3.35	16.25
644	Tennessee Valley Authority	Gallatin	2	10,749	3.35	16.25
645	Tennessee Valley Authority	Gallatin	3	10,749	3.35	16.25
646	Tennessee Valley Authority	Gallatin	4	10,749	3.35	16.25
647	Tennessee Valley Authority	John Sevier	1	11,517	1.88	15.10
648	Tennessee Valley Authority	John Sevier	2	11,517	1.88	15.10
649	Tennessee Valley Authority	John Sevier	3	11,517	1.88	15.10
650	Tennessee Valley Authority	John Sevier	4	11,517	1.88	15.10
651	Tennessee Valley Authority	Johnsonville	7	10,970	3.63	14.28
652	Tennessee Valley Authority	Johnsonville	8	10,970	3.63	14.28
653	Tennessee Valley Authority	Johnsonville	9	10,970	3.63	14.28
654	Tennessee Valley Authority	Johnsonville	10	10,970	3.63	14.28
655	Tennessee Valley Authority	Kingston	1	10,688	2.21	20.35
656	Tennessee Valley Authority	Kingston	2	10,688	2.21	20.35
657	Tennessee Valley Authority	Kingston	3	10,688	2.21	20.35
658	Tennessee Valley Authority	Kingston	4	10,688	2.21	20.35
659	Tennessee Valley Authority	Kingston	5	10,688	2.21	20.35
660	Tennessee Valley Authority	Kingston	6	10,688	2.21	20.35
661	Tennessee Valley Authority	Kingston	7	10,688	2.21	20.35
662	Tennessee Valley Authority	Kingston	8	10,688	2.21	20.35
663	Tennessee Valley Authority	Kingston	9	10,688	2.21	20.35
664	Tennessee Valley Authority	Paradise	1	10,268	4.18	18.66
665	Tennessee Valley Authority	Paradise	2	10,268	4.18	18.66
666	Tennessee Valley Authority	Paradise	3	10,268	4.18	18.66
667	Tennessee Valley Authority	Shawnee	1	10,500	2.87	15.51
668	Tennessee Valley Authority	Shawnee	2	10,500	2.87	15.51
669	Tennessee Valley Authority	Shawnee	3	10,500	2.87	15.51
670	Tennessee Valley Authority	Shawnee	4	10,500	2.87	15.51
671	Tennessee Valley Authority	Shawnee	5	10,500	2.87	15.51
672	Tennessee Valley Authority	Shawnee	6	10,500	2.87	15.51
673	Tennessee Valley Authority	Shawnee	7	10,500	2.87	15.51
674	Tennessee Valley Authority	Shawnee	8	10,500	2.87	15.51
675	Tennessee Valley Authority	Shawnee	9	10,500	2.87	15.51
676	Tennessee Valley Authority	Shawnee	10	10,500	2.87	15.51
677	Tennessee Valley Authority	Watts Bar	A	11,142	3.75	15.62
678	Tennessee Valley Authority	Watts Bar	B	11,142	3.75	15.62
679	Tennessee Valley Authority	Watts Bar	C	11,142	3.75	15.62
680	Tennessee Valley Authority	Watts Bar	D	11,142	3.75	15.62
681	Tennessee Valley Authority	Widows Creek "B"	7	11,234	3.90	15.01
682	Tennessee Valley Authority	Widows Creek "B"	8	11,234	3.90	15.01

(Continued)

Appendix A—Power Plant and Air Quality Data   433

TABLE 44. (Continued)

	Company Name*	Plant Name	Boiler Number	Average Heat Content of Coal, Btu/lb	Average Sulfur Content, %	Average Ash Content, %
683	Toledo Edison	Acme	13	11,410	2.68	15.19
684	Toledo Edison	Acme	14	11,410	2.68	15.19
685	Toledo Edison	Acme	15	11,410	2.68	15.19
686	Toledo Edison	Acme	16	11,410	2.68	15.19
687	Toledo Edison	Acme	91	11,410	2.68	15.19
688	Toledo Edison	Acme	92	11,410	2.68	15.19
689	Toledo Edison	Bay Shore	1	12,143	1.93	10.78
690	Toledo Edison	Bay Shore	2	12,143	1.93	10.78
691	Toledo Edison	Bay Shore	3	12,143	1.93	10.78
692	Toledo Edison	Bay Shore	4	12,143	1.93	10.78
693	UGI Corp. Luzerne Electric	Hunlock Creek	2	8,732	0.70	23.55
694	UGI Corp. Luzerne Electric	Hunlock Creek	6	8,732	0.70	23.55
695	Union Electric	Labadie	1	11,134	3.07	9.80
696	Union Electric	Labadie	2	11,134	3.07	9.80
697	Union Electric	Labadie	3	11,134	3.07	9.80
698	Union Electric	Labadie	4	11,134	3.07	9.80
699	Union Electric	Meramec	1	11,810	1.53	9.39
700	Union Electric	Meramec	2	11,810	1.53	9.39
701	Union Electric	Meramec	3	11,810	1.53	9.39
702	Union Electric	Meramec	4	11,810	1.53	9.39
703	Union Electric	Sioux	1	10,939	2.99	15.27
704	Union Electric	Sioux	2	10,939	2.99	15.27
705	Union Electric	Venice	7	11,912	1.31	7.90
706	Union Electric	Venice	8	11,912	1.31	7.90
707	Upper Peninsula Generating	Presque Isle	1	12,415	1.30	8.20
708	Upper Peninsula Generating	Presque Isle	2	12,415	1.30	8.20
709	Upper Peninsula Generating	Presque Isle	3	12,415	1.30	8.20
710	Upper Peninsula Generating	Presque Isle	4	12,415	1.30	8.20
711	Upper Peninsula Generating	Presque Isle	5	12,415	1.30	8.20
712	Utah Power & Light	Gadsby	2	12,072	0.50	8.51
713	Utah Power & Light	Gadsby	3	12,072	0.50	8.51
714	Utah Power & Light	Hale	2	12,013	0.53	9.80
715	Utah Power & Light	Huntington No. 2	2	---	--	---
716	Utah Power & Light	Naughton	1	9,509	0.50	4.50
717	Utah Power & Light	Naughton	3	9,509	0.50	4.50
718	Virginia Electric & Power	Bremo	3	12,391	0.89	10.99
719	Virginia Electric & Power	Bremo	4	12,391	0.89	10.99
720	Virginia Electric & Power	Mt. Storm	1	11,276	1.95	18.77
721	Virginia Electric & Power	Mt. Storm	2	11,276	1.95	18.77
722	Virginia Electric & Power	Mt. Storm	3	11,276	1.95	18.77
723	Western Massachusetts	West Springfield	1	---	--	---
724	Western Massachusetts	West Springfield	2	---	--	---
725	Western Massachusetts	West Springfield	3	---	--	---
726	Wisconsin Electric Power	North Oak Creek	1	11,457	2.09	10.71
727	Wisconsin Electric Power	North Oak Creek	2	11,457	2.09	10.71
728	Wisconsin Electric Power	North Oak Creek	3	11,457	2.09	10.71
729	Wisconsin Electric Power	North Oak Creek	4	11,457	2.09	10.71
730	Wisconsin Electric Power	Port Washington	1	12,118	3.43	10.56
731	Wisconsin Electric Power	Port Washington	2	12,118	3.43	10.56
732	Wisconsin Electric Power	Port Washington	3	12,118	3.43	10.56
733	Wisconsin Electric Power	Port Washington	4	12,118	3.43	10.56
734	Wisconsin Electric Power	Port Washington	5	12,118	3.43	10.56
735	Wisconsin Electric Power	Valley	1	11,848	3.22	10.39
736	Wisconsin Electric Power	Valley	2	11,848	3.22	10.39
737	Wisconsin Electric Power	Valley	3	11,848	3.22	10.39
738	Wisconsin Electric Power	Valley	4	11,848	3.22	10.39
739	Wisconsin Power & Light	Edgewater	1	10,930	2.53	8.94
740	Wisconsin Power & Light	Edgewater	2	10,930	2.53	8.94
741	Wisconsin Power & Light	Edgewater	3	10,930	2.53	8.94
742	Wisconsin Power & Light	Edgewater	4	10,930	2.53	8.94
743	Wisconsin Power & Light	Nelson Dewey	1	10,837	3.62	10.37
744	Wisconsin Power & Light	Nelson Dewey	2	10,837	3.62	10.37
745	Wisconsin Power & Light	Rock River	1	11,107	2.82	10.06
746	Wisconsin Power & Light	Rock River	2	11,107	2.82	10.06
747	Wisconsin Public Service	J. P. Pulliam	3	11,863	2.80	10.89
748	Wisconsin Public Service	J. P. Pulliam	4	11,863	2.80	10.89
749	Wisconsin Public Service	J. P. Pulliam	5	11,863	2.80	10.89
750	Wisconsin Public Service	J. P. Pulliam	6	11,863	2.80	10.89
751	Wisconsin Public Service	J. P. Pulliam	7	11,863	2.80	10.89
752	Wisconsin Public Service	J. P. Pulliam	8	11,863	2.80	10.89
753	Wisconsin Public Service	Weston	1	11,786	2.93	9.43
754	Wisconsin Public Service	Weston	2	11,786	2.93	9.43

TABLE 45. POWER PLANT AND AIR QUALITY DATA FOR THOSE PLANTS WITH ELECTROSTATIC PRECIPITATORS

*Co. Name	Year Boiler Placed in Service	Generating Capacity, MW	Design Coal Consumption, tons/hour	Air Flow at 100% Load, scf/min	Type of Firing	Boiler Manufacturer	Boiler Efficiency at 100% Load	% Excess Air Used
1	1971	788.8	250.	1,275,000	Pul. coal/Tangential	Combustion Eng.	89.1	18.
2	1956	187.5	66.7	274,000	Pul. coal/Tangential	Combustion Eng.	88.52	15.
3	1958	190.4	73.0	285,000	Pul. coal/Tangential	Combustion Eng.	88.52	15.
4	1972	788.8	250.0	1,275,000	Pul. coal/Tangential	Combustion Eng.	89.10	18.
5	1949	69.	28.5	140,000	Pul. coal/Tangential	Combustion Eng.	86.5	20.
6	1949	69.	28.5	140,000	Pul. coal/Tangential	Combustion Eng.	86.5	20.
7	1954	145.	80.	440,000	Pul. coal/Tangential	Combustion Eng.	87.0	20.
8	1967	552.	186.	1,250,000	Pul. coal/Tangential	Combustion Eng.	88.6	20.
9	1968	550.	179.	1,250,000	Pul. coal/Opposed	B & W	90.7	20.
10	1972	650.	250.	1,500,000	Pul. coal/Opposed	Foster Wheeler	88.99	25.
11	1973	650.	250.	1,500,000	Pul. coal/Opposed	Foster Wheeler	88.99	25.
12	1974	650.	250.	1,500,000	Pul. coal/Opposed	Foster Wheeler	88.99	25.
13	1960	186.	95.	590,000	Cyclone	B & W	88.8	20.
14	1958	183.	80.	450,000	Pul. coal/Opposed	Foster Wheeler	88.	20.
15	1959	178.	80.	450,000	Pul. coal/Opposed	Foster Wheeler	88.	20.
16	1969	576.	200.	1,169,000	Pul. coal/Opposed	B & W	91.0	30.
17	1970	576.	200.	1,169,000	Pul. coal/Opposed	B & W	91.0	30.
18	1971	576.	200.	1,169,000	Pul. coal/Opposed	B & W	91.0	30.
19	1963	294.	100.	630,000	Pul. coal/Tangential	Combustion Eng.	89.8	30.
20	1954	142.	46.	265,000	Pul. coal/Front	B & W	89.9	20.

*The numbers in the first column correspond to the same plant names in Tables 45 and 46 as they do in Table 44.

(Continued)

434 Electrostatic Precipitator Manual

TABLE 45. (Continued)

*Co. Name	Year Boiler Placed in Service	Generating Capacity, MW	Design Coal Consumption, tons/hour	Air Flow at 100% Load, scf/min	Type of Firing	Boiler Manufacturer	Boiler Efficiency at 100% Load	% Excess Air Used
21	1942	--	19.5 ea.	93,750	Pul. coal/Front	Foster Wheeler	87.7	20.
22	1943	--	19.5 ea.	93,750	Pul. coal/Front	B & W	87.7	20.
23	1958	223.	83.	317,292	Pul. coal/Front	B & W	89.8	20.
24	1958	223.	83.	317,292	Pul. coal/Front	B & W	89.8	20.
25	1961	223.	83.	317,292	Pul. coal/Front	B & W	89.8	20.
26	1969	818.1	421.	1,893,000	Pul. coal/Opposed	B & W	88.68	16.
27	1970	818.1	421	1,893,000	Pul. coal/Opposed	B & W	88.68	16.
28	1969	170.	71.5	319,222	Pul. coal/Front	Foster Wheeler	88.02	18.
29	1970	170.	71.5	319,222	Pul. coal/Front	Foster Wheeler	88.02	18.
30	1972	173.	70.0	365,333	Pul. coal/Front	Riley Stoker	87.92	18.
31	1965	80.	35.9	169,666	Pul. coal/Front	Riley Stoker	86.9	22.
32	1967	590.	247.5	800,000	Pul. coal/Front	B & W	--	20.
33	1967	590.	247.5	800,000	Pul. coal/Front	B & W	--	20.
34	1964	206.635	70.	600,000	Pul. coal/Front	Riley Stoker	90.00	20.
35	1971	207.00	72.	576,000	Pul. coal/Front	B & W	88.60	18.
36	1956	140.625	48.2	222,000	Pul. coal/Tangential	Combustion Eng.	89.9	20
37	1958	187.85	59.2	280,000	Pul. coal/Tangential	Combustion Eng.	90.0	20.
38	1960	206.635	68.42	286,000	Pul. coal/Tangential	Combustion Eng.	90.00	20.
39	1952	75.	26.3	121,000	Pul. coal/Tangential	Combustion Eng.	89.48	20.
40	1951	75.	31.6	127,000	Pul. coal/Front	B & W	87.8	25.
41	1954	112.5	48.	231,000	Pul. coal/Tangential	Combustion Eng.	89.95	4.0
42	1955	112.5	40.	210,000	Pul. coal/Front	B & W	89.1	4.0
43	1972	446.616	197.4	868,000	Pul. coal/Front	Riley Stoker	88.58	5.0
44	1966	410.85	130.7	700,000	Pul. coal/Opposed	Riley Stoker	90.0	20.
45	1968	657.00	223.0	1,130,000	Pul. coal/Tangential	Combustion Eng.	90.0	20.
46	1973	745.20	151.8	733,049	---	Riley Stoker	88.71	20.
47	1973	---	151.8	733,049	---	Riley Stoker	88.71	20.
48	1949	46.	19.5	95,500	Pul. coal/Front	B & W	88.	20.
49	1950	46.	19.5	95,500	Pul. coal/Front	B & W	88.	20.
50	1952	73.5	26.3	121,000	Pul. coal/Tangential	Combustion Eng.	90.	20.
51	1973	37.8	18.0	92,000	Pul. coal/Front	B & W	88.1	20.
52	1960	125.	60.9	316,100	Pul. coal/Front	Riley Stoker	87.2	20.
53	1968	250.	130.9	572,700	Pul. coal/Front	Riley Stoker	87.4	20.
54	1972	322.	153.5	827,894	Pul. coal/Front	Foster Wheeler	88.1	18.
55	1949	35.	17.	100,000	Pul. coal/Front	Riley Stoker	85.	20.
56	1949	35.	17.	100,000	Pul. coal/Front	Riley Stoker	85.	20.
57	1952	84.	39.	222,000	Pul. coal/Front	Riley Stoker	85.	20.
58	1958	100.	50.	270,000	Pul. coal/Front	Riley Stoker	89.	20.
59	1965	388.9	200.	875,760	Cyclone	B & W	87.93	16.
60	1972	616.	277.7	1,310,082	Cyclone	B & W	87.58	16.
61	1950	38.	18.75	86,890	Pul. coal/Front	B & W	86.6	25.
62	1950	38.	18.75	86,890	Pul. coal/Front	B & W	86.6	25.
63	1958	110.	47.1	201,333	Pul. coal/Front	B & W	89.05	25.
64	1948	34.	12.5	67,555	Pul. coal/Tangential	Combustion Eng.	85.2	25.
65	1948	34.	12.5	67,555	Pul. coal/Tangential	Combustion Eng.	85.2	25.
66	1949	34.	12.5	67,555	Pul. coal/Tangential	Combustion Eng.	85.2	25.
67	1949	34.	12.5	67,555	Pul. coal/Tangential	Combustion Eng.	85.2	25.
68	1960	239.4	91.	397,777	Pul. coal/Tangential	Combustion Eng.	87.24	25.
69	1960	450.	149.	608,958	Pul. coal/Opposed	B & W	89.7	20.
70	--	300.	131.	621,664	Pul. coal/Opposed	B & W	88.88	20.
71	1960	168.	62.	240,500	Pul. coal/Front	Combustion Eng.	90.05	20.
72	1952	100.	38.2	161,000	Pul. coal/Tangential	Combustion Eng.	89.4	24.
73	1953	100.	37.5	158,300	Pul. coal/Tangential	Combustion Eng.	89.4	24.
74	1954	125.	47.6	196,900	Pul. coal/Front	B & W	89.33	25.
75	1958	163.	62.0	240,500	Pul. coal/Tangential	Combustion Eng.	90.05	20.
76	1962	240.	86.8	351,900	Pul. coal/Tangential	Combustion Eng.	89.99	20.
77	1969	434.	173.0	669,200	Pul. coal/Tangential	Combustion Eng.	89.01	20.
78	1962	50.0	31.0	93,985	Pul. coal/Front	Riley Stoker	82.7	20.
79	1968	80.0	39.6	143,163	Pul. coal/Front	B & W	88.1	18.
80	1974	132.	61.5	228,109	Pul. coal/Front	B & W	87.7	20.
81	1959	22.	12.	---	Pul. coal/Front	---	87.0	25.
82	1947	20.	16.6	64,700	Pul. coal/Front	---	83.1	15.
83	1951	20.	16.6	64,700	Pul. coal/Front	---	83.3	15.
84	1961	37.5	18.2	73,000	Cyclone	B & W	88.1	15.
85	1965	37.5	18.2	73,000	Cyclone	B & W	87.3	15.
86	1968	80.	37.4	155,800	Cyclone	B & W	87.47	15.
87	1972	80.	37.4	155,800	Cyclone	B & W	88.67	15.
88	1970	105.	40.2	316,000	Pul. coal/Front	Riley Stoker	87.7	22.
89	1958	256.	92.	482,000	Pul. coal/Tangential	Combustion Eng.	90.2	24.
90	1948	46.	21.5	228,000	Pul. coal/Front	B & W	---	22.
91	1948	46.	21.5	228,000	Pul. coal/Front	B & W	---	22.
92	1948	46.	21.5	228,000	Pul. coal/Front	B & W	---	22.
93	1948	46.	21.5	228,000	Pul. coal/Front	B & W	---	22.
94	1949	86.	46.	242,620	Pul. coal/Tangential	Combustion Eng.	87.9	22.
95	1949	86.	46.	242,620	Pul. coal/Tangential	Combustion Eng.	87.9	22.
96	1959	233.	85.	471,520	Pul. coal/Tangential	Combustion Eng.	90.3	22.
97	1970	680.	230.	1,106,700	Pul. coal/Front	B & W	89.9	18.
98	1972	680.	230.	1,106,700	Pul. coal/Opposed	B & W	89.9	18.
99	1941	60.	25.2	167,120	Pul. coal/Front	B & W	88.7	20.
100	1941	60.	25.2	167,120	Pul. coal/Front	B & W	88.7	20.
101	1951	69.	33.3	195,150	Pul. coal/Front	B & W	88.5	23.
102	1951	69.	33.3	195,150	Pul. coal/Front	B & W	88.5	23.
103	1962	256.	88.4	487,340	Pul. coal/Tangential	Combustion Eng.	89.7	22.
104	1973	787.	332.5	2,257,232	Pul. coal/Tangential	Combustion Eng.	88.01	25.
105	1958	239.	100.	433,000	Pul. coal/Tangential	Combustion Eng.	89.4	15.
106	1961	358.	145.	545,000	Pul. coal/Tangential	Combustion Eng.	89.4	14.6
107	1945	59.	30.	110,000	Pul. coal/Front	B & W	86.4	20.
108	1953	69.	37.	120,000	Pul. coal/Front	B & W	87.0	25.
109	1949	173.	44.5	152,000	Cyclone	B & W	86.6	22.
110	1949	--	44.5	152,000	Cyclone	B & W	86.6	22.
111	1959	374.	139.	464,000	Pul. coal/Tangential	Combustion Eng.	89.4	18
112	1950	107.	37.	128,000	Cyclone	B & W	87.0	20.
113	1950	--	37.	128,000	Cyclone	B & W	87.0	20.
114	1959	360.	144.	470,000	Cyclone	B & W	89.4	16.
115	1965	660.	145.	578,000	Pul. coal/Tangential	Combustion Eng.	89.3	14.
116	1965	--	145.	578,000	Pul. coal/Tangential	Combustion Eng.	89.3	14.
117	1966	660.	145.	578,000	Pul. coal/Tangential	Combustion Eng.	89.3	14.
118	1966	--	145.	578,000	Pul. coal/Tangential	Combustion Eng.	89.3	14.
119	1967	660.	282.5	1,100,000	Cyclone	B & W	88.2	16.
120	1968	660.	282.5	1,100,000	Cyclone	B & W	88.2	16.
121	1972	892.8	176.5	1,217,000	Cyclone	B & W	89.03	15.
122	1972	--	176.5	1,217,000	Cyclone	B & W	89.03	15.
123	1961	54.	26.	86,800	Pul. coal/Tangential	Foster Wheeler	84.5	22.
124	1931	115.	20.	92,000	Pul. coal/Front	B & W	83.3	--
125	1931	--	28.	116,000	Pul. coal/Front	B & W	83.8	--
126	1931	--	28.	116,000	Pul. coal/Front	B & W	83.8	--
127	1952	121.	56.	191,000	Cyclone	B & W	88.6	20.
128	1958	326.	138.	463,000	Pul. coal/Tangential	Combustion Eng.	89.4	18.
129	1962	355.	150.	685,000	Pul. coal/Tangential	Combustion Eng.	89.4	24.
130	1955	188.	80.8	276,000	Cyclone	B & W	89.1	10.
131	1955	184.	80.8	276,000	Cyclone	B & W	89.1	10.
132	1957	299.	125.	467,000	Pul. coal/Tangential	Combustion Eng.	89.5	15.
133	1963	598.	224.	798,000	Pul. coal/Tangential	Combustion Eng.	89.0	20.
134	1929	208.	24.	100,000	Pul. coal/Front	B & W	75.6	25.
135	1929	--	24.	100,000	Pul. coal/Front	B & W	75.6	25.

(Continued)

Appendix A—Power Plant and Air Quality Data   435

TABLE 45. (Continued)

*Co. Name	Year Boiler Placed in Service	Generating Capacity, MW	Design Coal Consumption, tons/hour	Air Flow at 100% Load, scf/min	Type of Firing	Boiler Manufacturer	Boiler Efficiency at 100% Load	% Excess Air Used
136	1929	--	24.	100,000	Pul. coal/Front	B & W	75.6	25.
137	1929	--	24.	100,000	Pul. coal/Front	B & W	75.6	25.
138	1929	--	24.	100,000	Pul. coal/Front	B & W	75.6	25.
139	1929	--	24.	100,000	Pul. coal/Front	B & W	75.6	25.
140	1938	150.	25.	122,200	Pul. coal/Front	B & W	82.8	20.
141	1938	--	25.	111,000	Pul. coal/Front	B & W	82.8	20.
142	1938	--	25.	111,100	Pul. coal/Front	B & W	82.8	20.
143	1955	225.	84.	310,300	Pul. coal/Tangential	Combustion Eng.	89.3	18.
144	1962	389.	124.	470,000	Cyclone	B & W	89.4	16.
145	1953	200.	64.	360,500	Pul. coal/Front	B & W	90.3	25.
146	1954	200.	64.	360,500	Pul. coal/Front	B & W	90.3	25.
147	1958	376.	130.4	734,500	Pul. coal/Front	B & W	90.3	25.
148	1961	387.	134.2	755,900	Pul. coal/Tangential	Combustion Eng.	90.6	25.
149	1961	387.	134.2	755,900	Pul. coal/Tangential	Combustion Eng.	90.6	25.
150	1965	1028.	321.0	1,808,100	Pul. coal/Tangential	Combustion Eng.	90.8	25.
151	1948	66.	31.	190,000	Pul. coal/Tangential	Combustion Eng.	88.05	15.
152	1948	66.	31.	190,000	Pul. coal/Tangential	Combustion Eng.	88.05	15.
153	1950	66.	31.	190,000	Pul. coal/Tangential	Combustion Eng.	88.05	15.
154	1956	156.25	80.	340,000	Pul. coal/Tangential	Combustion Eng.	89.33	18.
155	1957	156.25	80.	340,000	Pul. coal/Tangential	Combustion Eng.	89.33	18.
156	1959	265.	92.	645,576	Pul. coal/Tangential	Combustion Eng.	88.9	17.
157	1961	265.	89.	625,071	Pul. coal/Front	B & W	88.9	17.
158	1955	156.25	88.	340,000	Pul. coal/Tangential	Combustion Eng.	88.	--
159	1958	156.25	88.	340,000	Pul. coal/Tangential	Combustion Eng.	88.	--
160	1962	265.	100.	630,000	Pul. coal/Tangential	Combustion Eng.	90.26	18.
161	1967	385.	150.	907,400	Pul. coal/Opposed	B & W	90.47	18.
162	1947	18.7	10.0	65,830	Pul. coal/Front	B & W	85.6	37.3
163	1947	18.7	10.0	69,700	Pul. coal/Front	B & W	85.6	36.8
164	1950	18.7	12.0	86,590	Pul. coal/Front	Riley Stoker	86.0	17.7
165	1957	54.4	25.0	124,570	Pul. coal/Front	Riley Stoker	86.5	20.0
166	1959	81.6	37.9	230,630	Pul. coal/Front	Riley Stoker	86.4	23.9
167	1969	345.6	137.	611,000	Pul. coal/Tangential	Combustion Eng.	58.35	20.0
168	1951	19.	13.95	75,600	Pul. coal/Front	Riley Stoker	86.0	17.7
169	1952	33.	13.95	63,290	Pul. coal/Front	Riley Stoker	86.0	17.7
170	1971	593.4	375.	1,189,000	Pul. coal/Tangential	Combustion Eng.	82.58	20.
171	1972	593.4	375.	1,189,000	Pul. coal/Tangential	Combustion Eng.	82.58	20.
172	1974	593.4	418.	2,600,000	Pul. coal/Tangential	Combustion Eng.	81.5	20.
173	1958	117.	54.	240,000	Pul. coal/Tangential	Combustion Eng.	88.6	20.
174	1959	147.1	54.	240,000	Pul. coal/Tangential	Combustion Eng.	88.6	20.
175	1937	37.5	21.	94,500	Pul. coal/Front	B & W	85.2	25.
176	1937	37.5	21.	94,500	Pul. coal/Front	B & W	85.2	25.
177	1940	37.5	21.	94,500	Pul. coal/Front	B & W	85.2	25.
178	1940	37.5	21.	94,500	Pul. coal/Front	B & W	85.2	25.
179	1971	610.2	250.	1,311,656	Pul. coal/Opposed	B & W	89.9	18.
180	1970	610.2	250.	1,311,656	Pul. coal/Opposed	B & W	89.9	18.
181	1972	610.2	250.	1,311,656	Pul. coal/Opposed	B & W	89.9	18.
182	1974	610.2	250.	1,311,656	Pul. coal/Opposed	B & W	89.9	18.
183	1948	69.	24.0	105,777	Pul. coal/Tangential	Combustion Eng.	87.9	20.
184	1949	69.	24.0	105,777	Pul. coal/Tangential	Combustion Eng.	87.9	20.
185	1950	69.	23.4	103,333	Pul. coal/Tangential	Combustion Eng.	87.7	20.
186	1951	69.	23.4	103,333	Pul. coal/Tangential	Combustion Eng.	87.7	20.
187	1952	69.	23.4	103,333	Pul. coal/Tangential	Combustion Eng.	87.7	20.
188	1953	69.	23.4	103,333	Pul. coal/Tangential	Combustion Eng.	87.7	20.
189	1961	75.	23.5	160,000	Pul. coal/Opposed	Foster Wheeler	87.	20.
190	1957	81.6	32.	228,000	Pul. coal/Front	B & W	90.	20.
191	1959	81.6	32.	228,000	Pul. coal/Front	B & W	90.	20.
192	1970	176.8	65.	360,000	Pul. coal/Front	B & W	90.	20.
193	1951	67.5	31.5	130,000	Pul. coal/Front	B & W	87.3	24.5
194	1951	67.5	31.5	130,000	Pul. coal/Front	B & W	87.3	24.5
195	1951	67.5	31.5	130,000	Pul. coal/Front	B & W	87.3	24.5
196	1951	67.5	31.5	130,000	Pul. coal/Front	B & W	87.3	24.5
197	1968	121.	45.	214,000	Pul. coal/Front	Riley Stoker	88.4	20.
198	1942	50.	21.	97,000	Pul. coal/Tangential	Combustion Eng.	87.7	22.
199	1943	50.	21.	97,000	Pul. coal/Tangential	Combustion Eng.	87.7	22.
200	1947	50.	21.	97,000	Pul. coal/Tangential	Combustion Eng.	87.7	22.
201	1943	50.	21.	97,000	Pul. coal/Tangential	Combustion Eng.	87.7	22.
202	1971	817.2	281.	1,530,000	Pul. coal/Opposed	B & W	90.92	18.
203	1973	822.6	281.	1,530,000	Pul. coal/Opposed	B & W	90.92	18.
204	1973	822.6	281.	1,530,000	Pul. coal/Opposed	B & W	90.92	18.
205	1974	817.2	281.	1,530,000	Pul. coal/Opposed	B & W	90.92	18.
206	1948	9.25	11.55	59,000	Pul. coal/Front	Combustion Eng.	87.6	22.
207	1949	9.25	11.55	59,000	Pul. coal/Front	Combustion Eng.	87.6	22.
208	1957	292.	99.	426,250	Pul. coal/Tangential	Combustion Eng.	89.17	18.
209	1953	169.	61.	258,000	Pul. coal/Front	B & W	88.4	23.
210	1953	156.	61.	258,000	Pul. coal/Front	B & W	88.4	23.
211	1954	156.	61.	258,000	Pul. coal/Front	B & W	88.4	23.
212	1954	169.	61.	258,000	Pul. coal/Front	B & W	88.4	23.
213	1961	353.	120.	542,000	Pul. coal/Front	Combustion Eng.	90.16	18.
214	1969	544.	198.	862,000	Pul. coal/Tangential	Combustion Eng.	90.62	18.
215	1948	11.4	15.4	74,000	Pul. coal/Front	B & W	87.4	26.
216	1948	11.4	15.4	74,000	Pul. coal/Front	B & W	87.4	26.
217	1968	11.4	15.3	69,000	Pul. coal/Front	Foster Wheeler	87.78	18.
218	1968	11.4	15.3	69,000	Pul. coal/Front	Foster Wheeler	87.78	18.
219	1957	165.	56.	292,520	Pul. coal/Tangential	Combustion Eng.	88.99	20.
220	1957	165.	56.	292,520	Pul. coal/Tangential	Combustion Eng.	88.99	20.
221	1959	275.	91.	487,640	Pul. coal/Tangential	Combustion Eng.	89.59	20.
222	1960	275.	91.	487,640	Pul. coal/Tangential	Combustion Eng.	89.59	20.
223	1961	275.	91.	487,640	Pul. coal/Tangential	Combustion Eng.	89.59	20.
224	1974	1080.	360.	1,874,400	Pul. coal/Opposed	B & W	90.23	20.
225	1941	40.	17.4	85,077	Pul. coal/Tangential	Combustion Eng.	86.3	19.
226	1941	40.	17.4	85,077	Pul. coal/Tangential	Combustion Eng.	86.3	19.
227	1942	40.	17.4	85,077	Pul. coal/Tangential	Combustion Eng.	86.3	19.
228	1953	125.	48.	233,587	Pul. coal/Tangential	Combustion Eng.	88.82	23.
229	1953	125.	48.	233,587	Pul. coal/Tangential	Combustion Eng.	88.82	23.
230	1940	40.	17.4	82,077	Pul. coal/Tangential	Combustion Eng.	86.8	23.
231	1940	40.	17.4	82,077	Pul. coal/Tangential	Combustion Eng.	86.8	23.
232	1948	65.	27.9	123,265	Pul. coal/Tangential	Combustion Eng.	87.5	23.
233	1948	65.	27.9	123,265	Pul. coal/Tangential	Combustion Eng.	87.5	23.
234	1949	70.	30.	144,867	Pul. coal/Tangential	Combustion Eng.	88.2	22.
235	1950	70.	30.	144,867	Pul. coal/Tangential	Combustion Eng.	88.2	22.
236	1955	150.	55.	276,447	Pul. coal/Tangential	Combustion Eng.	88.75	19.
237	1951	90.	40.	185,800	Pul. coal/Tangential	Combustion Eng.	88.66	22.
238	1951	90.	40.	185,800	Pul. coal/Tangential	Combustion Eng.	88.66	22.
239	1958	165.	58.9	292,520	Pul. coal/Tangential	Combustion Eng.	88.95	22.
240	1965	350.	117.	561,037	Pul. coal/Tangential	Combustion Eng.	89.74	18.
241	1966	350.	117	561,037	Pul. coal/Tangential	Combustion Eng.	89.74	18.
242	1969	650.	208.	973,350	Pul. coal/Tangential	Combustion Eng.	90.12	18.
243	1970	650.	208.	973,350	Pul. coal/Tangential	Combustion Eng.	90.12	18.
244	1952	100.	40.6	198,442	Pul. coal/Tangential	Combustion Eng.	88.8	23.
245	1952	100.	40.6	198,442	Pul. coal/Tangential	Combustion Eng.	88.8	23.
246	1954	133.	52.	253,946	Pul. coal/Tangential	Combustion Eng.	89.2	20.
247	1954	133.	52.	253,946	Pul. coal/Tangential	Combustion Eng.	89.2	20.
248	1970	525.	224.5	841,323	Pul. coal/Tangential	Combustion Eng.	89.4	18.
249	1952	80.	41.5	385,000	Pul. coal/Front	B & W	88.6	25.
250	1953	80.	41.5	385,000	Pul. coal/Front	B & W	88.6	25.

(Continued)

TABLE 45. (Continued)

*Co. Name	Year Boiler Placed in Service	Generating Capacity, MW	Design Coal Consumption, tons/hour	Air Flow at 100% Load, scf/min	Type of Firing	Boiler Manufacturer	Boiler Efficiency at 100% Load	% Excess Air Used
251	1954	100.	47.8	430,000	Pul. coal/Front	B & W	88.9	21.
252	1960	165.	75.0	636,000	Pul. coal/Front	B & W	88.4	18.
253	1942	--	23.8	240,000	Pul. coal/Front	Foster Wheeler	85.5	26.
254	1942	--	23.8	240,000	Pul. coal/Front	Foster Wheeler	85.5	26.
255	1949	--	37.6	342,000	Pul. coal/Front	Foster Wheeler	85.5	26.
256	1950	--	37.6	342,000	Pul. coal/Front	Foster Wheeler	85.5	26.
257	1950	--	37.6	342,000	Pul. coal/Front	Foster Wheeler	85.5	26.
258	1956	125.	64.0	400,000	Pul. coal/Front	Foster Wheeler	88.0	26.
259	1964	100.	42.5	280,000	Pul. coal/Front	B & W	89.07	20.0
260	1969	220.85	92.5	492,000	Pul. coal/Front	B & W	87.14	20.0
261	1957	74.	40.	368,000	Pul. coal/Front	Riley Stoker	87.27	20.0
262	1960	74.	39.	353,000	Pul. coal/Front	Riley Stoker	89.21	20.0
263	1953	180.876	75.85	368,501	Pul. coal/Tangential	Combustion Eng.	88.22	18.0
264	1954	180.876	75.85	368,501	Pul. coal/Tangential	Combustion Eng.	88.22	18.0
265	1955	180.876	75.85	368,501	Pul. coal/Tangential	Combustion Eng.	88.22	18.0
266	1970	200.	100.	404,440	Cyclone	B & W	87.15	11.0
267	1941	46.	25.	225,000	Pul. coal/Tangential	Combustion Eng.	85.0	22.0
268	1942	46.	25.	225,000	Pul. coal/Tangential	Combustion Eng.	85.0	22.0
269	1943	40.	27.	225,000	Pul. coal/Tangential	B & W	86.0	22.0
270	1948	40.	27.	225,000	Pul. coal/Tangential	B & W	87.0	22.0
271	1954	125.	41.	187,159	Pul. coal/Front	B & W	88.6	23.0
272	1954	125.	41.	187,159	Pul. coal/Front	B & W	88.6	23.0
273	1955	125.	41.	187,159	Pul. coal/Front	B & W	88.6	23.0
274	1970	578.	195.	1,006,011	Pul. coal/Opposed	Foster Wheeler	89.01	18.0
275	1971	806.	269.	1,382,440	Pul. coal/Tangential	Combustion Eng.	89.10	18.0
276	1972	789.	269.	1,382,440	Pul. coal/Tangential	Combustion Eng.	89.10	18.0
277	1974	952.	375.	1,775,706	Pul. coal/Tangential	Combustion Eng.	88.70	18.0
278	1963	245.	94.25	1,050,000	Pul. coal/Tangential	Combustion Eng.	89.3	18.0
279	1964	245.	94.25	1,050,000	Pul. coal/Tangential	Combustion Eng.	89.3	18.0
280	1965	250.0	97.1	401,600	Pul. coal/Opposed	B & W	89.08	18.0
281	1967	319.0	122.9	563,500	Pul. coal/Opposed	Riley Stoker	89.10	20.0
282	1968	480.7	185.7	714,200	Pul. coal/Opposed	B & W	89.09	18.0
283	1969	490.0	185.7	714,200	Pul. coal/Opposed	B & W	89.09	18.0
284	1948	22.	12.0	80,000	Pul. coal/Front	B & W	80.0	20.0
285	1948	22.	12.0	80,000	Pul. coal/Front	B & W	80.0	20.0
286	1964	125.	89.0	500,000	Pul. coal/Tangential	Combustion Eng.	89.0	18.0
287	1950	100.	50.0	195,041	Pul. coal/Tangential	Combustion Eng.	88.5	20.0
288	1950	100.	50.0	195,041	Pul. coal/Tangential	Combustion Eng.	88.5	20.0
289	1953	100.	50.0	195,041	Pul. coal/Tangential	Combustion Eng.	88.5	20.0
290	1957	125.	55.0	259,426	Pul. coal/Tangential	Combustion Eng.	88.3	20.0
291	1958	125.	55.0	269,436	Pul. coal/Tangential	Combustion Eng.	88.1	20.0
292	1974	350.	139.0	587,087	Pul. coal/Tangential	Combustion Eng.	89.1	18.0
293	1974	350.	139.0	587,087	Pul. coal/Tangential	Combustion Eng.	89.1	18.0
294	1965	149.6	56.4	277,300	Pul. coal/Tangential	Combustion Eng.	89.2	18.0
295	1967	190.4	71.3	334,200	Pul. coal/Tangential	Combustion Eng.	89.1	18.0
296	1959	93.75	32.1	152,705	Pul. coal/Tangential	Combustion Eng.	89.4	17.0
297	1961	93.75	32.15	152,705	Pul. coal/Tangential	Combustion Eng.	89.4	17.0
298	1970	370.	125.	597,294	Pul. coal/Front	Foster Wheeler	88.8	18.0
299	1973	578.	197.1	934,100	Pul. coal/Opposed	Foster Wheeler	89.01	18.0
300	1953	49.	19.6	91,405	Pul. coal/Front	B & W	87.2	25.0
301	1953	49.	19.6	91,405	Pul. coal/Front	B & W	87.2	25.0
302	1954	69.	26.9	151,120	Pul. coal/Front	B & W	88.9	15.
303	1958	113.636	40.3	224,221	Pul. coal/Front	Riley Stoker	90.0	18.
304	1973	175.	70.	365,333	Pul. coal/Front	Riley Stoker	87.92	18.
305	1974	175.	70.	365,333	Pul. coal/Front	Riley Stoker	87.92	18.
306	1969	28.7	14.5	75,000	Pul. coal/Front	General Electric	90.	1.5
307	1970	623.	267.	1,730,000	Cyclone	B & W	89.1	16.
308	1973	634.5	267.	1,730,000	Cyclone	B & W	89.1	16.
309	1953	75.	34.	140,900	Pul. coal/Tangential	Combustion Eng.	87.0	25.
310	1959	231.25	93.	389,900	Pul. coal/Tangential	Combustion Eng.	87.3	31.
311	1955	73.5	30.8	141,100	Pul. coal/Tangential	Combustion Eng.	87.2	24.
312	1956	108.8	44.0	201,900	Pul. coal/Tangential	Combustion Eng.	87.2	24.
313	1954	103.	42.7	202,800	Pul. coal/Tangential	Combustion Eng.	87.0	25.
314	1964	387.	151.	647,600	Pul. coal/Tangential	Combustion Eng.	88.8	20.
315	1955	217.26	89.	400,000	Pul. coal/Front	B & W	88.8	17.-18.
316	1955	217.26	89.	400,000	Pul. coal/Front	B & W	88.8	17.-18.
317	1955	217.26	89.	400,000	Pul. coal/Front	B & W	88.8	17.-18.
318	1955	217.26	89.	400,000	Pul. coal/Front	B & W	88.8	17.-18.
319	1955	217.26	89.	400,000	Pul. coal/Front	B & W	88.8	17.-18.
320	1956	217.26	89.	400,000	Pul. coal/Front	B & W	88.8	17.-18.
321	1964	---	232.	800,000	Cyclone	B & W	90.1	20.
322	1938	---	18.	180,000	Pul. coal/Front	Foster Wheeler	85.5	20.
323	1938	---	18.	180,000	Pul. coal/Front	Foster Wheeler	85.5	20.
324	1946	---	18.3	170,000	Pul. coal/Front	B & W	86.2	13.
325	1947	---	18.3	170,000	Pul. coal/Front	B & W	86.2	13.
326	1953	---	---	150,000	Spreader Stoker	B & W	79.	33.
327	1953	---	---	150,000	Spreader Stoker	B & W	79.	33.
328	1958	113.64	42.	387,000	Pul. coal/Tangential	Combustion Eng.	87.1	--
329	1961	113.64	42.	387,000	Pul. coal/Tangential	Combustion Eng.	87.1	--
330	1951	50.	18.6	---	Pul. coal/Tangential	Combustion Eng.	85.44	--
331	1953	69.	37.9	---	Pul. coal/Tangential	Combustion Eng.	85.82	--
332	1953	690.	37.9	---	Pul. coal/Tangential	Combustion Eng.	85.82	--
333	1956	113.64	46.5	---	Pul. coal/Tangential	Combustion Eng.	87.15	--
334	1967	253.44	96.	482,000	Pul. coal/Tangential	Combustion Eng.	89.06	--
335	1969	471.00	199.	963,000	Pul. coal/Tangential	Combustion Eng.	89.06	--
336	1959	37.5	25.15	114,000	Pul. coal/Front	Riley Stoker	85.5	25.
337	1967	218.45	92.	410,000	Pul. coal/Tangential	Combustion Eng.	87.0	18.
338	1958	50.0	25.5	104,200	Pul. coal/Front	Riley Stoker	85.0	22.
339	1941	---	15.0	59,200	Pul. coal/Front	---	80.56	20.
340	1944	---	15.0	59,900	Pul. coal/Front	---	81.04	20.
341	1945	---	15.0	59,900	Pul. coal/Front	---	82.89	20.
342	1950	---	19.5	79,800	Pul. coal/Front	B & W	85.52	25.
343	1937	12.15	13.1	47,000	Pul. coal/Tangential	Combustion Eng.	84.2	5.
344	1944	12.15	14.75	52,000	Pul. coal/Front	Riley Stoker	81.8	5.
345	1949	24.15	14.35	51,000	Pul. coal/Front	Riley Stoker	85.0	5.
346	1949	24.15	14.35	50,600	Pul. coal/Front	Riley Stoker	85.0	5.
347	1961	125.	61.5	221,000	Pul. coal/Tangential	Combustion Eng.	87.4	5.
348	1954	49.	24.8	151,209	Pul. coal/Front	B & W	88.5	23.
349	1958	81.6	40.9	246,957	Pul. coal/Front	Combustion Eng.	89.22	22.
350	1954	70.	42.	79,500	Pul. coal/Front	B & W	86.5	23.
351	1964	110.	50.	109,000	Pul. coal/Front	B & W	88.4	19.
352	1958	50.0	27.0	135,321	Pul. coal/Front	Riley Stoker	87.5	25.
353	1964	138.7	55.	421,670	Cyclone	B & W	89.0	10.4
354	1972	349.2	145.	500,000	Pul. coal/Front	Foster Wheeler	87.99	20.
355	1968	212.	89.22	504,148	Pul. coal/Tangential	Combustion Eng.	86.39	20.
356	1962	65.28	24.5	114,000	Cyclone	B & W	88.7	16.
357	1966	81.6	33.	178,000	Cyclone	B & W	87.61	16.
358	1971	157.5	51.1	287,000	Pul. coal/Front	Riley Stoker	88.5	20.
359	1950	---	21.	---	Pul. coal/Tangential	Combustion Eng.	85.7	12.5
360	1969	514.8	204.8	1,196,073	Pul. coal/Tangential	Combustion Eng.	89.1	20.
361	1958	187.5	84.4	581,600	Pul. coal/Tangential	Combustion Eng.	88.68	20.
362	1960	187.5	84.4	581,600	Pul. coal/Tangential	Combustion Eng.	88.6	20.
363	1964	188.1	83.0	573,000	Pul. coal/Tangential	Combustion Eng.	88.68	20.
364	1963	265.	100.	195,833	Pul. coal/Front	B & W	89.4	20.
365	1969	737.6	291.5	1,100,000	Pul. coal/Front	Foster Wheeler	89.3	20.

(Continued)

Appendix A—Power Plant and Air Quality Data   437

TABLE 45. (Continued)

*Co. Name	Year Boiler Placed in Service	Generating Capacity, MW	Design Coal Consumption, tons/hour	Air Flow at 100% Load, scf/min	Type of Firing	Boiler Manufacturer	Boiler Efficiency at 100% Load	% Excess Air Used
366	1957	100.	38.7	203,111	Pul. coal/Front	B & W	88.8	25.
367	1971	438.	167.0	810,810	Pul. coal/Front	Combustion Eng.	89.1	20.
368	1973	511.	219.5	1,640,000	Pul. coal/Tangential	Combustion Eng.	88.67	20.
369	1954	75.	36.85	282,000	Pul. coal/Front	B & W	88.2	25.
370	1953	75.	33.8	274,000	Pul. coal/Front	B & W	90.	25.
371	1954	50.	17.9	77,700	Pul. coal/Front	B & W	88.1	18.
372	1958	46.	20.15	91,200	Pul. coal/Tangential	Combustion Eng.	88.0	18.
373	1961	50.	20.15	91,200	Pul. coal/Tangential	Combustion Eng.	88.0	18.
374	1964	75.	31.35	144,700	Pul. coal/Front	B & W	88.4	18.
375	1968	80.	31.35	144,700	Pul. coal/Front	B & W	88.4	18.
376	1970	80.	31.35	144,700	Pul. coal/Front	B & W	88.4	18.
377	1973	165.	78.4	383,597	Pul. coal/Front	B & W	89.61	25.0
378	1939	81.5	10.25	47,600	Pul. coal/Front	B & W	88.2	18.
379	1939	---	10.25	47,600	Pul. coal/Front	B & W	88.2	18.
380	1949	---	12.60	50,800	Pul. coal/Front	B & W	88.0	18.
381	1951	---	12.60	50,800	Pul. coal/Front	B & W	88.0	18.
382	1951	---	12.60	50,800	Pul. coal/Front	B & W	88.0	18.
383	1954	112.5	55.0	221,221	Pul. coal/Front	Foster Wheeler	86.3	25.
384	1956	112.5	56.0	230,384	Pul. coal/Front	Foster Wheeler	86.2	21.
385	1958	147.1	65.0	265,727	Pul. coal/Front	Foster Wheeler	86.1	21.
386	1962	163.2	78.8	295,616	Pul. coal/Front	Foster Wheeler	86.8	25.
387	1966	209.44	91.0	378,083	Pul. coal/Front	Riley Stoker	87.2	21.
388	1969	272.0	105.75	467,749	Pul. coal/Tangential	Combustion Eng.	88.2	21.
389	1972	355.5	136.5	609,000	Pul. coal/Tangential	Combustion Eng	88.28	21.
390	1974	355.5	136.5	609,000	Pul. coal/Tangential	Combustion Eng.	88.28	21.
391	1942	25.	18.6	72,050	Pul. coal/Front	Combustion Eng.	85.2	25.
392	1942	25.	18.6	72,050	Pul. coal/Front	Combustion Eng.	85.2	25.
393	1947	69.	39.7	158,000	Pul. coal/Front	Foster Wheeler	86.0	25.
394	1949	69.	39.7	158,000	Pul. coal/Front	Foster Wheeler	86.0	25.
395	1950	74.75	39.7	158,000	Pul. coal/Front	Foster Wheeler	86.0	25.
396	1952	74.75	39.7	158,000	Pul. coal/Front	Foster Wheeler	86.0	25.
397	1961	44.	20.7	82,400	Pul. coal/Front	B & W	87.4	25.
398	1947	26.	12.	76,527	Pul. coal/Opposed	Foster Wheeler	87.	26.
399	1947	26.	12.	76,527	Pul. coal/Opposed	Foster Wheeler	87.	26.
400	1958	171.7	55.	330,000	Pul. coal/Tangential	Combustion Eng.	89.47	22.
401	1962	255.	79.	382,000	Pul. coal/Tangential	Combustion Eng.	90.57	22.
402	1951	75.	26.8	114,500	Pul. coal/Tangential	Combustion Eng.	89.0	23.
403	1951	75.	26.8	114,500	Pul. coal/Tangential	Combustion Eng.	89.0	23.
404	1953	75.	26.8	114,500	Pul. coal/Tangential	Combustion Eng.	89.0	23.
405	1965	12.5	12.	73,000	Pul. coal/Front	---	86.0	19.-22.
406	1966	12.5	12.	73,000	Pul. coal/Front	---	86.0	19.-22.
407	1974	15.0	19.6	83,400	Pul. coal/Front	Erie City Iron	86.9	15.-31.
408	1968	299.2	97.	474,000	Pul. coal/Opposed	Riley Stoker	88.9	20.
409	1973	578.	197.1	951,556	Pul. coal/Opposed	Foster Wheeler	89.0	18.
410	1968	172.8	91.	318,000	Pul. coal/Tangential	Combustion Eng.	86.46	21.
411	1968	26.	18.1	164,300	Spreader Stoker	Erie City Iron	82.88	40.
412	1969	66.	45.	150,192	Cyclone	B & W	88.	12.
413	1943	21.875	9.03	55,000	Pul. coal/Opposed	Foster Wheeler	86.67	31.
414	1943	21.875	9.03	55,000	Pul. coal/Opposed	Foster Wheeler	86.67	31.
415	1951	60.	29.4	200,000	Pul. coal/Tangential	Combustion Eng.	89.10	--
416	1950	29.4	11.7	103,500	Pul. coal/Front	B & W	87.5	25.
417	1950	29.4	11.7	103,500	Pul. coal/Front	B & W	87.5	25.
418	1953	100.	33.1	250,000	Pul. coal/Tangential	Combustion Eng.	89.6	22.
419	1948	15.	10.45	131,000	Stoker	Combustion Eng.	83.9	25.
420	1948	15.	10.45	131,000	Stoker	Combustion Eng.	83.9	25.
421	1952	20.	12.75	155,000	Stoker	B & W	84.6	28.
422	1952	20.	12.75	155,000	Stoker	B & W	84.6	28.
423	1955	135.	43.9	312,800	Pul. coal/Tangential	Combustion Eng.	89.34	24.
424	1958	135.	43.9	312,800	Pul. coal/Tangential	Combustion Eng.	89.34	24.
425	1962	194.	90.	317,778	Cyclone	B & W	88.6	17.
426	1968	421.6	182.	646,667	Cyclone	B & W	88.4	16.
427	1956	138.1	55.8	242,200	Pul. coal/Tangential	Combustion Eng.	87.74	18.
428	1959	138.1	55.8	242,200	Pul. coal/Tangential	Combustion Eng.	87.74	18.
429	1959	138.1	55.8	242,200	Pul. coal/Tangential	Combustion Eng.	87.74	18.
430	1970	115.1	49.	220,000	Pul. coal/Front	B & W	88.3	19.
431	1950	140.03	20.	398,000	Cyclone	B & W	87.3	20.
432	1950	---	20.	398,000	Cyclone	B & W	87.3	20.
433	1951	---	20.	398,000	Cyclone	B & W	87.3	20.
434	1974	520.968	226.	1,214,000	Pul. coal/Front	B & W	88.3	15.
435	1953	7.5	4.6	21,700	Pul. coal/Front	---	86.98	20.
436	1953	7.5	4.6	21,700	Pul. coal/Front	---	86.98	20.
437	1967	22.	10.3	45,960	Pul. coal/Front	B & W	88.5	19.
438	1968	598.4	246.	910,000	---	B & W	88.	16.
439	1952	81.0	51.	124,480	Pul. coal/Tangential	Combustion Eng.	85.	23.
440	1954	112.5	56.	169,750	Pul. coal/Front	Foster Wheeler	87.	25.
441	1955	113.635	58.	183,330	Pul. coal/Front	B & W	87.	23.
442	1960	179.52	94.	304,530	Pul. coal/Front	B & W	87.	25.
443	1942	57.5	40.	102,000	Pul. coal/Front	B & W	86.	25.
444	1944	62.5	40.	102,000	Pul. coal/Front	B & W	86.	25.
445	1956	113.635	67.	195,470	Pul. coal/Front	B & W	87.	20.
446	1959	163.2	93.	277,160	Pul. coal/Front	B & W	84.	20.
447	1953	46.	30.	83,000	Pul. coal/Front	Riley Stoker	86.	20.
448	1949	87.87	17.8	84,204	Pul. coal/Front	B & W	86.98	25.
449	1949	87.87	17.8	84,204	Pul. coal/Front	B & W	86.98	25.
450	1957	105.	42.5	161,010	Pul. coal/Front	B & W	89.1	27.
451	1943	43.75	26.5	94,604	Pul. coal/Front	B & W	85.9	25.
452	1948	43.75	26.4	97,650	Pul. coal/Front	B & W	86.2	30.
453	1969	18.328	8.93	40,100	Pul. coal/Front	B & W	87.0	18.
454	1944	31.25	17.85	74,064	Pul. coal/Front	B & W	85.05	25.
455	1944	31.25	17.85	74,064	Pul. coal/Front	B & W	85.05	25.
456	1947	31.25	17.85	74,064	Pul. coal/Front	B & W	85.05	25.
457	1947	31.25	17.85	74,064	Pul. coal/Front	B & W	85.05	25.
458	1950	50.0	26.1	107,920	Pul. coal/Front	B & W	87.48	25.
459	1950	50.0	26.1	107,920	Pul. coal/Front	B & W	87.48	25.
460	1955	159.5	62.5	257,833	Pul. coal/Front	B & W	89.1	25.
461	1955	159.5	62.5	257,033	Pul. coal/Front	B & W	89.1	25.
462	1959	185.	72.5	350,087	Pul. coal/Front	Foster Wheeler	88.88	20.
463	1960	185.	72.5	350,087	Pul. coal/Front	Foster Wheeler	88.88	20.
464	1961	185.	72.5	350,087	Pul. coal/Front	Foster Wheeler	88.88	20.
465	1962	185.	72.5	350,087	Pul. coal/Front	Foster Wheeler	88.88	20.
466	1967	317.5	117.95	548,520	Pul. coal/Opposed	B & W	89.13	18.
467	1969	623.	234.5	1,092,254	Pul. coal/Opposed	B & W	88.99	18.
468	1971	623.	234.5	1,092,254	Pul. coal/Opposed	B & W	88.99	18.
469	1974	1300.	480.	2,500,000	Pul. coal/Opposed	B & W	88.45	20.
470	1971	816.3	291.6	975,000	Pul. coal/Opposed	Foster Wheeler	88.8	18.
471	1971	816.3	291.6	975,000	Pul. coal/Opposed	Foster Wheeler	88.8	18.
472	1953	213.	77.	278,125	Pul. coal/Front	B & W	88.8	15.
473	1954	213.	77.	278,125	Pul. coal/Front	B & W	88.8	15.
474	1957	225.	81.4	317,292	Cyclone	B & W	89.3	17.
475	1958	225.	81.4	317,292	Cyclone	B & W	89.3	17.
476	1968	590.8	247.5	800,000	Pul. coal/Front	B & W	87.4	20.
477	1955	217.26	89.	400,000	Pul. coal/Front	B & W	88.8	17.-18.
478	1955	217.26	89.	400,000	Pul. coal/Front	B & W	88.8	17.-18.
479	1955	217.26	89.	400,000	Pul. coal/Front	B & W	88.8	17.-18.
480	1955	217.26	89.	400,000	Pul. coal/Front	B & W	88.8	17.-18.

(Continued)

TABLE 45. (Continued)

*Co. Name	Year Boiler Placed in Service	Generating Capacity, MW	Design Coal Consumption, tons/hour	Air Flow at 100% Load, scf/min	Type of Firing	Boiler Manufacturer	Boiler Efficiency at 100% Load	% Excess Air Used
481	1955	217.26	89.	400,000	Pul. coal/Front	B & W	88.8	17.-18.
482	1954	77.4	34.5	150,000	Pul. coal/Tangential	Combustion Eng.	88.66	25.
483	1957	104.5	44.85	199,000	Pul. coal/Tangential	Combustion Eng.	88.18	22.
484	1959	104.5	44.85	199,000	Pul. coal/Tangential	Combustion Eng.	88.18	22.
485	1963	134.0	57.	246,000	Pul. coal/Tangential	Combustion Eng.	88.20	22.
486	1968	225.6	84.75	440,000	Pul. coal/Front	Foster Wheeler	88.79	20.
487	1959	54.4	41.1	167,000	Pul. coal/Tangential	Combustion Eng.	82.69	23.
488	1964	75.	60.5	275,000	Pul. coal/Front	B & W	83.28	17.
489	1964	151.	70.	279,200	Cyclone	B & W	90.2	16.
490	1974	265.	116.	2,325,500	Pul. coal/Tangential	Combustion Eng.	87.4	20.
491	1939	7.5	7.75	26,100	Pul. coal/Front	B & W	77.6	20.
492	1939	7.5	7.75	26,100	Pul. coal/Front	B & W	77.6	20.
493	1948	7.5	6.92	26,000	Pul. coal/Front	B & W	84.1	20.
494	1954	30.	20.5	78,200	Pul. coal/Front	Riley Stoker	86.7	22.
495	1971	665.	400.	1,524,000	Pul. coal/Tangential	Combustion Eng.	85.5	20.
496	1972	665.	400.	1,524,000	Pul. coal/Tangential	Combustion Eng.	85.5	20.
497	1974	508.6	255.	2,313,300	Pul. coal/Tangential	Combustion Eng.	88.39	20.
498	1964	38.	9.25	38,000	Spreader Stoker	Erie City Iron	79.	30.
499	1972	38.	11.75	46,000	Spreader Stoker	Erie City Iron	78.	35.
500	1969	660.	255.	1,114,012	Pul. coal/Opposed	Foster Wheeler	89.69	20.
501	1969	660.	255.	1,114,012	Pul. coal/Opposed	Foster Wheeler	89.69	20.
502	1970	936.	325.	1,412,472	Pul. coal/Tangential	Combustion Eng.	90.41	20.
503	1971	936.	325.	1,412,472	Pul. coal/Tangential	Combustion Eng.	90.41	20.
504	1944	12.5	8.85	46,363	Pul. coal/Front	Erie City Iron	83.41	15.5
505	1944	12.5	8.85	46,364	Pul. coal/Front	Erie City Iron	83.41	15.5
506	1952	46.9	21.1	94,685	Pul. coal/Tangential	Erie City Iron	88.0	15.0
507	1952	46.9	21.1	94,685	Pul. coal/Tangential	Erie City Iron	88.0	15.0
508	1967	936.	316.	1,412,472	Pul. coal/Tangential	Combustion Eng.	90.41	20.
509	1968	936.	316.	1,412,472	Pul. coal/Tangential	Combustion Eng.	90.41	20.
510	1942	35.	15.4	57,680	Pul. coal/Front	B & W	87.3	25.
511	1950	50.	15.4	57,680	Pul. coal/Front	B & W	87.1	25.
512	1957	156.2	56.	230,737	Pul. coal/Tangential	Combustion Eng.	88.59	22.
513	1954	133.	47.	215,665	Pul. coal/Front	B & W	89.88	15.
514	1954	133.	47.	215,665	Pul. coal/Front	B & W	89.88	15.
515	1959	187.	62.8	328,951	Pul. coal/Tangential	Combustion Eng.	89.73	15.
516	1960	187.	62.8	328,951	Pul. coal/Tangential	Combustion Eng.	89.73	15.
517	1948	21.2	8.7	45,300	Pul. coal/Front	Erie City Iron	85.9	15.
518	1948	21.1	8.7	45,300	Pul. coal/Front	Erie City Iron	85.9	15.
519	1949	21.2	8.7	45,300	Pul. coal/Front	Erie City Iron	85.9	15.
520	1949	21.1	8.7	45,300	Pul. coal/Front	Erie City Iron	85.9	15.
521	1967	42.5	20.5	107,550	Pul. coal/Front	Foster Wheeler	88.6	20.
522	1967	42.5	26.4	139,050	Pul. coal/Front	B & W	86.5	20.
523	1966	103.0	42.5	223,610	Pul. coal/Front	B & W	89.1	20.
524	1958	105.0	42.5	223,610	Pul. coal/Front	B & W	89.1	20.
525	1964	132.8	58.4	325,060	Pul. coal/Front	B & W	89.1	20.
526	1961	363.	135.	625,000	Pul. coal/Tangential	Combustion Eng.	88.9	20.
527	1965	405.	150.	711,000	Pul. coal/Tangential	Combustion Eng.	89.	20.
528	1969	790.	281.	1,463,000	Pul. coal/Tangential	Combustion Eng.	90.	20.
529	1954	75.	44.	248,000	Pul. coal/Front	Foster Wheeler	84.2	40.
530	1954	156.25	57.5	417,780	Pul. coal/Front	Foster Wheeler	88.	20.
531	1956	156.25	57.5	414,670	Pul. coal/Front	Foster Wheeler	88.	20.
532	1971	734.	279.	1,540,000	Pul. coal/Tangential	Combustion Eng.	90.	20.
533	1973	800.	279.	1,510,000	Pul. coal/Tangential	Combustion Eng.	90.	20.
534	1949	40.	20.7	125,000	Pul. coal/Opposed	Foster Wheeler	83.4	40.
535	1949	40.	20.7	125,000	Pul. coal/Opposed	Foster Wheeler	83.4	40.
536	1949	40.	20.7	125,000	Pul. coal/Opposed	Foster Wheeler	83.4	40.
537	1949	40.	20.7	125,000	Pul. coal/Opposed	Foster Wheeler	83.4	40.
538	1951	107.	38.7	243,000	Pul. coal/Front	Foster Wheeler	88.0	20.
539	1953	156.	48.6	376,000	Pul. coal/Front	Foster Wheeler	88.0	20.
540	1959	353.6	100.8	480,000	Pul. coal/Tangential	Combustion Eng.	89.77	15.
541	1960	353.6	104.5	500,000	Pul. coal/Tangential	Combustion Eng.	89.91	15.
542	1947	55.	23.	126,447	Pul. coal/Front	B & W	88.6	20.
543	1952	82.	31.	144,076	Pul. coal/Tangential	Combustion Eng.	89.0	23.
544	1964	364.	116.	466,000	Pul. coal/Opposed	B & W	91.2	18.
545	1965	364.	116.	466,000	Pul. coal/Opposed	B & W	91.2	18.
546	1959	190.	55.	318,400	Pul. coal/Tangential	Combustion Eng.	92.1	20.
547	1960	190.	55.	318,400	Pul. coal/Tangential	Combustion Eng.	92.1	20.
548	1962	190.	55.	318,400	Pul. coal/Tangential	Combustion Eng.	92.1	20.
549	1970	573.	186.	1,000,000	Pul. coal/Tangential	Combustion Eng.	91.8	18.
550	1971	575.	186.	1,000,000	Pul. coal/Tangential	Combustion Eng.	91.8	18.
551	1949	95.	55.	220,000	Pul. coal/Tangential	Combustion Eng.	88.9	18.
552	1950	95.	55.	220,000	Pul. coal/Tangential	Combustion Eng.	88.9	18.
553	1954	108.	55.	222,000	Pul. coal/Tangential	Combustion Eng.	91.18	18.
554	1956	108.	55.	222,000	Pul. coal/Tangential	Combustion Eng.	91.18	18.
555	1951	44.	30.85	160,000	---	B & W	84.0	28.5
556	1951	44.	30.85	155,500	---	B & W	84.0	28.5
557	1955	100.	60.25	312,000	---	B & W	85.97	26.3
558	1960	44.	20.9	132,000	Pul. coal/Front	B & W	87.78	23.
559	1957	100.	61.35	288,000	---	B & W	86.33	27.5
560	1959	110.	61.35	292,000	---	B & W	86.39	26.5
561	1962	150.	62.4	330,000	Pul. coal/Front	B & W	87.69	18.
562	1968	350.	151.1	810,000	Pul. coal/Tangential	Combustion Eng.	88.29	27.
563	1973	382.5	214.	791,000	Pul. coal/Tangential	Combustion Eng.	84.65	27.
564	1964	166.	94.25	240,000	Pul. coal/Tangential	Combustion Eng.	86.66	27.
565	1954	66.	45.9	151,000	---	B & W	84.4	28.5
566	1970	531.	237.	1,292,000	Pul. coal/Tangential	Combustion Eng.	88.85	20.
567	1972	531.	237.	1,292,000	Pul. coal/Tangential	Combustion Eng.	88.85	20.
568	1949	43.3	25.82	100,000	Pul. coal/Front	Riley Stoker	87.1	20.
569	1949	43.3	25.82	100,000	Pul. coal/Front	Riley Stoker	87.1	20.
570	1951	43.3	25.82	100,000	Pul. coal/Front	Riley Stoker	87.1	20.
571	1959	150.	64.	266,000	Pul. coal/Front	Riley Stoker	88.9	20.
572	1959	150.	64.	266,000	Pul. coal/Front	Riley Stoker	88.9	20.
573	1960	150.	64.	266,000	Pul. coal/Front	Riley Stoker	88.9	20.
574	1961	150.	64.	266,000	Pul. coal/Front	Riley Stoker	88.9	20.
575	1950	33.3	17.9	130,000	Pul. coal/Front	Riley Stoker	87.4	25.
576	1950	33.3	17.9	130,000	Pul. coal/Front	Riley Stoker	87.4	25.
577	1950	33.3	17.9	130,000	Pul. coal/Front	Riley Stoker	87.4	25.
578	1956	125.	49.6	226,000	Pul. coal/Front	Riley Stoker	88.	24.
579	1968	387.	160.0	610,000	Pul. coal/Tangential	Combustion Eng.	89.	24.
580	1953	99.	42.5	191,000	Pul. coal/Front	Foster Wheeler	88.11	24.
581	1953	99.	42.5	191,000	Pul. coal/Front	Foster Wheeler	88.11	24.
582	1954	99.	42.5	191,000	Pul. coal/Front	Foster Wheeler	88.11	24.
583	1954	99.	42.5	191,000	Pul. coal/Front	Foster Wheeler	88.11	24.
584	1960	113.636	43.5	221,000	Pul. coal/Front	Foster Wheeler	90.45	16.
585	1968	345.6	112.5	573,000	Cyclone	B & W	89.66	16.
586	1973	330.0	200.	824,827	Pul. coal/Front	Foster Wheeler	88.05	18.
587	1955	33.	17.5	102,500	Pul. coal/Front	Riley Stoker	87.	17.
588	1973	60.	28.0	196,000	Pul. coal/Tangential	Combustion Eng.	87.5	30.
589	1969	54.4	23.9	103,652	Pul. coal/Front	B & W	91.7	18.
590	1959	81.6	29.3	222,000	Pul. coal/Tangential	Combustion Eng.	88.0	25.
591	1949	46.0	18.7	137,000	Pul. coal/Tangential	Combustion Eng.	87.7	25.
592	1951	62.5	23.0	155,000	Pul. coal/Tangential	Combustion Eng.	88.0	25.
593	1953	62.5	73.0	155,000	Pul. coal/Tangential	Combustion Eng.	88.0	25.
594	1957	81.6	28.0	222,000	Pul. coal/Tangential	Combustion Eng.	89.0	25.
595	1974	750.	326.	1,317,000	Pul. coal/Tangential	Combustion Eng.	88.77	18.

(Continued)

Appendix A—Power Plant and Air Quality Data

TABLE 45. (Continued)

*Co. Name	Year Boiler Placed in Service	Generating Capacity, MW	Design Coal Consumption, tons/hour	Air Flow at 100% Load, scf/min	Type of Firing	Boiler Manufacturer	Boiler Efficiency at 100% Load	% Excess Air Used
596	1974	750.	326.	1,317,000	Pul. coal/Tangential	Combustion Eng.	88.77	18.
597	1962	139.	43.	336,000	Pul. coal/Tangential	Combustion Eng.	89.6	22.5
598	1964	139.	43.	336,000	Pul. coal/Tangential	Combustion Eng.	89.6	22.5
599	1967	220.	70.5	550,000	Pul. coal/Opposed	Foster Wheeler	89.2	22.5
600	1958	125.	43.1	336,000	Pul. coal/Tangential	Combustion Eng.	89.55	22.5
601	1958	125.	43.1	336,000	Pul. coal/Tangential	Combustion Eng.	89.55	22.5
602	1953	75.	26.8	164,500	Pul. coal/Tangential	Combustion Eng.	88.99	22.5
603	1954	75.	26.8	164,500	Pul. coal/Tangential	Combustion Eng.	88.99	22.5
604	1955	100.	36.75	229,000	Pul. coal/Tangential	Combustion Eng.	89.25	22.5
605	1970	355.8	120.	738,933	Pul. coal/Opposed	Riley Stoker	89.8	20.
606	1971	355.8	120.	738,933	Pul. coal/Opposed	Riley Stoker	89.8	20.
607	1974	315.	120.5	524,000	Pul. coal/Front	Riley Stoker	89.1	20.
608	1966	81.6	31.15	182,900	Pul. coal/Front	Riley Stoker	88.5	23.
609	1966	81.6	31.15	182,900	Pul. coal/Front	Riley Stoker	88.5	23.
610	1969	172.8	55.4	312,000	Pul. coal/Front	Riley Stoker	88.4	23.
611	1970	172.8	55.4	312,000	Pul. coal/Front	Riley Stoker	88.4	23.
612	1955	46.	23.85	183,816	Pul. coal/Front	B & W	86.4	24.
613	1966	103.7	47.5	403,774	Pul. coal/Front	B & W	86.96	26.
614	1973	265.23	110.	1,020,633	Pul. coal/Front	B & W	88.04	20.
615	1970	718.1	392.5	1,492,583	Pul. coal/Tangential	Combustion Eng.	87.46	18.
616	1971	718.1	392.5	1,492,583	Pul. coal/Tangential	Combustion Eng.	87.46	18.
617	1960	272.	100.	550,000	Pul. coal/Opposed	B & W	89.66	23.
618	1960	272.	100.	550,000	Pul. coal/Opposed	B & W	89.66	23.
619	1961	272.	100.	550,000	Pul. coal/Opposed	B & W	89.66	23.
620	1962	244.8	100.	550,000	Pul. coal/Opposed	B & W	89.66	23.
621	1963	33.0	19.0	121,560	Cyclone	B & W	88.6	10.
622	1963	33.0	19.0	121,560	Cyclone	B & W	88.6	10.
623	1963	33.0	19.0	121,560	Cyclone	B & W	88.6	10.
624	1970	335.	182.3	680,000	Pul. coal/Opposed	Riley Stoker	88.3	15.
625	1973	325.	182.1	680,000	Pul. coal/Opposed	Riley Stoker	88.3	15.
626	1957	125.	49.7	222,000	Cyclone	B & W	88.7	13.
627	1958	125.	49.7	222,000	Cyclone	B & W	88.7	13.
628	1960	179.52	64.9	296,500	Cyclone	B & W	89.6	16.
629	1963	187.5	71.3	325,000	Cyclone	B & W	89.2	16.
630	1965	239.36	93.4	423,500	Pul. coal/Opposed	Riley Stoker	88.7	15.
631	1967	414.0	151.4	696,500	Pul. coal/Opposed	Riley Stoker	88.7	15.
632	1958	330.	98.	---	Cyclone	B & W	--	13.
633	1959	330.	98.	---	Cyclone	B & W	--	13.
634	1959	330.	98.	---	Cyclone	B & W	--	13.
635	1966	950.	316.5	1,512,770	Pul. coal/Tangential	Combustion Eng.	90.08	20.
636	1954	200.	76.1	305,610	---	B & W	88.5	20.
637	1955	200.	76.1	305,610	---	B & W	88.5	20.
638	1955	223.25	76.1	305,610	---	B & W	88.5	20.
639	1955	223.25	76.1	305,610	---	B & W	88.5	20.
640	1962	550.	213.5	780,800	---	B & W	89.59	20.
641	1972	1300.	509.	2,234,600	Pul. coal/Opposed	B & W	88.87	20.
642	1973	1300.	509.	2,234,600	Pul. coal/Opposed	B & W	88.87	20.
643	1956	300.	99.5	392,927	Pul. coal/Tangential	Combustion Eng.	88.5	20.
644	1957	300.	99.5	392,927	Pul. coal/Tangential	Combustion Eng.	88.5	20.
645	1959	327.6	111.5	468,238	Pul. coal/Tangential	Combustion Eng.	89.8	20.
646	1959	327.6	111.5	468,238	Pul. coal/Tangential	Combustion Eng.	89.8	20.
647	1955	223.25	69.85	308,230	Pul. coal/Tangential	Combustion Eng.	88.85	20.
648	1955	223.25	69.85	308,230	Pul. coal/Tangential	Combustion Eng.	88.85	20.
649	1956	200.	69.85	308,230	Pul. coal/Tangential	Combustion Eng.	88.85	20.
650	1957	200.	69.85	308,230	Pul. coal/Tangential	Combustion Eng.	88.85	20.
651	1958	172.8	61.75	264,575	---	Foster Wheeler	89.66	20.
652	1959	172.8	61.75	264,575	---	Foster Wheeler	89.66	20.
653	1959	172.8	61.75	264,571	---	Foster Wheeler	89.66	20.
654	1959	172.8	61.75	264,571	---	Foster Wheeler	89.66	20.
655	1954	175.	57.9	220,476	Pul. coal/Tangential	Combustion Eng.	88.64	16.
656	1954	175.	57.9	220,476	Pul. coal/Tangential	Combustion Eng.	88.64	16.
657	1954	175.	57.9	220,476	Pul. coal/Tangential	Combustion Eng.	88.64	16.
658	1954	175.	57.9	220,476	Pul. coal/Tangential	Combustion Eng.	88.64	16.
659	1954	200.	76.5	308,230	Pul. coal/Tangential	Combustion Eng.	88.64	20.
660	1955	200.	76.5	308,230	Pul. coal/Tangential	Combustion Eng.	88.64	20.
661	1955	200.	76.5	308,230	Pul. coal/Tangential	Combustion Eng.	88.64	20.
662	1955	200.	76.5	308,230	Pul. coal/Tangential	Combustion Eng.	88.64	20.
663	1955	200.	76.5	308,230	Pul. coal/Tangential	Combustion Eng.	88.64	20.
664	1963	704.	306.	1,166,120	Cyclone	B & W	89.66	16.
665	1963	704.	306.	1,166,120	Cyclone	B & W	89.66	16.
666	1969	1150.2	434.5	1,829,000	Cyclone	B & W	89.22	20.
667	1953	175.	58.15	238,158	---	B & W	88.33	20.
668	1953	175.	58.15	238,158	---	B & W	88.33	20.
669	1953	175.	58.15	238,158	---	B & W	88.33	20.
670	1954	175.	58.15	238,158	---	B & W	88.33	20.
671	1954	175.	58.15	238,158	---	B & W	88.33	20.
672	1954	175.	58.15	238,158	---	B & W	88.33	20.
673	1954	175.	58.15	238,158	---	B & W	88.33	20.
674	1955	175.	58.15	238,158	---	B & W	88.33	20.
675	1955	175.	58.15	238,158	---	B & W	88.33	20.
676	1956	175.	58.15	238,158	---	B & W	88.33	20.
677	1942	60.	26.2	117,442	---	B & W	88.03	20.
678	1942	60.	26.2	117,442	---	B & W	88.03	20.
679	1943	60.	26.2	117,442	---	B & W	88.03	20.
680	1945	60.	26.2	117,442	---	B & W	88.03	20.
681	1960	575.01	206.	877,538	Pul. coal/Tangential	Combustion Eng.	89.62	20.
682	1964	550.	226.25	846,977	Pul. coal/Tangential	Combustion Eng.	89.83	20.
683	1938	71.	11.	89,000	---	B & W	86.1	25.
684	1941	71.	15.	126,000	---	B & W	84.2	25.
685	1941	71.	15.	126,000	---	B & W	84.2	25.
686	1951	72.	31.	262,000	---	B & W	87.6	25.
687	1949	112.5	23.	192,000	---	B & W	87.3	25.
688	1949	112.5	23.	192,000	---	B & W	87.3	25.
689	1955	140.	49.	249,400	---	B & W	89.51	23.
690	1959	140.	49.	249,400	---	B & W	89.51	23.
691	1963	140.	50.	250,000	Pul. coal/Front	B & W	90.38	18.
692	1968	218.	75.	380,500	Pul. coal/Front	B & W	90.43	17.
693	1947	15.	12.	78,000	Pul. coal/Front	Foster Wheeler	73	40.
694	1959	50.	31.	160,000	Pul. coal/Front	Foster Wheeler	79.	40.
695	1970	555.	238.	1,023,530	Pul. coal/Tangential	Combustion Eng.	88.44	23.
696	1971	555.	238.	1,023,530	Pul. coal/Tangential	Combustion Eng.	88.44	23.
697	1972	555.	238.	1,023,530	Pul. coal/Tangential	Combustion Eng.	88.44	23.
698	1973	555.	238.	1,023,530	Pul. coal/Tangential	Combustion Eng.	88.44	23.
699	1953	137.5	54.	230,000	Pul. coal/Tangential	Combustion Eng.	88.1	23.
700	1954	137.5	54.	230,000	Pul. coal/Tangential	Combustion Eng.	88.1	23.
701	1959	289.	109.7	471,000	Pul. coal/Front	Foster Wheeler	88.82	23.
702	1961	359.	133.5	574,000	Pul. coal/Front	Foster Wheeler	86.77	23.
703	1967	549.8	193.5	830,000	Cyclone	B & W	89.12	23.
704	1968	549.8	193.5	830,000	Cyclone	B & W	89.12	23.
705	1950	---	50.4	217,000	---	B & W	86.05	23.
706	1950	---	50.4	217,000	---	B & W	86.05	23.
707	1955	25.0	12.7	59,800	Pul. coal/Front	Riley Stoker	85.6	25.
708	1962	37.5	19.6	91,900	Pul. coal/Tangential	Combustion Eng.	86.7	21.
709	1964	57.8	23.0	97,500	Pul. coal/Tangential	Combustion Eng.	88.7	18.
710	1966	57.8	23.0	97,500	Pul. coal/Tangential	Combustion Eng.	88.7	18.

(Continued)

440 Electrostatic Precipitator Manual

TABLE 45. (Continued)

*Co. Name	Year Boiler Placed in Service	Generating Capacity, MW	Design Coal Consumption, tons/hour	Air Flow at 100% Load, scfm	Type of Firing	Boiler Manufacturer	Boiler Efficiency at 100% Load	% Excess Air Used
711	1974	90.	37.0	162,000	Pul. coal/Front	Riley Stoker	87.3	20.
712	1952	69.	29.8	149,500	---	Riley Stoker	87.5	18.
713	1955	113.636	38.9	191,000	---	Combustion Eng.	88.6	18.
714	1950	44.	21.3	112,800	Pul. coal/Front	Riley Stoker	81.4	20.
715	1974	411.	175.	---	Pul. coal/Tangential	Combustion Eng.	89.18	--
716	1958	152.6	62.	258,000	Pul. coal/Tangential	Combustion Eng.	87.9	21.0
717	1971	306.	175.	690,000	Pul. coal/Tangential	Combustion Eng.	86.63	21.0
718	1950	69.	30.	134,114	Pul. coal/Front	B & W	87.6	25.
719	1958	185.277	55.8	265,826	Pul. coal/Front	B & W	89.3	18.
720	1965	570.24	215.	953,872	Pul. coal/Tangential	Combustion Eng.	90.04	23.
721	1966	570.24	215.	953,872	Pul. coal/Tangential	Combustion Eng.	90.04	23.
722	1973	522.0	214.	1,278,000	Pul. coal/Tangential	Combustion Eng.	89.72	23.
723	1949	46.	18.75	94,806	Pul. coal/Tangential	Combustion Eng.	88.	20.
724	1952	50.	18.75	94,806	Pul. coal/Tangential	Combustion Eng.	88.	20.
725	1957	113.636	45.4	225,000	Pul. coal/Tangential	Combustion Eng.	88.2	20.
726	1953	120.	44.5	222,500	---	Combustion Eng.	88.76	20.
727	1954	120.	44.5	222,500	---	Combustion Eng.	88.76	20.
728	1955	130.	45.25	228,100	---	Combustion Eng.	88.68	20.
729	1957	130.	45.25	228,100	---	Combustion Eng.	88.68	20.
730	1935	80.	39.3	209,600	---	Combustion Eng.	88.08	28.
731	1943	80.	39.6	214,500	---	Combustion Eng.	87.71	30.
732	1948	80.	37.9	205,200	---	Combustion Eng.	87.49	30.
733	1949	80.	38.3	207,500	---	Combustion Eng.	88.78	30.
734	1950	80.	36.3	207,500	---	Combustion Eng.	90.23	30.
735	1968	70.	32.89	164,500	Pul. coal/Front	Riley Stoker	87.85	20.
736	1968	70.	32.89	164,500	Pul. coal/Front	Riley Stoker	87.85	20.
737	1969	70.	32.89	164,500	Pul. coal/Front	Riley Stoker	87.85	20.
738	1969	70.	32.89	164,500	Pul. coal/Front	Riley Stoker	87.85	20.
739	1931	30.	21.6	150,000	Pul. coal/Front	B & W	84	12.
740	1941	30.	21.6	150,000	Pul. coal/Front	B & W	84.	12.
741	1951	60.	48.	240,000	Cyclone	B & W	89.7	12.
742	1969	330.	133.	650,000	Cyclone	B & W	88.8	12.
743	1960	113.6	45.	220,000	Cyclone	B & W	90.0	15.
744	1962	113.6	45.	220,000	Cyclone	B & W	90.0	15.
745	1954	79.6	49.	250,000	Cyclone	B & W	89.8	15.
746	1955	79.6	49.	250,000	Cyclone	B & W	89.8	15.
747	1943	30.	18.5	130,000	Pul. coal/Front	B & W	86.3	24.
748	1947	30.	20.5	142,000	Pul. coal/Front	B & W	86.2	22.
749	1949	50.	34.5	227,000	Pul. coal/Front	B & W	86.2	25.
750	1951	60.	42.0	227,000	Pul. coal/Front	B & W	85.8	25.
751	1958	75.	45.6	223,000	Pul. coal/Front	B & W	88.05	18.
752	1964	125.	68.0	371,000	Pul. coal/Front	B & W	87.6	22.
753	1954	60.	38.7	258,000	Pul. coal/Front	B & W	86.2	23.
754	1960	75.	36.0	252,000	Pul. coal/Front	Combustion Eng.	87.7	23.

TABLE 46. POWER PLANT AND AIR QUALITY DATA FOR THOSE PLANTS WITH ELECTROSTATIC PRECIPITATORS

*Co. Name	Type Fly Ash Collector**	ESP Manufacturer	Year ESP Placed in Service	ESP Design Efficiency, %	ESP Tested Efficiency, %	Mass Emission Rate, lbs/hr	Installed Cost, $1,000's***
1	E	Buell	1971	98.5	---	634.	2,550.
2	E	Western	1972	99.0	---	---	2,301.
3	E	Western	1972	99.0	---	---	2,301.
4	E	Buell	1972	99.5	---	---	3,728.
5	E	---	--	97.0	---	---	---
6	E	---	--	97.0	---	---	---
7	C	American Standard/UOP	1954/1957	97.5	93.00	---	673.
8	E	UOP	1967	99.	99.4	440.	1,668.
9	E	UOP	1968	99.	99.4	440.	2,453.
10	E	American Standard	1972	99.5	---	313.	2,025.
11	E	American Standard	1973	99.5	---	313.	2,700.
12	E	American Standard	1974	99.5	---	313.	2,700.
13	E	Koppers	1960	90.00	73.00	350.	407.
14	E	Koppers	1958	95.	96.3-96.4	900.	632.
15	E	Buell	1959	95.	96.3-96.4	900.	706.
16	E	Research Cottrell	1969	99.	---	450.	1,958.
17	E	Research Cottrell	1970	99.	---	450.	1,401.
18	E	Research Cottrell	1971	99.	---	450.	1,417.
19	E	Buell/American Standard	1963/1973	99.5	99.5	53.	2,167.
20	E	---	1954/1968	98.	83.00	650.	934.
21	E	---	1970	97.5	97.-98.4	950. each	650.
22	E	---	1970	97.5	97.-98.4	971. each	650.
23	E	Koppers	1975	99.7	---	73.6	---
24	E	Koppers	1974	99.7	---	73.6	---
25	E	Koppers	1974	99.7	---	73.6	---
26	E	Research Cottrell	1969	97.	---	4080.	4,172.
27	E	Research Cottrell	1970	97.	---	4080.	4,172.
28	E	Research Cottrell	1969	99.	97.6-99.9	183.	579.7
29	E	Research Cottrell	1970	99.	97.6-99.9	183.	597.9
30	E	Research Cottrell	1972	99.	97.6-99.9	168.	489.0
31	C	---	1972	99.9	98.40	51.7	196.7
32	E	Western	1967	95.	---	2424.	1,351.
33	E	Western	1967	95.	---	2401.	1,351.
34	E	Research Cottrell	1973	99.21	---	180.	2,389.
35	E	Research Cottrell	1971	99.0	---	181.	525.
36	E	Buell	1973	99.43	---	91.9	1,832.
37	E	Buell	1974	99.48	---	104.	2,178.
38	E	Buell	1974	99.43	---	920.	884.
39	E	Research Cottrell	1974	99.33	---	49.6	---
40	C	Buell	1974	99.24	---	65.	---
41	E	Research Cottrell	1973	99.24	---	75.	3,050.
42	C	Buell	1975	99.46	---	87.	---
43	C	UOP	1973	99.00	---	---	2,580.
44	E	Buell	1974	99.59	---	208.	5,100.8
45	E	Buell	1974	99.53	---	380.	7,788.4
46	E	UOP	1973	99.	---	---	---
47	E	UOP	1973	99.	---	---	---
48	E	Buell	1975	99.27	---	39.	---
49	E	Buell	1975	99.27	---	39.	---

*The numbers in the first column correspond to the same plant names in Tables 45 and 46 as they do in Table 44.
**Some plants have a combination mechanical collector - electrostatic precipitator. Those with an ESP only are designated as "E" under this heading, and those with a combination collector are designated as "C".
***Costs are reported as the original costs recorded on the utility's books of accounts and unitized as prescribed in the FPC List of Units of Property effective January 1, 1961. Certain items called for in this report are not specifically unitized in the referenced list of property units. In this case the most accurate figure available is desired. In the case of stacks without foundation, the stack cost plus those added costs essential to the stack operation and support are included.

(Continued)

Appendix A—Power Plant and Air Quality Data   441

TABLE 46. (Continued)

*Co. Name	Type Fly Ash Collector**	ESP Manufacturer	Year ESP Placed in Service	ESP Design Efficiency, %	ESP Tested Efficiency, %	Mass Emission Rate, lbs/hr	Installed Cost, $1,000's***
50	E	Research Cottrell	1974	99.25	---	51.	1,448.4
51	E	Research Cottrell	1973	99.5	99.5	12.0	575.
52	E	---	--	---	---	---	---
53	E	---	--	---	---	---	---
54	E	---	--	---	---	---	---
55	E	Koppers	1949	95.	---	107.	66,250.
56	E	Koppers	1949	95.	---	107.	66,250.
57	E	Koppers	1952	96.	---	244.	140,700.
58	E	Koppers	1958	97.	---	125.	163,500.
59	E	Western	1973	99.0	---	444.	7,991.
60	E	Buell	1972	99.0	---	569.	5,031.
61	E	Western	1969	97.1	97.7-98.7	77.5	640.5
62	E	Western	1969	97.1	97.7-98.7	77.5	640.5
63	E	Research Cottrell	1970	97.1	97.7-98.7	239.6	992.
64	E	Western	1972	98.0	---	36.0	569.
65	E	Western	1972	98.0	---	30.2	569.
66	E	Western	1972	98.0	---	40.3	569.
67	E	Western	1972	98.0	---	41.1	569.
68	E	Koppers	1961	97.0	99.6	749.5	1,396.
69	E	Research Cottrell	1960	95.	83.2	2423.	---
70	E	UOP	1976	99.5	---	290.	2,987.
71	C	Research Cottrell	1960	96.00	---	400.	535.
72	E	Research Cottrell	1974	99.5	---	---	3,129.
73	E	Research Cottrell	1974	99.5	---	---	3,257.
74	E	Research Cottrell	1973	99.5	---	---	3,647.
75	C	Research Cottrell	1958	96.	---	437.7	627.
76	E	Buell	1962	96.	---	670.1	454.
77	E	Buell	1969	98.	---	246.1	1,006.
78	C	American Standard	1971	96.0	---	76.5	500.
79	C	UOP	1968	99.5	---	256.0	224.1
80	E	American Standard	1974	99.35	---	133.7	1,942.
81	E	---	--	---	---	---	---
82	E	American Standard	1971	97.5	93.00	84.	259.1
83	E	American Standard	1971	97.5	93.00	84.	259.1
84	E	American Standard	1971	97.5	93.00	84.	259.1
85	E	American Standard	1971	97.5	93.00	84.	259.1
86	C	American Standard	1968	97.5	---	13.8	128.9
87	E	American Standard	--	97.5	---	27.55	94.6
88	E	Research Cottrell	1970	98.0	---	81.	200.
89	E	Buell	1958	95.	---	576.	424.
90	C	Research Cottrell/Koppers	1958	96.	93.2	200.1	---
91	C	---	1958	96.	93.2	200.3	---
92	C	Research Cottrell/Koppers	1958	96.	93.2	200.3	---
93	C	Research Cottrell/Koppers	1958	96.	93.2	200.3	---
94	E	Research Cottrell	1949	95.	---	355.	480.
95	E	Research Cottrell	1949	95.	---	355.	468.
96	E	Buell	1959	97.	---	608.	543.
97	E	Research Cottrell	1970	99.5	97.1	224.	888.
98	E	Research Cottrell	1972	99.5	99.5	224.	1,311.4
99	E	Research Cottrell	1941	90.	---	292.	151.
100	E	Research Cottrell	1941	90.	---	292.	149.
101	E	Research Cottrell/Koppers	1951	95.	---	238.	403.
102	E	Research Cottrell/Koppers	1951	95.	---	238.	402.
103	E	Koppers	1962	99.4	98.00	110.	705.
104	E	Research Cottrell	1973	99.3	99.3	408.6	2,752.
105	E	Research Cottrell	1958	99.0	97.5	374.5	1,041.
106	E	Research Cottrell	1961	98.0	95.8	500.4	1,359.
107	E	Research Cottrell	1945	92.	93.4	221.1	122.
108	E	Research Cottrell	1953	95.	95.8	171.6	306.
109	C	Western	1949	98.	---	32.0	715.
110	C	Western	1949	98.	---	32.0	715.
111	E	Research Cottrell	1959	98.5	90.9	558.3	1,588.
112	E	Research Cottrell	1971	98.0	---	33.3	365.
113	E	Research Cottrell	1971	98.0	---	33.3	365.
114	E	Research Cottrell	1966	98.0	---	147.5	2,327.
115	E	Research Cottrell	1965	99.	---	309.2	2,135.
116	E	Research Cottrell	1965	99.	---	309.4	---
117	E	Research Cottrell	1966	99.	---	283.5	1,992.
118	E	-	1966	99.	---	283.6	---
119	E	Research Cottrell	1967	98.	94.3	235	2,251.
120	E	Research Cottrell	1968	98.	94.3	235	2,089.
121	E	Research Cottrell	1972	99.5	---	74.	---
122	E	Research Cottrell	1972	99.5	---	74.	---
123	E	Western	1961	98.	93.9	141.1	979.
124	E	Western	1955	95.	---	120.3	977.
125	E	Western	1955	95.	---	120.3	949.
126	E	Western	1955	95.	---	120.3	---
127	E	Research Cottrell	1971	98.	---	39.2	1,800.
128	E	Koppers	1958	98.	---	405.	1,205.
129	E	Koppers	1962	98.	---	401.7	1,438.
130	E	Western	1955	90.	---	380.5	394.
131	E	Research Cottrell	1973	99.0	---	35.79	3,500.
132	E	Research Cottrell	1973	98.5	---	316.25	7,000
133	E	Research Cottrell	1963	98.	---	840.8	1,560.
134	E	Research Cottrell	1929	96.	---	100.5	---
135	E	Research Cottrell	1929	96.	---	100.5	---
136	E	Research Cottrell	1929	96.	---	100.5	---
137	E	Research Cottrell	1929	97.	---	72.9	---
138	E	Research Cottrell	1929	97.	---	72.9	---
139	E	Research Cottrell	1929	97.	---	72.9	---
140	E	Research Cottrell	1938	96.6	---	74.5	---
141	E	Research Cottrell	1938	96.6	---	74.5	---
142	E	Research Cottrell	1938	96.6	---	74.5	---
143	E	Research Cottrell	1955	98.	---	297.6	519.
144	E	Research Cottrell	1962	98.	---	124.1	1,807.
145	-	Western	1953	97.	---	286.6	962.
146	-	Western	1953	97.	---	286.6	962.
147	-	Research Cottrell	1955	99.	---	184.6	2,187.
148	-	Research Cottrell	1957	99.	---	192.0	2,022.
149	-	Research Cottrell	1957	99.	---	192.0	2,079.
150	-	Research Cottrell	1965	99.	---	4944.	10,300.
151	E	Western	1969	99.	95.2-98.6	63.87	470.
152	E	Western	1969	99.	95.2-98.6	63.87	470.
153	E	Western	1969	99.	95.2-98.6	63.87	470.
154	E	Western	1968	99.	95.2-98.6	127.75	1,115.
155	E	Western	1969	99.	95.2-98.6	127.75	1,115.
156	E	Koppers	1959	95.	59-88.	1328.26	486.5
157	E	Koppers	1961	95.	59-88.	1307.32	527.5
158	E	Western	1971	99.	96.9-97.6	168.	1,311.
159	E	Western	1971	99.	96.9-97.6	168.	1,311.
160	E	Koppers	1967	95.	89.9	670.8	374.9
161	E	Buell	1967	98.	95.5	981.3	595.7
162	E	UOP	1974	99.5	---	3.27	983.5
163	E	UOP	1974	99.5	---	11.95	983.5
164	E	UOP	1974	99.5	---	1.39	983.5
165	E	UOP	1974	99.5	---	5.63	1,610.

(Continued)

TABLE 46. (Continued)

*Co. Name	Type Fly Ash Collector**	ESP Manufacturer	Year ESP Placed in Service	ESP Design Efficiency, %	ESP Tested Efficiency, %	Mass Emission Rate, lbs/hr	Installed Cost, $1,000's***
166	E	UOP	1974	99.5	---	8.08	2,212.8
167	E	Research Cottrell	1969	99.0	---	224.	811.
168	E	UOP	1974	99.5	---	3.763	929.0
169	E	UOP	1974	99.5	---	5.825	929.0
170	E	Research Cottrell	1971	97.3	97.5-97.9	1707.	1,726.
171	E	Research Cottrell	1972	97.3	97.5-97.9	1707.	1,725.
172	E	Research Cottrell	1974	98.6	---	1570.	1,887.
173	C	Buell	1958	97.5	67.9-93.2	393.	412.
174	C	Buell	1958	97.5	67.9-93.2	393.	384.
175	C	Research Cottrell	1937	95.0	67.9-93.2	873.	135.
176	C	Research Cottrell	1954	95.0	67.9-93.2	873.	135.
177	C	Research Cottrell	1940	97.0	67.9-93.2	873.	134.
178	C	Research Cottrell	1954	97.0	67.9-93.2	873.	134.
179	E	Buell	1971	98.	---	1286.	1,251.
180	E	Buell	1970	98.	---	1286.	1,486.
181	E	Buell	1972	99.5	---	321.	4,259.
182	E	Buell	1974	99.5	---	321.	2,720.
183	E	Research Cottrell	1973	99.5	99.8-99.9	---	803.
184	E	Research Cottrell	1973	99.5	99.8-99.9	---	838.
185	E	Research Cottrell	1973	99.5	99.8-99.9	---	857.
186	E	Research Cottrell	1972	99.5	99.8-99.9	321.	862.
187	E	Research Cottrell	1972	99.5	99.8-99.9	321.	827.
188	E	Research Cottrell	1972	99.5	99.8-99.9	321.	996.
189	E	Research Cottrell	1969	97.5	---	5.	---
190	C	Research Cottrell	1957	98.	---	129.2	260.
191	C	Research Cottrell	1959	98.	---	129.2	275.
192	C	UOP	1970	99.5	98.6	65.0	461.
193	C	Western	1951	98.	98.6-98.9	88.	566.
194	C	Western	1951	98.	98.6-98.9	88.	566.
195	C	Western	1951	98.	98.6-98.9	88.	565.
196	C	Western	1951	98.	98.6-98.9	88.	565.
197	E	American Standard	1968	99.6	99.4	33.	455.
198	E	Research Cottrell	1942	---	---	237.	204.
199	E	Research Cottrell	1943	---	---	237.	203.
200	E	Research Cottrell	1947	---	---	237.	258.
201	E	Research Cottrell	1948	---	---	237.	235.
202	E	Research Cottrell	1971	99.6	98.-99.	224.	4,564.
203	E	Research Cottrell	1973	99.6	98.-99.	---	4,431.
204	E	Research Cottrell	1973	99.6	98.-99.	---	4,826.
205	E	Research Cottrell	1974	99.6	---	---	4,377.
206	C	Western	1967	97.66	96.5-98.5	41.	548.
207	C	Western	1967	97.66	96.5-98.5	41.	548.
208	C	Western	1957	97.6	---	390.	1,746.
209	C	Western	1953	98.1	---	200.	789.
210	C	Western	1953	98.1	---	200.	789.
211	C	Western	1954	98.1	---	200.	789.
212	C	Western	1954	98.1	---	200.	789.
213	C	Research Cottrell	1961	98.3	---	330.	1,943.
214	E	American Standard	1969	99.6	98.7	140.	3,910.
215	C	Research Cottrell	1968	99.6	98.5-98.9	12.	497.
216	C	Research Cottrell	1968	99.6	98.5-98.9	12.	497.
217	E	Research Cottrell	1967	99.0	98.5-98.9	30.	585.
218	E	Research Cottrell	1967	99.0	98.5-98.9	30.	585.
219	C	Research Cottrell	1971	99.0	---	91.	681.
220	C	Research Cottrell	1971	99.0	---	91.	582.
221	E	Research Cottrell/Western	1973/1959	99.5	---	107.	4,359./458.
222	E	Research Cottrell/Western	1972/1960	99.5	---	107.	4,359./458.
223	E	Research Cottrell/Western	1973/1961	99.5	---	107.	4,359./458.
224	E	Research Cottrell	1974	99.7	---	280.2	---
225	E	Buell	1972	99.0	---	43.3	2,369.
226	E	Buell	1972	99.0	---	43.3	2,369.
227	E	Buell	1972	99.0	---	43.3	2,369.
228	E	Buell	1973	99.0	---	110.	2,175.
229	E	Buell	1973	99.0	---	110.	2,175.
230	E	Buell	1972	99.0	---	43.3	2,517.
231	E	Buell	1972	99.0	---	43.3	2,517.
232	E	Research Cottrell	1973	99.0	---	65.0	1,382.
233	E	Research Cottrell	1973	99.0	---	65.0	1,382.
234	E	Research Cottrell	1971	99.	98.73	69.	1,008.
235	E	Research Cottrell	1971	99.	98.73	69.	1,008.
236	E	Buell	1972	99.2	99.55	91.7	2,322.
237	E	Research Cottrell	1970	99.	---	93.	1,191.
238	E	Research Cottrell	1970	99.	---	93.	1,191.
239	E	Buell	1973	99.	---	141.	3,508.
240	C	UOP/Buell	1972	99.0	---	187.	222./2,114.
241	C	UOP/Buell	1971	99.0	---	187.	222./2,114.
242	E	Research Cottrell	1972	99.7	---	124.5	4,298.
243	E	Research Cottrell	1972	99.7	---	124.5	4,298.
244	E	Buell	1973	99.03	---	82.7	2,619.
245	E	Buell	1972	99.03	---	82.7	2,619.
246	E	Buell	1972	99.06	---	92.5	2,619.
247	E	Buell	1973	99.06	---	92.5	2,619.
248	E	Research Cottrell	1970	99.5	99.4	285.7	1,473.
249	C	Research Cottrell	1952	98.1	---	215.7	378.6
250	C	Research Cottrell	1953	97.9	---	238.3	370.6
251	C	Research Cottrell	1954	97.9	---	282.	501.7
252	C	---	1960	98.3	---	351.	552.6
253	C	Research Cottrell	1942	95.0	---	235.	185.4
254	C	Research Cottrell	1942	95.0	---	235.	175.1
255	C	Research Cottrell	1949	97.5	---	300.	442.9
256	C	Research Cottrell	1950	97.5	---	300.	477.6
257	C	Research Cottrell	1950	97.5	---	300.	493.3
258	C	Research Cottrell	1956	98.2	---	450.	711.3
259	E	American Standard	1973	98.0	---	50.1	820.
260	E	American Standard	1973	98.0	---	94.5	1,396.
261	E	American Standard	1974	96.0	---	---	1,100.
262	E	American Standard	1974	96.0	---	---	1,100.
263	E	Research Cottrell	1971/1972	98.6	98.7-98.9	146./157.	---
264	E	Research Cottrell	1972/1972	98.6	98.7-98.9	151./151.	---
265	E	Research Cottrell	1972/1972	98.6	98.7-98.9	151./151.	---
266	E	UOP	1970	98.2	---	3240.	561.962
267	E	Research Cottrell	1945	90.0	---	220.	59.
268	E	Research Cottrell	1945	90.0	---	220.	59.
269	E	Western	1973	99.4	---	26.	1,063.
270	E	Research Cottrell	1948	90.0	---	220.	51.
271	E	Buell	1971	98.7	98.0	---	1,032.
272	E	Western	1969	98.0	---	72.	510.
273	E	Western	1969	98.0	---	72.	535.
274	E	Buell	1970	98.4	---	---	1,535.
275	E	Research Cottrell	1971	98.0	98.12	---	1,270.
276	E	Research Cottrell	1972	98.0	98.12	---	1,167.
277	E	Buell	1974	99.0	---	---	4,103.
278	E	Buell	1972	99.0	99.3-99.5	260.	3,496.
279	E	Buell	1972	99.0	99.3-99.5	260.	3,496.
280	E	American Standard	1965	98.0	---	300.	331.

(Continued)

Appendix A—Power Plant and Air Quality Data   443

TABLE 46. (Continued)

*Co. Name	Type Fly Ash Collector**	ESP Manufacturer	Year ESP Placed in Service	ESP Design Efficiency, %	ESP Tested Efficiency, %	Mass Emission Rate, lbs/hr	Installed Cost, $1,000's***
281	E	Research Cottrell	1967	98.0	---	360.	510.
282	E	Research Cottrell	1968	98.5	---	540.	650.
283	E	Buell	1969	98.3	---	540.	658.
284	E	Research Cottrell	1948	98.0	78.0-94.0	143.	80.
285	E	Research Cottrell	1948	98.0	78.0-94.0	144.	80.
286	E	American Standard	1964	98.0	78.0-94.0	266.	212.
287	E	Buell	1971	98.3	---	340.	1,006.
288	E	Buell	1968	98.3	---	340.	665.
289	E	Buell	1969	98.3	---	340.	902.
290	E	Buell	1970	98.3	---	374.	824.
291	E	Buell	1968	98.3	---	374.	816.
292	E	Buell	1974	99.0	---	622.	2,247.
293	E	Buell	1974	99.0	---	622.	2,241.
294	E	American Standard	1965	98.0	---	153.	281.
295	E	American Standard	1967	98.0	---	170.	306.
296	E	Buell	1968	98.2	94.5	106.9	522.
297	E	Buell	1969	98.2	94.5	106.9	462.
298	E	Buell	1970	98.0	93.06	430.	690.
299	E	Buell	1973	98.2	94.5	544.	2,191.
300	E	Buell	1974	99.5	---	14.0	1,472.7
301	E	Buell	1974	99.5	---	14.0	1,327.6
302	C	Western	1954	97.	---	100.	---
303	E	Research Cottrell	1958	98.5	---	53.	---
304	E	Research Cottrell	1973	99.	---	112.	1,500.
305	E	Research Cottrell	1974	99.	---	112.	1,500.
306	C	UOP	1969	97.	97.	---	150.
307	E	Western	1970	99.	96.63	304.	2,900.
308	E	Western	1973	99.	96.63	69.	1,752.7
309	E	Buell	1974	99.5	---	19.	1,697.
310	E	Buell	1972	99.	99.16	175.	2,690.
311	E	Buell	1973	99.5	99.50	15.5	1,418.
312	E	Buell	1974	99.5	---	22.38	1,528.
313	C	Research Cottrell	1972	99.67	99.30	---	980.
314	E	Buell	1970	99.	99.10	285.	2,200.
315	C	Western	1955	96.1	---	642.8	565.
316	C	Western	1955	96.1	---	642.8	565.
317	C	Western	1955	96.1	---	642.8	565.
318	C	Western	1955	96.1	---	642.8	565.
319	C	Western	1956	96.1	---	642.8	565.
320	C	Western	1964	90.	65.30	3564.	10,004.
321	E	Research Cottrell	1969	97.	---	72.2	308.8
322	E	Research Cottrell	1968	97.	---	72.2	214.2
323	E	Research Cottrell/UOP	1974	90./98.93	---	109./21.86	139.2/2,522.6
324	E	Research Cottrell/UOP	1974	90./98.93	---	108./21.86	---
325	E	American Standard/Research Cottrell	1973	99.78	---	51.43	2,228.7
326	C	American Standard/Research Cottrell	1973	99.78	---	51.43	---
327	C	Western/Research Cottrell	1969	98.9	99.2	51.7	415.5
328	C	Western/Buell	1971	99.5	99.4	168.	840.89
329	E	Buell	1974	99.25	99.-99.5	36.3	1,168.5
330	C	Western/Research Cottrell	1973	99.0	99.-99.5	57.6	706.6
331	C	Western/Research Cottrell	1972	99.0	99.-99.5	57.6	606.
332	E	Buell	1971	99.0	99.-99.5	---	633.
333	E	Research Cottrell/UOP	1968/1974	98.4	97.00	167.5	718.1/1,559.
334	E	Western	1969	97.00	92.00	801.0	836.0
335	E	American Standard	1973	99.	99.70	26.7	1,314.7
336	E	American Standard	1967	98.	99.10	326.	365.
337	C	Western	1972	99.	---	32.17	602.
338	E	UOP	1969	98.	---	96.	323.
339	E	UOP	1970	98.	---	107.	386.
340	E	UOP	1970	98.	---	60.	381.
341	E	UOP	1974	99.3	---	41.	1,126.
342	E	Buell	1973	99.1	99.4	20.67	850.
343	E	Buell	1973	99.1	99.4	22.29	850.
344	E	Buell	1973	99.2	99.8	23.17	970.
345	E	Buell	1973	99.2	99.8	23.17	970.
346	E	Buell	1972	99.2	99.8	87.5	1,960.
347	E	UOP	1973	99.3	---	27.	1,324.4
348	E	UOP	1973	99.3	---	43.	1,603.8
349	E	UOP	1974	99.3	---	30.9/73.5	1,596.8
350	E	UOP	1974	99.3	---	48.3/115.5	2,036.1
351	E	Western	1973	99.0	99.51	38.2	1,375.
352	E	Research Cottrell	1972	99.	---	47.4	2,454.
353	E	Research Cottrell	1972	99.	---	183.0	1,597.
354	E	UOP	1968	98.	---	257.	367.
355	E	American Standard	1968	97.	---	50.2	247.
356	E	Research Cottrell	1966	97.	---	150.8	245.
357	E	UOP	1971	99.35	98.82	63.8	350.
358	C	Koppers	1969	97.	---	77.	852.
359	E	Buell	1969	99.	---	290.	1,294.
360	E	Research Cottrell	1964	95.	---	1527.	440.
361	E	Research Cottrell	1972	99.5	99.5	219.	---
362	E	Research Cottrell	1972	99.5	99.5	219.	---
363	E	Koppers	1970	98.5	97.40	477.	1,453.
364	E	Research Cottrell	1969	98.5	97.70	1306.	1,641.
365	E	Buell	1973	98.5	97.70	166.	1,675.1
366	E	Western	1971	98.0	97.40	1036.	759.9
367	E	Buell	1973	98.5	---	345.	756.
368	E	Research Cottrell	1973	98.5	98.5	43.12	1,687.36
369	E	UOP	1974	99.5	---	49.4	1,460.5
370	E	Western	1954	97.5	96.9-98.0	57.5	105.0
371	E	American Standard	1958	97.5	96.9-98.0	60.5	144.1
372	E	American Standard	1961	97.5	96.9-98.0	60.5	116.3
373	E	UOP	1964	97.5	96.9-98.0	94.5	108.0
374	E	UOP	1968	97.5	96.9-98.0	94.5	132.0
375	E	Research Cottrell	1970	97.5	96.9-98.0	94.5	162.0
376	E	Research Cottrell	1973	99.5	99.5	75.0	326.21
377	E	Research Cottrell	1939	97.5	---	20.1	66.3
378	E	Research Cottrell	1939	97.5	---	20.1	66.1
379	E	Research Cottrell	1949	97.5	99.5	26.8	64.2
380	E	Research Cottrell	1951	97.5	99.5	26.8	82.5
381	E	Research Cottrell	1951	97.5	99.5	26.8	82.5
382	E	Research Cottrell	1954	97.5	97.5	282.	140.
383	E	Research Cottrell	1956	97.5	97.5	286.	339.
384	E	Research Cottrell	1958	97.5	97.5	333.	433.
385	E	Research Cottrell	1962	98.5	97.5-99.7	248.	489.
386	E	Research Cottrell	1966	98.5	97.5-99.7	303.	503.
387	E	Research Cottrell	1969	99.40	99.70	103.	899.
388	E	Buell	1972	99.4	99.5	192.	1,486.
389	E	Western	1974	99.4	99.5	192.	1,709.
390	E	Western	1942	96.0	98.1	91.	56.
391	E	Research Cottrell	1942	96.0	98.1	91.	56.
392	E	Research Cottrell	1947	97.5	99.5	169.	188.
393	E	Research Cottrell	1949	97.5	99.5	169.	213.
394	E	Research Cottrell	1950	97.5	99.5	169.	230.

(Continued)

TABLE 46. (Continued)

*Co. Name	Type Fly Ash Collector**	ESP Manufacturer	Year ESP Placed in Service	ESP Design Efficiency, %	ESP Tested Efficiency, %	Mass Emission Rate, lbs/hr	Installed Cost, $1,000's***
396	E	Research Cottrell	1952	97.5	99.5	169.	247.
397	C	American Standard	1961	99.5	---	17.7	810.
398	E	Research Cottrell	1947	94.00	85.90	1350.	60.
399	E	Research Cottrell	1947	94.00	93.50	1350.	60.
400	C	Buell	1958	99.00	97.3-98.2	185.	637.
401	C	Buell	1962	99.00	97.3-98.2	299.6	794.
402	E	Buell	1974	98.5	---	60.	1280.
403	E	Buell	1965	99.0	93.0	121.	280.
404	E	Buell	1974	98.7	---	60.	1280.
405	C	American Standard	1965	97.	---	89.	75.
406	C	American Standard	1966	97.	---	89.	75.
407	E	American Standard	1975	99.8	---	12.	555.
408	E	Western	1968	98.	---	202.	393.5
409	E	Western	1973	99.0	97.3	257.2	1728.2
410	E	Research Cottrell	1968	97.	95.2	426.14	1030.
411	C	Western	1968	98.5	---	75.98	75.
412	E	Research Cottrell	1969	94.5	95.14	149.	136.
413	E	Western	1973	99.8	---	28.0	1500.
414	E	Western	1973	99.8	---	28.0	1500.
415	E	Western	1973	99.8	---	17.0	3000.
416	E	Buell	1971	98.	---	93.3	218.2
417	E	Buell	1971	98.	---	93.3	218.2
418	E	Research Cottrell	1953	98.	---	78.5	218.2
419	E	Koppers	1974	99.5	---	8.	---
420	E	Koppers	1974	99.5	---	9.	---
421	E	Koppers	1974	99.5	---	---	---
422	E	Koppers	1974	99.5	---	---	---
423	E	Research Cottrell	1955	98.	---	166.	357.5
424	E	Research Cottrell	1958	98.	---	99.5	419.8
425	E	Western	1962	99.0	---	60.8	650.
426	E	Western	1968	99.0	---	128.5	860.
427	E	American Standard	1968	98.0	---	191.	650.
428	E	American Standard	1969	98.0	---	200.	712.
429	E	American Standard	1969	98.0	---	193.	665.
430	E	American Standard	1970	98.0	---	166.	401.
431	E	Western	1969	98.0	---	44.9	620.
432	E	Western	1970	98.0	---	44.9	583.
433	E	Western	1970	98.0	---	44.9	584.
434	E	Koppers	1974	99.75	---	52.3	1,630.
435	E	Research Cottrell	1971	94.	---	42.	185.
436	E	Research Cottrell	1971	94.	---	42.	185.
437	E	Research Cottrell	1972	94.	---	56.	372.
438	E	Research Cottrell	1968	99.	98.00	599.	1,370.
439	E	Research Cottrell	1952	97.	89.1	265.	233.
440	E	Research Cottrell	1954	97.	89.1	580.	385.
441	E	Research Cottrell	1955	97.8	95.2	713.	342.
442	E	Research Cottrell	1960	97.	89.1	991.	346.
443	E	Research Cottrell	1972	99.	---	30.	1,444.
444	E	Research Cottrell	1972	99.	---	28.	1,444.
445	E	Research Cottrell	1972	99.	---	45.	1,789.
446	E	Research Cottrell	1972	99.	---	88.	2,257.
447	E	UOP	1972	99.	99.00	66.	772.
448	E	Western	1970	99.	---	43.	473.
449	E	Western	1970	99.	---	43.	473.
450	E	Western	1957	99.	---	193.	335.
451	E	Western	1969	98.	94.4-97.5	130.	512.
452	E	Western	1969	98.	94.4-97.5	129.	463.
453	E	UOP	1969	98.	---	30.	---
454	E	Western	1971	99.	99.8	48.3	1,198.
455	E	Western	1972	99.	99.8	48.3	1,198.
456	E	Western	1971	99.	99.8	48.3	1,198.
457	E	Western	1971	99.	99.8	48.3	1,198.
458	E	Western	1971	99.	99.8	70.6	1,869.
459	E	Western	1971	99.	99.8	70.6	1,869.
460	C	---	1955	97.	67.8	506.8	362.
461	C	---	1955	97.	67.8	506.8	362.
462	E	Buell	1959	97.	81.3	617.	396.
463	E	Buell	1960	97.	81.3	617.	390.
464	E	Buell	1961	97.	81.3	617.	414.
465	E	Buell	1962	97.	81.3	617.	435.
466	E	American Standard	1967	99.	87.6	333.	671.
467	E	Western	1969	99.	87.6	666.	1,239.
468	E	Western	1971	99.	87.6	666.	1,238.
469	E	Koppers	1974	99.75	---	370.	6,000.
470	E	Research Cottrell	1971	98.5	---	1511.	3,426.
471	E	Research Cottrell	1971	98.5	---	1497.	3,426.
472	E	Western	1972	99.5	94.7-97.7	95.	4,213.
473	E	Western	1972	99.5	94.7-97.7	95.	4,213.
474	E	Western	1972	98.5	94.7-97.7	293.	4,213.
475	E	Western	1971	98.5	94.7-97.7	295.	4,213.
476	E	Western	1968	96.5	94.7-97.7	2969.	1,885.
477	C	Western	1955	96.1	---	642.8	535.
478	C	Western	1955	96.1	---	642.8	535.
479	C	Western	1955	96.1	---	642.8	535.
480	C	Western	1955	96.1	---	642.8	535.
481	C	Western	1955	96.1	---	642.8	535.
482	C	Buell	1954	96.	95.6	156.8	226.
483	C	Buell	1957	96.	95.6	204.6	218.
484	C	Buell	1959	96.	95.6	204.6	332.
485	E	Buell	1963	96.	95.6	259.1	259.
486	E	Buell	1968	98.	96.10	204.8	388.
487	E	Research Cottrell	1972	99.	99.4-99.7	41.7	890.
488	E	Research Cottrell	1972	99.	99.4-99.7	26.5	1,060.
489	E	UOP	1964	97.0	97.4	214.	610.
490	E	American Standard	1974	98.0	---	1196.6	363.
491	E	Buell	1955	90.	---	---	43.0
492	E	Buell	1955	90.	---	---	43.0
493	E	Buell	1955	90.	---	---	43.0
494	E	Buell	1955	90.	---	---	43.0
495	E/E	Koppers/	1971/1974	99.4/95.	---	866./577.	---
496	E/E	Koppers/	1972/1974	99.4/95.	---	866./577.	---
497	E	---	1974	99.33	---	400.	16,720.
498	E	American Standard	1972	95.	---	23.77	115.
499	E	American Standard	1972	95.	---	23.96	115.
500	E	Buell	1969	99.5	99.5	509.6	2,579.8
501	E	Buell	1969	99.5	99.5	509.6	2,579.8
502	E	Buell	1970	99.5	99.6	430.	3,406.8
503	E	Buell	1971	99.5	99.6	430.	3,417.4
504	E	Research Cottrell	1944	83.	89.-99.1	434.	68.3
505	E	Research Cottrell	1944	83.	89.-99.1	434.	68.3
506	E	Research Cottrell	1952	98.	89.-99.1	150.	331.3
507	E	Research Cottrell	1952	98.	89.-99.1	150.	331.3
508	E	Research Cottrell	1967	99.5	93.3	410.	2,754.1
509	E	Research Cottrell	1968	99.5	93.3	430.	2,738.6
510	C	Research Cottrell	1950	98.	97.6-98.4	112.8	379.3

(Continued)

Appendix A—Power Plant and Air Quality Data   445

TABLE 46. (Continued)

*Co. Name	Type Fly Ash Collector**	ESP Manufacturer	Year ESP Placed in Service	ESP Design Efficiency, %	ESP Tested Efficiency, %	Mass Emission Rate, lbs/hr	Installed Cost, $1,000's***
511	C	Research Cottrell	1950	98.	97.6-98.4	112.8	379.3
512	C	Research Cottrell	1957	98.	97.6-98.4	358.	565.1
513	C	Research Cottrell	1954	95.	89.6-96.4	805.	895.4
514	C	Research Cottrell	1954	95.	89.6-96.4	805.	895.4
515	C	Research Cottrell	1959	95.	89.6-96.4	1170.	627.2
516	C	Research Cottrell	1959	95.	89.6-96.4	1170	627.2
517	E	Research Cottrell	1948	94.	80.9-94.	131.2	61.6
518	E	Research Cottrell	1948	94.	80.9-94.	131.2	61.6
519	E	Research Cottrell	1949	94.	80.9-94.	191.2	61.6
520	E	Research Cottrell	1949	94.	80.9-94.	131.2	61.6
521	E	Research Cottrell	1967	98.	---	143.8	322.
522	E	Research Cottrell	1967	98.	---	185.9	293.
523	E	Buell	1966	98.	---	298.8	434.
524	E	Buell	1958	95.	---	747.4	162.
525	E	Buell	1954	98.	---	385.	300.
526	E	Research Cottrell/Buell	1961/1966	99.	---	387.	1,426.
527	E	Research Cottrell	1965	98.	---	2438.	1,052.
528	E	Western	1969	99.5	---	811.	1,100.
529	C	Research Cottrell	1954	94.3	---	277.	1,035.
530	E	Buell	1970	99.5	---	118.	---
531	E	Buell	1970	99.5	---	67.	---
532	E	Western	1971	99.5	---	2129.	2,388.
533	E	Western	1973	99.5	---	388.	2,337.
534	C	Western	1949	70.	---	---	439.6
535	C	Western	1949	70.	---	---	439.6
536	C	Western	1949	70.	---	---	439.6
537	C	Western	1949	70.	---	---	439.6
538	E	Research Cottrell	1951	96.	---	442.	910.
539	E	Research Cottrell	1953	96.	---	777.	910.
540	C	American Standard/Western	1959	95.	99.-99.4	173.	990.
541	C	American Standard/Western	1960	95.	99.-99.4	179.	990.
542	E	Research Cottrell	1947	96.0	95.6	166.3	116.
543	C	Research Cottrell	1952	98.4	96.1	56.0	177.
544	E	Research Cottrell	1964	97.5	94.9-96.6	847.	456.
545	E	Research Cottrell	1965	97.5	94.9-96.6	847.	451.
546	E	Research Cottrell	1959	97.5	90.9-95.9	1759.	465.
547	E	Research Cottrell	1960	97.5	90.9-95.9	1759.	466.
548	E	Research Cottrell	1962	97.5	90.9-95.9	1759.	378.
549	E	Research Cottrell	1970	99.5	98.84	---	695.
550	E	Research Cottrell	1971	99.5	98.84	---	695.
551	C	Research Cottrell	1949	99.3	87.6	1096.	215.
552	C	Research Cottrell	1950	99.3	87.6	1096.	215.
553	C	Research Cottrell	1954	99.7	98.1	1096.	250.
554	C	Research Cottrell	1956	99.7	98.1	1096.	243.
555	C	UOP	1969	97.5	93.2-97.5	270.	416.5
556	C	UOP	1966	90.0	93.2-97.5	778.	244.6
557	C	UOP	1965	87.0	93.2-97.5	1200.	425.6
558	E	UOP	1965	87.0	---	315.	237.12
559	E	Western	1965	90.0	---	1260.	349.73
560	E	Research Cottrell	1968	94.2	---	1135.	720.01
561	E	Western	1964	87.0	---	900.	340.53
562	E	Koppers	1968	87.0	---	1615.	744.01
563	E	Research Cottrell	1973	---	---	3457.	7,000.
564	E	Research Cottrell	1964	87.0	61.85	567.0	518.98
565	E	Buell	1962	97.0	---	842.	198.5
566	E	Western	1970	99.	91.15-93.78	7031.8	1,035.
567	E	Western	1972	99.	91.15-93.78	7229.8	1,000.
568	E	Buell	1973	98.6	98.8	64.94	874.
569	E	Buell	1973	98.6	98.8	64.94	874.
570	E	Buell	1973	98.6	98.8	64.94	874.
571	E	Western	1969	99.	99.	172.	1,493.
572	E	Western	1969	99.	99.	172.	1,493.
573	E	Western	1968	99.	99.	172.	1,493.
574	E	Western	1968	99.	99.	172.	1,493.
575	E	Western	1972	98.	98.-98.6	42.82	---
576	E	Western	1972	98.	98.-98.6	42.82	---
577	E	Western	1972	98.	98.-98.6	42.82	---
578	E	Research Cottrell	1969	98.5	98.5	147.	1,354.
579	E	Research Cottrell	1968	98.	98.0	407.	819.
580	E	Research Cottrell	1971	98.5	98.5	121.	1,133.
581	E	Research Cottrell	1970	98.5	98.5	121.	1,181.
582	E	Research Cottrell	1971	98.5	98.5	121.	1,133.
583	E	Research Cottrell	1969	98.5	98.5	121.	1,130.
584	E	UOP	1960	90.0	88.9	537.	259.
585	E	UOP	1968	92.4	97.5	194.	1,172.
586	E	Western	1973	99.5	---	47.0	4,185.
587	E	---	1974	99.8	---	---	200.
588	E	---	1973	99.8	---	---	300.
589	E	Research Cottrell	1969	99.	---	27.5	192.
590	E	Research Cottrell	1959	97.5	95.0	405.	358.8
591	E	Research Cottrell	1949	97.5	90.4-95.3	39.4	142.5
592	E	Research Cottrell	1951	97.5	90.4-95.3	42.3	169.7
593	E	Research Cottrell	1953	97.5	90.4-95.3	42.3	182.4
594	E	Research Cottrell	1957	97.5	90.4-95.3	102.2	217.9
595	E	Western	1974	99.5	---	420.	13,000.
596	E	Western	1974	99.5	---	420.	13,000.
597	E	Research Cottrell	1972	99.6	---	2055.	---
598	E	Research Cottrell	1972	99.6	---	3263.	---
599	E	Research Cottrell	1970	99.6	---	11801.	---
600	E	Research Cottrell	1971	99.9	---	8.2	494.8
601	E	Research Cottrell	1970	99.9	---	8.9	476.8
602	E	Research Cottrell	1968	99.6	99.0-99.12	31.5	360.15
603	E	Research Cottrell	1968	99.6	99.0-99.12	48.1	388.69
604	E	Research Cottrell	1969	99.6	99.0-99.12	66.5	447.91
605	E	Research Cottrell	1970	99.	---	234.39	736.60
606	E	Research Cottrell	1971	99.	---	234.39	620.62
607	E	Research Cottrell	1974	99.	---	330.	1,298.
608	E	American Standard/PCW	1966/1975	98.	39.-50./---	170.	211.5/600.
609	E	American Standard/PCW	1966/1975	98.	39.-50./---	170.	211.5/600.
610	E	Buell	1969	95.	92.1-94.5	662.	177.
611	E	Buell	1970	95.	92.1-94.5	662.	177.
612	C	Western	1973	97.6	99.0	50.1	733.
613	E	Western	1974	99.28	---	101.6	1,611.
614	E	---	1973	99.0	---	300.1	1,535.
615	E	Research Cottrell	1970	98.6	99.-99.8	600.	2,317.
616	E	Research Cottrell	1971	98.6	99.-99.8	600.	2,317.
617	E	Western	1960	95.	---	800.	---
618	E	Western	1960	95.	---	800.	---
619	E	Research Cottrell	1974	99.	99.6	714.	---
620	E	Research Cottrell	1973	99.	99.6	714.	---
621	C	American Standard	1974	96.	---	9.5	720.34
622	C	American Standard	1974	96.	---	11.0	720.34
623	C	American Standard	1974	96.	---	10.6	720.34
624	E	Western	1970	99.0	---	504.	540.
625	E	Western	1973	99.78	98.4	219.	---
626	E	Research Cottrell	1957	90.	---	354.	250.

(Continued)

TABLE 46. (Continued)

*Co. Name	Type Fly Ash Collector**	ESP Manufacturer	Year ESP Placed in Service	ESP Design Efficiency, %	ESP Tested Efficiency, %	Mass Emission Rate, lbs/hr	Installed Cost, $1,000's***
627	E	Research Cottrell	1958	90.	---	354.	216.
628	E	Research Cottrell	1960	93.	---	220.2	325.
629	E	American Standard	1963	95.5	---	174.7	249.
630	E	Research Cottrell	1965	98.5	97.2	291.	494.
631	E	American Standard/Research Cottrell	1967/1974	98.5/99.78	---	484./183.	567./7,113.
632	E	---	1972	99.0	97.5	216.9	3,261.
633	E	---	1971	99.0	97.5	216.9	3,261.
634	E	---	1972	99.0	97.5	216.9	3,261.
635	E	American Standard	1966	99.0	81.00	616.	1,606.
636	E	---	1971	97.0	---	155.3	1,982.5
637	E	---	1972	97.0	---	155.3	1,982.5
638	E	---	1972	97.0	---	155.3	1,982.5
639	E	---	1972	97.0	---	155.3	1,982.5
640	E	Koppers	1962	90.0	80.00	4804.	625.
641	E	American Standard	1972	99.0	99.06-99.1	805.4	3,887.5
642	E	American Standard	1973	99.0	99.06-99.1	805.4	3,887.5
643	E	Research Cottrell	1969	95.	---	150.	1,323.8
644	E	Research Cottrell	1970	95.	---	150.	1,323.8
645	E	Research Cottrell	1970	95.	---	295.	1,323.8
646	E	Research Cottrell	1969	95.	---	295.	1,323.8
647	E	---	1973	98.5	99.2-99.3	157.7	3,269.4
648	E	---	1973	98.5	99.2-99.3	157.7	3,269.4
649	E	---	1974	98.5	---	157.7	3,269.4
650	E	---	1974	98.5	---	157.7	3,269.4
651	E	---	1974	98.5	---	180.	1,327.3
652	E	---	1974	98.5	---	180.	1,327.3
653	E	---	1974	98.5	---	180.	1,327.3
654	E	---	1974	98.5	---	180.	1,327.3
655	E	Research Cottrell	1960	95.	---	84.6	155.2
656	E	Research Cottrell	1960	95.	---	84.6	155.2
657	E	Research Cottrell	1960	95.	---	84.6	155.2
658	E	Research Cottrell	1960	95.	---	84.6	155.2
659	E	Research Cottrell	1959	95.	---	112.	217.6
660	E	Research Cottrell	1960	95.	---	112.	217.6
661	E	Research Cottrell	1960	95.	---	112.	217.6
662	E	Research Cottrell	1960	95.	---	112.	217.6
663	E	Research Cottrell	1960	95.	---	112.	217.6
664	E	American Standard	1967	98.	---	672.	1,441.8
665	E	American Standard	1967	98.	---	672.	1,428.2
666	E	American Standard	1969	98.	---	829.	2,901.3
667	E	Research Cottrell	1970	90.	---	170.	710.6
668	E	Research Cottrell	1969	90.	---	170.	710.6
669	E	Research Cottrell	1969	90.	---	170.	710.6
670	E	Research Cottrell	1969	90.	---	170.	710.6
671	E	Research Cottrell	1970	90.	---	170.	710.6
672	E	Research Cottrell	1969	90.	---	170.	710.6
673	E	Research Cottrell	1970	90.	---	170.	710.6
674	E	Research Cottrell	1970	90.	---	170.	710.6
675	E	Research Cottrell	1969	90.	---	170.	710.6
676	E	Research Cottrell	1969	90.	---	170.	710.6
677	E	Research Cottrell	1969	95.00	95.00	117.	460.
678	E	Research Cottrell	1969	95.00	95.00	117.	460.
679	E	Research Cottrell	1969	95.00	95.00	117.	460.
680	E	Research Cottrell	1969	95.00	95.00	117.	460.
681	E	Western	1960	95.00	50.00	4017.	1,809.
682	E	Koppers	1964	90.0	99.00	5119.	648.
683	E	Western	1970	99.5	97.4-98.7	12.	457.
684	E	Western	1971	99.5	97.4-98.7	18.	641.
685	E	Western	1970	99.5	97.4-98.7	18.	421.
686	C	Research Cottrell	1951	98.7	97.4-98.7	151.	310.
687	C	Research Cottrell	1950	97.4	97.4-98.7	248.	245.
688	C	Research Cottrell	1950	97.4	97.4-98.7	248.	245.
689	C	UOP	1955	98.5	---	156.	578.
690	C	UOP	1959	98.8	---	161.	642.
691	E	Western	1963	98.5	---	101.	419.
692	E	Western	1968	99.5	95.1	101.	688.
693	C	Research Cottrell	1969	93.8	---	1250.	1,325.
694	C	Research Cottrell	1959	95.0	91.0	1700.	353.
695	E	Research Cottrell	1970	99.5	96.	161.6	2,169.
696	E	Research Cottrell	1971	99.5	96.	161.6	2,669.
697	E	Research Cottrell	1972	99.5	96.	161.6	2,900.
698	E	Research Cottrell	1973	99.5	96.	161.6	4,538.
699	E	Research Cottrell	1953	97.5	---	200.	425.
700	E	Research Cottrell	1954	97.5	---	200.	470.
701	E	Research Cottrell	1959	98.0	97.4	233.	981.
702	E	Research Cottrell	1961	97.5	---	465.	1,097.
703	E	Research Cottrell	1973	99.6	---	52.	2,973.
704	E	Research Cottrell	1973	99.6	---	52.	2,912.
705	E	Research Cottrell	1950	95.	88.5-90.5	502.	263.
706	E	Research Cottrell	1950	95.	88.5-90.5	502.	263.
707	E	Buell	1972	95.	---	27.	666.
708	E	Buell	1972	95.	---	46.	1,038.
709	E	Buell	1972	95.	---	47.	1,026.
710	E	Buell	1972	95.	---	47.	1,017.
711	E	Buell	1974	99.6	---	72.	1,045.
712	C	Western	1952	97.0	---	---	298.6
713	C	Western	1955	97.0	---	---	382.2
714	C	Western	1950	97.0	97.0	446.	148.399
715	E	Buell	1974	99.5	---	226.9	5,000.
716	E	Lodge Cottrell	1974	97.0	---	74.2	7,000.
717	E	Buell	1971	96.0	---	158.	814.
718	E	Western	1973	99.38	99.7-99.75	47.6	1,595.
719	E	Western	1973	99.38	99.7-99.75	79.0	2,193.
720	E	Research Cottrell	1973	99.83	---	110.5	9,748.
721	E	Research Cottrell	1973	99.83	---	110.5	9,374.
722	E	Research Cottrell	1973	99.2	85.00	660.0	4,608.
723	E	Research Cottrell	1949	95.0	---	9.88	184.
724	E	Research Cottrell	1952	95.0	---	9.88	140.
725	E	Research Cottrell	1957	97.5	---	19.76	310.
726	E	Research Cottrell	1970	99.5	98.73	72.	---
727	E	Research Cottrell	1970	99.5	98.73	72.	1,575.
728	E	Research Cottrell	1967	99.0	98.4	72.	583.
729	E	Research Cottrell	1967	99.0	98.4	72.	600.
730	E	Research Cottrell	1967	99.2	---	42.	534.
731	E	Research Cottrell	1968	99.2	---	42.	486.
732	E	Research Cottrell	1967	99.2	---	42.	513.
733	E	Research Cottrell	1965	99.2	---	42.	474.
734	E	Research Cottrell	1966	99.2	---	42.	510.
735	E	Research Cottrell	1968	99.0	98.43-99.36	47.	---
736	E	Research Cottrell	1968	99.0	98.43-99.36	47.	643.
737	E	Research Cottrell	1969	99.0	98.43-99.36	47.	---
738	E	Research Cottrell	1969	99.0	98.43-99.36	47.	586.
739	E	Research Cottrell	1951	90.0	---	150.	126.
740	E	Research Cottrell	1951	90.0	---	150.	126.
741	E	Research Cottrell	1973	99.5	---	9.3	2,000.

(Continued)

TABLE 46. (Continued)

*Co. Name	Type Fly Ash Collector**	ESP Manufacturer	Year ESP Placed in Service	ESP Design Efficiency, %	ESP Tested Efficiency, %	Mass Emission Rate, lbs/hr	Installed Cost, $1,000's***
742	E	Buell	1969	99.0	---	118.	854.
743	E	Research Cottrell	1974	99.5	---	114.0	2,400.
744	E	Research Cottrell	1974	99.5	---	114.0	2,400.
745	E	Research Cottrell	1971	99.5	---	8.6	1,350.
746	E	Research Cottrell	1972	99.5	---	8.6	1,350.
747	E	Research Cottrell	1973	98.0	---	76.1	713.
748	E	Research Cottrell	1974	98.0	---	37.9	760.
749	E	Research Cottrell	1974	97.0	---	205.7	970.
750	E	Research Cottrell	1951	90.0	88.3	258.6	237.
751	E	Western	1958	92.0	---	98.5	296.
752	E	Western	1974	97.0	---	220.1	1,632.2
753	E	Western	1972	99.0	99.7-99.8	10.8	1,857.
754	E	Western	1972	99.0	99.7-99.8	9.9	1,220.

Appendix B

Cascade Impactor Stage Parameters

TABLE 47. CASCADE IMPACTOR STAGE PARAMETERS
ANDERSEN MARK III STACK SAMPLER

Stage No.	No. of Jets	D_j-Jet Diameter (cm)	S-Jet to Plate Distance (cm)	$\frac{S}{D_j}$	Reynolds Number	Jet Velocity (m/sec)	Cumulative Fraction of Impactor Pressure Drop at each stage
1	264	.1638	.254	1.55	45	0.4	0.0
2	264	.1253	.254	2.03	59	0.8	0.0
3	264	.0948	.254	2.68	78	1.3	0.0
4	264	.0759	.254	3.35	98	2.0	0.0
5	264	.0567	.254	4.48	131	3.6	0.0
6	264	.0359	.254	7.08	206	9.0	0.2
7	264	.0261	.254	9.73	284	17.1	0.3
8	156	.0251	.254	10.12	500	31.5	1.0

MODIFIED BRINK MODEL B CASCADE IMPACTOR

Stage No.	No. of Jets	D_j-Jet Diameter (cm)	S-Jet to Plate Distance (cm)	$\frac{S}{D_j}$	Reynolds Number	Jet Velocity (m/sec)	Cumulative Fraction of Impactor Pressure Drop at each stage
0	1	.3598	1.016	2.82	326	1.4	0.0
1	1	.2439	0.749	3.07	481	3.0	0.0
2	1	.1755	0.544	3.10	669	6.0	0.0
3	1	.1375	0.424	3.08	853	9.7	0.0
4	1	.0930	0.277	2.98	1263	21.2	0.065
5	1	.0726	0.213	2.93	1617	35.3	0.255
6	1	.0573	0.191	3.33	2049	58.8	1.000

MRI MODEL 1502 INERTIAL CASCADE IMPACTORS

Stage No.	No. of Jets	D_j-Jet Diameter (cm)	S-Jet to Plate Distance (cm)	$\frac{S}{D_j}$	Reynolds Number	Jet Velocity (m/sec)	Cumulative Fraction of Impactor Pressure Drop at each stage
1	8	0.870	0.767	.88	281	0.5	0.0
2	12	0.476	0.419	.88	341	1.1	0.0
3	24	0.205	0.191	.96	411	3.2	0.0
4	24	0.118	0.191	1.61	684	8.9	0.0
5	24	0.084	0.191	2.27	973	18.2	0.045
6	24	0.052	0.191	3.60	1530	45.9	0.216
7	12	0.052	0.191	3.60	3059	102.3	1.000

(Continued)

TABLE 47. (Continued)

SIERRA MODEL 226 SOURCE SAMPLER

Stage No.	W-Jet Slit Width (cm)	Jet Slit Length (cm)	S-Jet to Plate Distance (cm)	$\frac{S}{W}$	Reynolds Number (@14.16 lpm)	Jet Velocity (m/sec) (@14.16 lpm)	Cumulative Fraction of Impactor Pressure Drop at each Stage
1	0.3590	5.156	0.635	1.77	602	1.3	0.0
2	0.1988	5.152	0.318	1.60	602	2.3	0.0
3	0.1147	3.882	0.239	2.08	800	5.4	0.0
4	0.0627	3.844	0.239	3.81	808	10.0	0.154
5	0.0358	3.869	0.239	6.68	802	17.4	0.308
6	0.0288	2.301	0.239	8.30	1348	36.9	1.000

UNIVERSITY OF WASHINGTON MARK III SOURCE TEST CASCADE IMPACTOR

Stage No.	No. of Jets	D_j-Jet Diameter (cm)	S-Jet to Plate Distance (cm)	$\frac{S}{D_j}$	Reynolds Number	Jet Velocity (m/sec)	Cumulative Fraction of Impactor Pressure Drop at each Stage
1	1	1.842	1.422	.78	1073	0.9	0.0
2	6	0.577	0.648	1.12	565	1.5	0.0
3	12	0.250	0.318	1.27	653	4.1	0.0
4	90	0.0808	0.318	3.94	269	5.2	0.019
5	110	0.0524	0.318	6.07	340	10.2	0.057
6	110	0.0333	0.318	9.55	535	25.4	0.189
7	90	0.0245	0.318	12.98	929	60.0	1.000

Appendix C

Particulate Matter, SO_x and NO_x Emission Limits

TABLE 48. PARTICULATE MATTER, SULFUR OXIDE, AND NITROGEN OXIDE EMISSION LIMITS FOR COAL-FIRED POWER BOILERS IN THE UNITED STATES[1,2]

State	Particulate Matter	Sulfur Oxides	Nitrogen Oxides
Alabama	0.12 lb/10⁶ Btu for existing sources with input > 250 x 10⁶ Btu/hr 0.10 lb/10⁶ Btu for new sources with input > 250 x 10⁶ Btu/hr	Category I Counties/1.8 lb SO_2/10⁶ Btu heat input-existing, 1.2 lb-new Category II Counties/4.0 lb SO_2/10⁶ Btu heat input-existing, 1.2 lb-new	0.7 lb NO_x/10⁶ Btu for new > 250 x 10⁶ Btu/hr
Alaska	0.05 grains/scf except 0.10 grains/scf prior to 7/1/72	500 ppm as SO_2	No standards
Arizona	Emission Rate = $17.0Q^{0.433}$ for input > 4200x10⁶ Btu/hr Emission Rate = $1.02Q^{0.769}$ for input < 4200x10⁶ Btu/hr	0.80 lb SO_2/10⁶ Btu heat input - new 1.0 lb SO_2/10⁶ Btu heat input - existing	0.7 lb NO_x/10⁶ Btu heat input for new sources (maximum 2 hour average)
Arkansas	Emission Rate = $17.31P^{0.16}$ for input > 60,000 lbs/hr after July, 1973 where P = process weight, tons/hr	0.2 ppm SO_2 for any 30 min. avg. beyond source premises	No standards
California	Each county has own regulations. See Table 4a for summary from counties responding to SoRI survey.		
Colorado	0.10 lb/10⁶ Btu for units with input ≥ 500 x 10⁶ Btu/hr	500 ppm	0.7 lb NO_x/10⁶ Btu heat input
Connecticut	0.10 lb/10⁶ Btu heat input - new 0.20 lb/10⁶ Btu heat input - existing	Fuels restricted to maximum S Content of 0.5% by weight	0.7 lb NO_x/10⁶ Btu - new above 250 x 10⁶ Btu/hr input 0.9 lb NO_x/10⁶ Btu - existing above 250 x 10⁶ Btu/hr input
Delaware	0.10 lb/10⁶ Btu heat input for new sources > 250 x 10⁶ Btu/hr	Fuel restricted to 1% S by weight .8 lb SO_2/10⁶ Btu for sources > 250 x 10⁶ Btu/hr	0.7 lb NO_x/10⁶ Btu for new sources > 250 x 10⁶ Btu/hr
Florida	0.1 lb/10⁶ Btu for new sources > 250 x 10⁶ Btu/hr (maximum 2 hour average)	1.2 lb/10⁶ Btu for new sources > 250 x 10⁶ Btu/hr (maximum 2 hour average) 1.5 lb/10⁶ Btu for existing > 250 x 10⁶ Btu/hr	0.7 lb NO_x/10⁶ Btu heat input (maximum 2 hour average)
Georgia	0.1 lb/10⁶ Btu for new sources > 250 x 10⁶ Btu/hr	1.2 lb/10⁶ Btu for new sources > 250 x 10⁶ Btu/hr (maximum 2 hour average)	0.7 lb NO_x/10⁶ Btu heat input for sources > 250 x 10⁶ Btu/hr
Hawaii	No standards	No standards for coal	No standards
Idaho	0.10 lb/10⁶ Btu for new sources > 250 x 10⁶ Btu/hr 0.12 lb/10⁶ Btu for existing (before 12/5/74) sources > 10,000 x 10⁶ Btu/hr	Coal limited to 1% sulfur by weight - existing 1.2 lb SO_2/10⁶ Btu for sources > 250 x 10⁶ Btu/hr	0.7 lb NO_x/10⁶ Btu for new sources ≥ 250 x 10⁶ Btu/hr
Illinois	0.10 lb/10⁶ Btu for new and existing sources > 250 x 10⁶ Btu/hr in any one hour	1.8 lb SO_2/10⁶ Btu in any one hour for major metro areas - existing 1.2 lb SO_2/10⁶ Btu new sources > 250 x 10⁶ Btu/hr	0.7 lb/10⁶ Btu new sources > 250 x 10⁶ Btu/hr (maximum 1 hour period)
Indiana	0.10 lb/10⁶ Btu for new sources > 250 x 10⁶ Btu/hr 0.6 lb/10⁶ Btu for new sources ≤ 250 x 10⁶ Btu/hr	1.2 lb SO_2/10⁶ Btu for new sources > 250 x 10⁶ Btu/hr	0.7 lb/10⁶ Btu new sources ≥ 250 x 10⁶ Btu/hr
Iowa	0.6 lb/10⁶ Btu for new sources 0.8 lb/10⁶ Btu for existing outside SMSA* 0.6 lb/10⁶ Btu for existing inside SMSA*	1.2 lb SO_2/10⁶ Btu for new sources > 250 x 10⁶ Btu/hr	No standards
Kansas	0.12 lb/hr/10⁶ Btu for input ≥ 10,000 x 10⁶ Btu/hr	1.5 lb/10⁶ Btu/hr for input ≥ 250 x 10⁶ Btu/hr	0.90 lb/10⁶ Btu/hr for input ≥ 250 x 10⁶ Btu/hr
Kentucky	0.10 lb/10⁶ Btu for input ≥ 250 x 10⁶ Btu/hr	1.2 lb/10⁶ Btu/hr for input of 250 x 10⁶ Btu/hr	0.7 lb/10⁶ Btu/hr for input ≥ 250 x 10⁶ Btu/hr
Louisiana	0.6 lb/10⁶ Btu heat input for new and existing not subject to Federal Regulations	2000 ppm by volume	No standards
Maine	0.1 lb/10⁶ Btu for input > 250 x 10⁶ Btu/hr	0.8 lb/10⁶ Btu for input > 250 x 10⁶ Btu/hr	No standards

[1] The Electrostatic Precipitator Manual by the McIlvaine Co., Chapter XIII, Section 4.1, pp. 53.1-54.0, August, 1977.
[2] Survey of all state air pollution agencies by Southern Research Institute in 1978.
*Standard metropolitan statistical area.

(Continued)

TABLE 48. (Continued)

State	Particulate Matter	Sulfur Oxides	Nitrogen Oxides
Maryland	0.03 grains/scf for input > 250 x 10^6 Btu/hr (new and existing sources)	Fuel limited to 1% sulfur in Areas I, III, IV 3.5 lb/10^6 Btu for input of 100 x 10^6 Btu/hr for Areas II, V, VI	0.5 lb/10^6 Btu (maximum 2 hour average) for new sources > 250 x 10^6 Btu/hr
Massachusetts	0.05 lb/10^6 Btu for new sources > 250 x 10^6 Btu/hr 0.15 lb/10^6 Btu for existing	0.28 lb SO_x/10^6 Btu in some areas, 0.55 lb sulfur/10^6 Btu in others	0.3 lb NO_x/10^6 Btu for new sources > 250 x 10^6 Btu/hr
Michigan	For pulverized coal equipment rated larger than 10^6 lb steam/hr or other modes of firing coal rated larger than 3 x 10^5 lb steam/hr, one must apply to commission for specific limits	1% sulfur coal as of 7/1/78	No standards
Minnesota	0.1 lb/10^6 Btu for new sources > 250 x 10^6 Btu/hr for most of the state	1.2 lb SO_x/10^6 Btu for new sources > 250 x 10^6 Btu/hr for most of state	0.7 lb NO_x/10^6 Btu for new sources > 250 x 10^6 Btu/hr
Mississippi	0.19 lb/10^6 Btu for input > 10,000 x 10^6 Btu/hr	4.8 lb SO_x/10^6 Btu heat input for sources > 250 x 10^6 Btu/hr	No standards
Missouri	0.18 lb/10^6 Btu for sources > 10,000 x 10^6 Btu/hr	1000 lb SO_2/hr	No standards
Montana	0.1 lb/10^6 Btu for new sources (maximum 2 hour average)	1.2 lb SO_2/10^6 Btu (maximum 2 hour average)	0.7 lb NO_x/10^6 Btu
Nebraska	0.1 lb/10^6 Btu for new sources (maximum 2 hour average)	1.2 lb SO_2/10^6 Btu (maximum 2 hour average)	0.7 lb NO_x/10^6 Btu
Nevada	0.1 lb/10^6 Btu for new sources > 250 x 10^6 Btu/hr	0.6 lb sulfur/10^6 Btu for new sources > 250 x 10^6 Btu/hr	0.7 lb NO_x/10^6 Btu for new sources > 250 x 10^6 Btu/hr
New Hampshire	0.10 lb/10^6 Btu for new sources > 250 x 10^6 Btu/hr and 0.19 for existing sources > 10,000 x 10^6 Btu/hr	1.5 lb sulfur/10^6 Btu for new 2.8 lb sulfur/10^6 Btu for existing	No standards
New Jersey	1000 lb/hr for source of 10,000 x 10^6 Btu/hr	0.2% by weight of sulfur in coal with several exceptions	No standards
New Mexico	0.05 lb/10^6 Btu for new sources > 250 x 10^6 Btu/hr Fine particulate emissions (<2 microns) cannot exceed 0.02 lb/10^6 Btu	0.34 lb SO_2/10^6 Btu for new and 1.0 lb SO_2/10^6 Btu for existing > 250 x 10^6 Btu/hr input	0.45 lb NO_2/10^6 Btu for new sources > 250 x 10^6 Btu/hr and 0.7 lb NO_2/10^6 Btu for existing
New York	0.1 lb/10^6 Btu for new sources (maximum 2 hour average)	0.60 lb sulfur/10^6 Btu for new sources > 250 x 10^6 Btu/hr for most areas	0.7 lb NO_x/10^6 Btu for new sources > 250 x 10^6 Btu/hr
North Carolina	0.10 lb/10^6 Btu for sources > 10,000 x 10^6 Btu/hr	1.6 lb SO_2/10^6 Btu for new sources 2.3 lb SO_2/10^6 Btu for existing	1.3 lb NO_2/10^6 Btu for sources $\geq$ 250 x 10^6 Btu/hr
North Dakota	0.8 lb/10^6 Btu for existing 0.1 lb/10^6 Btu for new sources > 250 x 10^6 Btu/hr	1.2 lb SO_2/10^6 Btu for new sources > 250 x 10^6 Btu/hr	No standards
Ohio	0.1 lb/10^6 Btu for new and existing > 1,000 x 10^6 Btu/hr in Priority 1 regions, .15 lb/10^6 Btu in Priority 2 and 3 regions	1.0 lb SO_2/10^6 Btu, new and existing	0.9 lb NO_x/10^6 Btu for new sources $\geq$ 250 x 10^6 Btu/hr
Oklahoma	0.1 lb/10^6 Btu for new and existing > 1,000 x 10^6 Btu/hr	1.2 lb SO_x/10^6 Btu, new	0.7 lb NO_x/10^6 Btu for new sources $\geq$ 50 x 10^6 Btu/hr
Oregon	0.1 lb/10^6 Btu for new sources 0.2 lb/10^6 Btu for existing	1% by weight sulfur limit in fuel	No standards
Pennsylvania	0.1 lb/10^6 Btu for new sources $\geq$ 600 x 10^6 Btu/hr	1.8 lb SO_2/10^6 Btu for sources $\geq$ 2,000 x 10^6 Btu/hr (for most areas)	0.7 lb NO_x/10^6 Btu
Rhode Island	0.10 lb/10^6 Btu for sources > 250 x 10^6 Btu/hr	0.55 lb sulfur/10^6 Btu in fuel or emissions of 1.1 lb SO_x/10^6 Btu	No standards
South Carolina	For sources $\geq$ 1,300 x 10^6 Btu/hr $E = 57.84 P^{-0.43}$, where E = emission rate, P = 10^6 Btu heat input/hr	Most counties are 3.5 lb SO_2/10^6 Btu	No standards
South Dakota	0.1 lb/10^6 Btu for new sources > 250 x 10^6 Btu/hr	1.2 lb SO_2/10^6 Btu for sources > 250 x 10^6 Btu/hr	0.7 lb NO_x/10^6 Btu
Tennessee	0.1 lb/10^6 Btu for existing sources $\geq$ 10,000 x 10^6 Btu/hr 0.1 lb/10^6 Btu for new sources $\geq$ 250 x 10^6 Btu/hr	1.2 lb SO_2/10^6 Btu for new sources > 250 x 10^6	0.7 lb NO_x/10^6 Btu for new sources $\geq$ 250 x 10^6 Btu/hr
Texas	0.1 lb/10^6 Btu for new sources > 250 x 10^6 Btu/hr 0.3 lb/10^6 Btu for existing sources	3.0 lb/10^6 Btu for existing 1.2 lb SO_2/10^6 Btu for new	0.7 lb NO_x/10^6 Btu for opposed-fired units, 0.5 for front-fired, 0.25 for tangentially-fired
Utah	0.1 lb/10^6 Btu for new sources $\geq$ 250 x 10^6 Btu/hr 85% control and 40% opacity for existing	1.2 lb SO_2/10^6 Btu for new 1% sulfur coal by weight	0.7 lb NO_x/10^6 Btu for new sources $\geq$ 250 x 10^6 Btu/hr
Vermont	0.1 lb/10^6 Btu for new and existing sources $\geq$ 250 x 10^6 Btu/hr	1.2 lb SO_2/10^6 Btu for new sources > 250 x 10^6 Btu/hr	0.3 lb NO_x/10^6 Btu for new sources > 250 x 10^6 Btu/hr
Virginia	0.1 lb/10^6 Btu for new sources > 250 x 10^6 Btu/hr 0.1 lb/10^6 Btu for existing sources > 10,000 x 10^6 Btu/hr	1.2 lb SO_2/10^6 Btu for new sources > 250 x 10^6 Btu/hr	No standards
Washington	0.1 grains/scf for new and existing sources	1000 ppm SO_2 for new and existing sources	No standards
West Virginia	.05 lb/10^6 Btu	2.0 lb SO_2/10^6 Btu/hr as of 6/30/78	No standards
Wisconsin	0.10 lb/10^6 Btu for new sources > 250 x 10^6 Btu/hr	1.2 lb SO_2/10^6 Btu for new sources > 250 x 10^6 Btu/hr	0.7 lb NO_x/10^6 Btu for new sources > 250 x 10^6 Btu/hr
Wyoming	0.10 lb/10^6 Btu for all sizes of new units	0.2 lb SO_2/10^6 Btu for new sources $\geq$ 250 x 10^6 Btu/hr	0.7 lb NO_x/10^6 Btu for new sources $\geq$ 250 x 10^6 Btu/hr

TABLE 48a. COUNTIES OF CALIFORNIA - EMISSION REGULATIONS FOR POWER PLANTS*

County	Particulate Matter	Opacity	Sulfur Oxides	Nitrogen Oxides
Santa Barbara	0.2 grains/ft³ for 1,000 cfm source 0.0635 grains/ft³ for 20,000 cfm source 0.0122 grains/ft³ for 1,500,000 cfm source	20%		
Merced	0.1 grains/scf	20%	0.2% by volume SO_2	225 ppm for source input > 1,775 x 10⁶ Btu/hr
Tehama	0.3 grains/ft³	40% (no more than 3 minutes in any hour)		
Placer	0.3 grains/scf for existing 0.1 grains/scf for new and 10 lbs/hr of combustion contaminants	20% (no more than 3 minutes in any hour) for new sources 40% for existing	0.2% by volume and 200 lbs/hr of sulfur compounds	140 lbs/hr of nitrogen oxides
North Coast Air Basin -				
Del Norte Humboldt Trinity Mendocino Sonoma	.23 grains/SCM (.10 grains/scf)	40% opacity (no more than 3 minutes in any one hour except 20% in Mendocino County)		
Plumas		40% for existing, 20% for new sources		
Fresno	0.10 grains/scf (0.23 grains/SCM) 10 lbs/hr of combustion contaminants	20%	0.2% by volume SO_2 and 200 lbs/hr of sulfur compounds	140 lbs/hr of nitrogen oxides
Kings	0.1 grains/scf	20%		
Monterey	0.10 lb/10⁶ Btu for new sources > 250 x 10⁶ Btu/hr	20% except 40% allowed for 2 minutes in any one hour	1.2 lb/10⁶ Btu	0.70 lb/10⁶ Btu
Madera	0.1 grains/scf and 10 lbs/hr of combustion contaminants	20% except 40% for no more than 2 minutes in any one hour	1.2 lb/10⁶ Btu	0.70 lb/10⁶ Btu
Ventura	0.1 grains/scf	20%		
South Coast - San Bernadino Zone	combustion contaminants which exceed both 11 lbs/hr (5 kg/hr) and 0.01 grains/scf (23 mg/m³) for new sources > 50 x 10⁶ Btu/hr	20%	0.5% sulfur by weight	225 ppm NOx
Glenn	0.3 grains/scf	40%	0.2% by volume (2,000 ppm)	
Shasta	0.10 grains/scf for new sources (.05 gr/scf for particulate matter < 10 microns)		1,000 ppm for new sources	
Tulare	0.1 grains/scf	20%		
Sutter	0.3 grains/scf and 10 lbs/hr of combustion contaminants	40%	0.2% sulfur compounds and 200 lbs/hr sulfur compounds	140 lbs/hr nitrogen oxides
El Dorado	0.10 grains/scf and 10 lbs/hr of combustion contaminants		200 lbs/hr sulfur compounds	140 lbs/hr nitrogen oxides
Butte	0.30 grains/scf	40%		
Sacramento	0.30 grains/scf (expected to be changed to 0.10 grains/scf for new sources)	40% (expected to be changed to 20% for new sources)	0.2% by volume (expected to be changed to 0.5% sulfur in fuels)	
Imperial	0.3 grains/scf existing 0.2 grains/scf new (after July 1, 1972)	40% existing 20% new (after July 1, 1972)		
Siskiyou	0.3 grains/scf	40%		
San Diego	0.1 lbs/10⁶ Btu for sources > 250 x 10⁶ Btu/hr (after August 17, 1971)	20% except 40% for no more than 2 minutes in any hour	1.2 lb/10⁶ Btu	0.70 lb/10⁶ Btu
Lake	No fuel-fired power boilers or associated regulations.			
Yolo-Solano	0.3 grains/scf and 40 lbs/hr of combustion particulates - existing	40% - existing	0.2% SO_2 and 200 lbs/hr of sulfur compounds - existing	140 lbs/hr of nitrogen oxides - existing
Kern	existing - 0.1 grains/scf, Valley Basin existing - 0.2 grains/scf, Desert Basin new - 0.1 grains/scf (3/19/74)	20% for more than 3 minutes in any one hour	0.2% SO_2 by volume	140 lbs/hr NOx, Valley Basin 0.7 lb/10⁶ Btu, Desert Basin (after 8/17/71)
San Joaquin	0.1 grains/scf - existing 0.10 lb/10⁶ Btu - new	20% except 40% for 2 minutes in any one hour	0.2% SO_2 by volume - existing 1.2 lbs/10⁶ Btu - new	225 ppm - existing 0.7 lb/10⁶ Btu - new
Bay Area -				
Alameda Contra Costa Marin Napa San Francisco San Mateo Santa Clara Solano Sonoma	generally, 0.15 grains/scf - existing 0.10 lbs/10⁶ Btu for new sources > 250 x 10⁶ Btu/hr	20% except 40% for not more than 2 minutes in any one hour	1.2 lbs/10⁶ Btu for new sources	0.70 lb/10⁶ Btu for new sources
Amador	0.10 grains/scf and 10 lbs of combustion contaminants	20% for more than 3 minutes in any one hour	200 lbs of sulfur compounds (calculated as SO_2)	140 lbs of nitrogen oxides (calculated as NO_2)

*Regulations were obtained from most of the Air Pollution Control District in California and summarized in this table as an indication of the emission limits experienced across the state. Each district has its own regulations. This survey was conducted in 1978.

TABLE 49. REGULATIONS APPLICABLE TO VISIBLE EMISSION ALLOWED FOR FUEL-FIRED BOILERS

	Existing Sources - Limits
1. Alabama	20% opacity or No. 1 on the Ringelmann chart except 60% or No. 3 on the Ringelmann chart for not more than 3 minutes in any 60 minutes.
2. Alaska	may not exceed 20% opacity for a period or periods aggregating more than 3 minutes in any hour.
3. Arkansas	may not be equal to or exceed 40% except for not more than five minutes in a 60 minute period (3 times in 24 hour maximum).
4. Arizona	may not exceed No. 2 Ringelmann (equivalent to 40% opacity).
5. California	each county has its own regulations.
6. Colorado	may not exceed 20% except 40% for no more than 3 minutes in any one hour.
7. Connecticut	may not exceed 20% except 40% for a period aggregating not more than 5 minutes in any 60 minutes.
8. Delaware	may not exceed either No. 1 on the Ringelmann chart or 20% opacity for more than 3 minutes in any one hour or more than 15 minutes in any 24 hour period.
9. Florida	may not exceed No. 1 of the Ringelmann chart (20% opacity) except No. 2 of the Ringelmann chart (40%) shall be permissible for not more than 2 minutes in any hour.
10. Georgia	may not have emissions equal to or greater than Ringelmann chart (20% opacity) except for emissions up to Ringelmann No. 2 for two minutes in any one hour. This is for fuel burning equipment constructed after January 1, 1972. Opacity requirements for equipment constructed prior to January 1, 1972, is 40%.
11. Hawaii	may not exceed 40%.
12. Idaho	may not exceed No. 2 on the Ringelmann chart (40% opacity). The new source standard for Idaho does not allow the emission's aggregating more than 3 minutes in any one hour which is greater than 20% opacity.
13. Illinois	may not exceed 30% opacity except may have opacity greater than 30% but not greater than 60% for a period or periods aggregating 8 minutes in any 60 minute period (limit to 3 times in any 24 hours).
14. Indiana	may not exceed 40% opacity for more than a cumulative total of 15 minutes in a 24 hour period.
15. Iowa	may not exceed 40% opacity except for a period or periods aggregating not more than 6 minutes in any 60 minute period.
16. Kansas	may not be equal to or greater than 40% opacity.
17. Kentucky	may not be equal to or greater than 40% opacity except for 60% for 6 minutes in any 60 minute period.
18. Louisiana	may not exceed 20% opacity.
19. Maine	may not exceed 40% opacity except for periods of not exceeding 5 minutes in any one hour or 15 minutes in any continuous 3 hour period.
20. Maryland	may not exceed 20% opacity except for 40% for a period or periods aggregating no more than 4 minutes in any sixty minutes.
21. Massachusetts	emissions may not be equal to or greater than 20% except 40% for a period or aggregate period of time in excess of 6 minutes during any one hour.
22. Michigan	may not exceed 20% except 40% for not more than 3 minutes in any 60 minute period for no more than 3 occasions during any 24 hour period.
23. Minnesota	may not exceed 20% except 60% for 4 minutes in any 60 minute period and 40% for 4 additional minutes in any 60 minute period.
24. Mississippi	may not exceed 40% except 60% for no more than 10 minutes per billion Btu gross heating value of fuel in any one hour per 24 hours.

(continued)

TABLE 49. (continued)

25. Missouri	may not be equal to or greater than 40% except 60% for a period or periods aggregating not more than 6 minutes in any 60 minutes. Kansas City's opacity limit is 20% except 60% for 6 minutes in any 60 minutes.
26. Montana	for equipment built before 1969, may not exceed 40%; after 1969, may not exceed 20%. Exception - 60% for 4 minutes in any 60 minutes.
27. Nebraska	may not be equal to or exceed 20%.
28. Nevada	may not be equal to or exceed 20% for a period or periods aggregating more than 3 minutes in any one hour.
29. New Hampshire	may not exceed 40%.
30. New Jersey	may not exceed 20% except for smoke which is visible for a period of not longer than 3 minutes in any consecutive 30 minute period.
31. New Mexico	may not exceed 20%.
32. New York	may not exceed 20% except for 3 minutes during any continuous 60 minute period.
33. North Carolina	may not exceed 40% for an aggregate of more than 5 minutes in any one hour or more than 20 minutes in any 24 hour period.
34. North Dakota	maximum allowable is 40%.
35. Ohio	may not exceed 20% except 60% for no more than 3 minutes in any 60 minutes.
36. Oklahoma	may not exceed 20% except 60% for no more than 5 minutes in any 60 minutes or more than 20 minutes in any 24 hour period.
37. Oregon	may not be equal to or greater than 40% for a period aggregating more than 3 minutes in any one hour except more stringent for special control areas.
38. Pennsylvania	may not be equal to or greater than 20% for more than 3 minutes in any one hour or equal to or greater than 60% at any time.
39. Puerto Rico	may not be equal to or greater than 20% except 60% for not more than 4 minutes in any 30 minutes
40. Rhode Island	may not be equal to or exceed 20%.
41. South Carolina	may not be equal to or exceed 40% except 60% for 5 minutes in one hour or 20 minutes in a 24 hour period.
42. South Dakota	may not exceed 20% except 40% is permissible for not more than 2 minutes in any hour.
43. Tennessee	may not exceed 40% for more than 5 minutes aggregate in any one hour or more than 20 minutes in any 24 hour period
44. Texas	may not exceed an opacity of 30% averaged over a 5 minute period.
45. Utah	may not exceed 40%.
46. Vermont	may not exceed 40% for more than 6 minutes in any hour. Opacity may never exceed 60%.
47. Virgin Islands	may not be equal to or greater than 40%.
48. Virginia	may not exceed 20% except for brief periods when starting a new fire, blowing tubes, or cleaning a fire box.
49. Washington	may not exceed 40% except for 15 minutes in any consecutive 8 hours.
50. Washington, D.C.	no visible emissions except less than 20% for 2 minutes in any 60 minute period and for an aggregate of 12 minutes in any 24 hour period.
51. West Virginia	may not be equal to or exceed 20% except 10% after June 30, 1975.
52. Wisconsin	may not be equal to or exceed 40% except 20% in Milwaukee and Lake Micigan AQCR's. Also 80% for 5 minutes in any one hour for cleaning or starting new fire in combustion equipment (3 times a day maximum).
53. Wyoming	may not exceed 40%.

Appendix D

Low Temperature Corrosion and Fouling[292]

INTRODUCTION

Flue gas temperatures which are in the range of 104-121°C (220-250°F) may result in corrosion and fouling of air heater elements and corrosion of precipitator elements. Operation at such low temperatures has caused corrosion and fouling of air heater elements in some installations, while others have experienced no difficulty with air heater exit temperatures as low as 104°C (220°F). An understanding of the factors which cause corrosion and fouling problems is important when operating with flue gases at low temperatures. The purpose of this appendix is to relate corrosion and fouling to fly ash and flue gas composition, fly ash resistivity, and temperature.

SULFURIC ACID OCCURRENCE IN FLUE GAS

SOx, H_2O, and H_2SO_4 Equilibria

A knowledge of the SO_3 concentration in the air heater and precipitator region of power plant exhaust systems is important from a standpoint of both corrosion and fly ash resistivity. The principal cause of corrosion in air heaters, and the most important factor in determining fly ash resistivity, is sulfuric acid, which results from the reaction of SO_3 with water vapor.

Most of the sulfur in power plant flue gases appears as SO_2, with typical SO_3 levels ranging from 1 to 2.5% of the SO_2. However, as Figure 301 shows, the equilibrium constant for the reaction

$$SO_2(g) + \tfrac{1}{2}O_2(g) = SO_3(g)$$

strongly favors the formation of SO_3 at temperatures below 537°C (1000°F) with 3% oxygen. This graph was calculated from data cited by Hedley.[293] The kinetics of the reaction are, of course, unfavorable in the absence of a catalyst, but it is thermodynamically feasible for SO_3 concentrations to exist at levels much greater than those normally encountered. Ratios of SO_3 to SO_2 as high as 0.1 have been reported.[294] Thus, since the formation of SO_3 is controlled by catalytic effects as well as the amount of excess air present, the concentration of SO_3 resulting from the combustion of a particular fuel can only be estimated in the absence of direct measurements.

The reaction between water vapor and SO_3 is given by

$$H_2O(g) + SO_3(g) = H_2SO_4(g).$$

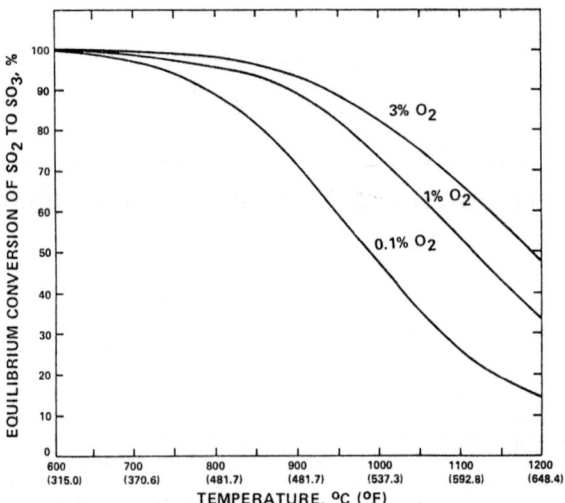

Figure 301. Equilibrium conversion of SO_2 to SO_3.

Figure 302 shows the equilibrium conversion of SO_3 to H_2SO_4 as a function of temperature for a typical flue gas water concentration of 8%. At temperatures below 204°C (400°F), essentially all of the SO_3 present is converted to H_2SO_4 at equilibrium. In contrast to the formation of SO_3, the formation of H_2SO_4 occurs rapidly in the thermodynamically feasible temperature range.[295] Thus, all SO_3 below the air heater in a power plant will exist as H_2SO_4, either in the vapor or liquid state. Since corrosion problems are associated with the presence of liquid phase sulfuric acid, the determination of the condensation characteristics of sulfuric acid from flue gas containing sulfuric acid and water vapor is a necessary step in evaluating the corrosion potential of a particular stack gas.

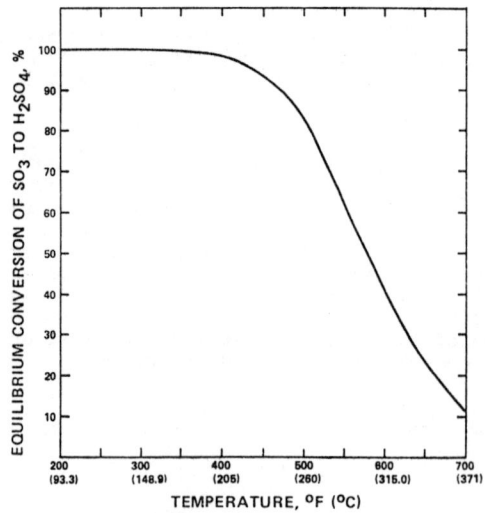

Figure 302. Equilibrium conversion of SO_3 to H_2SO_4 at 8.0 volume % H_2O in flue gas.

Determination Of The Sulfuric Acid Dew Point

Fly ash particles can influence the apparent dew point, or saturation temperature of H_2SO_4 in flue gas, but experience has shown that one commits practically no error by neglecting the presence of other gases and considering only the system sulfuric acid - water.[296] A thermodynamic analysis of the sulfuric acid - water - flue gas system, ignoring for the present the effect of fly ash, provides a theoretical basis for predicting acid dew points and condensate composition from vapor-liquid equilibria data.

For the case of ideal or quasi-ideal binary solutions, dew points of vapor mixtures composed of the binary solution vapor and noncondensable gases can easily be calculated from a knowledge of the pure component vapor pressures as a function of temperature. The H_2SO_4-H_2O system presents special problems because:

the H_2SO_4 and water undergo chemical reaction to form the various hydrates of sulfuric acid, and therefore the equilibrium relationships are strongly composition-dependent, and

H_2SO_4 has a very low pure component vapor pressure, thus making direct measurements extremely difficult.

The total vapor pressure of H_2SO_4 at low temperature is essentially the partial pressure of water above the acid solution, and this is available from the existing literature. In order to determine the dew point, however, the H_2SO_4 partial pressure at low temperature must be known, and the literature lacks such data.[296]

As a result of the experimental difficulties encountered in low temperature vapor pressure measurements, efforts have been made to calculate the partial pressure from liquid phase thermodynamic data. Abel[297] was the first to derive a relationship enabling the calculation of H_2SO_4, H_2O, and SO_3 partial pressures from standard state values of enthalpy, entropy, and heat capacity; and partial molal values of enthalpy, entropy, free energy, and heat capacity. Müller,[296] using Abel's calculated data, computed dew points of gases with low H_2SO_4 concentrations. Gmitro and Vermuelen[298] utilized thermodynamic data, which are claimed to be more recent and more complete, to calculate H_2SO_4, SO_3, and H_2O partial pressures from -50 to 400°C with solutions ranging from 10 to 100 weight percent H_2SO_4. Snowden and Ryan[295] have used Gmitro and Vermulen's partial pressure data to construct a chart which gives the dew point temperature of a gas as a function of H_2SO_4 and H_2O partial pressures. The composition of the acid condensate occurring at a given dew point is also provided.

The dew points predicted from Abel's data are about 30°F higher than those arrived at with Gmitro and Vermuelen's data. The difference in these two works lies mainly in the data available for the calculation of the partial pressures. Gmitro and Vermuelen had access to much more accurate data and should have obtained the more accurate results. However, their results do not agree with direct dew point measurements by the condensation technique, whereas Abel's partial pressures have been verified in part by use of this method.

A suspect assumption common to predictions of acid dew points based on both the Abel and Gmitro calculations is that the vapor state is an ideal gas, and that the vapor solution is also ideal. A gas mixture may behave nearly ideally volumetrically, but a component present in small amounts may exhibit significant departure from ideality if that component is associated in the vapor state.

Among the limitations of some presentations in the literature of the Müller correlation with 10% water vapor is that they do not indicate the effect of variations in the water vapor concentration on sulfuric acid dew points. The concentration of the condensate is also not provided. Figures 303 and 304 were prepared to present this information.

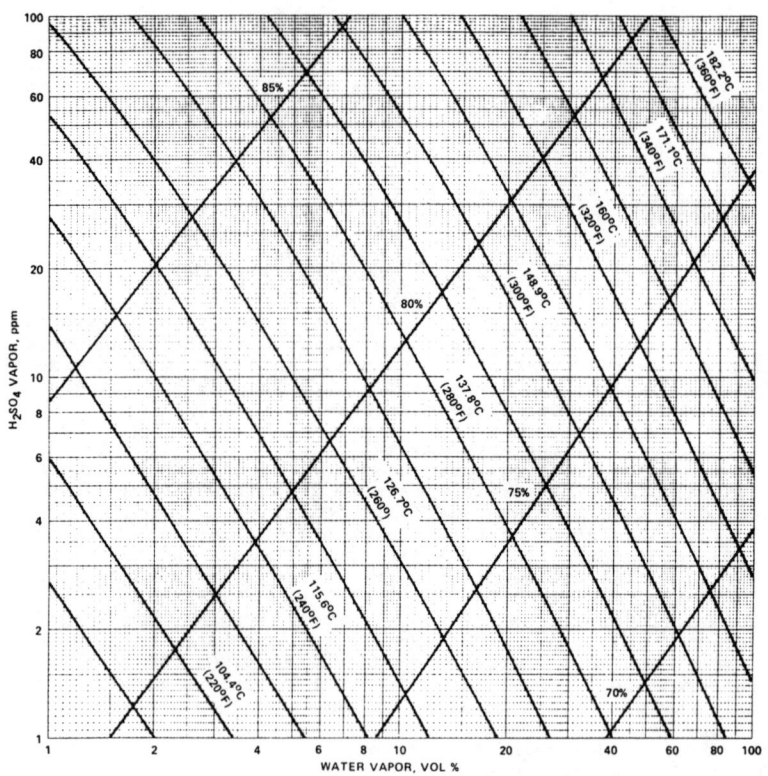

Figure 303. Dew point and condensate composition for vapor mixtures of H_2O and H_2SO_4 at 760 mm Hg total pressure (Abel and Greenewalt).[299]

Figure 303 is a sulfuric acid - water dew point chart prepared from Abel's H_2SO_4 partial pressures and Greenewalt's[299] water partial pressures above sulfuric acid solutions. The partial pressure data were calculated by computer from the following equations:

$$P_{H_2SO_4} = \exp\left[2.303\left(A + \frac{B}{T} + \frac{D(\ln T)}{2.303} + E \cdot T\right)\right] \quad (74)$$

and

$$P_{H_2O} = \exp\left[2.303\left(A' - \frac{B'}{T}\right)\right], \quad (75)$$

where T is in degrees Kelvin and partial pressures are in mm Hg. The constants in these equations are given by Abel and Greenewalt for various sulfuric acid concentrations. It should be noted that the range of uncertainty indicated by Abel for the constant B in equation 301 results in a dew point uncertainty of 4.45°C (24°F) at 10% water vapor.

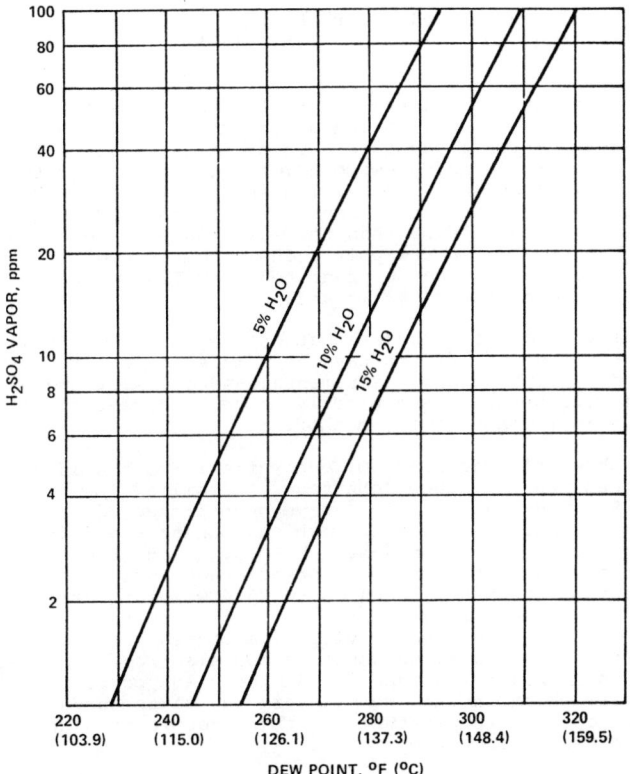

Figure 304. H_2SO_4 dew points for typical flue gas moisture concentrations.

The information contained in Figure 303, if it were accurate, would be of value in assessing the corrosion potential of a flue gas. The dew point temperature can be predicted from an analysis of H_2SO_4 and water vapor content, and if the gas is cooled to some temperature below the dew point, the equilibrium concentration of condensate and the amount condensed can be obtained. It should be pointed out, however, that the amount of condensate predicted from the use of a dew point chart such as Figure 303 is actually a prediction of the amount available for condensation. The amount of condensate depositing on a metal surface may differ from the chart prediction because of mass transfer considerations.

As an example of the use of the chart, consider a flue gas containing 10 ppm H_2SO_4 and 10% H_2O. Condensation would occur at about 275°F, and the condensate composition at that point would be about 79% H_2SO_4 by weight. If the gas were cooled to 250°F, 85% of the H_2SO_4 should be removed from the gas phase, and an insignificant amount of the water vapor would also be condensed. The condensation, therefore, follows the 10% water line, resulting in a condensate which would be the equilibrium composition of the condensate at 121°C (250°F), assuming the vapor phase is in equilibrium with the total liquid condensed. The composition change of the liquid is small over the temperature interval given as an example, ranging from 79% at 135°C (275°F) to 75% at 121°C (250°F).

It is apparent from Figure 303 that a knowledge of water vapor concentration is of fundamental importance. Appreciable changes in this variable can have a rather significant effect on the predicted sulfuric acid dew point, and if a gas is saturated with H_2SO_4, the condensate composition is determined by the water vapor content and temperature. Thus, if a surface is maintained at a known temperature lower than the sulfuric acid dew point, but higher than the water dew point, the concentration of acid condensate which occurs can be predicted from Figure 303 if the water vapor content of the gas is known.

In addition to the procedure based on calculated partial pressures, a number of efforts have been made to determine sulfuric acid dew points using instrumental and chemical procedures. Two methods will be discussed briefly: the condensation method and an electrical conductivity method.

The problem of measuring SO_3 concentration and acid dew point has been studied since Johnstone[300] examined the problem in 1929. Many papers[301-312] have been presented which employ the electrical conductivity method which Johnstone originated. The British Coal Utilization Research Association (BCURA) designed an instrument which has found widespread usage employing Johnstone's concept. This instrument, known as BCURA dew point meter, has been described in detail by Flint.[301] It is a portable instrument which measures the conductivity of a condensing film. The detector element is glass and contains two electrodes mounted flush with the surface. A tube inside the glass probe transports compressed air which is used to maintain the glass surface of the probe at the desired temperature. A thermocouple provides a read-out of the glass surface temperature.

If an electrically conductive film forms on the detector element, a current will flow that is proportional to the magnitude of the externally impressed voltage and the conductivity of the condensing film. The current flow is measured with a microammeter. A dew point is determined by inserting the detector element into a gas stream with the instrument temperature held at some value above the dew point. The element temperature is then alternately increased and decreased slowly to establish the exact temperature at which the increase in conductivity, and thus the dew point, occurs.

The condensation method is widely used for determinations of SO_3 in stack gases. The basic procedure employed consists of pumping the flue gas through a condenser coil maintained below the dew point of sulfuric acid, but above the normal water dew point. A heated sampling probe is used to obtain the flue gas samples, and a filter is inserted at the probe entrance to exclude particulate matter. A fritted glass filter follows the condenser to serve as a spray trap. When the sampling period is concluded, the H_2SO_4 is washed from the condenser, and the washings are collected and titrated.[313]

The condensation of a binary vapor mixture from a noncondensable gas is normally path-dependent, and the composition of the vapor leaving a condenser is not fixed merely by stating that the gas is saturated at a particular temperature. This is true because the degree of fractionation occurring during condensation depends on conditions which exist in the condenser. For the case of H_2SO_4-H_2O vapor mixtures in flue gas, however, the water vapor is in large excess, and no appreciable change in its concentration occurs until the water dew point is reached. The composition of the gas is, therefore, not path-dependent, and the state of the system is fixed if the gas is saturated with H_2SO_4 at a certain temperature and water vapor content. As a result, the condensation method can be used to obtain dew points of H_2SO_4-flue gas mixtures. Since the gas leaving the condenser is saturated with H_2SO_4 at the condenser exit temperature, the concentration of the exit vapor represents the dew point, or saturation temperature, of the gas.

Figure 305 presents the results obtained for flue gas dew points as a function of H_2SO_4(g) content by various investigators. To make an exact comparison, all of the curves should be for a gas of the same volume percent water vapor. However, reference to Figure 303 will indicate that a variation in water vapor concentrations from 7 to 10% can cause only about a 2.78 to 4.45°C (37 to 40°F) change in the dew point. Taylor's results were obtained with the BCURA dew point meter in a mixture of air, water vapor, and sulfuric acid.[312] Lisle's data were obtained using the condensation method, again with a mixture of air, water vapor, and sulfuric acid.[313] The dew point curves of Gmitro, Müller, and from Figure 303, are based on the previously discussed calculated partial pressures.

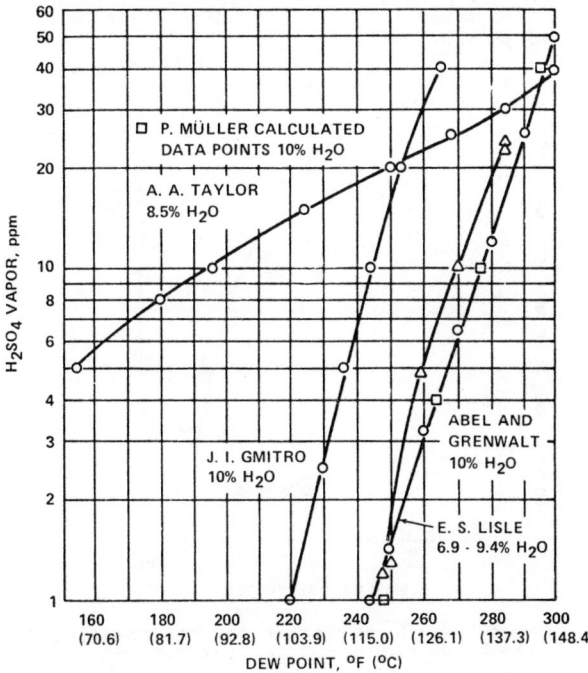

Figure 305. H_2SO_4 dew points obtained by various investigators.

It is obvious from Figure 305 that, except for Lisle and Sensenbaugh's checks of the data based on Abel's sulfuric acid partial pressures (Müller's data and Figures 303 and 304), there is little agreement between the results of the various investigators. The data obtained from calculated partial pressures agree in form, which is to be expected since the equations used to calculate the partial pressures are also of the same form. The nature of the disagreement between the calculated dew point and those obtained with the dew point meter suggest there is a sensitivity problem with the instrument at low sulfuric acid partial pressures.

In view of the difficulties with calculations based on liquid phase thermodynamic properties and the probable inaccuracy of dew point meters at low acid partial pressure, it can be concluded that the only reliable method of correlating sulfuric acid dew points with water and H_2SO_4 vapor concentration is a carefully planned experimental program based on the condensation method employed by Lisle and Sensenbaugh. In the absence of such data, the dew points based on Abel's partial pressure data can be used, since they have been verified in part by experiment and by the operational experience of several power plants.

Condensation Characteristics

As stated previously, the amount of acid condensate predicted from the use of a chart such as Figure 303 as a result of cooling to a temperature below the sulfuric acid dew point is a prediction of the amount available for condensation. Figure 306 shows that the predicted percentage of H_2SO_4 condensed increases and asymptotically approaches 100% as the temperature is lowered below the dew point. However, peak values of acid deposition rates at temperatures between the water and acid dew points have been observed by numerous investigators.

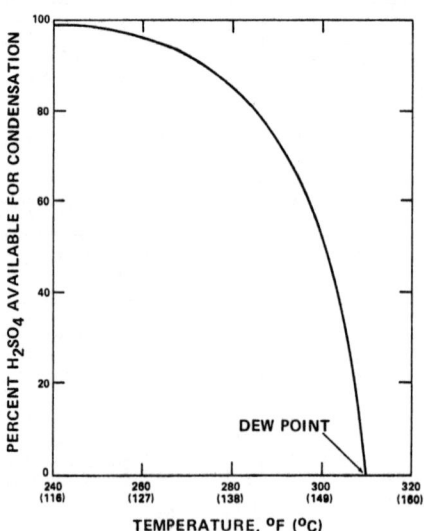

Figure 306. Percent H_2SO_4 available for condensation for flue gas of 100 ppm H_2SO_4 and 10% H_2O vapor (calculated from Figure 303).

The occurrence of such a peak in the condensation rate may be caused by a change in the diffusivity of the H_2SO_4 in the region close to the condensing surface. The rate of condensation is dependent on the diffusion rate of H_2SO_4 and water vapor to the surface. Small droplets of H_2SO_4 will form in the cooled gas adjacent to the surface, and the size of these droplets is likely to increase with decreasing temperature. The growth of the droplets would slow their diffusion to the surface and increase the probability that they would be carried forward in the gas stream. Thus, a temperature can be reached at which the slowed diffusion becomes dominant over the increased amount of condensate available for collision with the surface. This explanation is similar to one offered by Flint and Kear.[306] A typical condensate rate curve, obtained in a spiral condenser with a vapor mixture consisting of 7.5 vol % H_2O, 69 ppm H_2SO_4, and the balance air, is shown in Figure 307.[314]

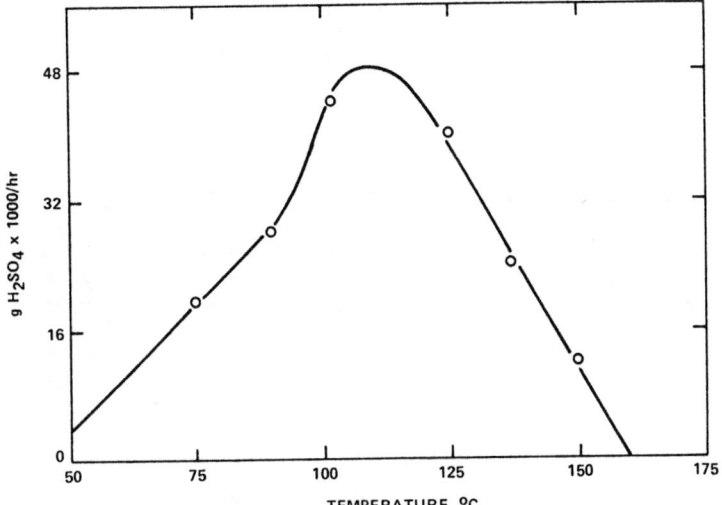

Figure 307. Variation in condensation rate with surface temperature (From H. D. Taylor).[314]

FACTORS INFLUENCING CORROSION RATES

Acid Strength

If a flue gas is known to be saturated with H_2SO_4 vapor at a temperature below the acid dew point, it is possible to predict the initial condensate composition as a function of the water vapor partial pressure and temperature. Since data are available in the literature concerning the corrosion rates of various materials as a function of acid concentrations, it is of interest to determine whether there is any relationship between corrosion rates measured in flue gas and the acid condensate strength predicted from a gas analysis.

A study of flue gas corrosion of low alloy steels by Piper and Van Vliet[315] provides data which illustrate the difficulty encountered in predicting corrosion rates of metals from acid condensate strength alone. The compositions of the low alloy steel specimens used in this study are given in Table 50. The corrosion tests were conducted by inserting specimens maintained at known temperatures into stack gas produced from a pulverized-fuel-fired steam generator. The average H_2SO_4 content of the stack gas was about 30 ppm. Figure 308 gives the predicted sulfuric acid condensate compositions for the range of stack gas water vapor concentrations experienced during the study.

TABLE 50. COMPOSITION, PERCENT BY WEIGHT, SPECTROGRAPHIC ANALYSIS OF SPECIMENS TESTED (from Piper and Van Vliet)[315]

Name	Mn	Si	Cu	Ni	Cr	Zr
Cor-ten	0.40	0.38	0.23	0.29	0.61	--
NAX-A	0.85	0.90	0.07	<0.1	0.59	Present
NAX-B	0.82	0.79	0.29	<0.1	0.60	Present
NAX-C	0.53	0.54	0.07	<0.1	<0.1	Present

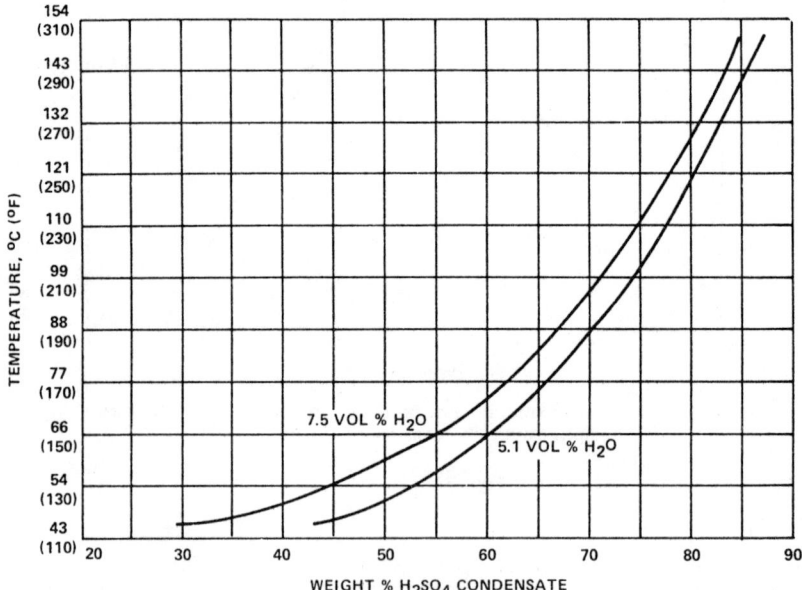

Figure 308. Equilibrium sulfuric acid condensate composition.

Figure 309 shows the average corrosion rate of selected steel specimens as a function of predicted H_2SO_4 condensate strength. The condensate strengths shown in Figures 308 and 309 were obtained from the computer printout of partial pressure for the H_2SO_4-H_2O system, using Greenewalt's equation (equation 75) for the partial pressure of water over sulfuric acid solutions. The widths of the surface in Figure 309 indicate the possible acid concentrations at each temperature over the range of water vapor partial pressures encountered in the stack gas.

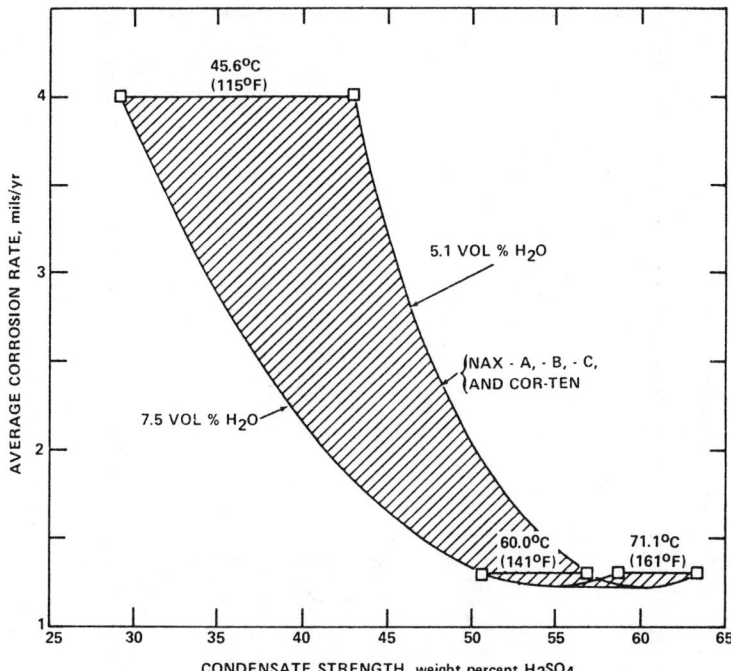

Figure 309. Corrosion of steel in flue gas as a function of calculated H_2SO_4 condensate strength (corrosion data from Piper and Van Vliet; H_2SO_4 data from Greenewalt).[315][299]

Figure 310 is a plot of corrosion rates of steel given by M. G. Fontana[316] at 23.4°C (75°F) as a function of acid concentration. The corrosion rates for steel specimens immersed in acid are orders of magnitude higher than those observed by Piper.

Since corrosion increases with temperature, the differences between the Fontana and Piper data are even greater than indicated because the latter's data were obtained at high temperatures.

The low alloy steels used in the Piper study would not be expected to exhibit greatly different corrosion rates in sulfuric acid solution than the ordinary carbon steel on which Fontana's data are based. Therefore, the orders of magnitude differences in corrosion rates indicated are largely a reflection of the differences in environment between the two situations.

Another contributing factor is the parabolic nature of the corrosion-time relationship usually found in corrosion work. Thus, because of the effects of fly ash and condensate deposition rates, it is not practical to predict or correlate corrosion rates of materials in flue gas solely on the basis of equilibrium condensate compositions.

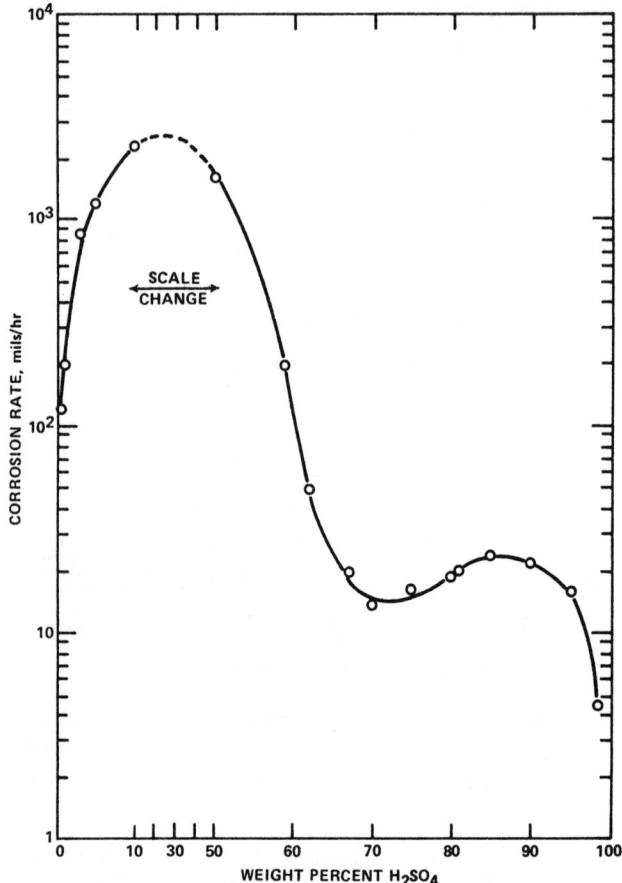

Figure 310. Corrosion of steel as a function of H_2SO_4 concentration at 23.4°C (75°F).[316]

Acid Deposition Rate

The corrosion rate of metal surfaces in flue gas at temperatures well above the water dew point is more strongly related to the amount of condensate deposited than to the concentration of the condensate. Consider, for example, a steel surface at 126.1°C (260°F) exposed to a flue gas with a bulk gas phase concentration of 10 ppm sulfuric acid vapor and 10% water vapor. A condensate strength of 77% H_2SO_4 would be expected, and if fly ash neutralizing ability is ignored, some nonzero rate of corrosion would be expected. If the same steel surface were exposed to a similar flue gas with 80 ppm sulfuric acid vapor, the predicted condensate strength would remain at 77% H_2SO_4, but the corrosion rate would be greater because of the increased quantity of acid condensate depositing on the metal. In both cases, decreasing the metal surface temperatures to a value approaching the water dew point [37.3 to 42.8°C (100 to 110°F)] of the flue gas would result in increased corrosion rates because of the highly corrosive dilute acid formed at these temperatures.

The temperature at which the maximum condensation rate of acid occurs has been correlated with the temperature of maximum corrosion in flue gases. Figure 311 was taken from a study by G. G. Thurlow, in which an air-cooled corrosion probe was exposed to flue gas produced from burning a 0.8% sulfur coal.[317] The rate of sulfate deposition shows a peak at the same surface temperature as the corrosion rate. This peak rate effect is often not observed with coal firing, but Black[310] and Clark[311] have found this phenomenon quite useful in correlating corrosion of air preheaters in oil fired units. The sulfur content of the fuel used in these studies ranged from 1.4 to 4.0%.

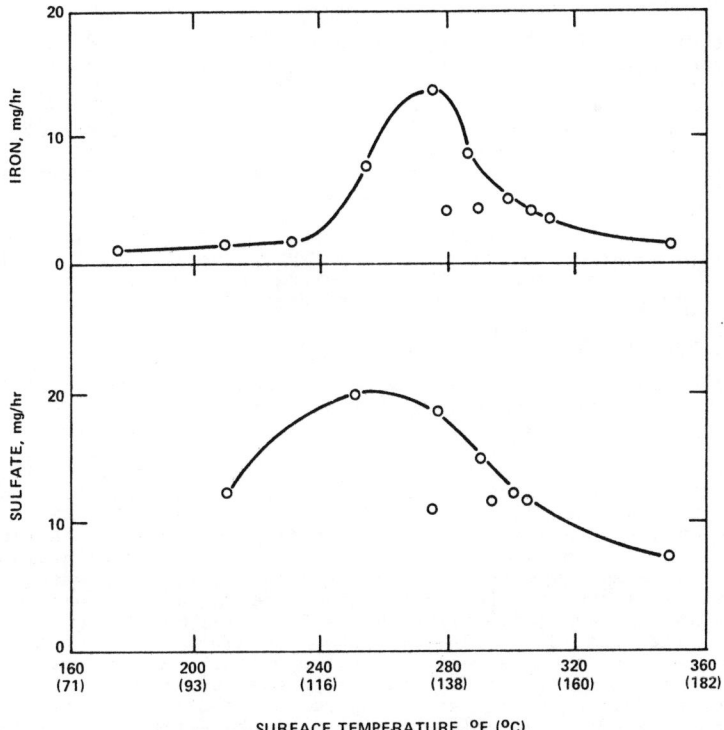

Figure 311. Variation of condensation and corrosion with surface temperature (data from Thurlow).[317]

Black and Clark's work was done with the BCURA dew point meter, and the peak rate of acid deposition was indicated by a peak rate of increase in current, measured as microamps per minute. The maximum corrosion rate is expected to occur in a regenerative air preheater at the point where the average metal temperature corresponds to the peak rate temperature indicated by the BCURA meter. By superimposing a plot of the dew point meter readings in the region of the peak over lines of average metal temperature, it was possible to match the peak rate temperatures with actual corrosion experience.

The above authors also found that the BCURA indication of the acid dew point was a poor indicator of flue gas corrosion potential, particularly when oil and gas mixtures are fired. This observation is not surprising since, as Figure 303 indicates, the dew point alone does not specify how much acid is available for condensation. The accuracy of the dew point meter may also be an important factor, because instructions for use of the meter state[318] that changes in dew point readings of less than 11°C (52°F) are not to be regarded as significant. Referring again to Figure 303, a change of dew point at 10% water vapor from 132 to 143°C (270 to 290°F) indicates a 370% increase in the H_2SO_4 vapor content of the flue gas.

Studies conducted by Lee, Freidrich and Mitchell,[308] in which the BCURA meter was employed with flue gas produced from burning low sulfur lignite, showed that the meter was unable to detect acid dew points with low sulfur coals. In one experiment, no acid dew point was detected by the meter in the presence of sulfuric acid vapor levels as high as 27 ppm. The author's explantation for this is that the condensed acid was completely neutralized by basic constituents in the fly ash.

Thus, since high fly ash resistivity is associated with low sulfuric acid vapor concentrations, the BCURA meter is not likely to be of value in assessing the low corrosion potential associated with a flue gas containing high resistivity fly ash.

Fly Ash Alkalinity

Although fly ash can cause severe plugging problems in air heaters, it is well established that alkaline ashes can neutralize a portion of the SO_3 and H_2SO_4 occurring in stack gases, thereby acting to reduce corrosion. Lee provides data which illustrate the interaction of acid condensate with fly ash. Figure 312 illustrates the effect of surface temperature on acid condensation rate when burning a 7% sulfur coal with 3% excess oxygen. The RBU plotted on the y axis in the upper graph is a measure of the rate of acid condensation when the BCURA dew point meter is maintained at the indicated temperatures. Data for the lower graph were obtained by isokinetically sampling the flue gas and collecting the fly ash and acid condensate in a Teflon vial maintained at 82, 100, 118, and 135°C (180, 212, 245, and 275°F). The contents of the vial were then extracted, and the extract was analyzed for acid or base content. If the extract pH was less than 7, the solution was titrated with sodium hydroxide, and the results were reported as a negative cation content. If the extract was basic, the solution was titrated with HCl, and the results were reported as an excess cation content, indicating that the condensed sulfuric acid had been completely neutralized.

The acid neutralizing ability of fly ash with various base contents is illustrated in Figure 313 for a flue gas with a typical dust loading of 11.4 gm/scm (5 gr/scf). The parallel lines each represent a base content of fly ash, expressed as milliequivalents reactive base per gram fly ash. Data obtained on Contract CPA 70-149 (A Study of Resistivity and Conditioning of Fly Ash) indicate that fly ash produced from burning a high sulfur coal has as much as 0.6 milliequivalents soluble base (1.7% CaO) per gram fly ash.[319] This quantity of base is capable of neutralizing 80 ppm H_2SO_4 in the gas phase, assuming that the flue gas has an ash concentration of 11.4 gm/scm (5 gr/scf). This is not to say that complete neutralization will occur, since the degree of neutralization obtained in the flue gas is a function of the rate of transfer of H_2SO_4 to the fly ash particles and the rate of reaction occurring on the particle surface.

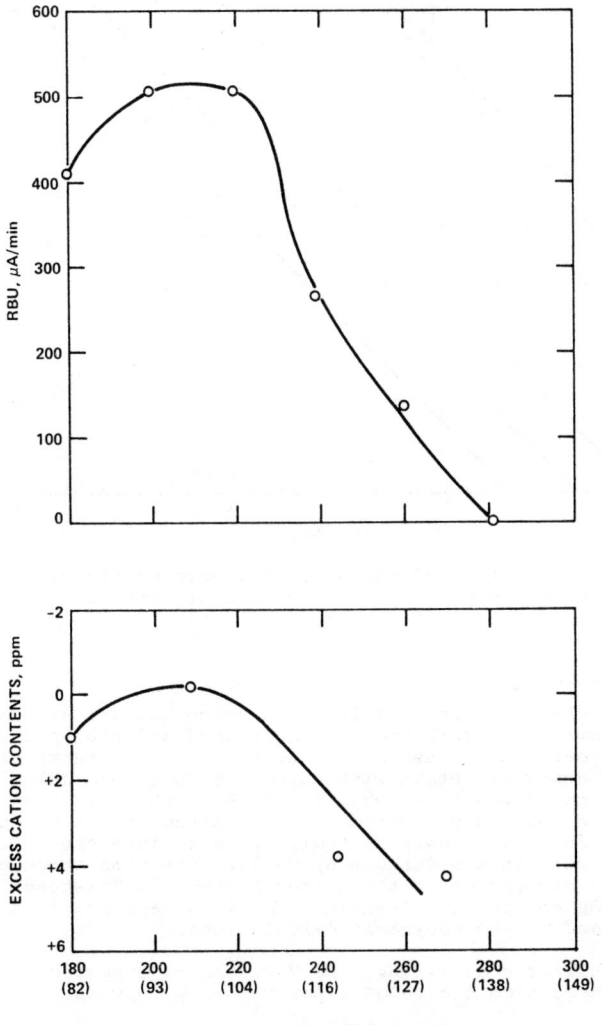

Figure 312. Variation in rate of acid buildup (RBU) and excess cation content of fly ash as a function of surface temperature. Coal contains 7% sulfur with 3% excess O_2 (data from Lee).[307]

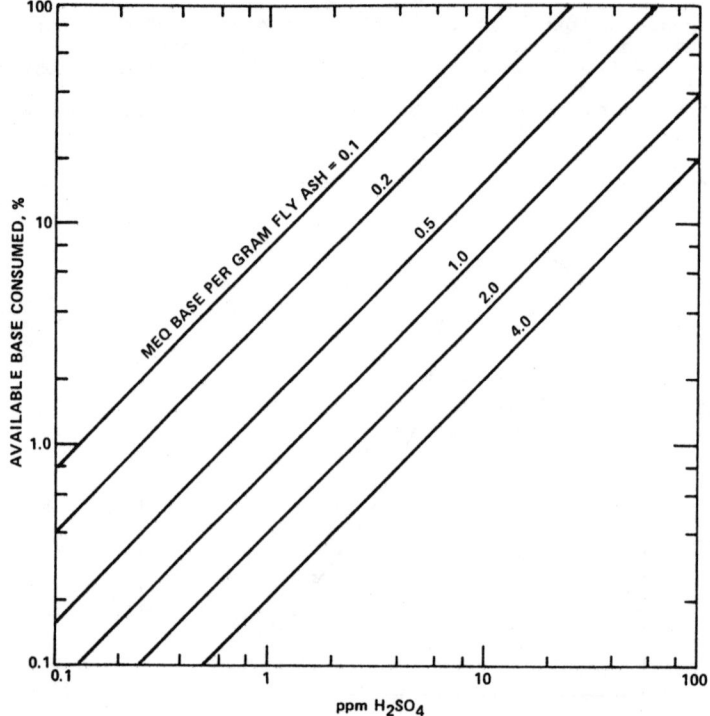

Figure 313. Consumption of the available base on fly ash as a function of the concentration of neutralizing acid in flue gas with 5 gr/scf fly ash.

Hydrochloric Acid

Sulfur, chlorine, and alkali metal compounds are associated with high temperature corrosion in coal-fired boilers, but low temperature corrosion is usually thought of only in terms of sulfuric acid. However, metals with surface temperature below the moisture dew point would be subjected to HCl attack if the chlorine content of the coal is converted to HCl. Although not all of the chlorine in coal appears as NaCl, it is of interest to examine the chemical reactions undergone by NaCl in the combustion process. The following discussion is taken from a study by Halstead[320] in which chloride and sulfate deposit formations were examined with probe tests and by thermodynamic calculations.

In pulverized coal firing, the NaCl can be expected to evaporate and undergo some degree of vapor phase hydrolysis.

$$NaCl(g) + H_2O(g) = NaOH(g) + HCl(g)$$

The reactions of the chloride and NaOH with SO_2 to form Na_2SO_4 are, however, of greater importance. They are

$$2NaCl(g) + H_2O(g) + SO_2(g) + \tfrac{1}{2}O_2(g) = Na_2SO_4(g) + 2HCl(g)$$

$$2NaOH(g) + SO_2(g) + \tfrac{1}{2}O_2(g) = Na_2SO_4(g) + H_2O(g)$$

Halstead calculated the equilibrium partial pressures of Na_2SO_4 and NaCl in flue gases produced from burning the coals listed in Table 51 at 5% O_2 excess, stoichiometric O_2, and 2% O_2 deficient. These calculations, together with deposition studies conducted with a cooled probe, indicate that almost total conversion of NaCl to Na_2SO_4 takes place with 3 to 5% excess oxygen in large boilers with good mixing of fuel and air. With lower oxygen levels, and when poor mixing and short residence times are encountered, the conversion of NaCl to Na_2SO_4 may be incomplete.

TABLE 51. SULFUR AND CHLORINE CONCENTRATIONS IN FLUE GAS (from Halstead)[320]

Sulfur in coal %	Chlorine in coal %	Sulfur compounds[a] in flue gas vol ppm	Chlorine compounds[a] in flue gas vol ppm
0.8	0.8	750	680
1.2	0.4	1100	340
1.8	0.07	1700	60

a. Calculated by assuming complete volatilization of all sulfur and chlorine in coal and one atom of sulfur or chlorine present in each gas molecule.

Thus, it can be seen that significant concentrations of HCl are likely to result from the combustion of chlorine-containing coal. The subject of HCl corrosion in flue gases has received comparatively little attention in the literature because it is not likely to occur unless temperatures near the water dew point are encountered. Air preheater elements, however, can drop below the moisture dew point if excessive water vapor, such as would occur from a steam leak, is present.

Figure 314, taken from a study by R. W. Kear,[321] illustrates the effect of HCl in a flue gas on corrosion of a test probe. This experiment was conducted using an apparatus which produced a synthetic flue gas by addition of SO_2 and Cl_2 to the fuel supply of a small laboratory burner. Analysis of the flue gas indicated that all chlorine was converted to HCl, resulting in 400 ppm HCl by volume. It should be noted, however, that corrosion could be caused by the presence of chlorine gas. The assumption that Figure 314 is an illustration of the effect of HCl gas is therefore dependent upon Kear's conclusion that all chlorine is converted to HCl in the burner flame. The SO_3, or H_2SO_4, content of this gas was reported as 36 ppm. The temperature at which the corrosion rate accelerates corresponds to the water dew point of the synthetic flue gas, which is about 7% by volume water vapor. When the metal surface temperature is above the water dew point, the presence of HCl has no effect on corrosion, but it can be seen from Figure 314 that drastic increases in corrosion occur due to HCl as the metal surface falls below the water dew point. The corrosion probe was exposed for a 30-minute period in each experiment.

Data obtained by Piper and Van Vliet[315] confirm Kear's results. Piper's data were obtained by exposing metal condensers, which could be cooled to selected temperatures, to flue gas produced from burning a 0.066% chloride coal. Analysis of the flue gas showed that HCl concentrations ranged from 16 to 82 ppm, and the sulfuric acid vapor concentration averaged 30 ppm. The relative rates of corrosion of low alloy steel specimens maintained at 71, 60, 46, and 30°C (161, 141, 115, and 87°F) for 2-month exposures were 1, 1, 3, and 66, respectively. The water dew point of the flue gas during the exposure period ranged from 32 to 40°C (91 to 104°F). It is thus apparent that the rate of attack greatly accelerated below the water dew point. This corrosion is a result of both H_2SO_4 and HCl, but the importance of the effect of HCl is indicated by the fact that at the water dew point, the chemical equivalents of chloride exceeded those of sulphate. Another important observation of the Piper study was that a vitreous enamel coating on Cor-Ten, which was used in a pilot-plant air preheater, was considerably attacked at temperatures below the water dew point.

Since high resistivity fly ash usually occurs in the absence of sulfuric acid vapor, it is of interest to consider such a situation in which appreciable concentrations of HCl exist. Piper analyzed the vapor-liquid equilibria data for the system HCl-H_2O, and concluded that, with an HCl vapor concentration of 82 ppm, the hydrochloric acid dew point would be 3.9°C (39°F) above the water dew point. A similar analysis of the water-SO_2 system indicated that the sulfurous acid dew point, for a stack gas with about 1900 ppm SO_2 and typical water vapor concentrations, would be the same as the water dew point.

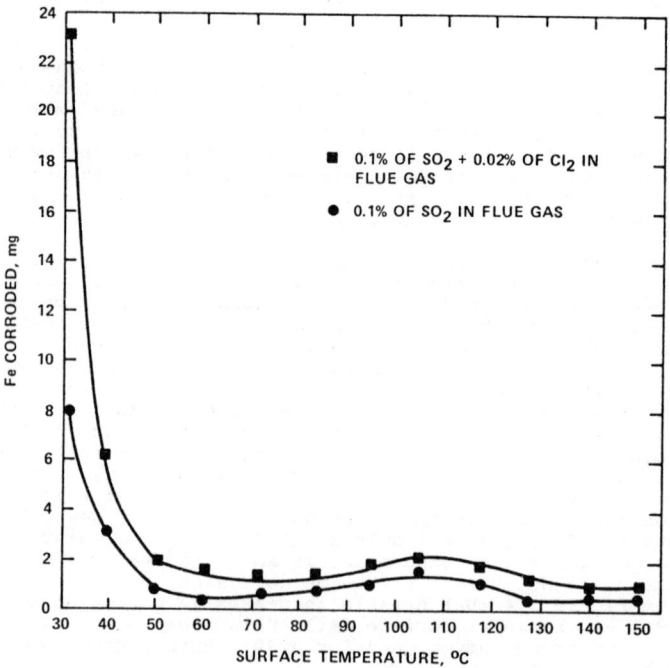

Figure 314. The effect of chlorine addition or corrosion of mild steel in a synthetic flue gas (from R. W. Kear).[321]

FOULING OF LOW TEMPERATURE SURFACES

Deposit formation, or fouling, in air heater elements is a combination of chemical and physical processes. At 600 to 700°F, which is the range of temperature normally encountered at the hot end of regenerative air heaters, the saturation partial pressure of the mineral components of fly ash is extremely low. Thus deposit formation in this region is not a result of condensation from the vapor phase, but is instead a mechanical process in which slag and refractory material are carried by the flue gas into the air heater elements. These particles can lodge within the passages of hot end elements and thereby accumulate additional deposits of finer dust particles.[322] Procedures are reported in the literature for removing such deposits.

If the flue gas contains appreciable amounts of H_2SO_4, corrosion and deposit buildup will occur simultaneously in the cooler regions of the air heater.[323] The following reaction will occur on steel surfaces which are below the H_2SO_4 dew point.

$$Fe + H_2SO_4 \rightarrow FeSO_4 + H_2$$

The ferrous sulfate can then oxidize to form ferric sulfate.

$$4FeSO_4 + 2H_2SO_4 + O_2 \rightarrow 2Fe_2(SO_4)_3 + 2H_2O$$

An extensive study of regenerative air heater deposits by the Bureau of Mines[324] found that deposits built up in thickness at the cold end of the air heater, and that this area was the principal region of corrosion and destruction of the element. All deposits found in this area exhibited the following characteristics: partial solubility in water, presence of sulfates, and acidity. The solubilities in water of these deposits varied over a wide range—13 to 98%. Deposits with highest solubilities were found on preheater test plates which were most severely attacked by acid. Some of the variations in deposit solubility were attributed to variations in the ability of the deposits to trap fly ash.

Reaction of the ferrous and ferric salts formed during corrosion with alkaline compounds sometimes used in washing air heaters can produce compounds that will result in additional fouling. Ferric sulfate, for example, can undergo the following reactions.[323]

$$Fe_2(SO_4)_3 + 3Ca(OH)_2 \text{ (lime)} \rightarrow 2Fe(OH)_3 + 3CaSO_4$$

$$Fe_2(SO_4)_3 + 6NaOH \rightarrow 2Fe(OH)_3 + 3Na_2SO_4$$

$$Fe_2(SO_4)_3 + 3Na_2CO_3 \text{ (soda ash)} + 3H_2O \rightarrow 2Fe(OH)_3 + 3Na_2SO_4 + 3CO_2$$

The $Fe(OH)_3$ (ferric hydroxide) is undesirable because it is a sticky, gelatinous precipitate which can cause severe fouling. The above reactions indicate that, in washing air heater elements or tubes, removing the soluble sulfates with a neutral water wash is desirable prior to a caustic wash.

It is important to note that deposit formation can occur in air heater elements in the absence of significant amounts of H_2SO_4. Chemical analysis of deposits from air heaters installed in some lignite-burning power stations has revealed no chemical evidence of deposition.[325] In one instance, moisture from steam cleaning action was found to be responsible for trapping ash deposits. Deposits formed in this manner are similar to cement and very difficult to remove.

In the absence of moisture and acid condensate problems, the nature of the fouling mechanisms discussed herein suggests that lowered cold end temperature would not result in increased deposit formation.

LABORATORY CORROSION STUDIES[319]

Samples of fly ash were obtained for corrosion studies from the precipitator hoppers of two plants with high dust resistivity problems. These ash samples have widely different soluble base contents, as can be seen from Table 52. Sulfur contents of the coal burned in the two plants range from 0.6 to 1.0%. Laboratory experiments were conducted to determine whether deposited layers of these ashes exhibit differing capabilities for neutralizing acid and inhibiting corrosion.

A schematic diagram of the apparatus used for the experiments is given in Figure 315, and the data obtained are presented in Tables 52 and 53. The corrosion specimen was a 2.54 cm (1 in.) diameter mild steel disc, and the amount of corrosion occurring as a result of exposure to H_2SO_4 was determined by measuring the weight loss.

TABLE 52. FLY ASH PROPERTIES

	pH of suspension	Soluble sulfate wt %	Soluble base as CaO		Mass median particle diameter, μ
			meq/g	wt %	
Neutral (from Plant 1)					
As received	6.70	0.31	0	0	38
Following experiment 4	1.69	23.4	0	0	-
Basic (from Plant 6)					
As received	12.25	1.2	2.7	7.6	18
Following experiment 3	8.72	23.1	Not determined		-

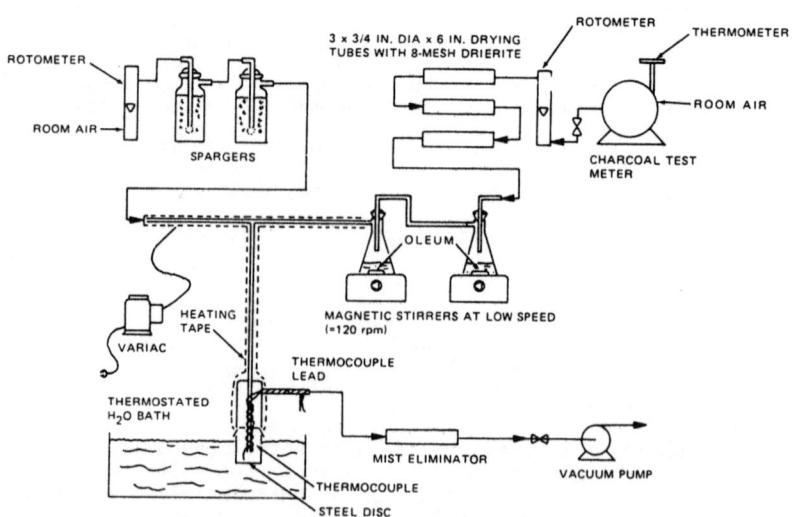

Figure 315. Schematic diagram of apparatus used in corrosion experiments.

TABLE 53. CORROSION RATE EXPERIMENTS

Experiment No.	H_2SO_4 Vapor Generator, % Acid Used	Duration Hr	Condensate Composition Wt % H_2SO_4	Temperature, °C (°F) Gas	Temperature, °C (°F) Water Bath	Temperature, °C (°F) Disc Surface	H_2SO_4 Condensate Rate meq/hr	Apparent Corrosion Rate mg/hr	Apparent Corrosion Rate meq/hr	Apparent Corrosion Rate[b] mils/yr	Ash Layer	H_2SO_4 Reacting With Disc Wt %
1	104	2.0	56	195 (383)	25 (78)	-- --	--	1.05	0.056	46	None	--
2	104	1.9	52	193 (379)	29 (84)	-- --	1.3	0.90	0.048	39	None	3.7
3	104	2.1	--	198 (388)	26 (79)	-- --	1.5	0.20	0.011	17[a]	Basic	0.7
4	104	2.0	--	212 (412)	26 (79)	-- --	1.6	0.40	0.022	34[a]	Neutral	1.4
5	107	1.0	36	176 (349)	2.8 (37)	32 (90)	5.0	32	1.7	1400	None	34
6	107	1.0	40	190 (374)	3.9 (39)	25 (77)	6.0	32	1.7	1400	None	28
7	107	1.0	--	198 (388)	2.8 (37)	35 (95)	10.1	18	0.97	790	Basic	10
8	107	1.0	--	200 (392)	2.8 (37)	30 (86)	8.6	17	0.91	740	Basic	11
9	107	1.0	--	199 (390)	2.2 (36)	30 (86)	12.4	53	2.8	2300	Neutral	23
10	107	1.0	--	198 (388)	3.9 (39)	27 (81)	12.4	41	2.2	1800	Neutral	18

a. Based on exposure of one side of disc to acid rather than both sides as in all other runs.
b. Assuming formation of $Fe_2(SO_4)_3$.

The experiments in Table 53 can be divided into two groups. In Experiments 1 through 4, the acid condensation rate on the disc was relatively low, but high condensation rates were achieved in Experiments 5 through 10 by increasing the strength of oleum used as an SO_3 vapor source and by lowering the temperature of the water bath. Water vapor concentrations of 2 - 2.5% by volume were provided by the water spargers. Since the air streams bearing H_2O and SO_3 vapor mix in the heated glass "T", a saturated mixture of air and H_2SO_4 is formed, and the condensation rate will depend on the temperature of the condensing surface and the concentration of H_2SO_4 in the gas phase. For both sets of experimental conditions, an examination of the corrosion rates (meq basis) and acid deposition rates in Table 53 shows that an excess of acid was present with respect to the amount of iron corroded in all experiments.

For the experiments with fly ash, the ash was deposited in the sample container in such a manner that the disc was covered to a thickness of approximately 0.2 mm. Acid did not sufficiently penetrate the ash to reach the underside of the disc in Experiments 3 and 4, and the penetration rates were calculated on the basis of one side only. Corrosion was observed on both sides in all other experiments; therefore, the total area of both disc surfaces was used as a basis of calculation.

A comparison of data from Experiment 3 with those from Experiment 4 indicates that the basic fly ash was more effective in reducing corrosion than the neutral ash. The equilibrium pH values of the ash samples prior to and following these experiments are given in Table 52. As would be expected, the neutral ash slurry is much more acidic than that of the basic ash after both have experienced an equivalent sulfate gain due to H_2SO_4 condensation. The fact that the basic ash produced a pH greater than 7 following the experiment shows that it was capable of neutralizing all of the condensed acid. Complete neutralization did not occur until the acid-ash mixture was slurried in water, however, as evidenced by the measurable degree of corrosion which occurred in Experiment 3.

For the experiments with low acid condensation rates, both the neutral and basic ash deposits reduced the weight loss rate of the disc, but the penetration rate calculated for Experiment 4 (neutral ash) is not significantly different from those of Experiments 1 and 2 (no ash). These results are to be expected, since the neutral character of the material from Plant 1 indicates that any corrosion inhibiting value which it exhibits is likely to be the result of physical rather than chemical factors.

High corrosion rates were obtained in Experiments 5 through 10 due to increased acid condensation rates and decreased condensate composition. The high percentage of H_2SO_4 reacting with the disc in these experiments is an indication of the greater corrosiveness of acid in the 36 - 40 wt % range. Some difficulty was encountered in maintaining constant experimental conditions, as indicated by variations in the disc surface temperatures and the acid condensation rates. Once again, the data suggest that the neutral ash has little corrosion inhibiting value, but significantly lower corrosion rates were obtained with the basic ash. In contrast to the conditions of Experiment 3, an excess of acid was present with respect to the base content of the ash layer for Experiments 7 and 8. If it is assumed that the same amount of base reacts per unit weight of basic ash in both sets of experiments, it can be shown that less than 30% of the condensing acid could have been neutralized in Experiments 7 and 8. The principal mechanism by which corrosion rates were reduced in Experiments 7 and 8 appeared to be the formation of a cement-like deposit which reduced the amount of acid reaching the metal surface. Such deposits would be likely to cause plugging of air heater elements in plant operation.

Generalizations concerning the direct effect of basic and neutral fly ashes on corrosion rates from these experiments are hazardous because of the complex nature of the corrosion process. However, it is possible to draw some conclusions regarding the interaction of the fly ash with condensing acid.

The reduced corrosion rate obtained in Experiment 3 indicates that the basic fly ash from Plant 6 neutralized a major portion of the acid as it condensed. This is an important observation because the data obtained has revealed the presence of unreacted acid on the surface of fly ash containing amounts of water soluble base substantially in excess of the apparent surface acidity. Thus, basic ash deposited on metal surfaces could conceivably present an acidic, and hence corrosive, environment to a metal surface and exhibit little or no neutralizing capability. A layer of $CaSO_4$, formed by reaction between H_2SO_4 and CaO, apparently can prevent the underlying soluble base from being utilized. The ash from Plant 6 contained appreciable sulfate when received from the precipitator hoppers (1.2%), but the experimental data presented here indicate that the sulfate did not present an impermeable barrier to the liquid condensate.

The neutral ash from Plant 1 would not be expected to provide a significant degree of protection from condensing acid, and the experimental data tend to confirm this. However, even a neutral ash can reduce the amount of acid available for corrosion in a flue gas by adsorbing SO_3. The small amount of sulfate (0.31%) present on the ash from Plant 1 when received indicates that some adsorption of SO_3 at high temperatures occurred. The operating temperature of the precipitator at Plant 1 is about 160°C (320°F), which is well above the H_2SO_4 dew point.

In conclusion, then, the data from these experiments indicate that a basic ash such as that from Plant 6 can be of significant value in neutralizing condensed acid and reducing air heater corrosion rates. However, in the presence of an excess of condensing acid, serious deposit formation problems could be expected. The neutral ash was of little or no apparent value in reducing corrosion rates, but it exhibited a lesser tendency to form cement-like deposits than did the basic material. The most important benefit to be expected from the presence of a basic fly ash from the standpoint of corrosion is the consumption of SO_3 by the basic material in the high temperature region prior to the air heater. Unfortunately, this also creates a high resistivity problem for precipitators operating in the 148°C (300°F) range.

SUMMARY OF FIELD EXPERIENCE AND PLANT DATA[292]

Table 54 is a compilation of available data from a number of power plants concerning fly ash, flue gas and coal composition, and fly ash resistivity. The data reported in this table were either obtained by SoRI personnel under Contracts CPA 70-149 and CPA 70-166 sponsored by the U.S. Environmental Protection Agency or made available to SoRI by the utility companies.

TABLE 54. PROPERTIES OF FLUE GAS AND FLY ASH FOR VARIOUS COAL-FIRED BOILERS

Plant Designation	Coal Analysis (Dry Basis)		Fly Ash Analysis							Flue Gas Analysis Precipitator Inlet			Typical Fly Ash Resistivity	
	Sulfur %	Ash %	pH	Water Slurry Sol base as CaO %	Sol SO_4 %	Ethanol Slurry pH	Free acid as H_2SO_4 %			SO_2 vol ppm	(Wet Basis) H_2SO_4 vapor vol ppm	H_2O vol %	Ω-cm	temp °C (°F)
6	0.7–1.0	8.5	12.2	7.6	1.2	>9.1	0			--	--	10.7	1.9×10^{11} 1.0×10^{10}	150 (302) 104 (220)
1	0.6	12	8.2	Negligible	0.23	4.6	0.008			375	<1	7.7	1.9×10^{12}	160 (320)
2[a]	0.5	5.9	11.1	2.10	1.50	8.1	0			387	<1	8.9	3.8×10^{12}	135 (275)
11	0.5	15–25	11.2	1.50	0.17	--	--			--	--	--	4.5×10^{11}	110 (230)
8-3[a]	0.5	8.6	9.4	0.35	0.77	--	--			365	<1	7.7	1.0×10^{12}	154 (309)
5	0.95–1.90	15.8–16.0	9.4	0.19	0.41–0.47	--	--			610–1030	0.8–4.4	7.0	3×10^{11} 1.5×10^{12}	124-(256-) 160 (319)
7[a]	2.1	21.9	5.1	0	0.36	--	--			1650	8.7	5.7	1.0×10^{12} 2.0×10^{11}	160 (319) 149 (300)
4	3.6	16.4	11.0	1.65	0.77	3.8	0.037			2680	15	8.0	1.0×10^{10}	142 (287)
9[a,b]	~3.5	~14	9.8	0.35	1.15	3.9	0.088			--	--	--	1.0×10^{9}	143 (290)
10[a,b]	3.2	11.2	6.4	0	0.40	4.4	0.02			--	--	--	--	--

a. Precipitator preceded by mechanical collector.
b. Corrosion of air heater has occurred.

Appendix D—Low Temperature Corrosion and Fouling 479

Of all the plants listed in Table 54, only Plants 10 and 9 have experienced significant air heater corrosion problems. As the following discussion will indicate, the factors that result in high resistivity fly ash usually indicate that no corrosion problems are to be expected.

The ash samples for which analyses are given in Table 54 were either collected from the precipitator hoppers or obtained with a resistivity apparatus at the precipitator inlet. The values of pH and free acid obtained in a 95% ethanol slurry, which are given for selected samples, are an indiation of acid present on the surface of the ash. Samples which show an acidic pH in 95% ethanol generally exhibit a minimum pH in water, followed by a rise to a basic equilibrium value as the water soluble base is dissolved. The presence of significant amounts of unreacted acid on the ash surface is thought to be an indication that the fly ash has been "conditioned" by sulfuric acid.

Data for SO_2 - SO_3 were obtained by SoRI personnel using procedures described elsewhere.[319] Resistivity data were also obtained by SoRI using either a point-plane or cyclone resistivity apparatus, with the exceptions of Plants 6 and 11. For these two plants, the data were given to SoRI by the operating utilities.

Plant 6 has successfully overcome a high dust resistivity problem by lowering the precipitator operating temperature to about 104°C (220°F) at full load. An inspection of the low temperature zone of this installation was conducted while the unit was off the line for routine maintenance. This plant had nine months of operation with low gas temperatures.

The areas examined for evidence of corrosion were the cold and intermediate zones of the air heater elements, the plates and wires in the precipitator, and the sides of the duct encompassing the precipitator assembly. No evidence of corrosion was found in the air heater elements. Thin deposits were noted in some areas of the cold-end elements, but these were insufficient to cause measurable draft losses. Minor corrosion was observed on the perforated plate distributors at the precipitator inlet. The rusted areas corresponded to regions of low gas velocity caused by duct geometry. The only significant corrosion in the entire assembly was found on the under side of the top plate of the precipitator housing. The top side of this plate is exposed to streams of low temperature bleed air from the plant exterior, and it is probable that temperatures below the water dew point were reached. The purpose of the bleed air is to maintain a positive pressure for prevention of dust buildup on the rapper bushings.

There as no direct measurement of SO_3 at Plant 6, but measurements from Plant 2, which uses a similar fuel, show that SO_3 levels above and below the air heater are less than 1 ppm. The soluble sulfate content of fly ash taken from the precipitator hoppers of Plant 6, if a dust concentration of 3.4 gm/m^3 (1.5 gr/ft^3) is assumed, is equivalent to an SO_3 concentration of 10 ppm. It is possible, however, that a portion of the sulfate originated from oxidation of SO_2 on the ash surface rather than from SO_3 in the bulk gas phase. Figure 303 shows that the dew point of a flue gas with 10 ppm SO_3 and 10.7% water vapor is estimated as 135°C (275°F). The minimum cold end average temperature of the air heater at Plant 6 is 60°C (140°F). It is therefore probable that some acid condensation, and possibly corrosion, would have occurred if the basic ash had not been present to combine with the SO_3 in the high temperature zone prior to the air heater, thus preventing the formation of H_2SO_4 vapor in the air heater region. Furthermore, the data in Tables 52 and 53 and the lack of surface acidity indicated in Table 54 show that any H_2SO_4 which may form in the air heater region is likely to be neutralized.

In view of the known dependence of fly ash resistivity on temperature and the presence of H_2SO_4 on the fly ash surface, the hypothesis of negligible H_2SO_4 in the low temperature zone at Plant 6 may seem inconsistent with the decrease in resistivity with temperature which occurs at this installation. This apparent inconsistency can be qualitatively resolved by attributing the resistivity behavior to increasing adsorption of water vapor on the fly ash surface with decreasing temperature. It is also possible that oxidation of SO_2 to SO_3 occurs on the ash surface, and provides surface H_2SO_4 for conditioning for a brief time period, after which the acid is neutralized. The following reaction sequence may be used to represent this hypothesis.

$$SO_2(g) + \tfrac{1}{2}O_2(g) \rightarrow SO_3(g)$$

$$SO_3(g) + H_2O(g) \rightarrow H_2SO_4(g) \leftrightarrow H_2SO_4(l)$$

$$H_2SO_4(g \text{ or } l) + CaO(s) \rightarrow CaSO_4(s) + H_2O(g)$$

Thus, by adsorption of water and/or surface formation of SO_3, it is possible to explain the lowering of resistivity with decreasing temperature in the absence of appreciable SO_3 concentrations in the bulk gas phase.

Plant 11 and Plant 10 are the other plants listed in Table 54 with lowered cold-end temperatures. Plant 11 operates with a low sulfur coal which produced a highly basic fly ash with a high resistivity. No corrosion problems have been experienced at this installation, as would be expected. Precipitator inlet temperatures range from 110-122°C (230-253°F).

Plant 10 has operated with precipitator inlet temperatures from 108-118°C (228-246°F). Excessive deterioration of air heater cold end elements occurred when gas temperatures were lowered to 108°C (228°F), and as a result, operating temperature has now been raised to 117-118°C (243-246°F). The reason for lowering the exit temperature was said to be a desire to increase boiler efficiency rather than a need to lower fly ash resistivity. Fly ash and coal samples supplied to SoRI were analyzed and are reported in Table 54. However, the sulfur content of the coal normally used was reported by the utility to be 1.2-1.35%. Analysis of the fly ash indicates a neutral ash similar to that from Plant 1, and little or no acid neutralizing ability would be expected. The low sulfate content indicates that, in spite of the high sulfur content of the coal and the relatively low temperature at which the ash was collected, a comparatively small amount of H_2SO_4 is collected by the ash. From the ash content of the coal, the mass loading at Plant 10 is estimated, prior to the mechanical collector, as 6.9 gm/scm (3 gr/scf),[326] and the sulfate content of the fly ash is equivalent to only 6.4 ppm H_2SO_4. It is therefore probable that most of the H_2SO_4 formed from the combustion of this relatively high sulfur coal remained in the gas phase and was available for condensation.

Although there are no resistivity measurements from Plant 10, it is possible to infer from the coal and fly ash analysis that a low resistivity fly ash (significantly less than 2×10^{10} Ω-cm) is probable at this installation at the precipitator operating temperatures. It has been shown from studies of H_2SO_4 conditioning under EPA Contract CPA 70-149 at Plant 1 that a sulfate gain of only 0.1-0.2% due to adsorption or condensation of H_2SO_4 is sufficient to lower resistivity by two orders of magnitude for a neutral fly ash.[326a]

Plants 9 and 4 normally operate with a high sulfur coal, and typical air heater exit temperatures for both units range from 135-140°C (275-285°F). These plants have low fly ash resistivities at normal operating temperatures, and at times the resistivity value at Plant 4 has been too low for proper precipitator operation with high gas velocity. The cold-end portion of the air heaters at both of these installations operates below the acid dew point, but the corrosion experience has been somewhat different. Plant 4 has an average cold-end temperature of about 93°C (200°F), and Figure 303 shows that most of the H_2SO_4 vapor is available for condensation at this temperature. Furthermore, measurements of SO_3 before and after the air heater have indicated, on at least one occasion, a significant drop in SO_3 concentration across the heater. It is therefore probable that significant amounts of H_2SO_4 are condensed, either on the ash in the cool boundary layer adjacent to the metal surface, or on the metal surface itself. In spite of this fact, the cold-end baskets (made of low-alloy steel) have been in service for at least ten years at Plant 4 without requiring replacement. Table 54 shows that the fly ash at this unit is highly basic, and would be expected to have significant acid neutralizing ability. However, the presence of surface acidity, as indicated by data obtained in a 95% ethanol slurry, suggests that a sulfate layer on the ash is preventing a portion of the water soluble base from being utilized.

Plant 9 has required some replacement of cold-end air heater elements, but not at an excessive rate. The data in Table 54 indicate that the fly ash from Plant 9 is less basic than that from Plant 4, but the presence of a mechanical collector at Plant 9 makes a direct comparison of the two fly ash analyses difficult because of the difference in particle size distribution. It is, however, reasonable to conclude that without the presence of the basic fly ashes at both installations, corrosion would have been more severe.

Plant 7 operates at high air heater exit temperatures with an intermediate sulfur coal. The resistivity values indicated in Table 54 for this plant would be classified as high, but the near-neutral character of the ash, together with the presence of appreciable concentrations of H_2SO_4 vapor in the gas phase and the slope of the resistivity temperature curve, suggest that acceptable resistivity values would occur at about 137°C (280°F). With an 26°C (80°F) inlet air side temperature, this would give a cold-end average of 82°C (180°F) for the air heater. The Air Preheater Company's cold-end temperature and material selection guide gives a suggested minimum average cold-end temperature of about 71°C (160°F) for a coal of 2.1% sulfur content and corrosion-resistant, low-alloy steel cold end elements.[327] Some degree of corrosion may occur because the cold-end metal temperatures fall appreciably below the acid dew point, and because the neutral ash at Plant 7 could be expected to have no significant acid neutralizing ability. However, the experience of the Air Preheater Company as represented by their materials and temperature guide, and the lack of excessive H_2SO_4 vapor concentrations found at 148-160°C (300-320°F) are indications that a severe corrosion problem should not occur at Plant 7 with the presently used fuel if air heater exit temperatures as low as 137°C (280°F) were employed.

The corrosion experience of Plant 5 (Unit 1) is of interest because the average air heater exit temperature is about 126°C (260°F). Sulfur content of the coal normally burned at this unit is approximately 1%, and a typical dust load would be 8.5 gm/scm (3.7 gr/scf). Coal composition varied during the time period in which resistivity data were taken, and possibly as a result, the resistivity data show considerable scatter and no strong variation with temperature. Nonetheless, the relatively high resistivity values are to be expected on the basis of the coal sulfur content and the moderately basic character of the fly ash. No corrosion problems have occurred at this unit, and none would be expected with the relatively low H_2SO_4 vapor concentrations which were measured.

Plants 8-3 and 2 are typical of installations burning very low sulfur coal; that is, no appreciable H_2SO_4 vapor concentrations are found in the bulk gas phase, the fly ash produces a basic water slurry, and the resistivity is unfavorably high in the normal operating temperature range of 135-148°C (275-300°F).

If the design of these plants were such that operation in the 104-115°C (220-240°F) range were possible, no corrosion problems would be expected because of the absence of H_2SO_4 vapor. Unfortunately, there is not a sufficient quantitative knowledge of the relationship between resistivity and temperature to predict with confidence that low temperature operation at these installations would produce resistivity below the critical value of 2×10^{10} Ω-cm. The fact that the flue gas water concentrations at Plants 2 and 8-3 are about 30% lower than that at Plant 6 is an unfavorable condition for achieving lowered resistivity. However, the fly ashes from Plants 2 and 8-3, and in particular, that from Plant 8-3, are less basic than the ash produced at Plant 6. Data obtained under Contract CPA 70-149 indicate that a highly basic ash requires a greater gain of H_2SO_4, either by condensation or adsorption, to lower resistivity than does a neutral ash. Thus, if lowering of resistivity is due to the combined effects of water adsorption and the formation of SO_3 on a fly ash surface discussed earlier, it could be argued that the resistivity of the extremely basic ash of Plant 6 would show less sensitivity to decreasing temperature than the fly ash at Plants 2 and 8-3. Since the variables of ash composition and flue gas water concentrations indicate opposing effects when comparing Plant 6 with Plants 2 and 8-3, it would be hazardous to equate the resistivity-temperature experience of Plant 6 with the other two installations.

METHODS OF ASSESSING CORROSION TENDENCIES OF FLUE GASES

Introduction

A comprehensive discussion of methods developed in England for assessing the corrosion and fouling potential of flue gases is given in a bulletin entitled, "Testing Techniques for Determining the Corrosive and Fouling Tendencies of Boiler Flue Gases" published by the Boiler Availability Committee.[318] The following discussion is a brief summary of the purpose and method of operation of those procedures which relate to low temperature corrosion and fouling.

Corrosion Probes

The purpose of corrosion probes is to measure the amount of corrosion produced by acid condensed on metal surfaces in a flue

gas environment. These probes provide a means of supporting a
prepared metal test specimen in flue gas streams at a selected
temperature. The BCURA probe is an air-cooled device in which
the surface temperature of the test specimen is monitored with a
thermocouple brazed to the body of the probe. Exposure periods
of 15-30 minutes are recommended, and the amount of corrosion is
determined by measuring weight loss of the specimen.

Probes designed for short term experiments are of value for
comparing relative effects of variations in operating parameters,
such as temperature and fuel composition. However, for prediction
of actual corrosion rates over extended periods, long term tests
of 100 hours or more are desirable. A liquid-cooled probe has
been designed by the Sheel Petroleum Company, Ltd., for such extended experiments.[328]

Acid Deposition Probes

An indirect measurement of the rate of acid deposition on a
cooled surface is given by the BCURA dew point meter, which has
been described previously. Since the conductivity readings of
the dew point meter can be influenced by substances other than
sulfuric acid, it is of interest to consider a direct means of
measuring acid deposition rates.

Alexander[329] has described an air-cooled deposition probe
which accomplishes this purpose. The probe consists of an air-
cooled, one-inch diameter stainless steel tube in which the cooling
air passes through the tube and discharges into the flue gas. The
amount of acid depositing on test areas of the probe, the surface
temperature of which is known, is determined by analysis of deposits obtained from the test surfaces.

Gas And Ash Analysis

An analysis of flue gas for SO_3, SO_2, H_2O, and dust loading,
along with analysis of the fly ash for soluble components, is
necessary for a qualitative assessment of the flue gas corrosion
potential. Procedures used by SoRI for these analyses are described
in the final report from Contract CPA 70-149.[319]

SUMMARY AND CONCLUSIONS

It has been established that the principal cause of corrosion
in the low temperature zone of power plant exhaust systems is
condensation of sulfuric acid, either directly onto metal surfaces or onto fly ash particles which subsequently come in contact
with the metal. Other acids, in particular hydrochloric acid, can
be responsible for corrosion at temperatures approaching the water
dew point of flue gas, but such temperatures are not normally encountered.

Fouling in the low temperature zone of air heaters is primarily
caused by reaction of sulfuric acid with fly ash and the metal
surfaces of the heat exchanger. A basic fly ash can neutralize
appreciable quantities of SO_3 upstream from the air heater region
but laboratory experiments suggest that reaction of highly basic
fly ashes with high concentrations of H_2SO_4 in the low temperature
zone can result in problems with deposit formation. This conclusion
is supported by the experience of the Central Electricity Generating
Board of England, in which medium sulfur coals with alkaline ashes
have produced fouling,[330] but little air heater wastage accompanied

the deposit formation. It is also possible to have deposit formation in the low temperature zone in the absence of sulfuric acid if excessive moisture from steam leaks or soot blowing is present.

Severe corrosion and fouling problems in regenerative air heaters are associated with the temperature at which peak rates in acid deposition occur. These peak rates often are not observed with coal firing due to the presence of fly ash, but in any case, the existence of such a peak is a manifestation of relatively high concentrations of free H_2SO_4 vapor. Thus, the resistivity of fly ash, due to the presence of excessive H_2SO_4, would be expected to be lower than desirable for proper precipitator operation with high gas velocity under these conditions. Resistivity data taken at plants burning high sulfur coals with alkaline fly ashes have demonstrated that resistivity values below the critical 2×10^{10} Ω-cm are obtained at temperatures above 137°C (280°F). Therefore, lowering precipitator operating temperatures is neither necessary nor desirable for the case of high sulfur coals, which produce relatively high concentrations of H_2SO_4 vapor.

An analysis of the factors which cause corrosion, and the operating experience of at least two power plants, have demonstrated that low temperature operation of precipitators [104-121°C (220-250°F)] will not cause low temperature corrosion and fouling problems with a flue gas containing a basic fly ash and no appreciable concentrations of sulfuric acid vapor. The occurrence of corrosion and high fly ash resistivity thus tend to be mutually exclusive phenomena. A possible exception to the rule would be a stack gas with high (over 100 ppm) HCl concentration.

For the case of a plant burning a low to medium sulfur coal which produces a near-neutral, high resistivity ash at approximately 148°C (300°F) and low concentrations of H_2SO_4 vapor, the occurrence of some degree of corrosion as a result of lowered cold-end temperatures cannot be rigorously excluded. However, data obtained have shown that amounts of sulfuric acid sufficient to "condition" a neutral ash can be adsorbed at temperatures well above the sulfuric acid dew point.[319] It is therefore probable that an acceptable fly ash resistivity could be obtained at a temperature sufficiently high to avoid appreciable condensation of sulfuric acid on the cold-end elements of an air preheater. A quantitative evaluation of resistivity and corrosion under such circumstances would require fly ash resistivity data and relative corrosion rates (obtained with a corrosion probe such as described earlier) as a function of temperature in the flue gas.